T0191783

Biotransformations in Organic Chemistry

Kurt Faber

Biotransformations in Organic Chemistry

A Textbook

Seventh extended
and corrected edition

 Springer

Kurt Faber
Department of Chemistry
University of Graz
Graz, Austria

ISBN 978-3-319-87116-5   ISBN 978-3-319-61590-5   (eBook)
DOI 10.1007/978-3-319-61590-5

Printed on acid-free paper

This Springer imprint is published by Springer Nature
The registered company is Springer International Publishing AG
The registered company address is: Gewerbestrasse 11, 6330 Cham, Switzerland

Preface

The use of natural catalysts – enzymes – for the transformation of nonnatural man-made organic compounds is not at all new: they have been used for more than 100 years, employed either as whole cells or isolated enzymes [1]. While the object of the early research was the elucidation of biochemical pathways and enzyme mechanisms, it was the steep rise of asymmetric synthesis during the 1980s that the enormous potential of enzymes for the synthesis of nonnatural organic compounds was recognized. What started as an academic curiosity in the late 1970s became a hot topic in synthetic organic chemistry in the 1990s. Driven by breathtaking developments in molecular biosciences, the search for novel enzymes, their production, and adaptation to industrial processes are continuously simplified, which is demonstrated by the wavelike appearance of novel biocatalytic principles. As a result of this extensive research, there have been an estimated 18,000 papers published on the subject to date. To collate these data as a kind of "super-review" would clearly be an impossible task, and, furthermore, such a hypothetical book would be unpalatable for the non-expert [2–5].

This textbook is written from an organic chemist's viewpoint to provide a *condensed* introduction into biocatalysis and to persuade synthetic chemists to think outside the box and to consider biocatalytic methods as an alternative tool for stereoselective synthesis. By this means, the wide repertoire of synthetic methods has been significantly widened and complemented, which is illustrated by the fact that the proportion of papers on asymmetric synthesis employing biocatalytic methods has constantly risen from zero in 1970 to about 8% at present. Certainly, biochemical methods are not superior in a general sense – they are no panacea – but they provide powerful tools to complement "chemical" methodology for a broad range of highly selective organic transformations. Synthetic chemists capable of using this potential have a clear advantage over those limited to nonbiological methods to tackle the new generation of synthetic problems at the interface between chemistry and biology, particularly in view of the necessity to use renewable feedstocks.

In this book, reliable biotransformations, which already had significant impact on organic chemistry, are put to the fore, including industrial-scale showcases.

Enzymes possessing great potential but still having to show their reliability are mentioned more briefly. The literature covered extends to spring 2017 and special credit is given to selected "very old" papers to acknowledge the appearance of novel concepts. The most useful references are selected from the pack, and special emphasis is placed on reviews and books, which are mentioned during the early paragraphs of each chapter to facilitate rapid access to a specific field if desired. After all, I tried to avoid writing a book with the charm of a telephone directory!

The first edition of this book appeared in 1992 and was composed as a monograph. It was not only well received by researchers in the field but also served as a basis for courses in biotransformations worldwide. In subsequent editions, emphasis was laid on didactic aspects in order to provide the first textbook on biocatalytic methods for organic synthesis, which served as a guide at several academic institutions for updating a dusty organic chemistry curriculum by incorporating biochemical methods. The need to account for the continuous emergence of novel enzymes and new protocols to use them prompted this updated and extended edition.

My growing experience in teaching biotransformations for organic synthesis at several universities and research institutions around the world has enabled me to modify the text of this seventh edition so as to facilitate a deeper understanding of the principles, not to mention the correction of errors, which escaped my attention during previous editions. I am grateful to numerous unnamed students for pointing them out and for raising tough questions.

I wish to express my deep gratitude to Stanley M. Roberts (UK) for undergoing the laborious task of correcting the first edition of this book. Special thanks go to Michael Müller, Martina Pohl, Wolf-Dieter (Woody) Fessner, (Germany), Nick Turner (UK), and Bernd Nidetzky (Graz) for their helpful hints and discussions. This revised edition would not have been possible without the great assistance of Wolfgang Kroutil for pointing out erroneous absolute configurations, Jörg Schrittwieser for discussing didactics, Georg Steinkellner for great enzyme pictures, and Melanie Hall for patiently answering my ignorant questions regarding molecular biology.

I shall certainly be pleased to receive comments, suggestions, and criticism from readers for incorporation in future editions.

Graz, Austria   Kurt Faber
Spring 2017

References

1. For the history of biotransformations, see:
Neidleman SG (1990) The archeology of enzymology. In: Abramowicz D (ed) Biocatalysis, Van Nostrand Reinhold, New York, pp 1–24
Roberts SM, Turner NJ, Willetts AJ, Turner MK (1995) Introduction to Biocatalysis Using Enzymes and Micro-organisms, Cambridge University Press, Cambridge, pp 1–33

2. For monographs, see:

Jones JB, Sih CJ, Perlman D (eds) (1976) Applications of Biochemical Systems in Organic Chemistry, part I and II, Wiley, New York

Davies HG, Green RH, Kelly DR, Roberts SM (1989) Biotransformations in Preparative Organic Chemistry, Academic Press, London

Halgas J (1992) Biocatalysts in Organic Synthesis, Studies in Organic Chemistry, vol 46, Elsevier, Amsterdam

Poppe L, Novak L (1992) Selective Biocatalysis, Verlag Chemie, Weinheim

Roberts SM, Turner NJ, Willetts AJ, Turner MK (1995) Introduction to Biocatalysis Using Enzymes and Micro-organisms, Cambridge University Press, Cambridge

Bornscheuer UT, Kazlauskas RJ (2006) Hydrolases for Organic Synthesis, 2nd ed., Wiley-VCH, Weinheim

Bommarius AS, Riebel B (2004) Biocatalysis, Fundamentals and Applications, Wiley-VCH, Weinheim

Grunwald P (2009) Biocatalysis, Biochemical Fundamentals and Applications, Imperial College Press, London

3. For reference books, see:

Drauz K, Gröger H, May O (eds) (2012) Enzyme Catalysis in Organic Synthesis, 3rd edn, 3 vols, Wiley-VCH, Weinheim

Liese A, Seelbach K, Wandrey C (eds) (2006) Industrial Biotransformations, 2nd edn, Wiley-VCH, Weinheim

Hilterhaus L, Liese A, Kettling U, Antranikian G (eds) (2016) Applied Biocatalysis, Wiley-VCH, Weinheim, Germany.

Riva S, Fessner W-D (eds) (2014) Cascade Biocatalysis, Wiley-VCH, Weinheim, Germany.

P Grunwald (ed) (2015) Industrial Biocatalysis, Stanford, Singapore.

4. For a collection of preparative procedures, see:

Whittall J, Sutton PW (eds) (2010) Practical Methods for Biocatalysis and Biotransformations, Wiley, Chichester

Jeromin GE, Bertau M (2005) Bioorganikum, Wiley-VCH, Weinheim

Faber K, Fessner W-D, Turner NJ (eds) (2015) Biocatalysis in Organic Synthesis, in: Science of Synthesis, 3 vols Thieme, Stuttgart.

5. For the application of biotransformations to stereoselective synthesis, see:

Patel (ed) (2016) Green Biocatalysis, Wiley, Hoboken, NJ.

Tao J, Lin G-Q, Liese A (eds) (2009) Biocatalysis for the Pharmaceutical Industry, Wiley, Singapore.

Tao J, Kazlauskas R (eds) (2011) Biocatalysis for Green Chemistry and Chemical Process Development, Wiley, Hoboken, NJ.

Goswami A, Stewart JD (eds) (2016) Organic Synthesis Using Biocatalysis, Elsevier, Amsterdam.

Turner N (2017) Biocatalysis in Organic Synthesis: The Retrosynthesis Approach, RSC, London.

Contents

Chapter 1
Introduction and Background Information

1.1 Introduction

Exponents of classical organic chemistry will probably hesitate to consider a biochemical solution for one of their synthetic problems due to the fact, that biological systems would have to be handled. Where the growth and maintenance of whole microorganisms is concerned, such hesitation is probably justified. In order to save endless frustrations, close collaboration with a microbiologist or a biochemist is highly recommended to set up fermentation systems [1, 2]. On the other hand, isolated enzymes or enzyme preparations are available in increasing numbers from commercial sources, that can be handled like any other chemical catalyst.[1] In addition, modern methods of molecular biology became simple and reliable enough to be operated by any organic chemist with a minimal background in biosciences. Hence, the cloning and overexpression of a desired enzyme is nowadays feasible within a short time and at modest cost, in particular when it is derived from bacterial sources. Due to the enormous complexity of biochemical reactions compared to the repertoire of classical organic reactions, many methods described in this book have a strong empirical aspect. This 'black box' approach may not entirely satisfy the scientific purists, but as organic chemists tend to be pragmatists, they accept that the understanding of a biochemical reaction mechanism is not a *conditio sine qua non* for the success of a biotransformation. After all, the exact structure of a Grignard-reagent is still unknown although it's an indispensable reagent for organic synthesis. Consequently, a lack of detailed understanding of a biochemical reaction should never deter us from using it, if its usefulness has been established.

Worldwide, about 85–90% of all chemical processes are performed catalytic [3, 4], leading to annual sales of chemicals around 3500 billion US$ [5]. In this context, biocatalytic methods, which stand for the application of Nature's toolset

---

[1]For a list of enzyme suppliers see the appendix (Chap. 5).

© Springer International Publishing AG 2018
K. Faber, *Biotransformations in Organic Chemistry*,
DOI 10.1007/978-3-319-61590-5_1

for the production of chemicals, represent an important pillar in the building of catalysis.[2] Within the area of biotechnology, biocatalysis has been coined as *White Biotechnology*[3] and more than 300 industrial-scale processes are documented worldwide [6].

1.2   Common Prejudices Against Enzymes

If one uses enzymes for the transformation of nonnatural organic compounds, the following prejudices are frequently encountered: [7]

- *'Enzymes are sensitive'.*

    This is certainly true for most enzymes if one thinks of boiling them in water, but that also holds for most organic reagents, e.g., butyl lithium. When certain precautions are met, enzymes can be remarkably stable.[4] Some candidates can even tolerate hostile environments such as temperatures greater than 100 °C and pressures beyond several hundred bars [8–10].

- *'Enzymes are expensive'.*

    Some are, but others can be very cheap if they are produced on large scale [11]. In bulk, prices of enzymes range from ~100,000 $ per kg for a diagnostic enzyme to ~100 $ for crude preparations, which are adequate for most chemical reactions. Due to the rapid advances in molecular biology, costs for enzyme production are constantly dropping and proteins can be reused if they are immobilized.

- *'Enzymes are only active on their natural substrates'.*

    This statement is certainly true for enzymes derived from the primary metabolism, which provides energy for the maintenance of life. However, it is definitely false for proteins involved in secondary metabolism, which ensures detoxification of xenobiotics, defence against offenders and adaption to a constantly changing environment. Much of the early research on biotransformations was impeded by a tacitly accepted dogma of traditional biochemistry which stated that 'enzymes are nature's own catalysts developed during evolution to enable metabolic pathways'. This narrow definition implied that man-made organic compounds could not be regarded as substrates. Once this scholastic problem was surmounted [12], it turned out that the fact that nature has developed its own peculiar catalysts over 3.9×10^9 years does not necessarily imply that they are designed to work only on their natural target molecules. Research during the past decades has shown that the substrate tolerance of many enzymes

---

[2]Other sectors of catalysis are heterogeneous, homogeneous and organo-catalysis.

[3]Other sectors of biotechnology have been defined as 'Red' (medical biotechnology), 'Green' (agricultural biotechnology), 'Blue' (marine biotechnology), and 'Grey' (environmental biotechnology).

[4]Basic rules for the handling of enzymes are described in Chap. 5.

is much wider than previously believed and that numerous biocatalysts are capable of accepting nonnatural substrates of an unrelated structural type by often exhibiting the same high specificities as for the natural counterparts. It seems to be a general trend, that, the more complex the enzyme's mechanism, the narrower the limit for the acceptability of 'foreign' substrates. After all, there are many enzymes whose natural substrates – if there are any – are unknown.

- *'Enzymes work only in their natural environment'.*
  It is generally true that an enzyme displays its highest catalytic power in water, which in turn represents something of a nightmare for the organic chemist if it is the solvent of choice. However, biocatalysts *can* function in nonaqueous media, such as organic solvents, ionic liquids, and supercritical fluids, as long as certain guidelines are followed. Although the catalytic activity is usually lower in nonaqueous environments, many other advantages can be accrued by enabling reactions which are impossible in water (Sect. 3.1) [13–17].

1.3 Advantages and Disadvantages of Biocatalysts

1.3.1 Advantages of Biocatalysts

- *Enzymes are very efficient catalysts.*
  Typically the rates of enzyme-mediated processes are 10^8–10^{10} times faster than those of the corresponding noncatalyzed reactions,[5] and are thus far above the values that chemical catalysts are capable of achieving [18–21]. As a consequence, chemical catalysts are generally employed at 0.1–1 mol% of catalyst loading, whereas most enzymatic reactions can be performed with a mole percentage of 0.01–0.001, which clearly makes them more effective by orders of magnitude (Table 1.1).

Table 1.1 Catalytic efficiency of representative enzymes

Enzyme	Reaction catalyzed	TOF [s^{-1}]
Carbonic anhydrase	Hydration of CO_2	600,000
Acetylcholine esterase	Ester hydrolysis	25,000
Penicillin acylase	Amide hydrolysis	2000
Lactate dehydrogenase	Carbonyl reduction	1000
Mandelate racemase	Racemisation	1000
α-Chymotrypsin	Amide hydrolysis	100

TOF turnover frequency

- *Enzymes are environmentally acceptable.*
  Unlike many (metal-dependent) chemical catalysts, biocatalysts are environmentally benign reagents since they are completely biodegradable.

---

[5]In exceptional cases rate accelerations can exceed a factor of 10^{17}.

- *Enzymes act under mild conditions.*

    Enzymes act within a range of about pH 5–8 (typically around pH 7) and in a temperature range of 20–40 °C (preferably at around 30 °C). This minimizes problems of undesired side-reactions such as decomposition, isomerization, racemization, and rearrangement, which often plague traditional methodology.
- *Enzymes are compatible with each other.*[6]

    Since enzymes generally function under the same or similar conditions, several biocatalytic reactions can be performed in a cascade-like fashion in a single flask. Such systems are particularly advantageous if unstable intermediates are involved and furthermore, an unfavorable equilibrium can be shifted towards the desired product by linking consecutive enzymatic steps. Multienzyme cascades are often denoted as 'artificial metabolism' (Sect. 3.2) [22].
- *Enzymes are not restricted to their natural role.*

    Many proteins exhibit a high substrate tolerance by accepting a large variety of man-made nonnatural substances. If advantageous for a process, the aqueous medium can often be replaced by an organic solvent (Sect. 3.1).
- *Enzymes can catalyze a broad spectrum of reactions.*

    Like catalysts in general, enzymes can only *accelerate* reactions but have no impact on the position of the thermodynamic equilibrium of the reaction. Thus, in principle, enzyme-catalyzed reactions can be run in both directions. The catalytic flexibility of enzymes is generally denoted as 'catalytic promiscuity' [23–29], which is divided into 'substrate promiscuity' (conversion of a nonnatural substrate), 'catalytic promiscuity' (a nonnatural reaction is catalyzed), and 'condition promiscuity' (catalysis occurring in a nonnatural environment).

There is an enzyme-catalyzed process equivalent to almost every type of organic reaction [30], for example:

- Hydrolysis-synthesis of esters [31], amides [32], lactones [33], lactams [34], ethers [35], acid anhydrides [36], epoxides [37], and nitriles [38].
- Oxidation of alkanes [39], alcohols [40], aldehydes, sulfides, sulfoxides [41], epoxidation of alkenes [42], hydroxylation and dihydroxylation aromatics [43], and the Baeyer-Villiger oxidation of ketones [44, 45].
- Reduction of aldehydes/ketones, alkenes, and reductive amination [46].
- Addition-elimination of water [47], ammonia [48], hydrogen cyanide [49].
- Halogenation and dehalogenation [50], electrophilic Friedel-Crafts-type alkylation [51] and acylation [52], nucleophilic aromatic substitution [53], *O*- and *N*-dealkylation [54], Kolbe-Schmitt carboxylation [55], and decarboxylation [56], isomerization [57], acyloin [58] and aldol reactions [59]. Even Michael additions [60], Stetter reactions [61], Nef reactions [62], Wittig-olefination [63], Mitsunobu-type inversions [64], the Cannizzaro-reaction [65], Prins-rearrangement [66], and Diels-Alder reactions [67–70] have been reported.

---

[6]Only proteases are exceptions to this rule for obvious reasons.

Some major exceptions, for which equivalent reaction types have not (yet) been found in nature, is the Cope rearrangement – although [3,3]-sigmatropic rearrangements such as the Claisen rearrangement are known [71, 72]. On the other hand, some biocatalysts can accomplish reactions extremely difficult to emulate in organic chemistry, such as the selective functionalization of nonactivated C-H bond of aliphatics.

Enzymes display three major types of selectivities:

- Chemoselectivity

    Usually an enzyme acts on a single type of functional group, leaving other sensitive functionalities, which would react under chemical catalysis, unchanged. As a result, reactions generally tend to be 'cleaner' so that laborious removal of impurities arising from side reactions, can largely be omitted.
- Regioselectivity and Diastereoselectivity

    Due to their complex three-dimensional structure, enzymes can distinguish between functional groups which are chemically identical but situated in different positions within the same substrate molecule [73, 74].
- Enantioselectivity

    Last but not least, all enzymes are made from L-amino acids and thus are chiral catalysts.[7] As a consequence, any type of chirality present in the substrate molecule is 'recognized' upon formation of the enzyme-substrate complex. Thus, a prochiral substrate may be transformed into an optically active product through a desymmetrization process and both enantiomers of a racemic substrate usually react at different rates, affording a kinetic resolution. These latter properties collectively constitute the 'stereoselectivity' (in desymmetrizations) or 'enantioselectivity' (in kinetic resolutions) of an enzyme and represent its most important feature for asymmetric exploitation [76]. It is remarkable that this key feature was already recognized by E. Fischer back in 1898 [77].

All the major biochemical events taking place within an organism are governed by enzymes. Since the majority of them are highly selective with respect to the chirality of a substrate, it is obvious that the enantiomers of a given bioactive compound such as a pharmaceutical or an agrochemical will cause different biological effects [78]. Consequently, in a biological context, enantiomers must be regarded as two distinct species. The isomer with the desired activity is denoted as the 'eutomer',[8] whereas its enantiomeric counterpart, possessing less or even undesired activities, is termed as the 'distomer'.[9] The range of effects derived from the distomer can extend from lower (although positive) activity, no response or toxic events. The ratio of the activities of both enantiomers is defined as the 'eudismic ratio'. Some representative examples of different biological effects are given in Scheme 1.1.

---

[7]For synthetically produced D-chiral proteins see [75].

[8]From Ancient Greek 'ευ', meaning 'good'.

[9]The Latin prefix 'dis' means 'apart', or having a negavite or reversing force.

6 1 Introduction and Background Information

Probably the most well-known and tragic example of a drug in which the distomer causes serious side effects is 'Thalidomide', which was administered as a racemate in the 1960s. At that time it was not known that the sedative effect resides in the (R)-enantiomer, but that the (S)-counterpart is highly teratogenic [79].[10]

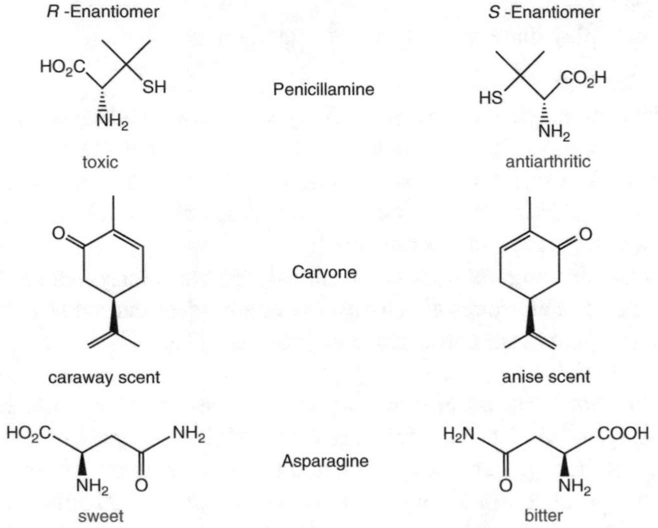

Scheme 1.1 Biological effects of enantiomers

As a consequence, racemates of pharmaceuticals and agrochemicals must be regarded with great caution. Quite astonishingly, 89% of the 537 chiral synthetic drugs on the market were sold in racemic form in 1990, while the respective situation in the field of pesticides was even worse (92% of 480 chiral agents were racemic) [80, 81]. Although at present many bioactive agents are still used as racemates for economic reasons, this situation is constantly changing due to increasing legislation pressure [82]. In 1992, the US Food and Drug Administration (FDA) adopted a long-awaited policy on the issue of whether chiral compounds may be applied as racemic mixtures or as single enantiomers [83–85]. According to these guidelines, the development of racemates is not prohibited a priori, but must undergo rigorous justification based on the separate testing of individual enantiomers. Consequently, single enantiomers are preferred over racemates, which is indicated by the fact that the number of new active pharmaceutical ingredients (APIs) in racemic form remained almost constant from 1992 to 1999, but they almost disappeared from 2001 onwards, going in hand with the doubling of numbers for single enantiomers [86, 87]. For agrochemicals the writing is on the wall: the current climate of 'environmentality' imposes a pressure for the development of enantiopure agents. Overall this has caused an increased need for enantiopure compounds [88, 89].

---

[10]According to a BBC-report, the sale of rac-thalidomide to third-world countries has been resumed in mid-1996!

Unfortunately, less than 10% of organic compounds crystallize as a conglomerate (the remainder form racemic crystals) largely denying the possibility of separating enantiomers by simple crystallization techniques – such as by seeding a supersaturated solution of the racemate with crystals of one pure enantiomer. Previously, enantiomerically pure auxiliary reagents were used in catalytic or in stoichiometric amounts [90]. They are often expensive and cannot always be recovered. Likewise, starting a synthesis with an enantiomerically pure compound from the stock of enantiopure natural compounds [91] such as carbohydrates, amino acids, terpenes or steroids – the so-called 'chiral pool' – has its limitations [92]. Considering the above-mentioned problems, it is obvious that enzymatic methods represent a valuable addition to the existing toolbox available for the asymmetric synthesis of fine chemicals [93].

1.3.2   Disadvantages of Biocatalysts

There are certainly some drawbacks worthy of mention for a chemist intent on using biocatalysts:

* *Enzymes are provided by nature in only one enantiomeric form.*
  Since there is no natural way of creating mirror-image enzymes from D-amino acids, it is impossible to invert the chiral induction of a given enzymatic reaction by choosing the 'other enantiomer' of the biocatalyst, a strategy which *is* possible with chiral chemical catalysts. To gain access to the other enantiomeric product, one has to follow a long and uncertain path in search for an enzyme with exactly the opposite stereochemical selectivity. However, this is sometimes possible, and strategies how nature transforms mirror-image substrates using stereo-complementary enzymes have recently been analyzed [94].
* *Enzymes require narrow operation parameters.*
  The obvious advantage of working under mild reaction conditions can sometimes turn into a drawback. If a reaction proceeds too slow under given parameters of temperature or pH, there is only a narrow operational window for alteration. Elevated temperatures as well as extreme pH lead to deactivation of the protein, as do high salt concentrations. The usual technique to increase selectivity by lowering the reaction temperature is of limited use with enzymatic transformations, although some enzymes remain catalytically active even in ice [95, 96]. Hence, the narrow temperature range for the operation of enzymes prevents radical changes, although positive effects from small changes have been reported [97].
* *Enzymes display their highest catalytic activity in water.*
  Due to its high boiling point, high heat of vaporization and its tendency to promote corrosion, water is usually the least suitable solvent for most organic reactions, although it is the 'greenest' of all solvents [98]. Furthermore, the

majority of organic compounds are only poorly soluble in aqueous media. Thus, shifting enzymatic reactions from an aqueous to an organic medium would be highly desired, but the unavoidable price one has to pay is usually some loss of catalytic activity, which is often in the order of one magnitude [99].

• *Enzymes are bound to their natural cofactors.*

It is a striking paradox, that although enzymes are extremely flexible for accepting nonnatural substrates, they are almost exclusively bound to their natural cofactors which serve as storage for chemical energy (ATP), as molecular shuttles of redox equivalents (heme, flavin, nicotinamide), or as carriers for toxic amines (pyridoxal phosphate) and highly reactive carbanion intermediates (thiamine diphosphate). These 'biological reagents' are prohibitively expensive to be used in stoichiometric amounts and some are relatively unstable molecules. With very few exceptions, they cannot be replaced by more economical man-made substitutes [100]. Although many cofactors can be efficiently recycled, others cannot (Sects. 2.1.4 and 2.2.1).

• *Enzymes are prone to inhibition phenomena.*

The tight binding of small molecules may lead to substrate and/or product inhibition, which causes a drop in reaction rate at elevated substrate and/or product concentrations, a factor which limits the efficiency of the process. Whereas substrate inhibition can be circumvented comparatively easily by keeping the substrate concentration at a low level through continuous addition, product inhibition is a more complicated problem. The gradual removal of product by physical means is usually difficult. Alternatively, a consecutive step may be linked to the reaction to effect in-situ chemical removal of the product.

• *Enzymes may cause allergies.*

Enzymes may cause allergic reactions. However, this may be minimized if enzymes are regarded as chemicals and handled with the same care.

1.3.3   Isolated Enzymes Versus Whole Cell Systems

The physical state of biocatalysts which are used for biotransformations can be very diverse. The final decision as to whether one should use isolated, more-or-less purified enzymes or whole microorganisms – either in a free or immobilized form – depends on many factors, such as (*i*) the type of reaction, (*ii*) whether there are cofactors to be recycled, and (*iii*) the scale in which the biotransformation has to be performed [101, 102],. The general pros and cons of using isolated enzymes versus whole (microbial) cells are outlined in Table 1.2.

A whole conglomeration of biochemistry, molecular biology, microbiology, biochemical and metabolic engineering – biotechnology – has led to the development of routes to a lot of speciality chemicals, starting from cheap carbon sources (usually carbohydrates, such as corn-steep liquor, molasses[11] or starch) and cocktails of salts, by using living (viable) whole cells. Such syntheses proceed through a multitude of steps and require fermentation, since they constitute de novo biosynthesis. Products from traditional fermentation are usually obtained with very high product titers because they are derived from the primary metabolism and thus have low toxicity. Typical products are organic acids (e.g. acetic, lactic, citric, succinic, itaconic, pyruvic acid), α-amino acids (e.g. L-Glu, L-Lys, L-Arg, L-Asp) and alcohols (e.g. 1-butanol, 1,3-propanediol).

Secondary metabolites, such as steroids (e.g. hydrocortisone), terpenoids (e.g. taxadiene, artemisinic acid) and antibiotics (e.g. penicillins, cephalosporins, tetracyclins) require more sophisticated pathway engineering and are obtained with reduced productivities due to their inherent toxicity [103–106]. Although many natural antibiotics were chemically synthesized, these processes are hopelessly uneconomic compared to the biotechnological routes. However, to date, many active pharmaceutical ingredients are derivatives of naturally occurring antibiotics, whose biologically synthesized backbone was chemically modified to reduce their toxicity and enhance their bioactivity and stability.

In contrast, the majority of microbially mediated biotransformations makes use of only a single (or a few) biochemical synthetic step(s) by using (or rather 'abusing'!) the microbe's enzymatic potential to convert a nonnatural organic compound into a desired product. Here, non-growing 'resting' cells are employed. Due to their constrained metabolism, less side reactions occur. The characteristics of processes using resting versus fermenting whole cells are outlined in Table 1.3.

Facilitated by rapid advances in molecular biology, the use of wild-type microorganisms from natural environments possessing >4000 genes[12] (which often show decreased yields and/or stereoselectivities due to competing enzyme activities) is constantly declining, while the application of recombinant cells (over)expressing the required protein(s) is rapidly increasing. Consequently, the catalytic protein becomes the dominant fraction in the cell's proteome and side reactions become negligible. If required, competing enzymes can be knocked out completely, as long as they are not of vital importance for the primary metabolism. Such taylor-made genetically engineered organisms for biotransformations are often called 'designer bugs'.

---

[11]Cheap forms of sucrose.

[12]E. coli has ~4500 genes and Saccharomyces cerevisiae (baker's yeast) ~6500 genes.

Table 1.2 Pros and cons of using isolated enzymes versus whole cell systems

Biocatalyst	Form	Pros	Cons
Isolated enzymes	Any	Simple apparatus, simple workup, better productivity due to higher concentration tolerance	Cofactor recycling necessary, limited enzyme stabilities
	Dissolved in water	High enzyme activities	Side reactions possible, lipophilic substrates insoluble, workup requires extraction
	Suspended in organic solvents	Easy to perform, easy workup, lipophilic substrates soluble, enzyme recovery easy	Reduced activities, redox reactions severely impeded
	Immobilized	Enzyme recovery easy	Loss of activity during immobilization
Whole cells	Any	No cofactor recycling necessary, no enzyme purification required	Expensive equipment, tedious workup due to large volumes, low productivity due to lower concentration tolerance, low tolerance of organic solvents, side reactions due to uncontrolled metabolism
	Growing culture	Higher activities	Large biomass, enhanced metabolism, more byproducts, process control difficult
	Resting cells	Workup easier, reduced metabolism, fewer byproducts	Lower activities
	Immobilized cells	Cell reuse possible	Lower activities

Table 1.3 Characteristics of resting versus fermenting cells

	Resting cells	Fermenting cells
Microbial cells	Resting	Growing
Reaction type	Short, catalytic	Long, biosynthetic
Number of reaction steps	Few	Many
Number of enzymes active	Few	Many
Starting material	Substrate	C + N source
Product	Natural or nonnatural	Only natural
Concentration tolerance	High	Low
Product isolation	Easy	Tedious
Byproducts	Few	Many

1.4   Enzyme Properties and Nomenclature

1.4.1   Structural Biology in a Nutshell

The polyamide chain of an enzyme is kept in a three-dimensional structure – the one with the lowest ΔG [107] – which is predominantly determined by its primary sequence.[13] For an organic chemist, an enzyme may be compared with a ball of yarn: Due to the natural aqueous environment, the hydrophilic polar groups (such as – COO^-, $-OH$, $-NH_3^+$, $-SH$, and $-CONH_2$) are mainly located on the outer surface of the enzyme in order to become hydrated, with the lipophilic substituents (the aryl and alkyl chains lacking affinity to water) being buried inside. As a consequence, the surface of an enzyme is covered by a tightly bound layer of water, which cannot be removed by lyophilization. This residual water, or 'structural water' (see Fig. 1.1), accounts for about 5–10% of the total dry weight of a freeze-dried enzyme [108]. It is tightly bound to the protein's surface by hydrogen bonds necessary to retain its three-dimensional structure and thus its catalytic activity. As a consequence, structural water is a distinctive part of the enzyme and differs significantly in its physical state from the 'bulk water' of the surrounding solution. There is restricted rotation of the 'bound water' and it cannot freely reorientate upon freezing, hence it's (formal) freezing point is about $-20\,°C$. Exhaustive drying of an enzyme (e.g., by chemical means) causes drastic changes in its conformation resulting in a loss of activity.

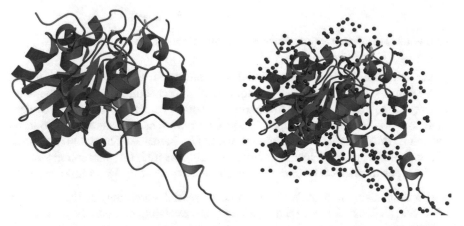

Fig. 1.1 Ribbon representation of the crystal structure of a *Candida antarctica* lipase B mutant bearing an inhibitor (*yellow*) bound to the active site (*left*). Structural water molecules are depicted as red dots (*right*) (PDB entry 3icw, courtesy of U. Wagner)

---

[13]The amino acid sequence of a protein is generally referred to as its 'primary structure', whereas the three-dimensional arrangement of the polyamide chain (the 'backbone') in space is called the 'secondary structure'. The 'tertiary structure' includes the arrangement of all atoms, i.e., the amino acid side chains are included, whereas the 'quaternary structure' describes the aggregation of several protein molecules to form oligomers.

The whole protein structure is stabilized by a large number of relatively weak binding forces such as van der Waals interactions[14] of aliphatic chains and π-π stacking of aromatic amino acids, which are predominantly located inside the protein core (Scheme 1.2). In contrast, stronger hydrogen bonds and salt bridges made up of Coulomb interactions are often close to the surface. As a consequence of the weak binding forces inside and strong bonds at the surface, in a rough approximation, enzymes have a soft core but a hard shell and thus represent delicate and soft (jellyfish-like) structures.

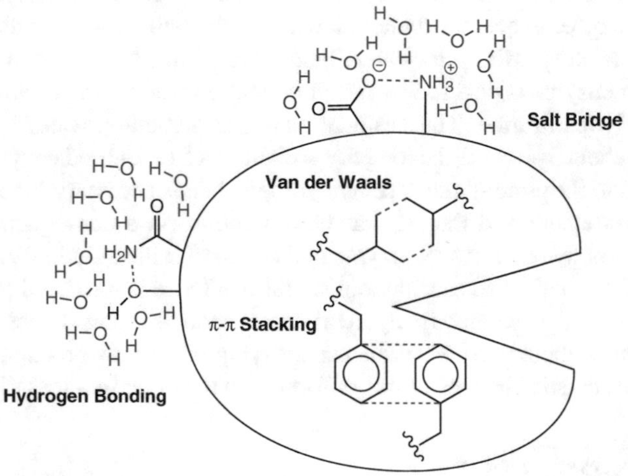

Scheme 1.2 Schematic representation of binding forces within a protein structure

The latter facilitates conformational movements during catalysis (such as the 'induced fit', see below) and is a prerequisite of the pronounced dynamic character of enzyme catalysis. Besides the main polyamide backbone, the only covalent bonds are –S–S– disulfide bridges. Enzymes are intrinsically unstable in solution and can be deactivated by denaturation, caused by increased temperature, extreme pH, or an unfavorable dielectric environment such as high salt concentrations.

The types of reaction leading to an enzyme's deactivation are as follows [109]:

– Rearrangement of peptide chains (due to partial unfolding) starts at around 40–50 °C. Most of these rearrangements are reversible and therefore relatively harmless.
– Hydrolysis of peptide bonds in the backbone, in particular adjacent to asparagine units, occurs at more elevated temperatures. Functional groups of amino acids (e.g. asparagine and glutamine residues) can be hydrolytically cleaved to furnish aspartic and glutamic acid, respectively. Both reaction mechanisms are favored by the presence of neighboring groups, such as glycine, which enable the formation of a cyclic intermediate. Thus, a negative charge (i.e., –COO⁻) is

---

[14]Also called London forces.

created from a neutral group ($-CONH_2$). In order to become hydrated, this newly generated carboxylate moiety pushes towards the protein's surface causing irreversible deactivation.

- Thiol groups may interchange the $-S-S-$ disulfide bridges, leading to a modification of covalent bonds within the enzyme.
- Elimination and oxidation reactions (often involving cysteine residues) cause the destruction of the protein.

Thermostable enzymes from thermophilic microorganisms show an astonishing upper operation limit of 60–80 °C and differ from their mesophilic counterparts by only small changes in primary structure [110]. The three-dimensional structure of such enzymes is often the same as those derived from mesophiles [111], but generally they possess fewer asparagine residues and more salt- or disulfide bridges. More recently, numerous (thermo)stable mutant enzymes have been obtained by genetic engineering. It is a common phenomenon, that an increased thermostability of proteins often goes in hand with an enhanced tolerance for organic solvents.

1.4.2 Mechanistic Aspects of Enzyme Catalysis

Generally, numerous groups – occasionally also coordinated metal ions and cofactors – in the active site of an enzyme cooperate to effect catalysis. Individual enzyme mechanisms have been elucidated in cases where the exact three-dimensional structure is known. For many enzymes used for the biotransformation of nonnatural organic compounds, assumptions are made about their molecular action. However, the logic of organic reaction mechanisms, which is based on thinking in terms of polarities – nucleophile/electrophile, acid/base, electron source/sink –, represents an excellent intellectual basis for explaining protein catalysis and organic chemists in particular will quickly see that there is nothing 'magic' about enzymes: they simply perform excellent organic chemistry.

The unparalleled catalytic power of enzymes has sparked numerous studies on mechanistic theories to provide a molecular understanding of enzyme catalysis for almost a century. Among the numerous theories and rationales, the most illustrative models for the organic chemist are discussed here [112–114].

'Lock-and-Key' Mechanism
The first proposal for a general mechanism of enzymatic action was developed by E. Fischer in 1894 [115, 116]. It assumes that an enzyme and its substrate mechanistically interact like a lock and key[15] (Fig. 1.2). Although this assumption was quite sophisticated at that time, it assumes a completely rigid enzyme structure.

---

[15]'To use a picture I want to say that enzyme and glucoside must go together like key and lock in order to exert a chemical effect upon each other', see [115] p. 2992.

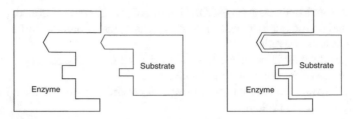

Fig. 1.2 Schematic representation of the 'lock-and-key' mechanism

Thus, it cannot explain why many enzymes do act on larger substrates, while they are inactive on smaller counterparts. Given Fischer's rationale, small substrates should be transformed at even higher rates than larger substrates since the access to the active site would be easier. Furthermore, the hypothesis cannot explain why many enzymes are able to convert not only their natural substrates but also numerous nonnatural compounds possessing different structural features. Consequently, a more sophisticated model had to be developed.

Induced-Fit Mechanism
This rationale, which takes into account that enzymes are not entirely rigid but rather represent delicate and soft structures, was developed by Koshland Jr. in the 1960s [117, 118].[16] It comprises that upon approach of a substrate during the formation of the enzyme-substrate complex, the enzyme can change its conformation under the influence of the substrate structure so as to wrap itself around its guest (Fig. 1.3). This phenomenon was denoted as the 'induced fit', which can be illustrated by the interaction of a hand (the substrate) and a glove (the enzyme). This advanced model can indeed explain why several structural features on a substrate are required in addition to the reactive group. These structural features may be located at quite a distance from the actual site of the reaction. The most typical 'induced-fit' enzymes are the lipases. They can convert an amazingly large variety of artificial substrates which possess structures which do not have much in common with the natural substrates – triglycerides.

A schematic representation of the 'induced-fit' mechanism is given in Fig. 1.3: Whereas A represents the reactive group of the substrate, X depicts the complementary reactive group(s) of the enzyme – the 'chemical operator'. Substrate part B forces the enzyme to adapt a different (active) conformation,[17] where the 'active' groups X of the enzyme are correctly positioned to effect catalysis. If part B is missing, no conformational change ('induced fit') takes place and thus the chemical operators stay in their inactive state.

---

[16]'A precise orientation of catalytic groups is required for enzyme action; the substrate may cause an appreciable change in the three-dimensional relationship of the amino acids at the active site, and the changes in protein structure caused by a substrate will bring the catalytic groups into proper orientation for reaction, whereas a non-substrate will not.' See [117].

[17]Conformational changes take place by hinge- and shear-type movements [119].

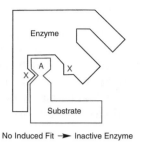
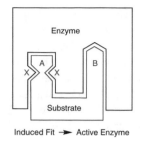

Fig. 1.3 Schematic representation of the 'induced-fit' mechanism

Desolvation and Solvation-Substitution Theory

In an attempt to explain the high catalytic rates of enzymes, M.J.S. Dewar developed a different rationale [120].[18] This so-called 'desolvation theory' assumes that the kinetics of enzyme reactions have much in common with those of gas-phase reactions. Hence, if a substrate enters the active site of the enzyme, it replaces *all* of the water molecules at the active site and a formal gas-phase reaction can take place which mimics two reaction partners interacting without a 'disturbing' solvent. In solution, the water molecules impede the approach of the partners, hence the reaction rate is reduced. This theory would, inter alia, explain why small substrate molecules are often slower converted than larger analogues, since the former are unable to replace *all* the water molecules at the active site.

The 'desolvation' theory has recently been extended by a 'solvation-substitution' theory [122]. It is based on the assumption that the enzyme cannot strip off the water surrounding the substrate to effect a 'desolvation', because this would be energetically unfavorable. Instead, the solvent is *displaced* by the environment of the active site thereby affecting 'solvation substitution'. Thus, the (often) hydrophobic substrate replaces the water with the (often) hydrophobic site of the enzyme which favors the formation of the enzyme-substrate complex. In addition, the replacement of water molecules within the active site during the substrate-approach decreases the dielectric constant within this area, which in turn enhances electrostatic enzyme-substrate interactions. The latter cause proper substrate-orientation thereby leading to an enhancement of catalytically productive events. This phenomenon is denoted as 'electrostatic catalysis' and was coined as 'Circe-effect' by W. P. Jencks.

Entropy Effects

A major reason for the exceptional catalytic efficiency of enzymes over small (chemical) catalysts is derived from the difference in size: The average (monomeric) protein used in biotransformations has a molecular weight of ~60,000 Da and a diameter of ~50 Å, while a typical chemical homogeneous catalyst weights ~600 Da and measures ~10 Å across. This allows the enzyme to enclose its substrate completely, with the catalytically active groups being positioned in

---

[18]A 'record' of rate acceleration factor of 10^{14} has been reported. See [121].

close proximity, while nonproductive movements are restricted. This rate-enhancing entropy effect is very similar to an *intra*molecular reaction, where the reacting groups are prearranged in close proximity. In contrast, *inter*molecular reactions are slower because the reacting partners have to find each other through diffusion [123–127]. In many proteins, the substrate has to approach the active site through a tunnel, which ensures its proper orientation, which enhances the number of productive catalytic events. In contrast, small chemical catalysts need many more collisions for successful catalysis.

Electrostatic and Covalent Catalysis
Most transition states involve charged intermediates, which are stabilized within the active site of an enzyme via ionic bonds in 'pockets' or 'holes' bearing a matching opposite charge. Such charges are provided by (Lewis acid-type) metal ions, typically Zn^{2+} or are located on acidic or basic amino acid side chains (such as His, Lys, Arg, Asp, or Glu). It is important to note that the (modest) pK_a of typical amino acid side chains, such as $-NH_3^+$ or $-CO_2^-$ can be substantially altered up to 2–3 pK_a-units through neighboring groups within the enzyme environment. As a consequence, the (approximately neutral) imidazole moiety of His can act as strong acid or base, depending on its molecular environment. Computer simulation studies suggested that in enzymes electrostatic effects provide the largest contribution to catalysis [128]. As a prominent example, the tetrahedral intermediate of carboxyl ester hydrolysis is stabilized in serine hydrolases by the so-called 'oxyanion hole' (Scheme 2.1).

Many enzymes form covalent bonds with their substrates during catalysis, such as the acyl-enzyme intermediate in carboxyl ester hydrolysis (Scheme 2.1) or the glycol monoester intermediate in epoxide hydrolysis (Scheme 2.84). Despite the covalent enzyme-substrate bond, such species are metastable and should be regarded as 'activated intermediates'. Some enzymes utilize cofactors, such as pyridoxal phosphate (PLP) or thiamine diphosphate (TPP), to form covalent intermediates during catalysis.

Since some enzymes can act faster than what would be predicted by the 'over-the-barrier' transition-state model ($\Delta\Delta G^{\neq}$), 'through-the-barrier' quantum tunneling of protons or electrons has been postulated [129, 130].

Three-Point Attachment Rule
This rationale to explain the enantioselectivity of enzymes was suggested by A.G. Ogston [131]. Since chirality is a quality of space, a substrate must be positioned firmly in three dimensions within the active site of an enzyme in order to ensure spatial recognition and to achieve a high degree of enantioselection. As a consequence, at least three different points of attachment of the substrate onto the active site are required.[19]

Although the majority of chiral molecules subjected to biotransformations possess central chirality located on an sp^3-carbon atom, all types of chiral molecules can be 'recognized', including compounds bearing a stereogenic sp^3-

---

[19]The following rationale was adapted from [132].

heteroatom such as phosphorus or sulfur [133] or possess axial or planar chirality involving sp^2- or sp-carbon atoms [134], respectively (Scheme 1.3). The use of the Three-Point Attachment Rule is described as follows (Figs. 1.4–1.6):

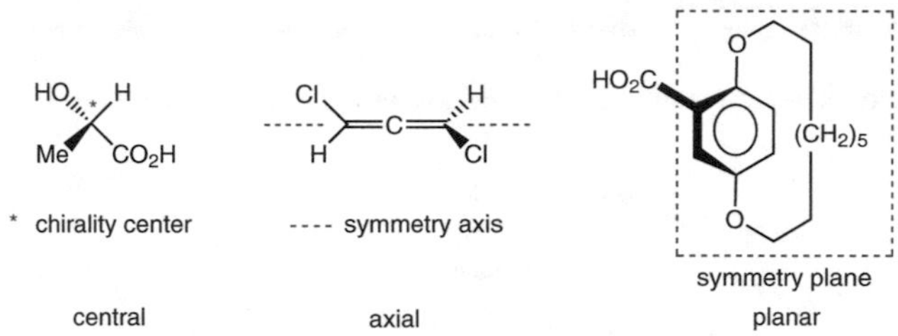

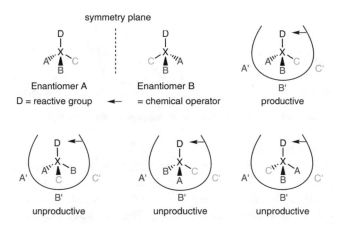

Scheme 1.3   Examples for central, axial, and planar chirality

Enantiomer differentiation (Fig. 1.4)   Substrate enantiomer A is a good substrate because it allows an optimal interaction of its groups (A, B, C) with the complementary binding sites of the enzyme (A′, B′, C′) to ensure productive binding by optimal orientation of the reactive group (D) towards the chemical operator (✓) which is required for a successful transformation. In contrast, substrate enantiomer B is a poor substrate, because optimal binding and orientation of the reactive group D is impossible regardless of its orientation in the active site. Thus, poor catalysis will be observed.

Fig. 1.4   Schematic representation of enzymatic enantiomer discrimination

Enantiotopos differentiation (Fig. 1.5)   If a prochiral substrate (C), bearing two chemically identical but stereochemically different enantiotopic groups (A), is involved, the model can be applied to rationalize the favored transformation of one of the two groups A leading to an 'enantiotopos differentiation'. Due to the preferred binding mode, group A on top (highlighted in bold) will react. Enantiotopos and -face nomenclature is depited in Scheme 1.4.

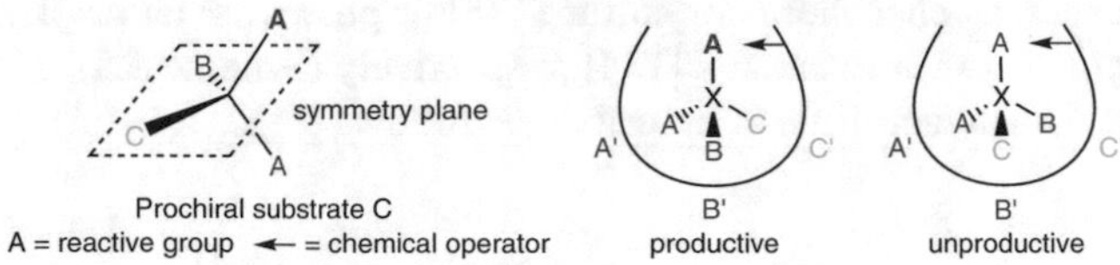

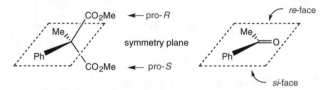

Prochiral substrate C
A = reactive group   ◄— = chemical operator

B'
productive

B'
unproductive

Fig. 1.5 Schematic representation of enzymatic enantiotopos discrimination

CO₂Me ◄— pro-R

symmetry plane

CO₂Me ◄— pro-S

re-face

si-face

Scheme 1.4 Enantiotopos and -face nomenclature

Enantioface discrimination The ability of enzymes to distinguish between two enantiomeric sides of a prochiral substrate (D) – an 'enantioface differentiation' – is illustrated in Fig. 1.6. An optimal match between the functional groups of prochiral substrate D leads to an attack of the chemical operator on the central atom X from the top-side. The mirror image orientation (also called 'alternative fit') of substrate D in the active site leads to a mismatch in binding, thus an attack by the chemical operator from the opposite side is disfavored.

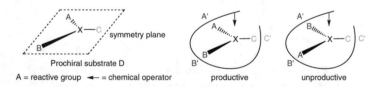

Prochiral substrate D
A = reactive group   ◄— = chemical operator

B'
productive

B'
unproductive

Fig. 1.6 Schematic representation of enzymatic enantioface discrimination

Kinetic Reasons for Selectivity

As every other catalyst, an enzyme (Enz) accelerates the reaction by lowering the energy barrier between substrate (S) and product (P) – the activation energy (E_a) [135]. The origin of this catalytic power – the rate acceleration – is attributed to the transition-state stabilization of the reaction [136], assuming that the catalyst binds more strongly to the transition state [S$^{\neq}$] than to the ground state of the substrate, by a factor approximately equal to the acceleration rate [137] (Fig. 1.7). In contrast to chemical catalysts, enzymes bind substrates very strong and the dissociation constant for an [EnzS]$^{\neq}$ complex has been estimated to be in the range of 10^{-20} molar [138]. In terms of reaction velocity, a $\Delta G^{\neq}$ of ~17 kcal/M translates into a catalytic rate of ~1 s^{-1}. Adding (or subtracting) 1.4 kcal/M from this value reduces (or enhances) the rate by about one order of magnitude.

Virtually all stereoselectivities of enzymes originate from the energy difference in enzyme-transition state complexes [EnzS]$^{\neq}$ (Fig. 1.8). In an enantioselective reaction, both of the enantiomeric substrates A and B (Fig. 1.4) or the two forms of mirror-image orientation of a prochiral substrate involving its enantiotopic groups or faces (Figs. 1.5 and 1.6) compete for the active site of the enzyme. Due to the chiral environment of the active site, *diastereomeric* enzyme-substrate complexes [EnzA] and [EnzB] are formed, which possess different values of free energy (ΔG) for their respective transition states [EnzA]$^{\neq}$ and [EnzB]$^{\neq}$. The result is a difference in activation energy ($\Delta\Delta G^{\neq}$) of enantiomers or the 'enantiomeric orientations' of prochiral compounds, respectively. As a consequence, one enantiomer (or orientation) will be transformed faster than the other. This process is generally referred to as 'chiral recognition'.

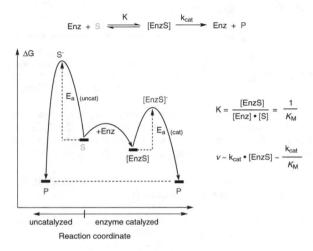

Fig. 1.7 Energy diagram of catalyzed versus uncatalyzed reaction. *Enz* enzyme, *S* substrate, *P* product, *[EnzS]* enzyme-substrate complex, *K* equilibrium constant for [EnzS] formation, k_{cat} reaction rate constant for [EnzS] $\rightarrow$ Enz + P, E_a activation energy, $\neq$ denotes a transition state, K_M Michaelis–Menten constant, v reaction velocity

The value of this difference in free energy, expressed as $\Delta\Delta G^{\neq}$, is a direct measure for the selectivity of the reaction which in turn determines the ratio of the individual reaction rates (v_A, v_B) of enantiomeric substrates A and B (or the two enantiotopic faces or groups competing for the active site of the enzyme, Fig. 1.8).[20] These values are of great importance since they determine the optical purity of the product. $\Delta\Delta G^{\neq}$ is composed of an enthalpy ($\Delta\Delta H^{\neq}$) and an entropy term ($\Delta\Delta S^{\neq}$). The enthalpy of activation is usually dominated by the breakage and formation of bonds when the substrate is transformed into the product. The entropy contribution includes the energy balance from the 'order' of the system, i.e., orienting the reactants, changes in conformational flexibility during the 'induced-fit', and various concentration and solvation effects.

---

[20]The individual reaction rates v_A and v_B correspond to $v_A = (k_{cat}/K_M)_A \cdot [\text{Enz}] \cdot [\text{A}]$ and $v_B = (k_{cat}/K_M)_B \cdot [\text{Enz}] \cdot [\text{B}]$, respectively, according to Michaelis–Menten kinetics. The ratio of the individual reaction rates of enantiomers is an important parameter for the description of the enantioselectivity of a reaction: $v_A/v_B = E$ ('Enantiomeric Ratio', see Sect. 2.1.1).

Table 1.4 lists some representative values of enantiomeric excess of product (e.e.) corresponding to a given $\Delta\Delta G^{\neq}$ of the reaction.

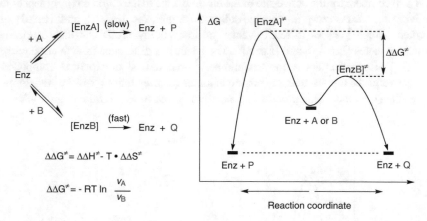

Fig. 1.8 Energy diagram for an enzyme-catalyzed enantioselective reaction. *Enz* enzyme, *A and B* enantiomeric substrates, *P and Q* enantiomeric products; *[EnzA] and [EnzB]* diastereomeric enzyme-substrate complexes; $^{\neq}$ denotes a transition state, $\Delta\Delta G$, $\Delta\Delta H$ and $\Delta\Delta S$ free energy, enthalpy, and entropy difference, resp., *R* gas constant, *T* temperature, v_A *and* v_B reaction velocities of A and B, resp.

Table 1.4   Free energy values $\Delta\Delta G^{\neq}$ for representative optical purities of product (e.e.) and the corresponding ratio of reaction rates of enantiomers (v_A, v_B)

$\Delta\Delta G^{\neq}$ [kcal/mol]	v_A/v_B	e.e. [%]
0.118	1.2	10
0.651	3	50
1.74	19	90
2.17	39	95
3.14	199	99
4.50	1999	99.9

$e.e.[\%] = \frac{P-Q}{P+Q} \times 100$

Due to the logarithmic dependence between $\Delta\Delta G^{\neq}$ and e.e. of the product (Fig. 1.8), even a very small difference in free energy (e.g., 0.65 kcal/mol) can lead to a considerable enantiomeric excess of product (50%) (Table 1.4) and only a modest 1.75 kcal/mol is required to reach ~90% e.e. However, the same amount of $\Delta\Delta G^{\neq}$ is roughly needed to push this value further to 99%! For virtually absolute selectivities, however, $\Delta\Delta G^{\neq}$ has to be considerably higher ($\geq$4.50 kcal/mol). Hence, the influence of $\Delta\Delta G^{\neq}$ on the stereoselectivity of the reaction is very sensitive, which makes accurate predictions virtually impossible.

1.4.3   Classification and Nomenclature

At present about 6500 enzymes have been recognized by the International Union of Biochemistry and Molecular Biology (IUBMB) [139–142] and if the prediction that there are about 25,000 enzymes existing in Nature is true [143], the bulk of this vast reservoir of biocatalysts still remains to be discovered and is waiting to be used. However, only a minor fraction of the enzymes already investigated (~10%) is commercially available. However, this number is steadily increasing.

For identification purposes, every enzyme has a four-digit number in the general form [EC A.B.C.D], where EC stands for 'Enzyme Commission'. The following properties are encoded:

A. denotes the main type of reaction (see Table 1.5);
B. stands for the subtype, indicating the substrate class or the type of transferred molecule;
C. indicates the nature of the co-substrate;
D. is the individual enzyme number.

As depicted in Table 1.5, enzymes have been classified into six categories according to the type of reaction they can catalyze. At first glance, it would seem advantageous to keep this classification throughout this book, since organic chemists are used to thinking in terms of reaction principles. Unfortunately, this does not work in practice for the following reasons: Due to the varying tolerance for nonnatural substrates, the importance for practical applications in organic synthesis is not at all evenly distributed amongst the different enzyme classes, as may be seen from the 'utility' column in Table 1.5 (compare Chap. 4). Furthermore, due to the widespread use of crude enzyme preparations (containing more than one active biocatalyst), one often does not know which enzyme is actually responsible for the biotransformation. Last but not least, there are many useful reactions which are performed with whole microbial cells, for which it can only be speculated as to which of the numerous enzymes in the cell is actually involved in the transformation.

Table 1.5 Classification of enzymes

Enzyme class	Number		Reaction type	Utility[a]
	Classified	Available		
1. Oxidoreductases	~2000	~100	Oxidation/reduction: oxygenation of C–H, C–C, C=C bonds, or overall removal or addition of hydrogen equivalents	+++
2. Transferases	~1900	~100	Transfer of groups: aldehydic, ketonic, acyl, sugar, phosphoryl, methyl, NH_3	++
3. Hydrolases	~1700	~200	Hydrolysis/formation of esters, amides, lactones, lactams, epoxides, nitriles, anhydrides, glycosides, organohalides	+++
4. Lyases	~700	~50	Addition/elimination of small molecules on C=C, C=N, C=O bonds	++
5. Isomerases	~250	~10	Isomerizations: rearrangement, epimerization, racemization, cyclization	+
6. Ligases	~200	~10	Formation/cleavage of C–O, C–S, C–N, C–C bonds with concomitant triphosphate cleavage	±

[a]The estimated utility of an enzyme class for the transformation of nonnatural substrates ranges from +++ (very useful) to ± (little use) [144]. (Based on the biotransformation database of Kroutil and Faber (2016) ~17,000 entries.)

One particular warning should be given concerning catalytic activities which are measured in several different systems:

According to the SI system, catalytic activity is defined by the "katal" (1 kat = 1 mol s^{-1} of substrate transformed). Since its magnitude is far too big for practical application, it has not been widely accepted. The transformation of one mole of an organic compound within one second resembles an industrial-scale reaction and is thus not suited to describe enzyme kinetics. As a consequence, a more appropriate standard – The 'International Unit' (1 I.U. = 1 μmol of substrate transformed per min) – has been defined. Unfortunately, other units such as nmol/min or nmol/h are also common, mainly to make the numbers of low catalytic activity look bigger. After all, it should be kept in mind that the activities using *nonnatural* substrates are often significantly below the values which were determined for *natural* substrates. As a rule of thumb, enzymes from the primary metabolism typically display k_{cat} values of ~100 s^{-1} because they are highly evolved and hence specific for a given substrate. On the other hand, proteins from the secondary metabolism show a broad substrate tolerance and are about one order of magnitude slower.

A comparison of the activity of different enzyme preparations is only possible if the assay procedure is performed exactly in the same way. Since most enzyme suppliers use their own experimental setup, an estimation of the cost/activity ratio of biocatalysts from various commercial sources is seldom possible by using published data, unless activity data were determined independently.

The *catalytic power* of a (bio)catalyst can be conveniently described by the so-called 'turnover frequency' (TOF), which has the dimension of [time^{-1}] (Table 1.1). It indicates the number of substrate molecules which are converted by a single (bio)catalyst molecule in a given period of time, (assuming that each catalyst molecule has a single active site) [145].[21] For biochemical reactions, this unit is the second, for the slower chemical catalysts, the minute (or the hour, for very slow catalysts) is usually preferred. For the majority of enzymes used in biotransformations, TOFs are within the range of 10–1000 s^{-1}, whereas the respective values for chemical catalysts are one to two orders of magnitude lower.

$$\text{Turnover Frequency (TOF)} = \frac{\text{Number of Substrates Converted}}{\text{Number of Catalyst Molecules} \times \text{Time}} \left[\frac{\text{Mol}}{\text{Mol} \times \text{Time}}\right]$$

The *catalytic productivity* of a (bio)catalyst is characterized by the dimensionless 'turnover number' (TON). It denotes the number of substrate molecules converted per number of catalyst molecules used within a given time, e.g. required to complete a reaction. TONs for enzymes typically range from 10^3 to 10^6 [146].

$$\text{Turnover Number (TON)} = \frac{\text{Number of Substrates Converted}}{\text{Number of Catalyst Molecules}} \left[\frac{\text{Mol}}{\text{Mol}}\right]$$

---

[21]For enzymes obeying Michaelis–Menten kinetics the TON is equal to $1/k_{cat}$.

The comparison of TOFs and TONs of different (bio)catalysts should be exercised with great caution [147], since TOFs only indicate how fast the catalyst acts at the onset of the reaction within a short time span, but it does not tell anything about its long-time performance.

The *operational stability* of a (bio)catalyst is described by the dimensionless 'total turnover number' (TTN), which is determined by the moles of product formed by the amount of catalyst spent. In other words, it stands for the amount of product which is produced by a given amount of catalyst during its whole lifetime. If the TONs of repetitive batches of a reaction are measured until the catalyst is dead, the sum of all TONs would equal to the TTN. TTNs are also commonly used to describe the efficiency of cofactor recycling systems.

$$\text{Total Turnover Number (TTN)} = \frac{\text{Number of Substrates Converted}}{\text{Number of Catalyst Molecules}} \left[\frac{\text{Mol}}{\text{Mol}}\right] \text{(Lifetime)}$$

The efficiency of microbial transformations (where the catalytic activity of enzymes involved cannot be measured) is characterized by the so-called 'productivity number' (PN) [148], defined as

$$\text{Productivity Number (PN)} = \frac{\text{Amount of Product Formed}}{\text{Biocatalyst (dry weight)} \times \text{Time}} \left[\frac{\text{Mol}}{\text{g} \times \text{Time}}\right]$$

which is the amount of product formed by a given amount of whole cells (dry weight) within a certain period of time. This number resembles the specific activity as defined for pure enzymes, but also includes several other important factors such as inhibition, transport phenomena, and concentration.

1.4.4   Coenzymes

A remarkable proportion of synthetically useful enzymes require cofactors (coenzymes),[22] which have a molecular weight of only few hundred Da, in contrast to the typical 50,000 Da of enzymes used in biotransformations. Coenzymes serve as molecular shuttles to stabilize and transfer sensitive 'chemical reagents', for instance redox-equivalents (e.g., complex hydrides, electrons, and activated oxygen species), toxic intermediates (ammonia) and water-sensitive carbanion species. Alternatively, 'chemical energy' is stored in energy-rich functional groups, such as acid anhydrides in ATP or PAPS. As a rule of thumb, enzymes are bound to their

---

[22] A 'cofactor' is tightly bound to an enzyme (e.g., FAD), whereas a 'coenzyme' can dissociate into the medium (e.g., NADH). In practice, however, this distinction is not always made in a consequent manner.

natural cofactors, which cannot be replaced by more economical man-made chemical substitutes.

Table 1.6 Common coenzymes required for biotransformations

Coenzyme	Reaction type	Recycling[a]
NAD$^+$/NADH	Carbonyl reduction &	(+) [+++]
NADP$^+$/NADPH	Alcohol oxidation	(+) [++]
ATP[b]	Phosphorylation	(+) [+]
SAM	C$_1$-alkylation	(+) [±]
Acetyl-CoA	C$_2$-alkylation	(+) [±]
Flavins	Baeyer-Villiger-, N- & S-oxidation, C=C reduction	(−)[c]
Pyridoxal-phosphate	Transamination, racemization	(−)
Thiamine diphosphate	C–C ligation	(−)
Metal-porphyrins	Peroxidation, oxygenation	(−)[c]
Biotin	Carboxylation	(−)

[a]Recycling of a cofactor is necessary (+) or not required (−), the feasibility of which is indicated in square brackets ranging from 'simple' [+++] to 'complicated' [±]
[b]For other triphosphates, such as GTP, CTP, and UTP, the situation is similar
[c]Many flavin- and metal porphyrin-dependent mono- or dioxygenases require additional NAD(P)H as an indirect reducing agent

Some cofactors are rather sensitive molecules and are gradually destroyed due to undesired side reactions occurring in the medium, in particular NAD(P)H and ATP. Cofactors are too expensive to be used in the stoichiometric amounts formally required. Accordingly, when coenzyme-dependent enzymes are employed, the corresponding coenzymes are used in catalytic amounts in conjunction with an efficient and inexpensive in-situ regeneration system. Some cofactor recycling methods are highly developed and are applicable to industrial scale, others are still problematic (Table 1.6). Fortunately, some coenzymes are tightly bound to their respective enzymes so that external recycling is not required. Whereas the redox potential of NADH and NADPH is largely independent of the type of enzyme, the potential of flavin cofactors is significantly determined by the protein, to which it is (covalently or noncovalently) attached. Consequently, the redox capabilities of flavoproteins encompass a broader range of reactions compared to nicotinamide-depending enzymes.

Many enzymes require coordinated metals such as (Lewis-acid type) Zn, Ca, Mg or redox-active Fe, Cu, Mn. Ni, Co, V, W and Mo are less common. In most cases, metals are tightly bound to the enzyme and are not an issue if they are supplied to the medium.

1.4.5 Enzyme Sources

The large majority of enzymes used for biotransformations in organic chemistry are employed in a crude form and are relatively inexpensive. Many preparations typically contain only about 1–30% of actual enzyme, the remainder being inactive proteins, stabilizers, buffer salts, or carbohydrates from the fermentation broth from which they have been isolated. This may even be an advantage, because crude preparations are often more stable than purified enzymes.

Pure enzymes are usually very expensive and are mostly sold by the unit, while crude preparations are often shipped in kg amounts. Since the techniques for protein purification through His- or Strep-tagging are becoming easier and more economic, the use of (partially) purified enzymes in biotransformations is rapidly increasing.

The main sources of enzymes for biotransformations are as follows [149, 150]:

– The detergent industry produces many proteases and lipases in huge amounts. These are used as additives for detergents to effect the hydrolysis of proteinogenic and fatty impurities in the laundry process at neutral pH and modest temperatures.
– The food industry uses proteases and lipases for meat and cheese processing and for the transesterification of fats and oils [151]. Glycosidases and decarboxylases are predominantly employed in the brewing and baking industries, respectively.
– Numerous enzymes can be isolated from cheap mammalian organs, such as kidney or liver, or from slaughter waste.
– Only a small proportion of enzymes used in biotransformations is obtained from plant sources, such as fruits (e.g., fig, papaya, pineapple) and vegetables (e.g., tomato, potato). While sensitive plant cell cultures were used in the past, they are nowadays cloned into a sturdy host microorganism (*Escherichia coli*, *Pichia*, *Aspergillus*) for their production.
– The richest and most convenient sources of enzymes are microorganisms. An impressive number of biocatalysts are derived from bacterial and (lower) fungal origin by cheap fermentation.

Primary and secondary metabolism The following guidelines help to enhance the hit-rate in search of a suitable (microbial) enzyme for the transformation of non-natural compounds: [152]

The central (primary) metabolism provides energy to sustain life and growth by breakdown of carbon sources, e.g. via glycolysis of a carbohydrate yielding acetyl-CoA, which is burnt in the Krebs-cycle to CO_2 and H_2O. In order to optimize the metabolic flux, enzymes from primary metabolic pathways always have been under heavy selection pressure to improve their catalytic efficiency and specificity for their substrates. Although they appear ideal for preparative-scale biotransformations regarding their catalytic rates ($\sim 100 \text{ s}^{-1}$), they possess a very narrow substrate spectrum. In other words, they are fast, but only on a single natural substrate. In contrast, for the biotransformation of non-natural compounds, catalytic allrounders are required, which are predominantly found in secondary metabolic pathways

where they take care of defence (through biosynthesis of antibiotics) and detoxification of xenobiotics. The utility of these enzymes is enhanced by a broad substrate
spectrum encompassing also non-natural compounds. The down-side is their modest degree of evolution, which makes them slower by about one order of magnitude.
However, this drawback makes them highly evolvable through genetic engineering.
As shown in Fig. 1.9, enzymes from primary metabolism (A, B) are highly
optimized for their substrate and mutation predominantly produces less active
variants. In contrast, the soft fitness landscape of enzymes from secondary metabolism (C, D) allows to modify their catalytic properties so as to adapt them for a
wider range of (non-natural) substrates. When proteins are within a certain range of
sequence-relationship, an overlap of catalytic activities – promiscuity (P) – can be
explored [153]. When working with wild-type organisms, it is advisable to harvest
them towards the late stage of the exponential growth phase, when the cheap carbon
source is consumed and the 'hungry' cells switch on their enzyme outfit from the
secondary metabolism to survive.

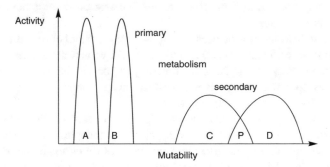

Fig. 1.9 Catalytic activity and evolvability of enzymes from primary and secondary metabolic
pathways.

References

1. Goodhue CT (1982) Microb. Transform. Bioact. Compd. 1: 9
2. Roberts SM, Turner NJ, Willetts AJ, Turner MK (1995) Introduction to Biocatalysis Using
   Enzymes and Micro-organisms. Cambridge University Press, Cambridge
3. Thomas JM, Harris KDM (2016) Energy Environ. Sci. 9: 687.
4. Wittcoff HA, Reuben BG, Plotkin JS (2013) Industrial Organic Chemicals, Wiley, 3rd ed.,
   p. 637.
5. http://www.essentialchemicalindustry.org/the-chemical-industry/the-chemical-industry.html
   accessed Nov 20, 2016.
6. OECD (2011) Future Prospects for Industrial Biotechnology, OECD Publishing, http://dx.
   doi.org/10.1787/9789264126633-en; ISBN: 9789264126633 (PDF); 9789264119567 (print).
   p. 24.
7. Rozzell JD (1999) Bioorg. Med. Chem. 7: 2253.
8. Baross JA, Deming JW (1983) Nature 303: 423
9. Hough DW, Danson MJ (1999) Curr. Opin. Chem. Biol. 3: 39

10. Prieur D (1997) Trends Biotechnol. 15: 242
11. Tufvesson P, Lima-Ramos J, Nordblad M, Woodley JM (2011) Org. Proc. Res. Dev. 15: 266.
12. Feyerabend P (1988) Against Method. Verso, London
13. Laane C, Boeren S, Vos K, Veeger C (1987) Biotechnol. Bioeng. 30: 81
14. Carrea G, Ottolina G, Riva S (1995) Trends Biotechnol. 13: 63
15. Bell G, Halling PJ, Moore BD, Partridge J, Rees DG (1995) Trends Biotechnol. 13: 468
16. Koskinen AMP, Klibanov AM (eds) (1996) Enzymatic Reactions in Organic Media. Blackie Academic & Professional, London
17. Gutman AL, Shapira M (1995) Synthetic Applications of Enzymatic Reactions in Organic Solvents. In: Fiechter A (ed) Adv. Biochem. Eng. Biotechnol., vol. 52, pp 87–128, Springer, Berlin Heidelberg New York
18. Wolfenden R, Snider MJ (2001) Acc. Chem. Res. 34: 938
19. Menger FM (1993) Acc. Chem. Res. 26: 206
20. Zechel DL, Withers SG (2000) Acc. Chem. Res. 33: 11
21. Williams DH, Stephens E, Zhou M (2003) Chem. Commun. 1973.
22. Garcia-Junceda E (2008) Multi-step Enzyme Catalysis. Wiley-VCH, Weinheim
23. Bornscheuer UT, Kazlauskas RJ (2004) Angew. Chem., Int. Ed. 43: 6032
24. Hult K, Berglund P (2007) Trends Biotechnol. 25: 231
25. Walsh C (2001) Nature 409: 226
26. Khersonsky O, Tawfik DS (2010) Ann. Rev. Biochem. 79: 471.
27. O'Brien PJ, Herschlag D (1999) Chem. Biol. 6: R91
28. Kazlauskas RJ (2005) Curr. Opin. Chem. Biol. 9: 195
29. Penning TM, Jez JM (2001) Chem. Rev. 101: 3027
30. Sih CJ, Abushanab E, Jones JB (1977) Ann. Rep. Med. Chem. 12: 298
31. Boland W, Frößl C, Lorenz M (1991) Synthesis 1049
32. Schmidt-Kastner G, Egerer P (1984) Amino Acids and Peptides. In: Kieslich K (ed) Biotechnology. Verlag Chemie, Weinheim, vol 6a, pp 387–419
33. Gutman AL, Zuobi K, Guibe-Jampel E (1990) Tetrahedron Lett. 31: 2037
34. Taylor SJC, Sutherland AG, Lee C, Wisdom R, Thomas S, Roberts SM, Evans C (1990) J. Chem. Soc., Chem. Commun. 1120
35. Zhang D, Poulter CD (1993) J. Am. Chem. Soc. 115: 1270
36. Yamamoto Y, Yamamoto K, Nishioka T, Oda J (1988) Agric. Biol. Chem. 52: 3087
37. Leak DJ, Aikens PJ, Seyed-Mahmoudian M (1992) Trends Biotechnol. 10: 256
38. Nagasawa T, Yamada H (1989) Trends Biotechnol. 7: 153
39. Mansuy D, Battoni P (1989) Alkane Functionalization by Cytochromes P450 and by Model Systems Using O_2 or H_2O_2. In: Hill CL (ed) Activation and Functionalization of Alkanes. Wiley, New York
40. Lemiere GL, Lepoivre JA, Alderweireldt FC (1985) Tetrahedron Lett. 26: 4527
41. Phillips RS, May SW (1981) Enzyme Microb. Technol. 3: 9
42. May SW (1979) Enzyme Microb. Technol. 1: 15
43. Boyd DR, Dorrity MRJ, Hand MV, Malone JF, Sharma ND, Dalton H, Gray DJ, Sheldrake GN (1991) J. Am. Chem. Soc. 113: 667
44. Walsh CT, Chen YCJ (1988) Angew. Chem., Int. Ed. 27: 333
45. Servi S (1990) Synthesis 1
46. Koszelewski D, Lavandera I, Clay D, Guebitz G, Rozzell D, Kroutil W (2010) Angew. Chem., Int. Ed. 47: 9337
47. Findeis MH, Whitesides GM (1987) J. Org. Chem. 52: 2838
48. Akhtar M, Botting NB, Cohen MA, Gani D (1987) Tetrahedron 43: 5899
49. Effenberger F, Ziegler T (1987) Angew. Chem., Int. Ed. 26: 458
50. Neidleman SL, Geigert J (1986) Biohalogenation: Principles, Basic Roles and Applications. Ellis Horwood, Chichester
51. Stecher H, Twengg M, Ueberbacher BJ, Remler P, Schwab H, Griengl H, Gruber-Khadjawi M (2009) Angew. Chem., Int. Ed. 48: 9546

52. Hayashi A, Saitou H, Mori T, Matano I, Sugisaki H, Maruyama K (2012) Biosci. Biotechnol. Biochem. 76: 559.
53. Maechling S, Lindell S (2006) Targets Heterocycl. Syst. 10: 66.
54. Buist PH, Dimnik GP (1986) Tetrahedron Lett. 27: 1457
55. Glueck SM, Gümüs S, Fabian WMF, Faber K (2010) Chem. Soc. Rev. 39: 313.
56. Ohta H (1999) Adv. Biochem. Eng. Biotechnol. 63: 1
57. Schwab JM, Henderson BS (1990) Chem. Rev. 90: 1203
58. Fuganti C, Grasselli P (1988) Baker's Yeast Mediated Synthesis of Natural Products. In: Whitaker JR, Sonnet PE (eds) Biocatalysis in Agricultural Biotechnology, ACS Symposium Series, vol 389, pp 359–370
59. Toone EJ, Simon ES, Bednarski MD, Whitesides GM (1989) Tetrahedron 45: 5365
60. Kitazume T, Ikeya T, Murata K (1986) J. Chem. Soc., Chem. Commun. 1331
61. Pohl M, Lingen B, Müller M (2002) Chem. Eur. J. 8: 5288
62. Durchschein K, Ferreira-da Silva B, Wallner S, Macheroux P, Kroutil W, Glueck SM, Faber K (2010) Green Chem. 12: 616
63. Tyagi V, Fasan R (2016) Angew. Chem. Int. Ed. 55: 2512.
64. Schober M, Toesch M, Knaus T, Strohmeier GA, van Loo B, Fuchs M, Hollfelder F, Macheroux P, Faber K (2013) Angew. Chem. Int. Ed. 52: 3277.
65. Wuensch C, Lechner H, Glueck SM, Zangger K, Hall M, Faber K (2013) ChemCatChem 5: 1744.
66. Dickschat JS (2016) Nat. Prod. Rep. 33: 87.
67. Williams RM (2002) Chem. Pharm. Bull. 50: 711
68. Oikawa H, Katayama K, Suzuki Y, Ichihara A (1995) J. Chem. Soc., Chem. Commun. 1321
69. Pohnert G (2001) ChemBioChem 2: 873
70. Klas K, Tsukamoto S, Sherman DH, Williams RM (2015) J. Org. Chem. 80: 11672.
71. Abe I, Rohmer M, Prestwich GD (1993) Chem. Rev. 93: 2189
72. Ganem B (1996) Angew. Chem., Int. Ed. 35: 936
73. Sweers HM, Wong CH (1986) J. Am. Chem. Soc. 108: 6421
74. Bashir NB, Phythian SJ, Reason AJ, Roberts SM (1995) J. Chem. Soc., Perkin Trans. 1, 2203
75. Jung G (1992) Angew. Chem., Int. Ed. 31: 1457
76. Sih CJ, Wu SH (1989) Topics Stereochem. 19: 63
77. Fischer E (1898) Zeitschr. physiol. Chem. 26: 60
78. Crossley R (1992) Tetrahedron 48: 8155
79. De Camp WH (1989) Chirality 1: 2
80. Ariens EJ (1988) Stereospecificity of Bioactive Agents. In: Ariens EJ, van Rensen JJS, Welling W (eds) Stereoselectivity of Pesticides. Elsevier, Amsterdam, pp. 39–108
81. Crosby J (1997) Introduction. In: Collins AN, Sheldrake GN, Crosby J (eds) Chirality in Industry II, pp 1–10, Wiley, Chichester
82. Millership JS, Fitzpatrick A (1993) Chirality 5: 573
83. Borman S (1992) Chem. Eng. News, June 15: 5
84. FDA (1992) Chirality 4: 338
85. US Food & Drug Administration (2004) Pharmaceutical Current Good Manufacturing Practices (cGMPs) for the 21st Century – a Risk-Based Approach: Final Report
86. Farina V, Reeves JT, Senanayake CH, Song JJ (2006) Chem. Rev. 106: 2734
87. Agranat H, Caner H, Caldwell J (2002) Nat. Rev. Drug Discov. 1: 753
88. Sheldon RA (1993) Chirotechnology. Marcel Dekker, New York
89. Collins AN, Sheldrake GN, Crosby J (eds) (1992, 1997) Chirality in Industry, 2 vols. Wiley, Chichester
90. Morrison JD (ed) (1985) Chiral catalysis. In: Asymmetric Synthesis, vol 5. Academic Press, London
91. Hanessian S (1983) Total Synthesis of Natural Products: the 'Chiron' Approach. Pergamon Press, Oxford

92. Scott JW (1984) Readily available chiral carbon fragments and their use in synthesis. In: Morrison JD, Scott JW (eds) Asymmetric Synthesis. Academic Press, New York, vol 4, pp 1–226
93. Margolin AL (1993) Enzyme Microb. Technol. 15: 266
94. Mugford P, Wagner U, Jiang Y, Faber K, Kazlauskas R (2008) Angew. Chem. Int. Ed. 47: 8782
95. Schuster M, Aaviksaar A, Jakubke HD (1990) Tetrahedron 46: 8093
96. Yeh Y, Feeney RE (1996) Chem. Rev. 96: 601
97. Phillips RS (1996) Trends Biotechnol. 14: 13
98. Li C-J (ed.) (2010) Handbook of Green Chemistry, vol. 5, Wiley VCH, Weinheim.
99. Klibanov AM (1990) Acc. Chem. Res. 23: 114
100. Paul CE, Hollmann F (2016) Appl. Microbiol. Biotechnol. 100: 4773.
101. Schrewe M, Julsing MK, Bühler B, Schmid A (2013) Chem. Soc. Rev. 42: 6346.
102. Willrodt C, Karande R, Schmid A, Julsing MK (2015) Corr. Opin. Biotechnol. 35: 52.
103. Sun J, Alper HS (2014) J. Ind. Microbiol. Biotechnol. 42: 423.
104. Porro D, Branduardi P, Sauer M, Mattanovich D (2014) Curr. Opin. Biotechnol. 30: 101.
105. Jones JA, Toparlak ÖD, Koffas MA (2015) Curr. Opin. Biotechnol. 33: 52.
106. Paddon CJ, Keasling JD (2014) Nat. Rev. Microbiol. 12: 355.
107. Anfinsen CB (1973) Science 181: 223
108. Cooke R, Kuntz ID (1974) Ann. Rev. Biophys. Bioeng. 3: 95
109. Ahern TJ, Klibanov AM (1985) Science 228: 1280
110. Adams MWW, Kelly RM (1998) Trends Biotechnol. 16: 329
111. Mozhaev VV, Martinek K (1984) Enzyme Microb. Technol. 6: 50
112. Jencks WP (1969) Catalysis in Chemistry and Enzymology. McGraw-Hill, New York
113. Fersht A (1985) Enzyme Structure and Mechanism, 2nd edn. Freeman, New York
114. Walsh C (ed) (1979) Enzymatic Reaction Mechanism. Freeman, San Francisco
115. Fischer E (1894) Ber. dtsch. chem. Ges. 27: 2985
116. Lichtenthaler FW (2003) Angew. Chem., Int. Ed. 33: 2364
117. Koshland DE (1958) Proc. Natl. Acad. Sci. USA 44: 98
118. Koshland DE, Neet KE (1968) Ann. Rev. Biochem. 37: 359
119. Gerstein M, Lesk AM, Chotia C (1994) Biochemistry 33: 6739
120. Dewar MJS (1986) Enzyme 36: 8
121. Lipscomb WN (1982) Acc. Chem. Res. 15: 232
122. Warshel A, Aqvist J, Creighton S (1989) Proc. Natl. Acad. Sci. USA 86: 5820
123. Page M I (1977) Angew. Chem. 89: 456
124. Ottosson J, Rotticci-Mulder JC, Rotticci D, Hult K (2001) Protein Sci. 10: 1769
125. Lipscomb WN (1982) Acc. Chem. Res. 15: 232
126. Ottosson J, Fransson L, Hult K (2002) Protein Sci. 11: 1462
127. Johnson LN (1984) Inclusion Compds. 3: 509
128. Warshel A, Sharma PK, Kato M, Xiang Y, Liu H, Olsson MHM (2006) Chem. Rev. 106: 3210
129. Garcia-Viloca M, Gao J, Karplus M, Truhlar DG (2004) Science 303: 186
130. Masgrau L, Roujeinikova A, Johanissen LO, Hothi P, Basran J, Ranaghan KE, Mulholland AJ, Sutcliffe MJ, Scrutton NS, Leys D (2006) Science 312: 237
131. Ogston AG (1948) Nature 162: 963
132. Jones JB (1976) Biochemical Systems in Organic Chemistry: Concepts, Principles and Opportunities. In: Jones JB, Sih CJ, Perlman D (eds) Applications of Biochemical Systems in Organic Chemistry, part I. Wiley, New York, pp 1–46
133. Kielbasinski P, Goralczyk P, Mikolajczyk M, Wieczorek MW, Majzner WR (1998) Tetrahedron: Asymmetry 9: 2641
134. Cipiciani A, Fringuelli F, Mancini V, Piermatti O, Scappini AM, Ruzziconi R (1997) Tetrahedron 53: 11853
135. Eyring H (1935) J. Chem. Phys. 3: 107

136. Kraut J (1988) Science 242: 533
137. Wong CH (1989) Science 244: 1145
138. Wolfenden R (1999) Bioorg. Med. Chem. 7: 647
139. International Union of Biochemistry and Molecular Biology (1992) Enzyme Nomenclature. Academic Press, New York
140. Schomburg D (ed) (2002) Enzyme Handbook. Springer, Heidelberg
141. Appel RD, Bairoch A, Hochstrasser DF (1994) Trends Biochem. Sci. 19: 258
142. Bairoch A (1999) Nucl. Acids Res. 27: 310; <http://www.expasy.ch/enzyme/>
143. Kindel S (1981) Technology 1: 62
144. Crout DHG, Christen M (1989) Biotransformations in Organic Synthesis. In: Scheffold R (ed) Modern Synthetic Methods, vol 5. pp. 1–114
145. Behr A (2007) Angewandte Homogene Katalyse. Wiley-VCH, Weinheim, p. 40
146. Mahler HR, Cordes HE (1971) Biological Chemistry, 2nd ed. Harper & Row, London
147. Farina V (2004) Adv. Synth. Catal. 346: 1553
148. Simon H, Bader J, Günther H, Neumann S, Thanos J (1985) Angew. Chem., Int. Ed. 24: 539
149. Chaplin MF, Bucke C (1990) Enzyme Technology. Cambridge University Press, New York
150. White JS, White DC (1997) Source Book of Enzymes. CRC Press, Boca Raton
151. Spradlin JE (1989) Tailoring Enzymes for Food Processing, in: Whitaker JR, Sonnet PE(eds) ACS Symposium Series, vol 389, p 24, J. Am. Chem. Soc., Washington
152. Bar-Even A, Noor E, Savir Y, Liebermeister W, Davidi D, Tawfik DS, Milo R (2011) Biochemistry 50: 4402.
153. Romero PA, Arnold FH (2009) Nat. Rev. Mol. Cell Biol. 10: 866.

Chapter 2
Biocatalytic Applications

2.1 Hydrolytic Reactions

Of all the types of enzyme-catalyzed reactions, hydrolytic transformations involving amide and ester bonds are the easiest to perform using proteases, esterases, or
lipases. The key features that have made hydrolases the favorite class of enzymes
for organic chemists during the past two decades are their lack of sensitive cofactors
(which otherwise would need to be recycled) and the large number of readily
available enzymes possessing relaxed substrate specificities to choose from.
About half of the total research in the field of biotransformations has been
performed using hydrolytic enzymes of this type [1, 2]. The reversal of the reaction,
giving rise to ester or amide *synthesis*, has been particularly well investigated using
enzymes in organic solvent systems. The special methodologies involved in this
latter type of reaction are described in Sect. 3.1.

Other applications of hydrolases, such as those involving the formation and/or
cleavage of phosphate esters, epoxides, nitriles, and organo-halides, are described
in separate chapters.

2.1.1 *Mechanistic and Kinetic Aspects*

The mechanism of amide- and ester-hydrolyzing enzymes is very similar to that
observed in the chemical hydrolysis by a base. A nucleophilic group from the active
site of the enzyme attacks the carbonyl group of the substrate ester or amide. This
nucleophilic 'chemical operator' can be either the hydroxy group of a serine (e.g.,
pig liver esterase, subtilisin, and the majority of microbial lipases), a carboxylate
group of an aspartic acid (e.g., pepsin) [3], or the thiol functionality of cysteine
(e.g., papain) [4–6].

© Springer International Publishing AG 2018
K. Faber, *Biotransformations in Organic Chemistry*,
DOI 10.1007/978-3-319-61590-5_2

The mechanism, which has been elucidated in greater detail, is that of the serine hydrolases [7, 8] (Scheme 2.1): Two additional groups (Asp and His) located close to the serine residue (which is the actual reacting chemical operator at the active site) form the so-called catalytic triad [9–12].[1] The special arrangement of these three groups effects a decrease of the pK_a of the serine hydroxy group thus enabling it to perform a nucleophilic attack on the carbonyl group of the substrate R^1–CO–OR^2 (step I). Thereby the acyl moiety of the substrate becomes covalently linked to the enzyme, forming the 'acyl-enzyme intermediate' by liberating the leaving group (R^2–OH). Then a nucleophile (Nu), usually water, can in turn attack the acyl-enzyme intermediate, regenerating the enzyme and releasing a carboxylic acid R^1–COOH (step II).

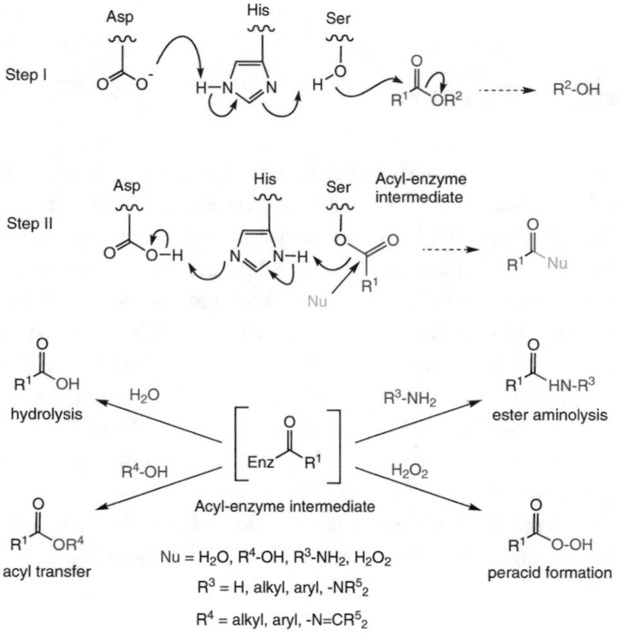

Scheme 2.1 The serine hydrolase mechanism

When the enzyme is operating in an organic solvent at low water concentrations – more precisely, at low water activity – any other nucleophile can compete with the water for the acyl-enzyme intermediate, thus leading to a number of synthetically useful transformations:

- Attack of another alcohol R^4–OH leads to a different ester R^1–CO–OR^4 via an interesterification reaction, called 'acyl transfer' [13, 14].
- The action of ammonia furnishes a carboxamide R^1–CO–NH$_2$ via an ammonolysis reaction [15, 16].

---

[1]In acetylcholine esterase from electric eel and lipase from *Geotrichum candidum* Asp within the catalytic triad is replaced by Glu [11, 12].

- An incoming amine R^3–NH_2 results in the formation of an N-substituted amide R^1–CO–NH–R^3, yielding an enzymatic aminolysis of esters [17, 18].
- Hydrazinolysis provides access to hydrazides [19, 20], and the action of hydroxylamine results in the formation of hydroxamic acid derivatives [21].
- Peracids of type R^1–CO–OOH are formed when hydrogen peroxide is acting as the nucleophile [22].
- Thiols (which would lead to thioesters) are unreactive [23].

During the course of all of these reactions, any type of chirality in the substrate is 'recognized' by the enzyme, which causes a preference for one of the two possible stereochemical pathways. The magnitude of this discrimination is governed by the kinetics and is a crucial parameter since it stands for the 'selectivity' of the reaction. It should be noted, that the following chapter is not an elaboration on enzyme kinetics, but rather a compilation of the most important conclusions needed for obtaining optimal results from stereoselective enzymatic transformations.

Since hydrolases nicely exemplify all different types of chiral recognition, we will discuss the underlying principles of these chiral recognition processes and the corresponding kinetic implications here [24]. Most of these types of transformations can be found within other groups of enzymes as well, and the corresponding rules can be applied accordingly.

Enantioface Differentiation
Hydrolases can distinguish between the two enantiomeric faces of achiral substrates such as enol esters possessing a plane of symmetry within the molecule [25]. The attack of the enzyme's nucleophilic chemical operator predominantly occurs from one side, leading to an unsymmetric enolization of the unstable free enol towards one preferred side within the chiral environment of the enzyme's active site [26]. During the course of the reaction a new center of chirality is created in the product (Scheme 2.2).

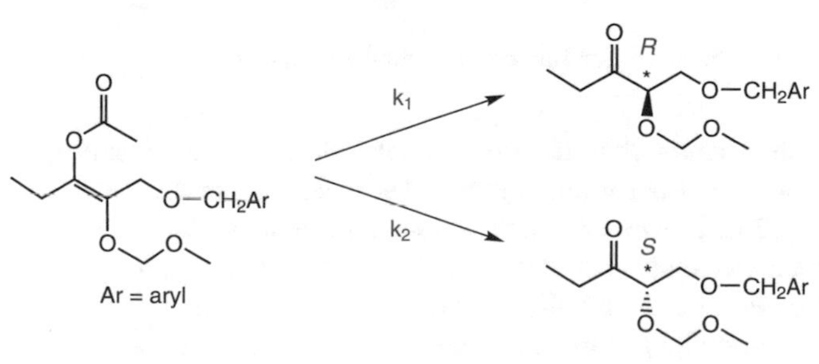

Scheme 2.2   Enantioface differentiation (achiral substrates)

Enantiotopos Differentiation

If prochiral substrates possessing two chemically identical but enantiotopic reactive groups X (designated pro-*R* and pro-*S*) are subjected to enzymatic hydrolysis, a chiral discrimination between them occurs during the transformation of group X into Y, thus leading to a chiral product (Scheme 2.3). During the course of the reaction the plane of symmetry within the substrate is broken. The single-step asymmetric hydrolysis of a prochiral α,α-disubstituted malonic diester by pig liver esterase or α-chymotrypsin is a representative example [27]. Here, the reaction terminates at the monoester stage since highly polar compounds of such type are heavily hydrated in an aqueous medium and are therefore generally not accepted by hydrolases [28].

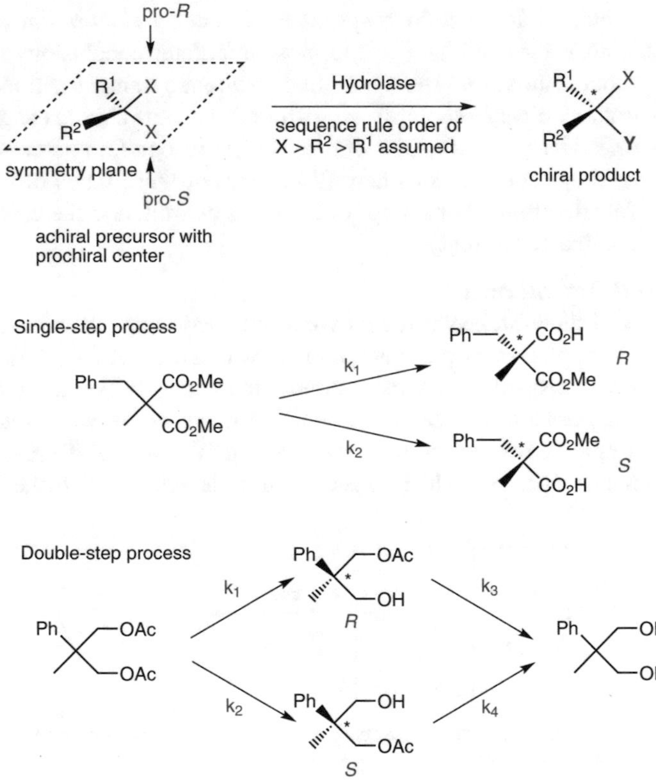

Scheme 2.3 Enantiotopos differentiation (prochiral substrates)

On the other hand, when the substrate is a diacetate, the resulting monoester is less polar and thus usually undergoes further cleavage in a second step to yield an achiral diol [29]. However, since the second step is usually slower, the chiral monoester can be trapped in fair yield if the reaction is carefully monitored.

Similarly, the two chemically identical groups X, positioned on carbon atoms of opposite (*R*,*S*)-configuration in a *meso*-substrate, will react at different rates in a hydrolase-catalyzed reaction (Scheme 2.4). In this way, the optically inactive *meso*-

substrate is transformed into an optically active product due to the transformation of one of the reactive groups from X into Y along with the destruction of the plane of symmetry within the substrate. Numerous open-chain or cyclic *cis-meso*-diesters have been transformed into chiral monoesters by this technique [30]. Again, for dicarboxylates the reaction usually stops after the first step at the carboxylate monoester stage, whereas two hydrolytic steps are usually observed with diacetate esters [31]. The theoretical yield of chiral product from single-step reactions based on an enantioface or enantiotopos differentiation or a desymmetrization of *meso*-compounds is always 100%.

If required, the interconversion of a given chiral hemiester product into its mirror-image enantiomer can be achieved by a simple two-step protection–deprotection sequence. Thus, regardless of the stereopreference of the enzyme which is used to perform the desymmetrization of the bifunctional prochiral or *meso*-substrate, both enantiomers of the product are available and no 'unwanted' enantiomer is produced. This technique is often referred to as the '*meso*-trick' [25].

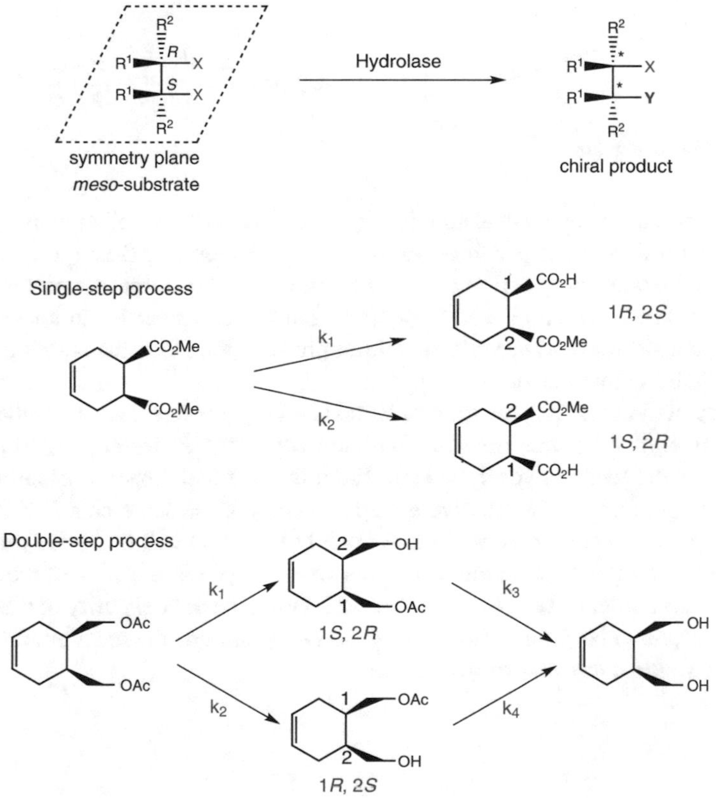

Scheme 2.4 Desymmetrization of *meso*-substrates

Since hydrolytic reactions are performed in an aqueous environment, where the molar concentration of water is ~55.5 mol/L, they are virtually completely irreversible. The kinetics of all of the single-step reactions described above is very simple (Fig. 2.1): a prochiral or a *meso*-substrate S is transformed into two enantiomeric products P and Q at different rates, determined by the apparent first-order rate constants k_1 and k_2, respectively (Schemes 2.2–2.4). The selectivity of the reaction (denoted α [32]) is only governed by the ratio of k_1/k_2, which is *independent of the conversion* and therefore remains constant throughout the reaction. Thus, the optical purity of the product (e.e.$_P$) is *not* dependent on the extent of the conversion. Consequently, the selectivity observed in such a reaction can*not* be improved by stopping the reaction at different extents of conversion, but only by changing the 'environment' of the system (e.g., via substrate modification, choice of another enzyme, the addition of organic cosolvents, and variations in temperature or pH). Different techniques for improving the selectivity of enzymatic reactions by variations in the 'environment' are presented on pp. 72–79 and 102–103.

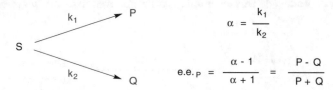

Fig. 2.1 Single-step kinetics

As mentioned above, occasionally a second successive reaction step cannot be avoided with diesters of prochiral or *meso*-diols (Schemes 2.3 and 2.4). For such types of substrates the reaction does not terminate at the chiral monoester stage to give the desired products P and Q (step 1), but rather proceeds via a second step (usually at a slower rate) to yield an achiral product (R). Here, the reaction kinetics become more complicated.

As depicted in Fig. 2.2, the ratio of P and Q – i.e., the optical purity of the desired product (e.e.$_P$) – depends now on four rate constants, k_1 through k_4, due to the presence of the second hydrolytic step. From the fact that enzymes usually show a continuous preference for reactive groups possessing the same chirality,[2] one can conclude that if S is transformed more quickly into P, Q will be hydrolyzed faster into diol R than P. Thus, the rate constants governing the selectivity of the reaction are often at an order of $k_1 > k_2$ and $k_4 > k_3$. Notably, the optical purity of the product monoester (e.e.$_P$) becomes a *function of the conversion* of the reaction, and generally follows the curve shown in Fig. 2.2.

---

[2]These groups are called homochiral.

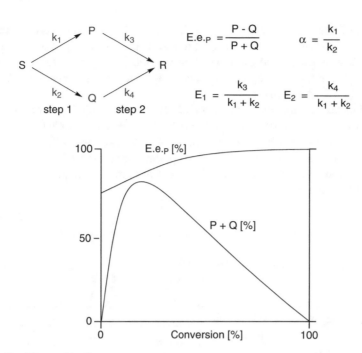

Fig. 2.2 Double-step kinetics

During early stages of the reaction, the optical purity of the product is mainly determined by the selectivity (α) of the first reaction step, which constitutes an enantiotopos or enantioface differentiation, depending on the type of substrate.

As the reaction proceeds, the second hydrolytic step, being a kinetic resolution, starts to take place to a more significant extent due to the increased formation of monoester P + Q. Its apparent 'opposite' selectivity compared to that of the first step (remember that $k_1 > k_2$, $k_4 > k_3$) leads to an enhancement of optical purity of the product (e.e.$_P$), because Q is hydrolysed faster than P. In contrast, the product concentration [P + Q] follows a bell-shaped curve: After having reached a maximum at a certain conversion (as long as the first step is faster than the second), the product concentration finally drops off again when most of the substrate S is consumed and the second hydrolytic step (forming R at the expense of P + Q) begins to dominate. The same analogous considerations are pertinent for the reverse situation – an esterification reaction.

In general, it can be stated that the ratio of reaction rates of the first versus the second step $(k_1 + k_2)/(k_3 + k_4)$ has a major impact on the *chemical yield* of P + Q, whereas the match or mismatch of the selectivities ($k_1 > k_2$, $k_3 < k_4$ or $k_1 > k_2$, $k_3 > k_4$, respectively) determines the *optical purity* of the product. In order to obtain a high chemical yield, the first step should be considerably faster than the second to ensure that the chiral product is accumulated, because then it is formed faster than it is further converted $[(k_1 + k_2) \gg (k_3 + k_4)]$. For a high e.e.$_P$, the selectivities of both steps should match each other ($k_1 > k_2$, $k_4 > k_3$), i.e., if P is formed predominantly in the first step from S, it should react at a slower rate than Q in the second step. Figure 2.2 shows a typical example of such a double-step process, where the first step is about ten times faster than the second, with selectivities matching ($k_1 = 100$, $k_2 = 10$, $k_3 = 1$, $k_4 = 10$).

In addition to trial-and-error experiments (i.e., by stopping such double-step reactions at various intervals and checking the yield and optical purity of the product), the e.e.-conversion dependence may also be calculated [33]. The validity of this method has been verified by the desymmetrization of a prochiral *meso*-diacetate using pig liver esterase (PLE) and porcine pancreatic lipase (PPL) as shown in Scheme 2.5 [34].

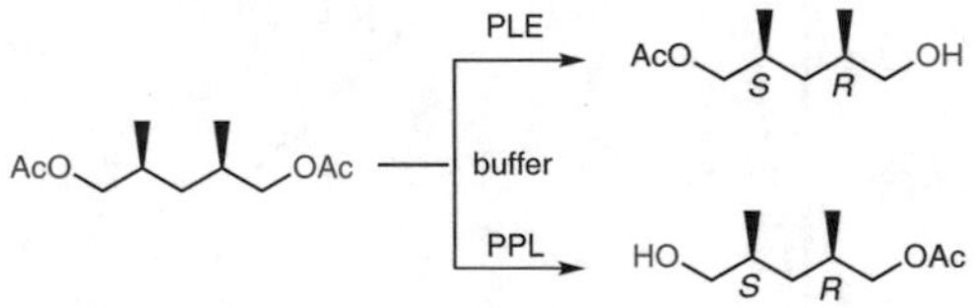

Enzyme	Stereochemical Preference	Kinetic Constants		
		α	E1	E2
PLE	pro-*R*	2.47	0.22	0.60
PPL	pro-*S*	15.6	0.04	0.18

Scheme 2.5 Desymmetrization of a *meso*-diacetate

Enantiomer Differentiation

When a racemic substrate is subject to enzymatic hydrolysis, chiral discrimination of the enantiomers occurs [35]. It should be noted that the chirality does not necessarily have to be of a central type, but can also be axial or planar to be 'recognized' by enzymes (Scheme 1.3). Due to the chirality of the active site of the enzyme, one enantiomer fits better into the active site than its mirror-image counterpart and is therefore converted at a higher rate, resulting in a kinetic resolution of the racemate. The vast majority of enzymatic transformations constitute kinetic resolutions and, interestingly, this potential of hydrolytic enzymes was realized as early as 1903 [36]! It is a remarkable observation that in biotransformations, kinetic resolutions outnumber desymmetrization reactions by about 1:4, which is presumably due to the fact that there are more racemic compounds possible as opposed to prochiral and *meso*-analogs. After all, prochiral and *meso*-compounds have only two functional groups (R^1, R^2) available for variation, whereas racemates have three (R^1, R^2, R^3) [37].

The most striking difference from the above-mentioned types of desymmetrization reactions, which show a theoretical yield of 100%, is that in kinetic resolution each of the enantiomers can be obtained in only 50% yield.

In some ideal cases, the difference in the reaction rates of both enantiomers is so extreme that the 'good' enantiomer is transformed quickly and the other is not converted at all. Then the enzymatic reaction will cease automatically at 50% conversion when there is nothing left of the more reactive enantiomer (Scheme 2.6) [38].

In practice, however, the enantioselectivity is not ideal, and the difference in – or more precisely the ratio of – the reaction rates of the enantiomers is not infinite, but measurable. The thermodynamic reasons for this have been discussed in Chap. 1 (Fig. 1.8). What one observes in these cases is not a complete standstill of the reaction at 50% conversion but a marked slowdown in reaction rate at around this point. In these numerous cases one encounters some crucial dependencies:

- The velocity of the transformation of each substrate enantiomer varies with the degree of conversion, since their ratio does not remain constant during the reaction.
- Therefore, the optical purity of both substrate (e.e.$_S$) and product (e.e.$_P$) becomes *a function of the conversion*.

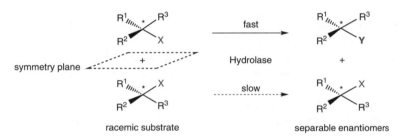

Scheme 2.6 Enantiomer differentiation

A very useful treatment of the kinetics of enzymatic resolution, describing the dependency of the conversion (c) and the enantiomeric excess of substrate (e.e.$_S$) and product (e.e.$_P$), was developed by C.J. Sih in 1982 [39] on a theoretical basis described by K.B. Sharpless [40] and K. Fajans [41]. The parameter describing the selectivity of a resolution was introduced as the dimensionless 'Enantiomeric Ratio' (*E*), which remains constant throughout the reaction and is only determined by the 'environment' of the system [42–45].[3] *E* corresponds to the ratio of the relative second-order rate constants (v_A, v_B) of the individual substrate enantiomers (A, B) and is related to the k_{cat} and K_M values of enantiomers A and B according to Michaelis–Menten kinetics as follows (for the thermodynamic background see Fig. 1.8):

---

[3]The Enantiomeric Ratio (*E*) is a synonym for the so-called selectivity factor (*s*). Whereas *E* is used more often in biocatalyzed kinetic resolutions, the *s*-factor is more common in chemocatalysis. In a mathematical sense, both are identical and describe the ratio of the relative (second-order) rate constants of enantiomers. For a comprehensive discussion see [45].

$$\text{Enantiomeric Ratio} \quad E = \frac{v_B}{v_A} = \frac{\left[\dfrac{k_{cat}}{K_M}\right]_A}{\left[\dfrac{k_{cat}}{K_M}\right]_B} \quad \Delta\Delta G^{\neq} = -RT \ \ln \ E$$

The 'Enantiomeric Ratio' is not to be confused with the term 'enantiomer ratio' (e.r.), which is used to quantify the enantiomeric composition of a mixture of enantiomers (e.r. = [A]/[B]) [46]. Related alternative methods for the experimental determination of E-values have been proposed [47–49].

Irreversible Reaction Hydrolytic reactions in aqueous solution can be regarded as completely irreversible due to the high 'concentration' of water present (55.5 mol/L). Assuming negligible enzyme inhibition, thus both enantiomers of the substrate are competing freely for the active site of the enzyme, Michaelis–Menten kinetics effectively describe the reaction in which two enantiomeric substrates (A and B) are transformed by an enzyme (Enz) into the corresponding enantiomeric products (P and Q, Fig. 2.3).

Instead of determining all individual rate constants (k_{cat}, K_M) for each of the enantiomers (a wearisome task for synthetic organic chemists, particularly when A and B are not available in enantiopure form), the ratio of the initial reaction rates of the substrate enantiomers ($E = v_A/v_B$) can be mathematically linked to the conversion (c) of the reaction, and the optical purities of substrate (e.e.$_S$) and product (e.e.$_P$). In practice, these parameters are usually much easier to determine and do not require the availability of pure enantiomers.

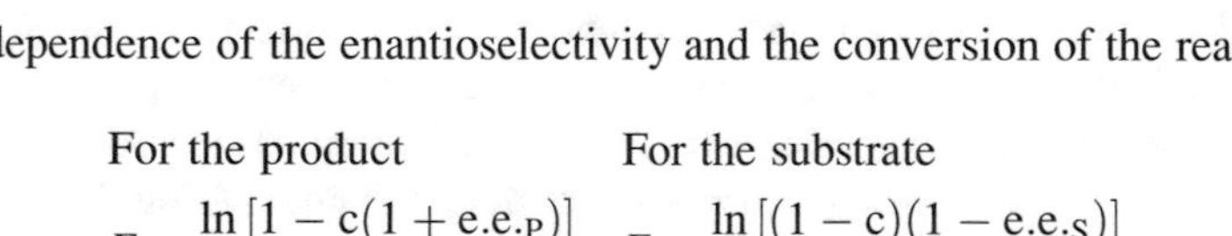

$$\text{Enz} + \text{A} \underset{}{\overset{(K_M)_A}{\rightleftharpoons}} [\text{Enz A}] \xrightarrow{(k_{cat})_A} \text{Enz} + \text{P}$$

$$\text{Enz} + \text{B} \underset{}{\overset{(K_M)_B}{\rightleftharpoons}} [\text{Enz B}] \xrightarrow{(k_{cat})_B} \text{Enz} + \text{Q}$$

Enz = enzyme
A, B = enantiomeric substrates
[EnzA], [EnzB] = diastereomeric enzyme-substrate complexes
P, Q = enantiomeric products

Fig. 2.3 Enzymatic kinetic resolution (irreversible reaction)

The dependence of the enantioselectivity and the conversion of the reaction is:

For the product   For the substrate

$$E = \frac{\ln\left[1 - c(1 + \text{e.e.}_P)\right]}{\ln\left[1 - c(1 - \text{e.e.}_P)\right]} \qquad E = \frac{\ln\left[(1 - c)(1 - \text{e.e.}_S)\right]}{\ln\left[(1 - c)(1 + \text{e.e.}_S)\right]}$$

c = conversion, e.e. = enantiomeric excess of substrate (S) or product (P),

E = Enantiomeric Ratio

The above-mentioned equations give reliable results except for very low and very high levels of conversion, where accurate measurement is impeded by errors derived from sample manipulation. In such cases, the following equation is recommended instead, because here only values for the optical purities of substrate and product need to be measured, which are *relative* quantities, in contrast to the conversion, which is an *absolute* quantity [50].

$$E = \frac{\ln \dfrac{[\text{e.e.}_P(1 - \text{e.e.}_S)]}{(\text{e.e.}_P + \text{e.e.}_S)}}{\ln \dfrac{[\text{e.e.}_P(1 + \text{e.e.}_S)]}{(\text{e.e.}_P + \text{e.e.}_S)}}$$

Two examples of enzymatic resolutions with selectivities of $E = 5$ and $E = 20$ are depicted in Fig. 2.4. The curves show that the product (P + Q) can be obtained in its highest optical purities before 50% conversion, where the enzyme can freely choose the 'well-fitting' enantiomer from the racemic mixture. So, the 'well-fitting' enantiomer is predominantly depleted from the reaction mixture during the course of the reaction, leaving behind the 'poor-fitting' counterpart. Beyond 50% conversion, the enhanced relative concentration of the 'poor-fitting' counterpart leads to its increased transformation by the enzyme. Thus, the e.e.$_P$ rapidly decreases beyond 50% conversion.

Analogous trends are seen for the optical purity of the residual slow-reacting enantiomer of the substrate (e.e.$_S$). Its optical purity remains low before 40%, then climbs significantly at around 50%, and reaches its maximum beyond the 60% conversion point.

Very high optical purity of substrate can be reached by extending the reaction beyond ~60% conversion, albeit at the price of reduced yield. Attractive optical purities for the substrate and product demand a very high enantioselectivity.

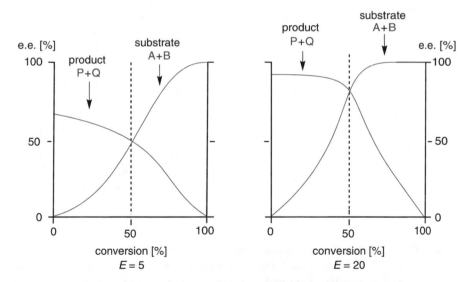

Fig. 2.4 Dependence of optical purities (e.e.$_S$/e.e.$_P$) on the conversion

Using the equations discussed above, the expected optical purity of substrate and product can be calculated for a chosen point of conversion and the enantiomeric ratio (E) can be determined as a convenient *conversion-independent* value for the 'enantioselectivity' of an enzymatic resolution. Free shareware programs for the calculation of the enantiomeric ratio for irreversible reactions can be obtained from the internet [51].[4] As a rule of thumb, enantiomeric ratios below 15 are inacceptable for practical purposes. They can be regarded as being moderate to good in the range of 15–30, and above this value they are excellent. However, values of $E > 200$ cannot be accurately determined due to the inaccuracies emerging from the determination of the enantiomeric excess (e.g., by NMR, HPLC, or GC), because in this range even an extremely small variation of e.e.$_S$ or e.e.$_P$ causes a significant change in the numerical value of E.

In order to obtain optimal results from resolutions of racemic substrates which exhibit moderate selectivities (E values ca. 20), one can proceed as follows (see Fig. 2.5): The reaction is terminated at a conversion of 40%, where the 'product' curve reaches its optimum in chemical and optical yield being closest to the 'ideal' point X (step 1). The product is isolated and the remaining substrate – showing a low optical purity at this stage of conversion – is subjected to a second hydrolytic step, until an overall conversion of about 60% is reached, where the 'substrate' curve is closest to X (step 2). Now, the substrate is harvested with an optimal chemical and optical yield and the 20% of product from the second step is sacrificed or recycled. This two-step process [52] can be used to allow practical use of numerous enzyme-catalyzed kinetic resolutions which show incomplete selectivities.

Fig. 2.5 Two-step enzymatic resolution

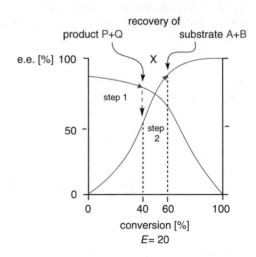

Reversible Reaction The situation becomes more complicated when the reaction is reversible [53, 54]. Then, the concentration of the nucleophile which attacks the acyl-enzyme intermediate is limited and is not in excess (like water in a hydrolytic reaction). In this situation, the equilibrium constant (K) of the reaction – neglected in the irreversible type of reaction – plays an important role and therefore has to be determined.

---

[4]http://biocatalysis.uni-graz.at/enantio/

The equations linking the enantioselectivity of the reaction (the Enantiomeric Ratio E), the conversion (c), the optical purities of substrate (e.e.$_S$) and product (e.e.$_P$), and the equilibrium constant K are as follows:

For the product   For the substrate

$$E = \frac{\ln\left[1 - (1 + K)c(1 + e.e._P)\right]}{\ln\left[1 - (1 + K)c(1 - e.e._P)\right]} \qquad E = \frac{\ln\left[1 - (1 + K)(c + e.e._S\{1 - c\})\right]}{\ln\left[1 - (1 + K)(c - e.e._S\{1 - c\})\right]}$$

c = conversion, e.e. = enantiomeric excess of substrate (S) or product (P),

E = Enantiomeric Ratio, K = equilibrium constant of the reaction

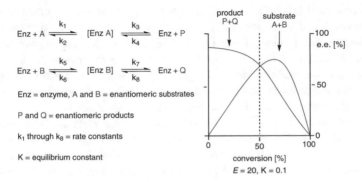

$$Enz + A \underset{k_2}{\overset{k_1}{\rightleftarrows}} [Enz\ A] \underset{k_4}{\overset{k_3}{\rightleftarrows}} Enz + P$$

$$Enz + B \underset{k_6}{\overset{k_5}{\rightleftarrows}} [Enz\ B] \underset{k_8}{\overset{k_7}{\rightleftarrows}} Enz + Q$$

Enz = enzyme, A and B = enantiomeric substrates

P and Q = enantiomeric products

k_1 through k_8 = rate constants

K = equilibrium constant

product P+Q   substrate A+B

conversion [%]

$E = 20,\ K = 0.1$

Fig. 2.6 Enzymatic kinetic resolution (reversible reaction)

As shown in Fig. 2.6, the product curve of an enzymatic resolution following a reversible reaction type remains almost the same as in the irreversible case. However, a significant difference is found in the substrate curve: particularly at higher levels of conversion (beyond 70%) the reverse reaction (i.e., esterification instead of a hydrolysis) starts to predominate. Since the enantiopreference of the substrate stays the same in both directions, it follows that the *same* enantiomer from the substrate and the product react preferentially in both the forward and the reverse reaction. Assuming that A is the better substrate than B, accumulation of product P and unreacted B will occur. For the reverse reaction, however, P is a better substrate than Q, because it is of the *same* chirality as A and therefore it will be transformed back into A at a faster rate than B into Q. As a result, the optical purity of the remaining substrate is depleted as the conversion increases. In other words, the reverse reaction, predominantly taking place at higher conversion levels, constitutes a second – and in this case an undesired – selection of chirality which causes a depletion of e.e. of the remaining substrate.

All attempts of improving the optical purity of substrate and product of reversible enzymatic resolutions are geared at shifting the reaction out of the equilibrium to obtain an irreversible type. The easiest way to achieve this is to use an excess of nucleophile: in order to obtain an equilibration constant of $K > 10$, about 20 M equivalents of nucleophile versus substrate are sufficient to obtain a virtually irreversible type of reaction. Other techniques, such as using special cosubstrates which cause an irreversible type of reaction, are discussed in Sect. 3.1.1.

Sequential Biocatalytic Resolutions For a racemic substrate bearing *two* chemically and stereochemically identical reactive groups, an enzymatic resolution proceeds through two consecutive steps via an intermediate monoester stage. During the course of such a reaction the substrate is forced to enter the active site of the enzyme twice – it is therefore 'double-selected'. Since each of the selectivities of both of the sequential steps determine the final optical purity of the product, exceptionally high selectivities can be achieved by using such a 'double-sieving' procedure.

As depicted in Fig. 2.7, a bifunctional racemic substrate consisting of its enantiomers A and B is enzymatically resolved via a first step to give the intermediate enantiomeric products P and Q. The selectivity of this step is governed by the constants k_1 and k_3. Then, both of the intermediate monoester products (P, Q) undergo a second reaction step, the selectivity of which is determined by k_2 and k_4, to form the enantiomeric final reaction products R and S. As a result, the optical purity of the substrate (A, B), the intermediate monoester (P, Q), and the final products (R, S) are a *function of the conversion* of the reaction, as shown by the curve in Fig. 2.7. The selectivities of each of the steps (E_1 and E_2) can be determined experimentally and the optical purities of the substrate e.e.$_{A/B}$, the intermediate e.e.$_{P/Q}$, and the final product e.e.$_{R/S}$ can be calculated [55, 56].

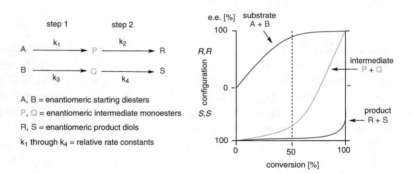

Fig. 2.7 Kinetis of sequential kinetic resolution of bifunctional substrates

It has been shown that the maximum overall selectivity (E_{tot}) of a sequential kinetic resolution can be related to the individual selectivities (E_1, E_2) of each of the steps [57]. E_{tot} represents the enantioselectivity that a hypothetical single-step resolution would need to yield the enantiomeric purity of the two-step resolution.

$$E_{tot} \sim \frac{E_1 \times E_2}{2}$$

This technique has been proven to be highly flexible. It was shown to work successfully not only in a hydrolytic reaction using cholesterol esterase [58] or microbial cells [59], but also in the reverse esterification direction in an organic solvent catalyzed by a *Pseudomonas* sp. lipase (Scheme 2.7). In a related fashion, a successful sequential resolution of a bifunctional 1,2-amine via ester aminolysis was reported [60].

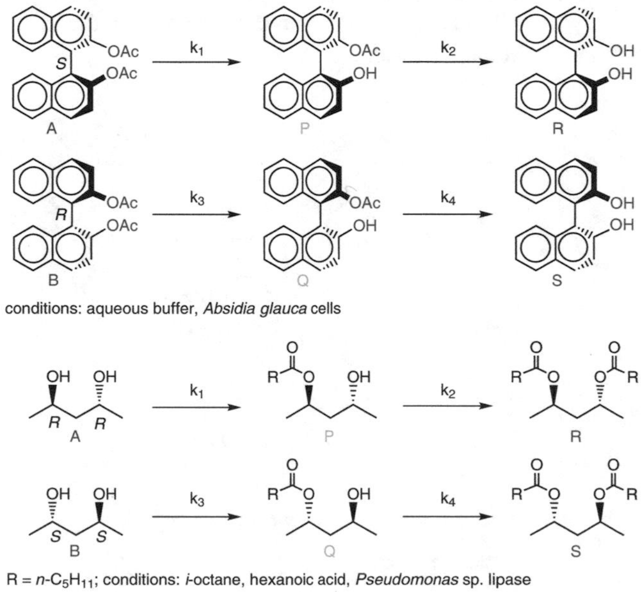

conditions: aqueous buffer, *Absidia glauca* cells

$R = n\text{-}C_5H_{11}$; conditions: *i*-octane, hexanoic acid, *Pseudomonas* sp. lipase

Scheme 2.7   Sequential enzymatic resolution of bifunctional substrate via hydrolysis or esterification

A special type of sequential enzymatic resolution involving a hydrolysis-esterification [61] or an alcoholysis-esterification sequence [62] is depicted in Fig. 2.8. In view of the mechanistic symmetry of enzymatic acyl transfer reactions (Scheme 3.6), the resolution of a racemic alcohol can be effected by enantioselective hydrolysis of the corresponding ester or by esterification of the alcohol. As the biocatalyst displays the same stereochemical preference in both reactions, the desired product can be obtained with higher optical yields, if the two steps are coupled sequentially. The basis of this approach parallels that of product recycling in hydrolytic reactions. However, tedious chromatographic separation of the intermediates and accompanying re-esterification is omitted.

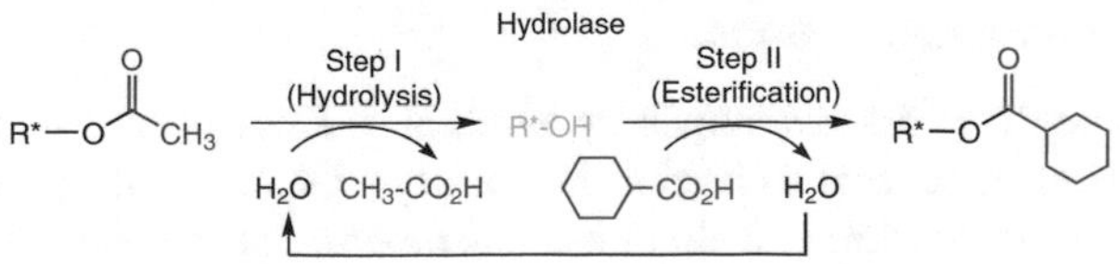

Fig. 2.8   Mechanism of sequential enzymatic kinetic resolution of monofunctional substrate via concurrent hydrolysis-esterification

As shown in Scheme 2.8, the racemic starting ester (A/B) is hydrolyzed to give alcohols (P/Q) in an organic medium containing a minimum amount of water, which in turn, by the action of the same lipase, are re-esterified with cyclohexanoic acid present in the mixture. Thus, the alcohol moiety of the substrate has to enter the active site of the lipase twice during the course of its transformation into the final product ester (R/S). An apparent selectivity of $E_{tot} = 400$ was achieved in this way, whereas the corresponding isolated single-step resolutions of this process were $E_1 = 8$ for the hydrolysis of acetate A/B, and $E_2 = 97$ for the esterification of alcohol P/Q with cyclohexanoic acid.

conditions: water-saturated hexane, cyclohexane carboxylic acid,
Mucor sp. lipase

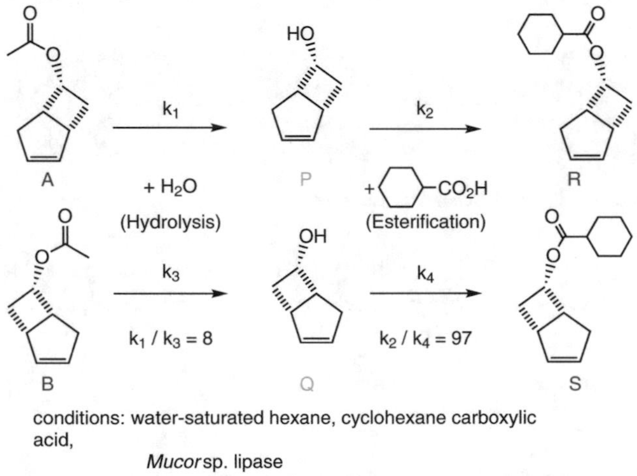

Scheme 2.8 Sequential enzymatic kinetic resolution of monofunctional substrate via concurrent hydrolysis-esterification in aqueous-organic solvent

Deracemization

Despite its widespread use, kinetic resolution has several disadvantages, particularly on an industrial scale. After all, an ideal process should lead to a single enantiomeric product in 100% chemical yield. The drawbacks of kinetic resolution are as follows:

- The theoretical yield of each enantiomer is limited to 50%. Furthermore, in general only one stereoisomer is desired and there is little or no use for the other.
- Separation of the product from the remaining substrate may be laborious, in particular when simple extraction or distillation fails [63].
- As explained above, the optical purity of substrate and/or product is often less than perfect for kinetic reasons.

To overcome these disadvantages by avoiding the occurrence of the undesired 'wrong' enantiomer, several strategies are possible [64, 65]. All of these processes which lead to the formation of a single stereoisomeric product from a racemate are called 'deracemizations' [66–68].

Repeated Resolution In order to avoid the loss of half of the material in kinetic resolution, it has been a common practice to racemize the unwanted enantiomer after separation from the desired product and to subject it again to kinetic resolution in a subsequent cycle, until virtually all of the racemic material has been converted into a single stereoisomeric product. For obvious reasons, this laborious procedure is not justified for laboratory-scale reactions, but it is a viable option for resolutions on an industrial scale, in particular for continuously operated processes, where the re-racemized material is simply fed back into the subsequent batch of the resolution process. At first sight, repeated resolution appears less than ideal and it certainly lacks synthetic elegance, bearing in mind that an infinite number of cycles are theoretically required to transform all of the racemic starting material into a single stereoisomer. Upon closer examination, though, re-racemization holds certain merits: a simple calculation shows that although only 50% of the desired enantiomer is obtained after a single cycle, the overall (theoretical) yield increases to ~94% after only four cycles [69].

In practice, however, deracemization via repeated resolution is often plagued by low overall yields due to the harsh reaction conditions required for (chemical) racemization [70]. In view of the mild reaction conditions displayed by enzymes, racemases of EC-class 5 are increasingly being employed [71, 72].

In-Situ Inversion The final outcome of a kinetic resolution of a racemate is a mixture of enantiomeric product and substrate. Separating them by physical or chemical means is often tedious and might pose a serious drawback to commercial applications, especially if the mixture comprises an alcohol and an ester. However, if the molecule has only a single center of chirality, the alcohol can be chemically inverted into its enantiomer *before* separating the products (Scheme 2.9) [73, 74]. Introduction of a good leaving group, LG (e.g., tosylate, triflate, nitrate, or Mitsunobu intermediate) yields an activated ester, which can be hydrolyzed with *inversion* of configuration, while the stereochemistry of the remaining carboxylic acid substrate ester is *retained* during hydrolysis. As a result, a *single* enantiomer is obtained as the final product. Since the e.e.$_S$ and e.e.$_P$ are a function of the conversion, it is obvious that the point where the kinetic resolution is terminated and the in-situ inversion is performed, has to be carefully chosen in order to obtain a maximum of the final e.e.$_P$. The optimal value for the conversion can be calculated as a function of the E value of the reaction, and it is usually at or slightly beyond a conversion of 50% [75, 76].

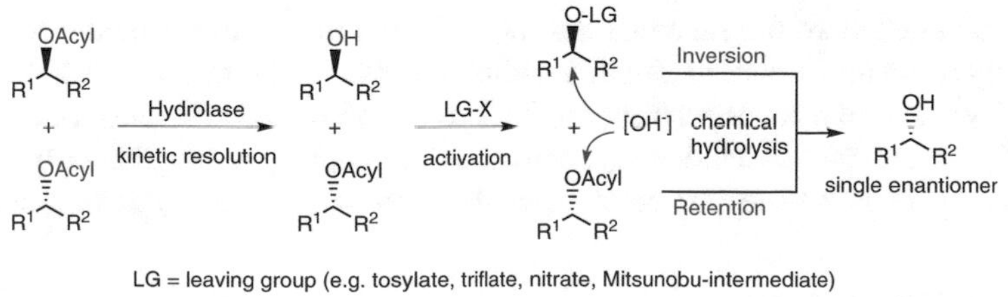

LG = leaving group (e.g. tosylate, triflate, nitrate, Mitsunobu-intermediate)

Scheme 2.9 Kinetic resolution followed by in-situ inversion

Dynamic resolution is a more elegant approach [77–82] This comprises a classic resolution with an additional feature, i.e., the resolution is carried out using conditions under which the substrate enantiomers are in a rapid equilibrium (racemizing). Thus, as the well-accepted substrate-enantiomer is depleted by the enzyme, the equilibrium is constantly adjusted by racemization of the poorly accepted counterpart. To indicate the nonstatic character of such processes, the term 'dynamic resolution' has been coined [83, 84].[5]

In this case, several reactions occur simultaneously and their relative rates determine the stereochemical outcome of the whole process (Fig. 2.9):

- The enzyme should display high specificity for the enantiomeric substrates A/B ($k_A \gg k_B$ or $k_B \gg k_A$).
- Spontaneous hydrolysis (k_{spont}) should be a minimum since it would yield racemic product.
- Racemization of the substrate should occur at an equal or higher rate compared to the biocatalytic reaction in order to provide a sufficient amount of the 'well-fitting' substrate enantiomer from the 'poor-fitting' counterpart ($k_{rac}^{Sub} \geq k_A$ or k_B, resp.).
- Racemization of the product (k_{rac}^{Prod}) should be minimal.

Although the above-mentioned criteria are difficult to meet experimentally, the benefits are impressive. Examples of this type of biotransformation have increased recently [85–91]; several examples are given in subsequent chapters.

The kinetics of a dynamic resolution is outlined in the following example [78, 92]. Figure 2.9 shows the e.e.$_S$ and e.e.$_P$ plotted for an enantiomeric ratio of $E \sim 10$. In a classic resolution process, the product is formed in ~83% e.e. at the very beginning of the reaction, but this value rapidly decreases when the reaction is run towards ~50% conversion as indicated by the symbol '*'. In a dynamic process, this depletion *does not* occur, because the enzyme always encounters racemic substrate throughout the reaction since the 'well-fitting' enantiomer is not depleted but constantly restored from the 'poor-fitting' counterpart via racemization. Thus, e.e.$_P$ remains constant throughout the reaction as indicated by the dashed arrow.

The e.e.$_P$ of dynamic processes is related to the enantioselectivity (E value) through the following formulas [93]:

$$\text{e.e.}_P = \frac{(E-1)}{(E+1)} \qquad E = \frac{(1+\text{e.e.}_P)}{(1-\text{e.e.}_P)}$$

In the case where the racemization (k_{rac}^{Sub}) is limited, the dynamic resolution gradually turns into a classic kinetic resolution pattern. Figure 2.9 shows the extent of the depletion of e.e.$_P$ depending on the conversion for several ratios of k_{rac}^{Sub}/k_A ($E \sim 10$). As can be expected, e.e.$_P$ decreases only slightly during the early stage of the reaction because the fast-reacting enantiomer is sufficiently available during

---

[5]Dynamic resolution is a type of second-order asymmetric transformation [79, 83]

this period. At higher levels of conversion, however, a serious drop in e.e.$_P$ will occur if the racemization cannot cope with the demand of the enzyme for the faster-reacting substrate enantiomer.

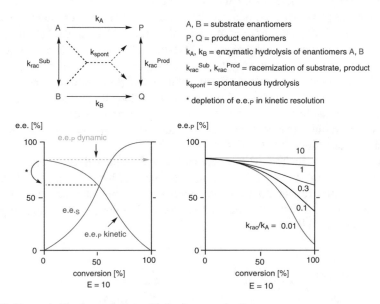

Fig. 2.9 Dynamic kinetic resolution with in-situ racemization

It is obvious that a high e.e.$_P$ for dynamic resolutions can only be achieved for reactions displaying excellent selectivities. For example, values for $E \sim 19$ and ~ 40 will lead to an e.e.$_P$ of 90% and 95%, respectively, but for an enantiomeric excess of 98% an enantiomeric ratio of ~ 100 is required.

2.1.2   Hydrolysis of the Amide Bond

The enzymatic hydrolysis of the carboxamide bond is associated to the biochemistry of amino acids and peptides [94]. The world production of enantiomerically pure amino acids was estimated to comprise a market of ca. US $ 11 billion per annum in 2015 [95]. The amino acids dominating this area with respect to output and value are produced by fermentation (L-lysine, L-phenylalanine, L-tryptophan, L-threonine, L-arginine, L-histidine, L-isoleucine, L-serine, L-valine) [96] and by synthesis (D,L-methionine) on industrial scale. However, a considerable number of optically pure D- and L-amino acids are prepared by using one of the enzymatic methods discussed below. L-Amino acids are used as additives for animal feed, for infusion solutions and as enantiopure starting materials for the synthesis of pharma- and agrochemicals or artificial sweeteners. Selected amino acids possessing the unnatural D-configuration have gained an increasing importance as bioactive compounds or components of such agents. For instance, D-phenylglycine and its p-hydroxy derivative are used for the synthesis of antibiotics such as ampicillin and amoxicillin, respectively, and D-valine is an essential component of the insecticidal synthetic pyrethroid fluvalinate (Scheme 2.208).

Among the principal methods for the enzymatic synthesis of enantiomerically pure amino acids depicted in Scheme 2.10, the most widely applied strategy is the resolution of racemic starting material (synthetically prepared from inexpensive bulk chemicals) employing easy-to-use hydrolytic enzymes such as proteases, esterases, and lipases. In contrast, more sophisticated procedures are the (1) reductive amination of α-keto acids using α-amino acid dehydrogenases (pp. 158–161), (2) asymmetric addition of ammonia onto α,β-unsaturated carboxylic acids catalyzed by ammonia lyases (Sect. 2.5.2), and (3) amino-group transfer using α-transaminases (Sect. 2.6.2) [97–99].

The hydrolytic methods discussed below were selected from the numerous strategies for amino acid synthesis [94, 100–106] for their flexibility, since they are not restricted to the 20 canonical amino acids, but also accept nonnatural analogs and give rise to D- or L-enantiomers. Several of these methods are employed on industrial scale [107].

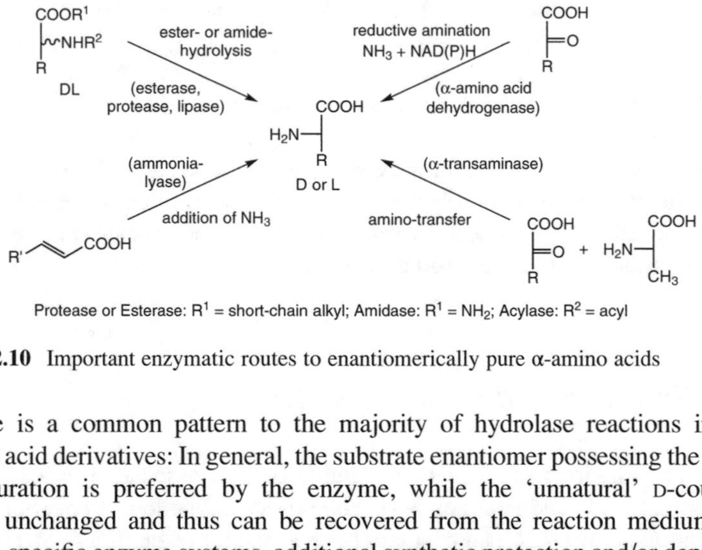

Scheme 2.10 Important enzymatic routes to enantiomerically pure α-amino acids

There is a common pattern to the majority of hydrolase reactions involving α-amino acid derivatives: In general, the substrate enantiomer possessing the 'natural' L-configuration is preferred by the enzyme, while the 'unnatural' D-counterpart remains unchanged and thus can be recovered from the reaction medium. Using strictly L-specific enzyme systems, additional synthetic protection and/or deprotection steps are required in those cases where the unnatural D-amino acid constitutes the desired product. However, enzymes with complementary enantiopreference are available for some processes such as the amidase, hydantoinase and acylase method (see below) to directly obtain the desired enantiomer. The work-up procedure is usually easy, because the difference in solubility of the product and the remaining substrate at different pH medium facilitates their separation by extraction.

However, there is a limitation to the majority of these methods: the α-carbon atom bearing the amino group must not be fully substituted, since such bulky substrates are generally not accepted by hydrolases. Thus, enantiopure α-methyl or α-ethyl amino acids are generally not accessible by these methods, although some exceptions are known [108, 109].

The recycling of the undesired enantiomer from the kinetic resolution is of crucial importance particularly on an industrial scale [110]. In the past, amino acid esters were thermally racemized at about 150–170 °C, milder conditions for the racemization of amino acid amides employed the formation of Schiff bases with aromatic aldehydes (such as benzaldehyde or salicylaldehyde) (Scheme 2.13). Nowadays, racemases [111] are used in dynamic resolution processes.

Esterase Method
A racemic amino acid ester can be enzymatically resolved by the action of a protease or (in selected cases) an esterase or a lipase. Remarkably, the first resolution of this type using a crude porcine pancreatic extract was reported in 1905 [112]! The catalytic activity of a protease on a carboxylic ester bond has frequently been denoted as 'esterase activity', although the mechanism of action does not differ from that of an amide hydrolysis. Bearing in mind the greater stability of an amide bond as compared to that of an ester, it is reasonable that a protease, which is able to cleave a much stronger amide bond, is capable of hydrolyzing a carboxylic ester. Esterases, on the other hand, are generally unable to cleave amide bonds, although they can catalyze their formation via ester aminolysis (Sect. 3.1.3, Scheme 2.1). This does not apply to highly strained β-lactams, which can be hydrolyzed by some esterases (pig liver esterase) or lipases (Scheme 2.19) [113].

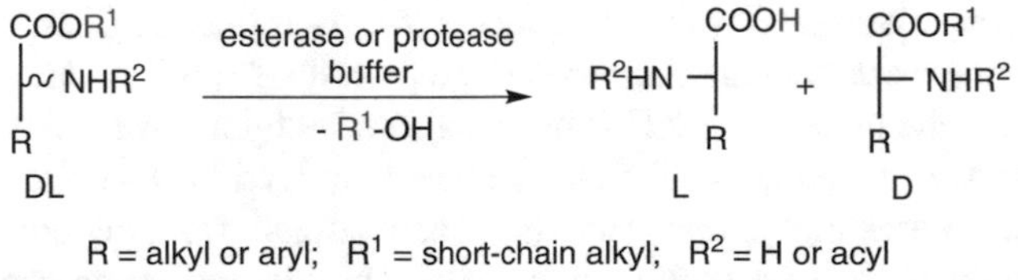

$$COOR^1 \quad \xrightarrow[\substack{\text{esterase or protease} \\ \text{buffer} \\ - R^1\text{-OH}}]{} \quad COOH \quad COOR^1$$

R = alkyl or aryl; R^1 = short-chain alkyl; R^2 = H or acyl

Scheme 2.11 Enzymatic resolution of α-amino acid esters via the esterase method

The amino group of the substrate may be either free or (better) protected by an acyl functionality, preferably an acetyl-, benzoyl-, or the *tert*-butyloxycarbonyl-(Boc)-group in order to avoid possible side reactions such as ring-closure going in hand with the formation of diketopiperazines. The ester moiety should be a short-chain aliphatic alcohol such as methyl or ethyl to ensure a reasonable reaction rate with esterases or proteases. When lipases are used, it is recommended to use more lipophilic alcohol residues (e.g., *n*-butyl, *n*-hexyl, *n*-octyl) or activated analogs bearing electron-withdrawing substituents, such as chloroethyl [114] or trifluoroethyl [115], to ensure high reaction rates.

Numerous enzymes have been used to hydrolyze *N*-acyl amino acid esters, the most versatile and thus very popular catalyst being α-chymotrypsin isolated from bovine pancreas (Scheme 2.12) [118–120]. Since it is one of the early examples of a pure enzyme which became available for biotransformations, its mode of action is well understood. A useful and quite reliable model of its active site has been proposed in order to rationalize the stereochemical outcome of resolutions performed with α-chymotrypsin [121, 122]. Alternatively, other proteases, such as subtilisin [123, 124], thermolysin [125], and alkaline protease [126] are also commonly used for the resolution of amino acid esters. Even whole microorganisms such as lyophilized cells of baker's yeast, possessing unspecific proteases, can be employed [127].

Carbonic anhydrase – an enzyme termed for its ability to catalyze the hydration of carbon dioxide forming hydrogen carbonate – can also be employed. In contrast to the above-mentioned enzymes, it exhibits the opposite enantiopreference by hydrolyzing the D-*N*-acylamino acid esters [116].

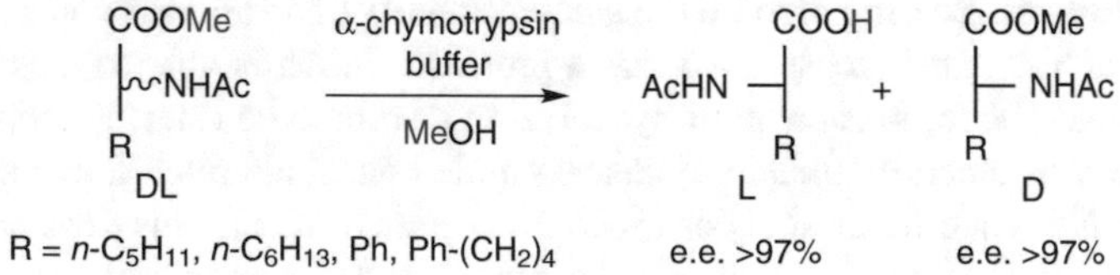

Scheme 2.12 Resolution of *N*-acetyl α-amino acid esters by α-chymotrypsin [116, 117]

An efficient dynamic resolution process for α-amino acid esters has been developed using a crude industrial protease preparation from *Bacillus licheniformis* ('alcalase')[6] (Scheme 2.13) [128]. The remaining unhydrolyzed D-enantiomer of the substrate was racemized in situ, catalyzed by pyridoxal-5-phosphate (PLP, vitamin B_6). Interestingly, this trick has been copied from nature, since pyridoxal-5-phosphate is an essential cofactor for biological amino-group transfer. PLP spontaneously forms a Schiff base with the amino acid ester (but not with the amino acid) which facilitates racemization through reversible proton migration. A range of racemic amino acid esters were dynamically resolved in excellent chemical and optical yield. As a more economical substitute for pyridoxal 5-phosphate, its nonphosphorylated analog (pyridoxal) or salicylaldehyde are preferable for large-scale applications.

---

[6]'Alcalase' is mainly used as additive in detergents for the degradation of proteinogenic impurities, its major enzyme component is subtilisin Carlsberg (alkaline protease A).

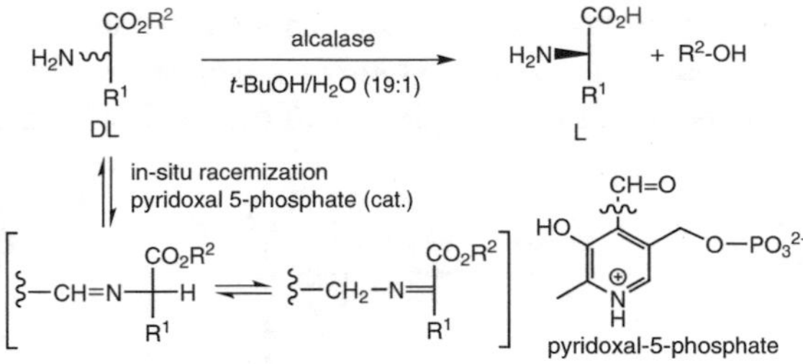

R^1	R^2		Product	
			yield [%]	e.e. [%]
Ph-CH$_2$-	Ph-CH$_2$-	L-Phe	92	98
Ph-CH$_2$-	n-Bu-	L-Phe	92	98
4-Hydroxyphenyl-CH$_2$-	Ph-CH$_2$-	L-Tyr	95	97
4-Hydroxyphenyl-CH$_2$-	n-Pr-	L-Tyr	95	97
(CH$_3$)$_2$CH-CH$_2$-	Ph-CH$_2$-	L-Leu	87	93
n-Bu-	Ph-CH$_2$-	L-NorLeu	87	90
Et-	Ph-CH$_2$-	L-NorVal	87	91

Scheme 2.13 Dynamic resolution of α-amino acid esters via the esterase method

Amidase Method

α-Amino acid amides are hydrolyzed enantioselectively by amino acid amidases (occasionally also termed aminopeptidases) obtained from various sources, such as kidney and pancreas [129] and from different microorganisms, in particular *Pseudomonas*, *Aspergillus*, or *Rhodococcus* spp. (Scheme 2.14, top) [130]. For industrial applications, special amidases (e.g., from *Mycobacterium neoaurum* and *Ochrobactrum anthropi*) have been developed [131, 132]. They are also accept α-substituted α-amino acid amides, which are otherwise not easily hydrolyzed due to steric hindrance [108]. Unreacted D-amino acid amides can be separated from the L-amino acids by extraction into organic solvents due to their different solubility at various pH. After separation, unreacted D-amino acid amides can be recycled ex-situ via base-catalyzed racemization of the corresponding Schiff-base intermediates in a separate step in analogy to the process depicted in Scheme 2.13 [133]. Since amino acid amides are less susceptible to spontaneous chemical hydrolysis in the aqueous environment than the corresponding esters, the products which are obtained by this method are often of higher optical purities compared to those obtained by the esterase method.

In order to avoid tedious separation and ex-situ racemization of the undesired enantiomer from kinetic resolution an elegant dynamic two-enzyme process was developed (Scheme 2.14, bottom) [134]. D-Amino acid amides were hydrolyzed enantioselectively using a thermostable mutant of D-amino acid amidase from *Ochrobactrum anthropi* SV3, while in-situ racemization of the racemic substrate was accomplished by a double mutant of α-amino-ε-caprolactam racemase (L19V/

L78T) from *Achromobacter obae*. Opposite L-amino acids were obtained by using an L-amino acid amidase from *Brevundimonas diminuta*. Both amidases and the racemase were co-expressed into a single *E. coli* host to facilitate handling.

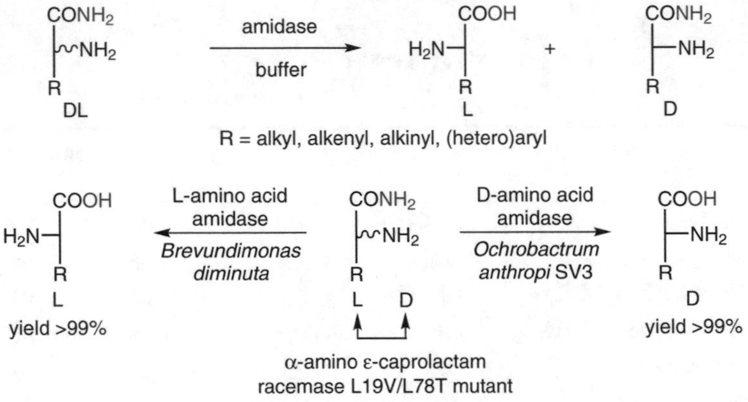

R	E.e. [%]	
	L	D
Ph-	>99	88
4-F-C$_6$H$_4$-	95	83
Ph-CH$_2$-	92	99
4-HO-C$_6$H$_4$-CH$_2$-	95	>99
2-, 3-, or 4-F-C$_6$H$_4$-CH$_2$-	91-98	93-97
4-Cl-C$_6$H$_4$-CH$_2$-	88	73
Ph-(CH$_2$)$_2$-	98	77

Scheme 2.14 Kinetic and dynamic resolution of amino acid amides via the amidase method

Acylase Method

Aminoacylases catalyze the hydrolysis of *N*-acyl amino acid derivatives, with the acyl groups preferably being acetyl, chloroacetyl, propionyl or benzoyl. Alternatively, the corresponding *N*-carbamoyl- and *N*-formyl derivatives can be used [135]. Enzymes of the amino acylase type have been isolated from hog kidney, and from *Aspergillus* or *Penicillium* spp. [136–138]. The versatility of this type of enzyme has been demonstrated by the resolution of racemic *N*-acetyl tryptophan, -phenylalanine, and -methionine on an industrial scale using column reactors (Scheme 2.15) [139, 140].

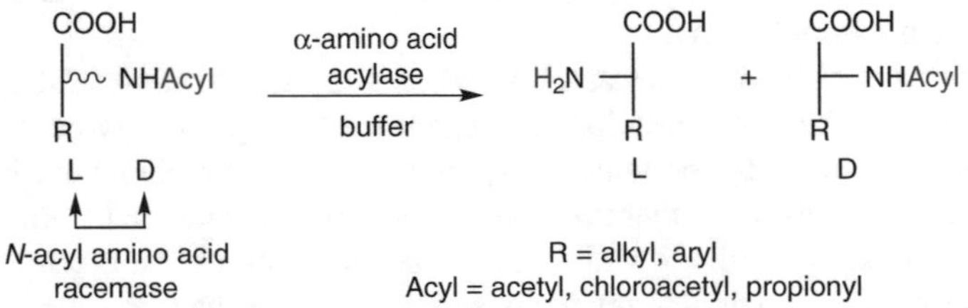

Scheme 2.15 Enzymatic resolution of *N*-acyl amino acids via the acylase method

On a laboratory scale, the readily available amino acylase from hog kidney is recommended [141]. It proved to be extremely substrate-tolerant, allowing variations of the alkyl- or aryl-moiety R within a wide structural range while retaining very high specificities for L-enantiomers, which made it a reliable tool for the synthesis of bioactive compounds [142–144]. Unwanted enantiomers of N-acetyl amino acids can be racemized ex-situ by heating with acetic anhydride, which involves activation of the acid moiety via a mixed anhydride, which undergoes cyclization to form an oxazolinone (azlactone). The latter is subject to racemization via an intermediate achiral enol. Like the amidase process, on large scale the acylase method was converted into a dynamic process by in-situ racemization of the nonreacting N-acylamino acid using an N-acylamino acid racemase [145–147]. In contrast to the majority of amino acid racemases, which are cofactor-dependent (usually pyridoxal-5-phosphate), an enzyme which was isolated from *Amycolatopsis* sp. requires a divalent metal ion such as Co, Mn, or Mg for catalytic activity [148].

Although the majority of N-acylamino acid acylases are L-selective, several stereo-complementary D-acylases were identified [149–152], which allow to access D-amino acids. Cyclic amino acids, such as piperidine-2-carboxylic acid are valuable building blocks for the synthesis of pharmaceuticals, such as the anticancer drug Incel, respectively. In order to access both enantiomers by choice of the appropriate enzyme, enantiocomplementary acylases from microbial sources were developed using classic enrichment techniques. An L-acylase from *Arthrobacter* sp. furnishes the free L-amino acid plus the unreacted D-N-acyl-substrate enantiomer, while opposite enantiomers were obtained using a D-specific acylase from *Arthrobacter xylosoxidans* (Scheme 2.16) [153–154].

Interestingly, even N-acyl α-amino*phosphonic* acid derivatives have been resolved using penicillin acylase [155].

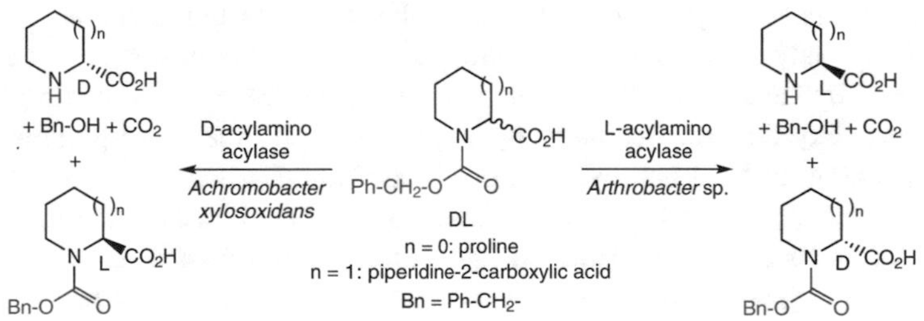

Scheme 2.16 Resolution of cyclic N-benzyloxycarbonyl amino acids using enantiocomplementary acylases

Hydantoinase Method
5-Substituted hydantoins are obtained in racemic form from cheap starting materials such as an aldehyde, hydrogen cyanide, and ammonium carbonate using the Bücherer–Bergs synthesis [156]. Hydantoinases from different microbial sources catalyze the hydrolytic ring-opening to form the corresponding N-carbamoyl–α-amino acids [157–159]. In nature, many (but not all) of these enzymes are

responsible for the cleavage of dihydropyrimidines occurring in pyrimidine catabolism, therefore they are often also called 'dihydro-pyrimidinases' (Scheme 2.17) [160–162].

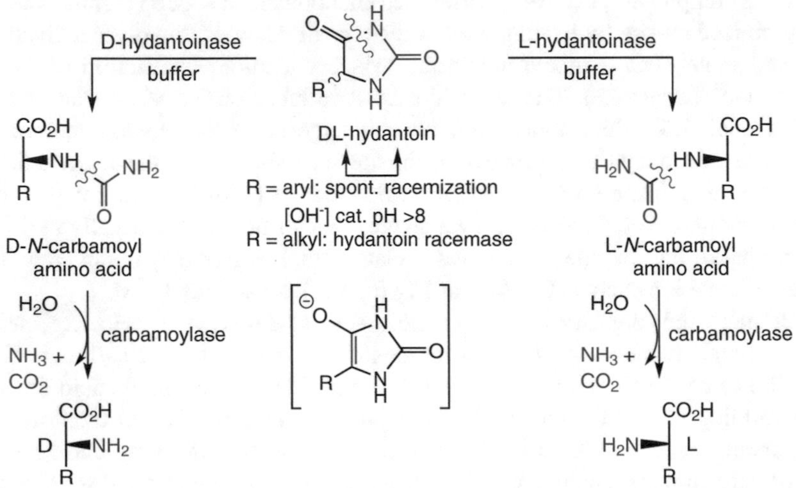

Scheme 2.17 Enzymatic resolution of hydantoins via the hydantoinase method

In contrast to the above-mentioned amino acid resolution methods involving amino acid esters, -amides, or N-acylamino acids where the natural L-enantiomer is preferably hydrolyzed, hydantoinases usually convert the opposite D-enantiomer [163–165], and L-hydantoinases are known to a lesser extent [166–168]. In addition, D-hydantoinases usually possess a broader substrate spectrum than their L-counterparts. Previously, N-carbamoyl amino acids thus obtained were chemically deprotected by treatment with nitrous acid or by exposure to an acidic pH (<4). Nowadays, they are enzymatically hydrolyzed to yield the corresponding amino acids by use of an N-carbamoyl amino acid amidohydrolase (carbamoylase) with matching enantiopreference, which is often produced by the same microbial species [169]. One property of 5-substituted hydantoins, which makes them particularly attractive for large-scale resolutions is their ease of racemization. When R contains an aromatic group, the enantiomers of the starting hydantoins are readily equilibrated at slightly alkaline pH (>8), which is facilitated by resonance stabilization of the corresponding enolate. In contrast, aliphatic substituted hydantoins racemize very slowly under the reaction conditions compatible with hydantoinases due to the lack of enolate stabilization. For such substrates the use of hydantoin racemases is required to render a dynamic resolution process, which ensures a theoretical yield of 100% [170, 171].

Lactamase Method

Due to their cyclic structure, cyclic amides (γ-, δ- and ε-lactams) are chemically considerably more stable and thus cannot be hydrolyzed by conventional proteases.

However, they can be resolved using a special group of proteases acting on cyclic amide bonds – lactamases [172].

The bicyclic γ-lactam shown in Scheme 2.18 is an important starting material for the production of antiviral agents, such as Carbovir and Abacavir. It can be efficiently resolved using enantiocomplementary γ-lactamases from microbial sources: an enzyme from *Rhodococcus equi* produced the (*S*)-configurated amino acid (plus enantiomeric non-converted lactam), and another lactamase isolated from *Pseudomonas solanacearum* acted in an enantiocomplementary fashion by providing the corresponding mirror-image products [173].

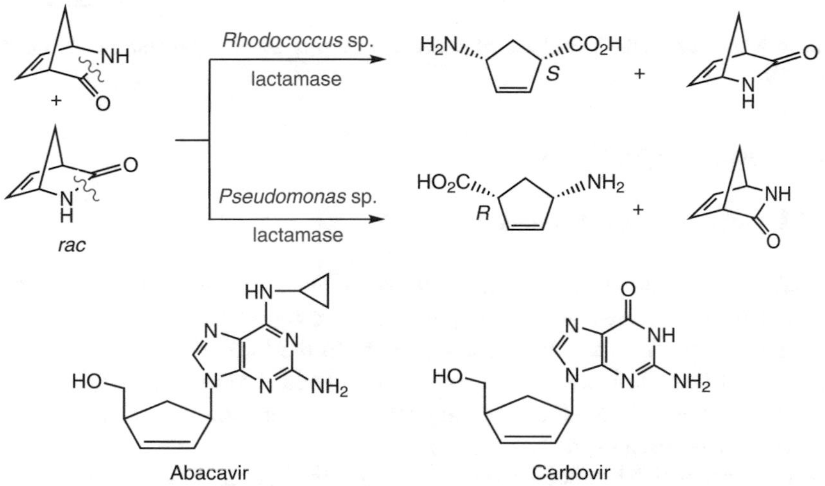

Scheme 2.18 Enzymatic resolution of bicyclic γ-lactams via the lactamase method

For the biocatalytic synthesis of lysine, an enantioselective lactamase is employed in the kinetic resolution of *rac*-α-amino-ε-caprolactam (Scheme 2.19, top) [174]. A suitable α-amino-ε-caprolactam racemase was found in several bacterial species, such as *Achromobacter*, *Alcaligenes* and *Flavobacterium*, detailed studies were performed with the enzyme from *Achromobacter obae* [175]. Quite remarkably, this racemase also accepts non-cyclic amino acid amides [176]. The racemase is used in combination with a suitable D- or L-α-amino-ε-caprolactamase in a dynamic process for the production of D- or L-lysine on an industrial scale in 100% yield at ~4000 t per annum from the racemic lactam [177].

In contrast to γ-, δ- and ε-lactams, highly strained β-lactams are more easily susceptible to enzymatic hydrolysis and thus can be (slowly) hydrolyzed by carboxyl ester hydrolases, such as esterases [178] and lipases [179, 180]. The bicyclic lactam shown in Scheme 2.19 (bottom), which serves as starting material for the synthesis of the antifungal agent (−)-cispentacin, was efficiently resolved using *Rhodococcus equi* lactamase [181].

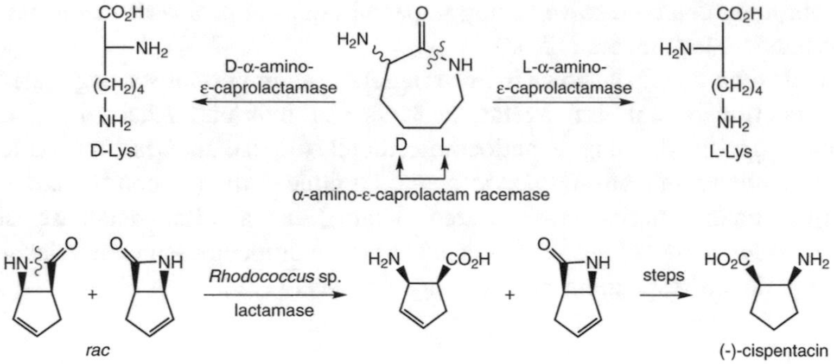

Scheme 2.19 Enzymatic hydrolysis of strained β-lactams and α-amino-ε-caprolactam using lactamases

2.1.3   Ester Hydrolysis

2.1.3.1   Esterases and Proteases

In contrast to the large number of readily available microbial lipases, less than a dozen of true 'esterases' – such as pig and horse liver esterases (PLE [182] and HLE, respectively) – have been used to perform the bulk of the large number of highly selective hydrolyses of carboxylic esters. Thus, the use of a different esterase is not easy in cases where the reaction proceeds with insufficient selectivity with a popular enzyme such as PLE.

An esterase which has been shown to catalyze the hydrolysis of nonnatural esters with exceptionally high selectivities is acetylcholine esterase (ACE). It would certainly be a valuable enzyme to add to the limited number of available esterases but it has a significant disadvantage since it is isolated from *Electrophorus electricus* – the electric eel. Comparing the natural abundance of this species with the occurrence of horses or pigs, its high price – which is prohibitive for large-scale applications – is probably justified. Thus, the number of ACE applications is limited [183–186]. Additionally, also cholesterol esterase is of limited use, since it seems to prefer bulky substrates which show structural similarities to the natural substrates of cholesterol esterase, i.e., steroid esters [58, 187].

To overcome this narrow range of readily available esterases, whole microbial cells are sometimes used instead of isolated enzyme preparations [188]. Although some highly selective conversions using whole-cell systems have been reported, it is clear that any optimization by controlling the reaction conditions is very complicated when whole cells are employed, because in most cases the nature of the actual active enzyme system remains unknown.

More recently, novel microbial esterases [189, 190] such as carboxyl-esterase NP [191] have been identified from an extensive screening in search for biocatalysts with high specificities for certain types of substrates. Since they have been made available in generous amounts by genetic engineering [192], they are now being used more widely. Despite numerous efforts directed towards the cloning and overexpression of microbial esterases, the number of synthetically useful enzymes – possessing a relaxed substrate specificity by retaining high enantioselectivity – are limited: many novel esterases showed disappointing selectivities [193, 194].

Fortunately, as mentioned in the foregoing chapter, a large number of proteases can also selectively hydrolyze carboxylic esters and this effectively compensates for the limited number of esterases [195]. The most frequently used members of this group are α-chymotrypsin [196], subtilisin [197] and, to a somewhat lesser extent, trypsin, pepsin [198], papain [199], penicillin acylase [200, 201] and a protease from *Aspergillus oryzae*. The latter enzyme seems to be particularly useful for the selective hydrolysis of bulky esters. As a rule of thumb, when acting on nonnatural carboxylic esters, most proteases seem to retain a preference for the hydrolysis of that enantiomer which mimics the configuration of an L-amino acid more closely [202].

Since many of the studies on the ester-hydrolysis catalyzed by α-chymotrypsin and subtilisin have been performed together with PLE in the same investigation, representative examples are not singled out in a separate section but are incorporated into the following chapter.

The structural features of more than 90% of the substrates which have been transformed by esterases and proteases can be reduced to the general formulas given in Scheme 2.20. The following general rules can be applied to the construction of substrates for esterases and proteases:

- For both esters of the general type I and II, the center of chirality (marked by an asterisk [*]) should be located as close as possible to the site of the reaction (that is, the carbonyl group of the ester) to ensure an optimal chiral recognition. Thus, α-substituted carboxylates and esters of secondary alcohols are usually more selectively hydrolyzed than their β-substituted counterparts and esters of chiral primary alcohols, respectively.
- Both substituents R^1 and R^2 can be alkyl or aryl groups, but they should differ in size and polarity to aid the chiral recognition process of the enzyme. They may also be joined together to form cyclic structures.
- Polar or charged functional groups located at R^1 and R^2, such as –OH, –COOH, –CONH$_2$, or –NH$_2$, which are heavily hydrated in an aqueous environment should be absent, since esterases (and in particular lipases) do not accept highly polar hydrophilic substrates. If such moieties are required, they should be masked with an appropriate lipophilic protective group.
- The alcohol moieties R^3 of type-I esters should be as short as possible, preferably methyl or ethyl. If neccessary, the reaction rate of ester hydrolysis may be enhanced by attaching electron-withdrawing groups to the alcohol moiety to give methoxymethyl or 2-haloethyl esters, respectively. In contrast, carboxylates bearing long-chain alcohols are usually hydrolyzed at reduced reaction rates with esterases and proteases.

- The same considerations are applicable to acylates of type II, where short-chain acetates or propionates are the preferred acyl moieties. Increasing the carbonyl reactivity of the substrate ester by adding electron-withdrawing substituents such as halogen or methoxy (leading to α-halo- or α-methoxyacetates) is a frequently used method to enhance the reaction rate in enzyme-catalyzed ester hydrolysis [203].
- One limitation in substrate construction is common for both types of sub-strates: the remaining hydrogen atom at the chiral center must not be replaced, since α,α,α-trisubstituted carboxylates and esters of tertiary alcohols are usually too bulky to be accepted by esterases and proteases, although there are some rare exceptions to this rule [204–207]. This limitation turns them into potential protective groups for carboxy- and alcoholic functionali-ties, such as t-butyl esters and pivalates, in case an enzymatic hydrolysis is not desired. For serine ester hydrolases, the rare ability to hydrolyze bulky esters was attributed to an atypical Gly-Gly-Gly-X-sequence motif (instead of the common Gly-X-motif) in the oxyanion cavity located within the active site, which was found in *Candida rugosa* and *Candida antarctica* lipase A [208–210].
- It is clear that both general substrate types (which themselves would constitute racemic substrates) may be further combined into suitable prochiral or *meso*-substrates (Scheme 2.20).

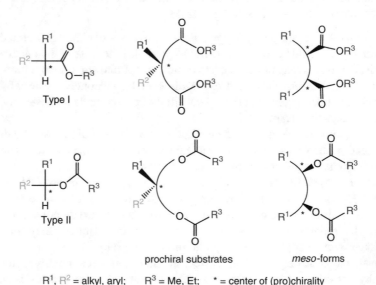

R¹, R² = alkyl, aryl;   R³ = Me, Et;   * = center of (pro)chirality

Scheme 2.20 Types of substrates for esterases and proteases

Pig Liver Esterase and α-Chymotrypsin
Amongst all the esterases, pig liver esterase (PLE) is clearly the champion considering its general versatility. This enzyme is constitutionally complex and consists of several so-called *iso*enzymes, which are associated as trimers of three individual proteins [211]. However, for many applications this crude mixture can be used without any problems although the isoenzyme subunits often possess similar (but not identical [212]) stereospecificities [213]. Thus, the selectivity of crude PLE may vary, depending on the source and the pretreatment of the enzyme preparation [214]. The biological role of PLE is the hydrolysis of various esters occurring in the porcine diet, which would explain its exceptionally wide substrate tolerance. For preparative reactions it is not absolutely necessary to use the expensive commercially available enzyme preparation because a crude acetone powder which can easily be prepared from pig liver is a cheap and efficient alternative [215].

In general, hepatic esterases from related sources such chickens, hamsters, guinea pigs, or rats were found to be less versatile when compared to PLE. In certain cases, however, esterases from rabbit [216, 217] and horse liver (HLE) [218, 219] proved to be useful substitutes for PLE.

Mild Hydrolysis Acetates of primary and secondary alcohols such as cyclopropyl acetate [220] and methyl or ethyl carboxylates (such as the labile cyclopentadiene ester [221]) can be selectively hydrolyzed under mild conditions using PLE, avoiding decomposition reactions which would occur during a chemical hydrolysis under acid or base catalysis (Scheme 2.21). For example, this strategy has been used for the final deprotection of the carboxyl moiety of prostaglandin E_1 avoiding the destruction of the delicate molecule [222, 223].

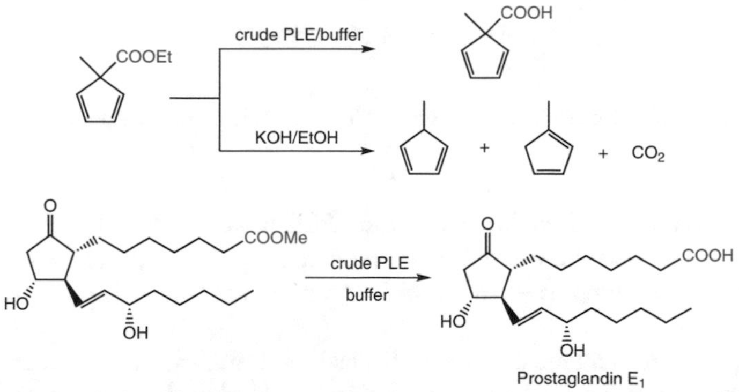

Scheme 2.21 Mild ester hydrolysis by porcine liver esterase

Regio- and Diastereoselective Hydrolysis Regiospecific hydrolysis of dimethyl malate at the 1-position could be effected with PLE as catalyst (Scheme 2.22) [224]. Similarly, hydrolysis of an *exo/endo*-mixture of diethyl dicarboxylates with a bicyclo[2.2.1]heptane framework occurred only on the less hindered *exo*-position

[225] leaving the *endo*-ester untouched, thus allowing a facile separation of the two positional isomers in a diastereomeric mixture.

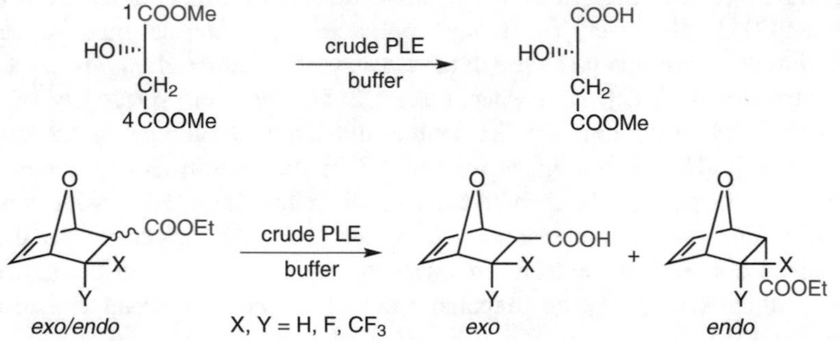

Scheme 2.22 Regio- and diastereoselective ester hydrolysis by porcine liver esterase

Separation of *E/Z*-Isomers With *E/Z*-diastereotopic diesters bearing an aromatic side chain, PLE selectively hydrolyzed the ester group in the more accessible (*E*)-*trans*-position to the phenyl ring, regardless of the *p*-substituent [226] (Scheme 2.23). In analogy to the hydrolysis of dicarboxylates (Scheme 2.3) the reaction stopped at the (*Z*)-monoester stage with no diacid being formed. Other hydrolytic enzymes (proteases and lipases) were less selective in this case.

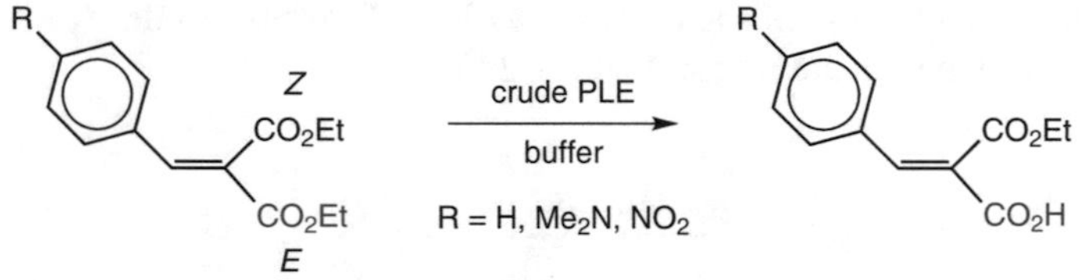

Scheme 2.23 Regioselective hydrolysis of *E/Z*-diastereotopic diesters by porcine liver esterase

Desymmetrization of Prochiral Diesters PLE has been used less frequently for the resolution of racemic esters (where α-chymotrypsin has played a more important role) but was employed more widely for the desymmetrization of prochiral diesters.

As depicted in Scheme 2.24, α,α-disubstituted malonic diesters can be selectively transformed by PLE or α-chymotrypsin to give the corresponding chiral monoesters [227, 228]. These transformations demonstrate an illustrative example for an 'alternative fit' of substrates with different steric requirements. While PLE preferentially hydrolyses the pro-S ester group on substrates possessing small α-substituents (R) ranging from ethyl through n-butyl to phenyl, an increase of the steric bulkiness of R forces the substrate to enter the enzyme's active site in an opposite (flipped) orientation. Thus, with the more bulky substituents the pro-R ester is preferentially cleaved.

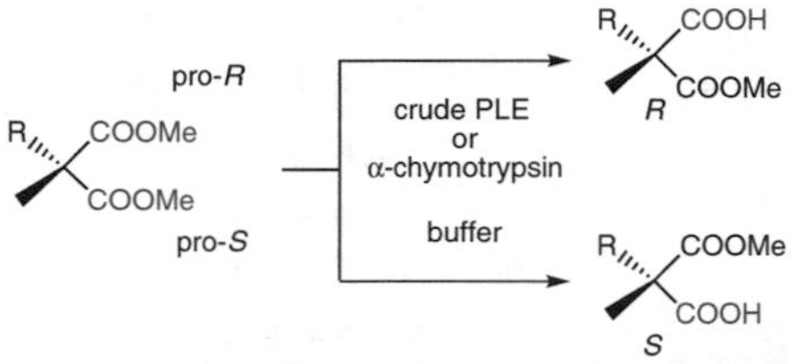

Enzyme	R	Configuration	e.e. [%]
PLE*	Ph-	S	86
PLE	C_2H_5-	S	73
PLE	n-C_3H_7-	S	52
PLE	n-C_4H_9-	S	58
PLE	n-C_5H_{11}-	R	46
PLE	n-C_6H_{13}-	R	87
PLE	n-C_7H_{15}-	R	88
PLE	p-MeO-C_6H_4-CH_2-	R	82
PLE	t-Bu-O-CH_2-	R	96
α-chymotrypsin	Ph-CH_2-	R	~100

* The ethyl ester was used.

Scheme 2.24 Desymmetrization of prochiral malonates by porcine liver esterase and α-chymotrypsin

As shown in Scheme 2.25, the prochiral center may be moved away from the ester moiety into the β-position. Thus, chiral recognition by PLE [229–233] and α-chymotrypsin [234–237] is retained during the desymmetrization of prochiral 3-substituted glutaric diesters. Whole cells of *Acinetobacter lowffii* and *Arthrobacter* spp. have also been used as a source for esterase activity [238] and, once again, depending on the substitutional pattern on carbon-3, the desymmetrization can lead to both enantiomeric products.

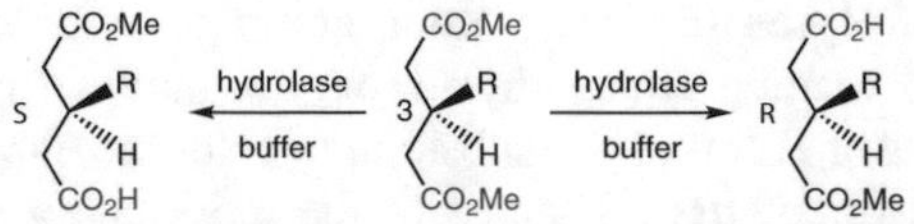

Hydrolase	R	Product	e.e. [%]
α-chymotrypsin*	AcNH-	R	79
α-chymotrypsin	Ph-CH₂-O-	R	84
α-chymotrypsin	CH₃OCH₂O-	R	93
PLE	AcNH-	R	93
PLE	CH₃-	R	90
PLE	Ph-CH₂-CH=CH-CH₂-	S	88
PLE	t-Bu-CO-NH-	S	93
PLE	HO-	S	12
α-chymotrypsin*	HO-	R	85
Acinetobacter sp.*	HO-	R	>95
Arthrobacter sp.*	HO-	S	>95

* The corresponding ethyl esters were used.

Scheme 2.25 Desymmetrization of prochiral glutarates

Acyclic *meso*-dicarboxylic esters with a glutaric acid backbone were also good substrates for PLE [239] and α-chymotrypsin (Scheme 2.26) [240]. Interestingly, an additional hydroxy group in the substrate led to an enhancement of the chiral recognition.

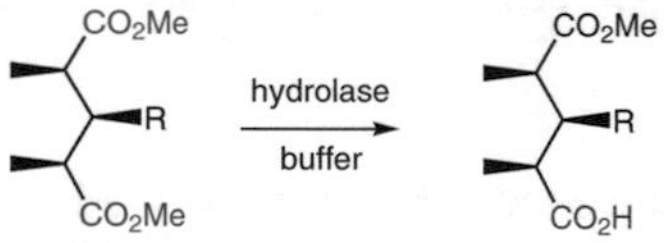

R	enzyme	e.e. [%]
H	crude PLE	64
OH	crude PLE	98
H	α-chymotrypsin	77

Scheme 2.26 Desymmetrization of acyclic *meso*-dicarboxylates by α-chymotrypsin and porcine liver esterase

The synthetic potential of the desymmetrization of cyclic *meso*-1,2-dicarboxylates by PLE is demonstrated in Scheme 2.27 [241]. A striking reversal of stereopreference was caused by variation of the ring size: when the rings are small ($n = 1, 2$), the (S)-carboxyl ester is selectively cleaved, whereas the (R)-counterpart preferentially reacts when the rings are larger ($n = 4$). The highly flexible cyclopentane derivative of moderate ring size is in the middle of the range and its chirality is not very well recognized. The fact that the nature of the alcohol moiety of such esters can have a significant impact in both the reaction rate and stereochemical outcome of the hydrolysis was shown by the poor chiral recognition of the corresponding diethyl ester of the cyclohexane derivative, which was slowly hydrolyzed to give the monoethyl ester of poor optical purity [242].

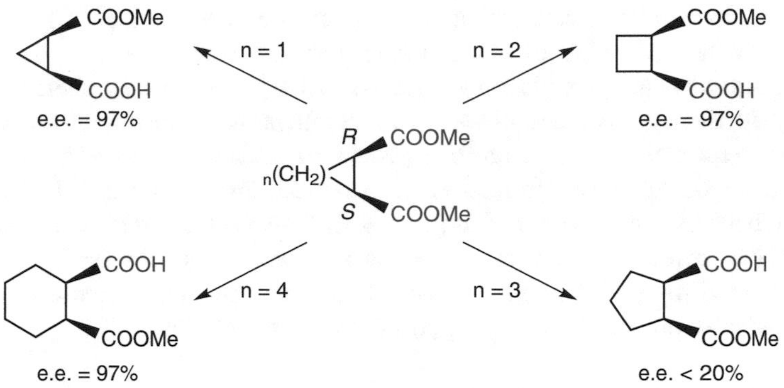

Scheme 2.27 Desymmetrization of cyclic *meso*-1,2-dicarboxylates by porcine liver esterase

Bulky bicyclic *meso*-dicarboxylates, which were extensively used as optically pure building blocks for the synthesis of bioactive products, are well accepted by PLE [243]. While the *exo*-configurated diester was a good substrate (Scheme 2.28, top), the corresponding more sterically hindered *endo*-counterpart was hydrolyzed at a significantly reduced reaction rate and stereoselectivity (e.e. 64%). The importance of the appropriate choice of the alcohol moiety is exemplified with unsaturated analogs [231] (Scheme 2.28, bottom): While the short-chain methyl and ethyl esters were hydrolyzed with high selectivities, the propyl ester was not.

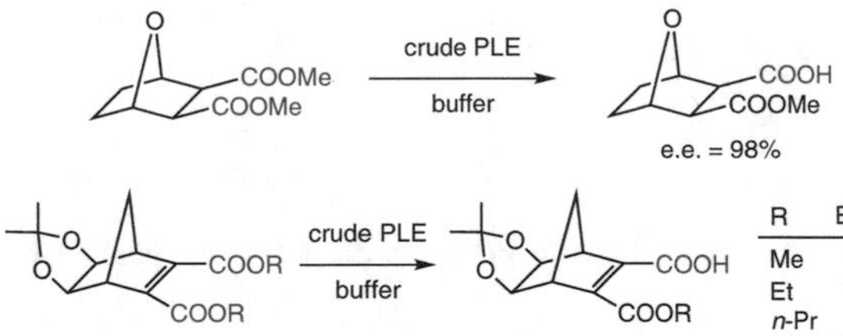

R	E.e. [%]
Me	85
Et	~100
n-Pr	45

Scheme 2.28 Desymmetrization of bicyclic *meso*-1,2-dicarboxylates by porcine liver esterase

Cyclic *meso*-diacetates can be hydrolyzed in a similar fashion. As shown in Scheme 2.29 (top), the cyclopentene *meso*-monoester [244], which constitutes one of the most important chiral synthons for prostaglandin synthesis [245], was obtained in an e.e. of 80–86% using crude PLE. In accordance with the above-mentioned hypotheses for the construction of esterase substrates, a significant influence of the acyl moiety of the ester was observed: the optical purity of the monoester gradually declined from 80–86% to 33% as the acyl chain of the starting substrate ester was extended from acetate to butanoate. A detailed study of the stereoselectivity of PLE isoenzymes revealed that isoenzymes PLE-1–3 gave almost identical results as the crude PLE preparation, whereas isoenzymes PLE-4 and PLE-5 showed lower stereoselectivities with a preference for the opposite enantiomer [246].

In order to avoid recrystallization of the optically enriched material (80–86% e.e.) obtained with crude PLE to enantiomeric purity, a search for a more selective esterase revealed that acetylcholine esterase (ACE) was the best choice [247]. It hydrolyzed the cyclopentene diester with excellent stereoselectivity but with the *opposite* stereopreference as with PLE (Scheme 2.29, bottom). Similar results were obtained by using lipases from porcine pancreas [248] and *Candida antarctica* [249]. When structural analogs of larger ring size were subjected to ACE hydrolysis, a dramatic effect on the stereochemical course was observed: while the six-membered *meso*-diester gave a racemic product, the seven-membered analog led to optically pure monoester of opposite configuration [250].

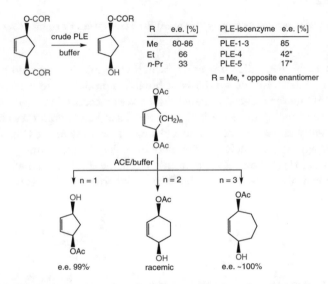

Scheme 2.29 Desymmetrization of cyclic *meso-sec*-diacetates by porcine liver esterase and acetylcholine esterase

Cyclic *meso*-diacetates containing protected nitrogen functionalities proved to be excellent substrates for PLE, although the chiral recognition is more difficult due to the fact that diastereotopic *prim*-alcohols have to be differentiated (Scheme 2.30). In the benzyl-protected 1,3-imidazolin-2-one system – which serves as a starting material for the synthesis of the vitamin (+)-biotin – the optical yield of PLE-catalyzed hydrolysis of the *cis*-diacetate [251] was much superior to that of the corresponding *cis*-dicarboxylate [252].

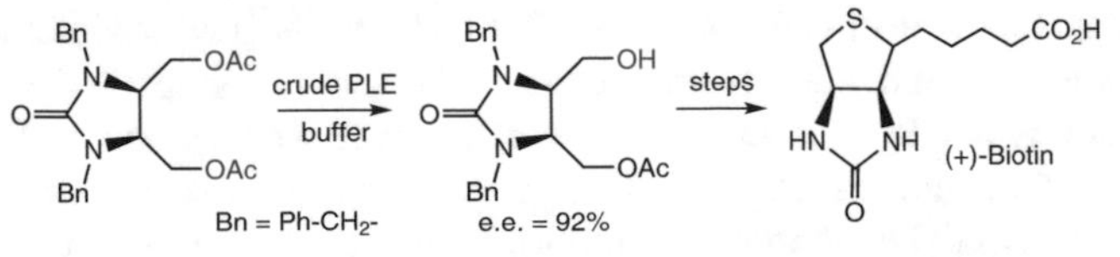

Scheme 2.30 Desymmetrization of *N*-protected cyclic *meso-prim*-diacetate by porcine liver esterase

Resolution of Racemic Esters Although PLE-catalyzed resolution of racemic esters have been performed less often as compared to the desymmetrization of prochiral and *meso*-diesters, it has been proven to be a valuable technique for the resolution of non-natural esters. Interestingly, chirality does not necessarily need to be located on a tetrahedral carbon atom, as in the case of the *trans*-epoxy dicarboxylate [253], also axial chirality of an allenic carboxylic ester [254] and an iron-tricarbonyl complex [255] were well recognized by PLE (Scheme 2.31, top).

Resolution of an *N*-acetylaminocyclopentene carboxylate shown in Scheme 2.31 (bottom) was used to access optically pure starting material for the synthesis of carbocyclic nucleoside analogs with antiviral activity [256]. Also a very bulky tricyclic monoester (required for natural product synthesis) was resolved with remarkably good selectivity [257].

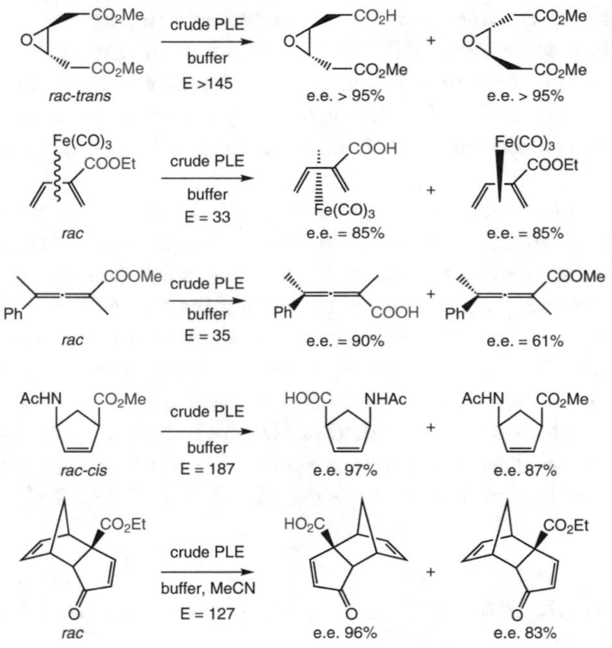

Scheme 2.31 Resolution of racemic carboxylic esters by porcine liver esterase

An example demonstrating the high stereospecificity of PLE is the kinetic resolution of the cyclic *trans*-1,2-diacetate shown in Scheme 2.32 [258]. The (R, R)-diacetate enantiomer – possessing two ester groups showing the matching (R)-configuration – was hydrolyzed from the racemic mixture via the monoester stage to yield the corresponding (R,R)-diol. The (S,S)-diacetate remained untouched, since it possesses only nonmatching (S)-ester groups. Again, as observed in the desymmetrization of *cis-meso*-1,2-dicarboxylates, the enantioselectivity strongly depended on the ring size: while the four- and six-membered substrates gave excellent results with opposite enantiopreference, the five-membered substrate analog was not suitable. It should be noted that a desymmetrization of the corresponding *cis-meso*-1,2-diacetates is impeded by nonenzymic acyl migration which leads to facile racemization of any chiral monoester that is formed.

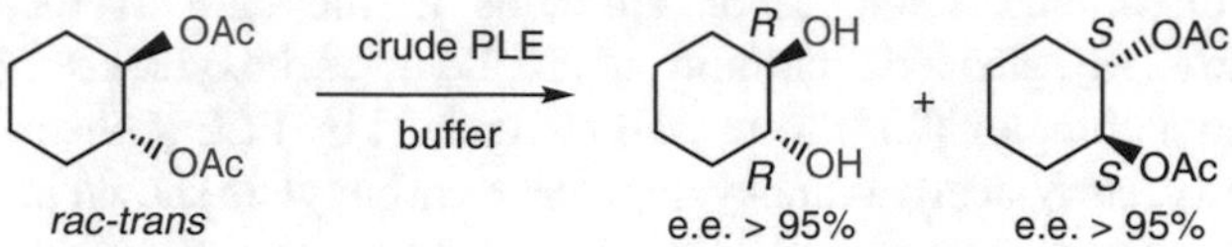

Scheme 2.32 Resolution of a cyclic *trans*-1,2-diacetate by porcine liver esterase

Inspired by the broad substrate range of porcine liver esterase, cloning and overexpression of PLE isoenzymes was persued over the past years in order to provide a reliable enzyme source and to overcome imperfect stereoselectivities of crude PLE preparations [259–261]. In addition, for the application of PLE-derived pharma products in humans, the use of enzymes from animal sources is undesirable due to the risk of contaminations by viruses and prions and due to the fact that products derived from pigs are considered impure by several world religions.

Analysis of the amino acid sequences of PLE isoenzymes revealed that the remarkably small differences of ca. 20 amino acids are not distributed randomly but are located within distinct conserved areas. Among the different isoenzymes, PLE-1 (also termed γ-PLE) and an isoenzyme termed A-PLE ('alternative pig liver esterase') [261, 262] were shown to be most useful for stereoselective ester hydrolysis. The latter enzyme, which was expressed at a high level in *Pichia pastoris*, is remarkably stable and showed perfect enantioselectivity for the industrial-scale resolution of methyl (4*E*)-5-chloro-2-isopropyl-4-pentenoate, which is a key building block for the synthesis of the renin inhibitor Aliskiren, which is used in the treatment of hypertension (Scheme 2.33) [263].

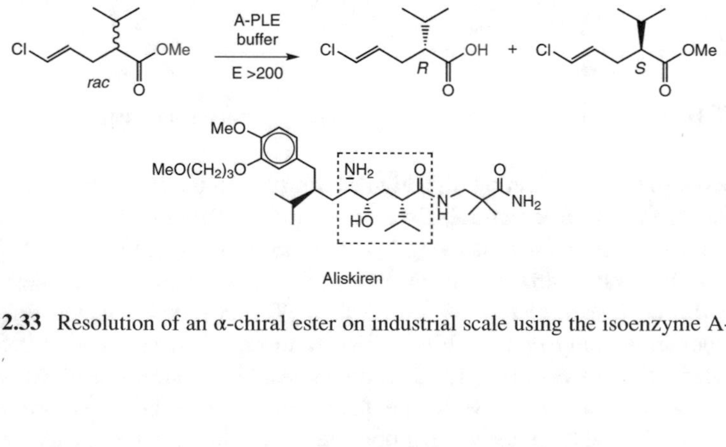

Scheme 2.33 Resolution of an α-chiral ester on industrial scale using the isoenzyme A-PLE

Microbial Esterases

Complementary to the use of isolated enzymes, whole microbial cells have also been used to catalyze esterolytic reactions. Interesting cases are reported from bacteria, yeasts, and fungi, such as *Bacillus subtilis* [264], *Brevibacterium ammoniagenes* [265], *Bacillus coagulans* [266], *Pichia miso* [26], and *Rhizopus nigricans* [267]. Although the reaction control becomes more complex on using whole microbial cells, the selectivities achieved are sometimes surprisingly high [268]. Since hydrolytic reactions do not require any cofactors, which are usually recycled by the metabolism of a 'living' fermenting organism, lyophilized 'resting' microbial cells can be used to minimize potential side reactions caused by competing enzymes. For instance, baker's yeast is a rich source of esterase activity, which was employed to resolve 1-alkyn-3-yl acetates with high selectivities [269] (Scheme 2.34).

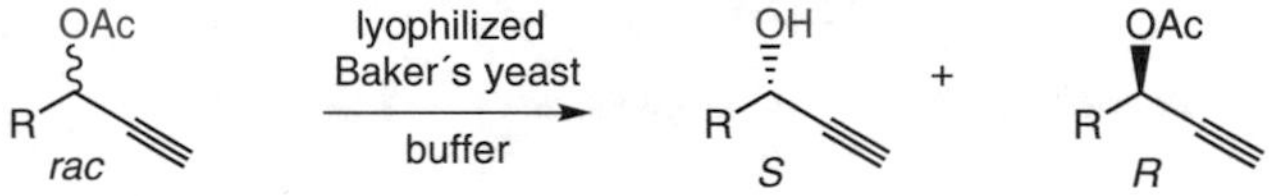

R	E	e.e. [%]	e.e. [%]
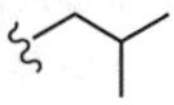 ξ–CH(CH₃)₂ chain	46	91	72
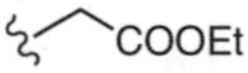 ξ–COOEt	~100	91	>97
 ξ–COOEt	89	96	59

Scheme 2.34 Hydrolytic resolution of *sec*-alcohols using whole (resting) cells of baker's yeast

Due to the importance of α-aryl- and α-aryloxy-substituted propionic acids as antiinflammatory agents (e.g., naproxen, ibuprofen) and agrochemicals (e.g., the herbicide diclofop), respectively, where the majority of the biological activity resides in only one enantiomer (*S* for α-aryl- and *R* for α-aryloxy derivatives,[7] a convenient way for the separation of their enantiomers was sought by biocatalytic methods. An extensive screening program carried out by the industry has led to isolation of an esterase from *Bacillus subtilis* [270] (Scheme 2.35). The enzyme, termed 'carboxyl esterase NP', accepts a variety of substrates esters, including naproxen [271, 272]. It exhibits highest activity and selectivity when the substrate has an aromatic side chain, as with α-aryl- and α-aryloxypropionic acids. With α-aryl derivatives the corresponding (*S*)-acids are obtained. Also α-aryloxy analogs are resolved with similar high specificities, but products have the opposite spatial configuration, taking into account that a switch in CIP sequence priority occurs when going from aryl to aryloxy. This means that the stereochemical preference of carboxyl esterase NP is reversed when an extra oxygen atom is introduced between the chiral center and the aromatic moiety.

---

[7]Be aware of the switch in the Cahn-Ingold-Prelog sequence priority.

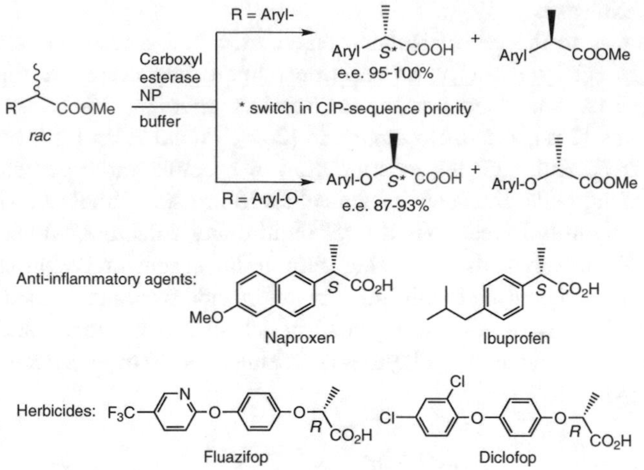

Scheme 2.35 Resolution of α-substituted propionates by carboxylesterase NP

Esterase Activity of Proteases

Numerous highly selective ester hydrolyses catalyzed by α-chymotrypsin [120] and papain have featured in excellent reviews [273] and the examples shown above illustrate their synthetic potential. On large scale, subtilisin (a protease which is widely used in detergent formulations) is a lost cost alternative [274–276].

The major requirements for substrates of type I (see Scheme 2.20) to be selectively hydrolyzed by proteases, such as α-chymotrypsin, pepsin, papain and subtilisin are the presence of a polar and a hydrophobic group on the α-center (R^1 and R^2, respectively) to mimic the natural substrates – amino acids.

Proteases are also useful for regioselective hydrolytic transformations (Scheme 2.36). For example, while regio-selective hydrolysis of a dehydroglutamate diester at the 1-position was achieved using α-chymotrypsin, the 5-ester was attacked by papain [277]. The latter is one of the few enzymes used for organic synthetic transformations originating from plant sources (papaya). Related protease preparations are derived from fig (ficin) and pineapple stem (bromelain) [278].

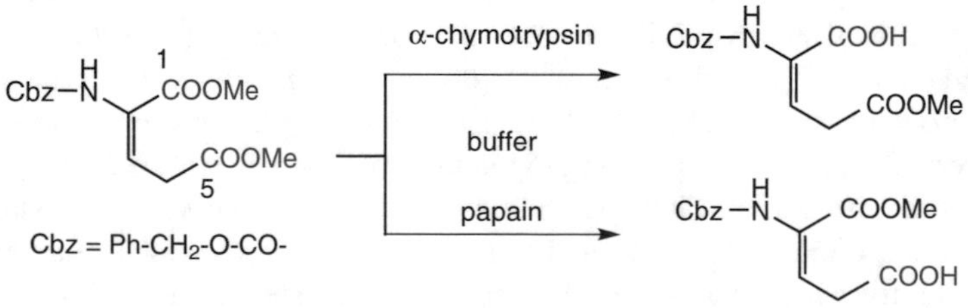

Scheme 2.36 Regio-complementary ester hydrolysis by proteases

In addition to the above mentioned enzymes, two proteases have emerged as highly selective biocatalysts for hydrolysis. Penicillin acylase is highly chemoselective for the cleavage of a phenylacetate group in its natural substrate, penicillin G, which is industrially used for the production of 6-aminopenicillanic acid (6-APA) (Scheme 2.37). This reaction has become a paradigm of how biocatalysis can contribute to make a chemical process more environmentally friendly. Chemical hydrolysis of 1 ton of Pen G requires ~10 tons of organic solvent (half of which is chlorinated) and problematic reagents, such as PCl_5 (600 kg) and amines (1 ton). The enzymatic hydrolysis using immobilized (reusable) Pen G acylase only requires a base to neutralize the acid formed [279]. Due to its specificity for a phenylacetate moiety, this enzyme can be employed in enzymatic protecting group chemistry [280, 281]. For instance, phenylacetyl groups can be removed in a highly chemoselective fashion in the presence of acetate esters (Scheme 2.37) [282, 283]. Furthermore, it can be used for the resolution of esters of primary [284] and secondary alcohols [285] as long as the acid moiety consists of a phenylacetyl group or a structurally closely related (heterocyclic) analog [286–288]. Some structural similarity of the alcohol moiety with that of the natural substrate penicillin G has been stated as being an advantage.

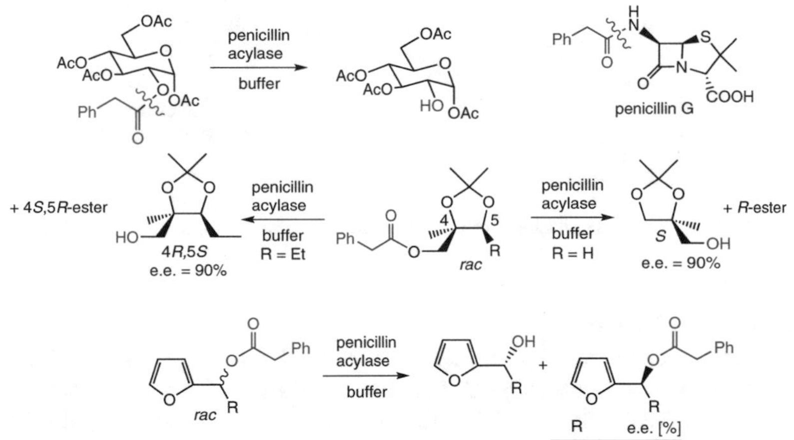

Scheme 2.37 Chemo- and enantioselective ester hydrolyses catalyzed by penicillin acylase

Along the same lines, a protease derived from *Aspergillus oryzae*, which has hitherto mainly been used for cheese processing, has proven useful for the resolution of sterically hindered substrates such as α,α,α-trisubstituted carboxylates [289] (Scheme 2.38). While 'traditional' proteases such as subtilisin were plagued by slow reaction rates and low selectivities, the α-trifluoromethyl mandelic ester (which constitutes a precursor of the chiral derivatization agent 'Mosher's acid' [290]) was successfully resolved by *Aspergillus oryzae* protease [291].

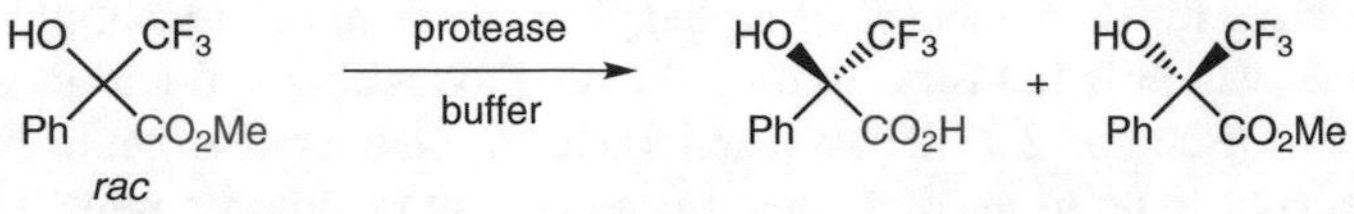

HO   CF₃   protease   HO   ⁗CF₃   HO,   CF₃
   ╳   ─────────→   ╳   +   ╳
Ph   CO₂Me   buffer   Ph   CO₂H   Ph   CO₂Me
   rac

protease	E	e.e. [%]	e.e. [%]
subtilisin	2	25	25
Aspergillus oryzae protease	46	88	88

Scheme 2.38 Resolution of bulky esters by subtilisin and *Aspergillus oryzae* protease

An elegant example of a protease-catalyzed hydrolysis of a carboxylic ester was demonstrated by the dynamic resolution of the antiinflammatory agent 'ketorolac' via hydrolysis of its ethyl ester by an alkali-stable protease derived from *Streptomyces griseus* (Scheme 2.39) [85]. When the hydrolysis was carried out at pH > 9, base-catalyzed in-situ racemization of the substrate ester provided more of the enzymatically hydrolyzed (*S*)-enantiomer from its (*R*)-counterpart, thereby raising the theoretical yield of this racemate resolution to 100%.

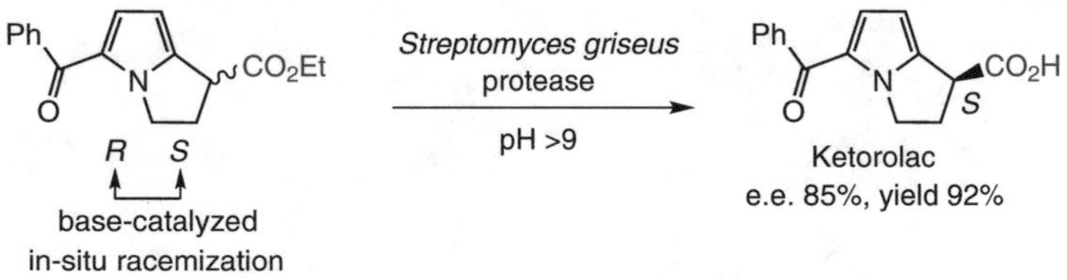

Ph   *Streptomyces griseus*   Ph
   ╲   protease   ╲
    ╲   ⁀CO₂Et   ─────────────────────→   ╲   ⁀CO₂H
   O   N   pH >9   O   N
                                                                              S
       R   S   Ketorolac
        └──┘   e.e. 85%, yield 92%
   base-catalyzed
   in-situ racemization

Scheme 2.39 Dynamic resolution with in-situ racemization by protease from *Streptomyces griseus*

Optimization of Selectivity

Stereoselective enzymatic hydrolysis of nonnatural esters often shows imperfect selectivities with moderate to good Enantiomeric Ratios of about $E = 3–20$, which translates into e.e.$_P$ values of 50–90%. In order to avoid tedious and material-consuming processes to enhance the optical purity of the product, e.g., by crystallization or via repeated kinetic resolution, several methods exist to improve the selectivity of an enzymatic transformation itself [24, 292]. In principle, they can be applied to all types of enzymes.

Since every catalytic system consists of three main components – (bio)catalyst, substrate, and medium – there are three possibilities for the tuning of the selectivity:

- Substrate modification is a straightforward and widely employed strategy.
- Altering the properties of the medium – pH, temperature, cosolvents – within certain limits is a simple and powerful technique to enhance enzyme selectivities.
- The ability to choose a different biocatalyst with a superior selectivity for a given substrate depends on the number of available candidates from the same

enzyme class. Enzyme screening is certainly a good option for proteases and lipases, but not within the relatively small group of esterases. The construction of enzyme mutants possessing altered stereospecificities by enzyme engineering is a laborious, but powerful strategy.

Substrate engineering is a promising technique, which is applicable to all types of enzymatic transformations. As may be concluded from the foregoing examples, the ability of an enzyme to 'recognize' the chirality of a given substrate predominantly depends on its steric shape. Although also electronic effects are involved, they are usually less important [293–296]. Thus, by variation of the substrate structure (most easily performed by chosing a protective group of different size and/or polarity) an improved fit of the substrate can be achieved, which leads to an enhanced selectivity of the enzyme.

Scheme 2.40 shows the optimization of a PLE-catalyzed desymmetrization of 3-aminoglutarate diesters using the 'substrate engineering' approach [231]. By varying the N-protecting group (X) in size and polarity, the optical purity of the monoester could be significantly enhanced as compared to the unprotected original substrate. In addition, a remarkable reversal of stereopreference was achieved upon the stepwise increase of the size of group X, which allowed to control the absolute configuration of the product.

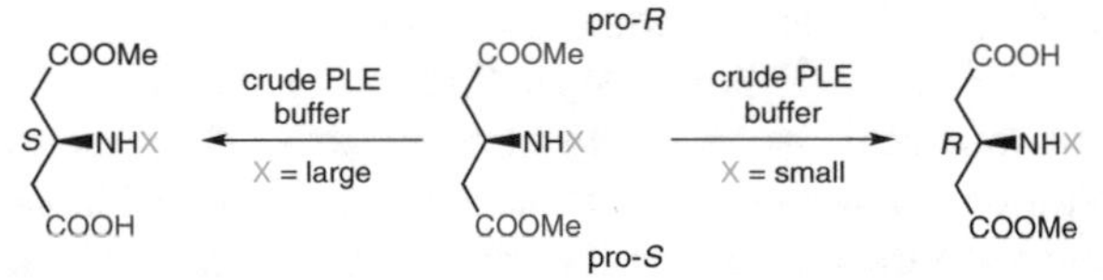

X	Configuration	e.e. [%]
H	R	41
CH$_3$-CO-	R	93
CH$_2$=CH-CO-	R	8
C$_2$H$_5$-CO-	R	6
n-C$_4$H$_9$-CO-	S	2
(CH$_3$)$_2$CH-CO-	S	54
c-C$_6$H$_{11}$-CO-	S	79
(CH$_3$)$_3$C-CO-	S	93
Ph-CH$_2$-O-CO-	S	93
(E)-CH$_3$-CH=CH-CO-	S	>97

Scheme 2.40 Optimization of porcine liver esterase-catalyzed hydrolysis by substrate modification

Another approach to substrate modification is based on the observation that enzyme selectivities are often enhanced with rigid substrate structures bearing π-electrons (Scheme 2.41). Thus, when a highly flexible aliphatic C-4 within a substrate (Sub) is stereochemically not well recognized, it can be 'chemically hidden' in the corresponding thiophene derivative, which is often transformed more selectively. Then, the enantioenriched heteroaromatic product is desulfurized by catalytic hydrogenation using Raney-Ni to yield the saturated desired product in high e.e. [297, 298].

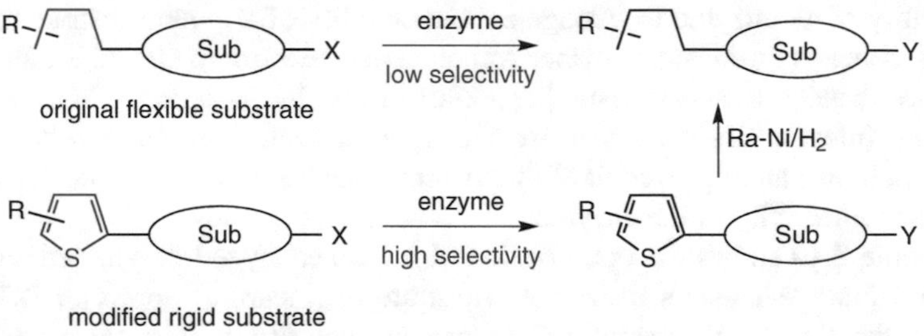

Scheme 2.41 Optimization of selectivity via introduction of a rigid thiophene unit

Medium Engineering Variation of the aqueous solvent system by the addition of water-miscible organic cosolvents such as methanol, *tert*-butanol, acetone, dioxane, acetonitrile, dimethyl formamide (DMF), and dimethyl sulfoxide (DMSO) is a promising and frequently used method to improve the selectivity of hydrolytic enzymes (Scheme 2.42) [299–301]. Depending on the stability of the enzyme, the concentration of cosolvent can be varied from ~10 to ~50% of the total volume. At higher concentrations, however, enzyme deactivation is unavoidable. Many studies have shown that a significant selectivity enhancement can be obtained, especially by addition of dimethyl sulfoxide or low-molecular-weight alcohols, such as *tert*-butanol. However, the price to pay on addition of water-miscible organic cosolvents to the aqueous reaction medium is a depletion in the reaction rate. The molecular reasons for enhanced enzyme selectivities in modified solvent systems is only partly understood and reliable predictions on the outcome of a medium engineering cannot be made (see Sect. 3.1). Consequently, this technique bears a strongly empirical character and requires trial and error experimentation, but in practice, however, the selectivity-enhancing effects are often dramatic.

The selectivity enhancement of PLE-mediated hydrolyses upon the addition of methanol, *tert*-butanol, and dimethyl sulfoxide to the reaction medium is exemplified in Scheme 2.42. The optical purities of products were in a range of ~20–50% when a pure aqueous buffer system was used, but the addition of methanol and/or DMSO led to a significant improvement [302].

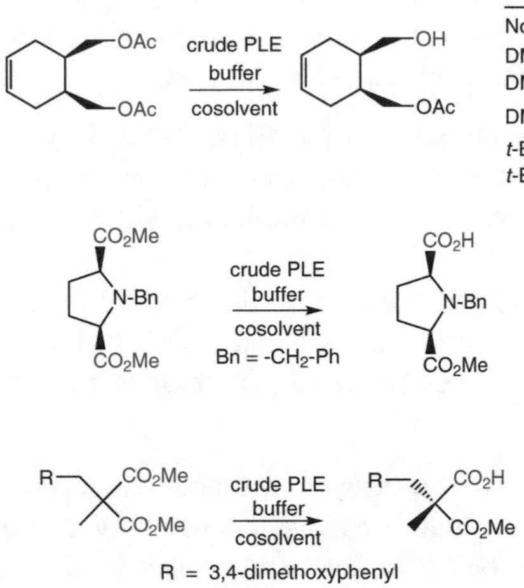

Organic Cosolvent	Relative Rate [%]	e.e. [%]
None	100	55
DMSO (20%)	70	59
DMSO (40%)	28	72
DMF (20%)	35	84
t-BuOH (5%)	70	94
t-BuOH (10%)	44	96

Organic Cosolvent	e.e. [%]
None	17
MeOH (10%)	>97
DMSO (25%)	>97

Organic Cosolvent	e.e. [%]
None	25
DMSO (50%)	93

Scheme 2.42 Selectivity enhancement of porcine liver esterase by addition of organic cosolvents

In biotransformations performed with crude enzyme preparations (e.g. lipases) or whole microbial cells (e.g. baker's yeast) stereoselectivities can be improved by addition of 'enhancers' (e.g. amines, alcohols), which act as noncompetitive inhibitors for competing (iso)enzymes possessing lower (or even opposite) selectivities [303]. This phenomenon is discussed on pp. 102–103 and p. 149, respectively.

Variation of pH Reactions catalyzed by hydrolases are usually performed in aqueous buffer systems with a pH close to that of the pH optimum of the enzyme. Because the conformation of an enzyme depends on its ionization state (among others), variation of the pH and the type of buffer will influence the selectivity of a given reaction. Such variations are facilitated by the fact that the pH activity profile of the more commonly used hydrolytic enzymes is rather broad and thus allows pH variations while maintaining an adequately high activity [304–307].

Variation of Temperature Enzymes, like other catalysts, generally are considered to exhibit their highest selectivity at low temperatures, as supported by experimental observations with hydrolases [308]. A rational understanding of temperature effects on enzyme stereoselectivity was proposed using dehydrogenases [309, 310]. It is based on the so-called 'racemic temperature' (T_{rac}) at which a given enzymatic reaction will proceed without stereochemical discrimination due to the fact that the activation energy of the reaction is the same for both stereochemical directions. In other words, there is no difference in free energy between $[EnzA]^{\neq}$ and $[EnzB]^{\neq}$, consequently $\Delta\Delta G^{\neq} = 0$ (Fig. 1.8).

$$\Delta\Delta G^{\neq} = \Delta\Delta H^{\neq} - T \cdot \Delta\Delta S^{\neq} \quad \text{If} \quad \Delta\Delta G^{\neq} = 0 \quad \text{then} \quad T = T_{rac} = \frac{\Delta\Delta H^{\neq}}{\Delta\Delta S^{\neq}}$$

$T_{rac} = $ 'Racemic Temperature'

From the Gibb's equation given above it follows that only the entropy term $\Delta\Delta S^{\neq}$ (but not the enthalpy $\Delta\Delta H^{\neq}$) is influenced by the temperature. Thus, the selectivity of an enzymatic reaction depends on the temperature as follows:

- At temperatures below T_{rac} the contribution of entropy is minimal and the stereochemical outcome of the reaction is mainly dominated by the activation enthalpy difference ($\Delta\Delta H^{\neq}$). The optical purity of product(s) will thus *decrease* with *increasing* temperature.
- On the other hand, at temperatures greater than T_{rac}, the reaction is controlled mainly by the activation entropy difference ($\Delta\Delta S^{\neq}$) and enthalpy plays a minor role. Therefore, the optical purity of product(s) will *increase* with *increasing* temperature.

However, the major product obtained at a temperature above T_{rac} will be the antipode to that below T_{rac}, thus a temperature-dependent *reversal* of stereochemistry is predicted. The validity of this rationale has been proven with the asymmetric reduction of ketones using a dehydrogenase from *Thermoanaerobium brockii* [311] (Sect. 2.2.2). In contrast to the above-mentioned dehydrogenases from thermophilic organisms, the majority of hydrolases used for biotransformations (except *Candida antarctica* lipase B) possess more restricted thermal operational limits, which narrows the possibility of a significant selectivity enhancement by variation of the reaction temperature. From the data available, it can be seen that upon lowering the temperature both an increase [312] or a decrease in the selectivity of hydrolase reactions may be observed [313], depending on whether the reaction was performed above or below the racemic temperature (T_{rac}) of the enzyme used. The modest upper temperature of about 50 °C for the majority of enzymes represents a serious limitation, while impressive effects have been observed upon cooling (-20 to -60 °C) [314–316]. In order to enhance reaction rates of organic-chemical reactions, microwave (MW) irradiation has become fashionable [317].[8]

While conventional heating is due to polychromatic infrared radiation, microwaves are generated in a monochromatic manner. The benefit of MW heating has been proven in numerous types of organic reactions, but the existence of special microwave-effects (the so-called hot-spot theory) is still heavily debated [318–321]. For enzyme-catalyzed reactions, MW heating has been shown to be superior to conventional heating by leading to reduced enzyme deactivation and enhanced selectivities [322–324].

Enzyme Engineering Molecular biology has enabled the redesign of enzymes possessing improved performance in terms of enhanced stability at extreme

---

[8]By definition, the range of microwave irradiation extends from 1 to 300 GHz; however, due to the resonance frequency of water (19.5 GHz), most of the applications are close to the latter range, i.e., 0.9 and 2.45 GHz.

temperatures and pH, and at high concentrations of reactants and organic (co)-solvents, which is crucial for the construction of process-stable proteins for biotechnological applications. In addition to improved stability, enzymes also can be engineered for enhanced (stereo)selectivities, which represents an equivalent to ligand tuning of homogeneous catalysts. Only some key issues are discussed below since this area requires special expertise in molecular biology – not necessarily a playground of synthetic organic chemists. For a deeper understanding, excellent introductory chapters can be found in recent books and reviews [325–332].

There are two distinct philosophies to enzyme engineering:

1. *Rational protein design* requires detailed knowledge of the three-dimensional structure of an enzyme, preferably from its high-resolution crystal structure or NMR measurements [333]. Alternatively, a computer-generated homology model may help, if the sequence identity is high enough. A sequence identity of ~70% translates into a reasonably well-defined model showing a root mean square deviation of 1–2 Å, which drops to a low value of 2–4 Å for proteins having only ~25% identity. In a first step, docking of the substrate to the active site allows to identify amino acid residues, which appear to interact closely with the structural features of the substrate during binding. Steric incompatibilities, such as collisional interference between residues, insufficient substrate binding in large pockets, or nonmatching polarities between hydrogen bonds or salt bridges can be identified and proposals for the replacement of (usually only few) amino acids can be made. The corresponding mutants are generated and tested for their catalytic properties. Sometimes, this rational approach yields impressive results, but quite often mutant enzymes tell us that the rational analysis of the substrate binding based on a *static* (crystal) structure is insufficient to explain the *dynamic* process of protein (re)folding upon formation of the enzyme–substrate complex, which is a prerequisite to support the *dynamics* of protein catalysis [334]. In addition, the tempting notion that mutations close to the active site are always better than distant ones is only a single aspect of a more complex story [335].

2. *Directed evolution* requires the availability of the gene(s) encoding the enzyme of interest, a suitable (microbial) expression system, a method to create mutant libraries, and an effective selection system – while structural information is irrelevant here.[9] Traditionally, mutant libraries are created by error-prone polymerase chain reaction (epPCR) with low mutation frequencies (1–3 mutations per 1000 base pairs). Since the possible number of mutants generated from a given protein exponentially increases by the number of mutations,[10] the crucial

---

[9]The principle of directed evolution was first described by M. Eigen, see [336].

[10]The possible number of mutants generated from a protein possessing 200 amino acids are 3800 variants for a single mutation, 7,183,900 for two mutations, and 8,429,807,368,950 for only four mutations. Complete ramdomization would result in 20^{200} enzyme variants, which is more than the mass of the universe, even if only one molecule of each enzyme were to be produced.

aspect lies in the selection problem [337]: In order to identify the one (or the few) mutant protein(s) with improved properties amongst the vast number of variants (typically 10^4–10^6), which are (more or less) randomly generated, an efficient screening method is required to find the tiny needle in the very big haystack. Adequate screening methods usually rely on spectral changes during catalysis. The drawback of this first-generation screening method is the requirement for a chromogenic or fluorogenic 'reporter group' in the substrate, which usually consists of a large (hetero)aromatic moiety which needs (at least) 10–14 π-electrons to be 'visible' by UV/VIS or fluorescence spectroscopy. Classic reporter groups are (colorless) p-nitrophenyl derivatives, such as esters or glycosides, which liberate the (yellow) p-nitrophenolate anion upon enzyme catalysis. The latter can be spectrophotometrically monitored at 410 nm (Scheme 2.43). Unfortunately, by introduction of the chromogenic reporter group, the original substrate (e.g., a methyl ester) is modified to a structurally very different p-nitrophenyl substrate ester analog. Since the mutants are screened for optimal activity/selectivity on the surrogate substrate, their performance with the 'real' (methyl ester) substrate will be less efficient. In order to create 'real' mutant enzymes for 'real' substrates, more sophisticated screening methods are recommended based either on a multienzyme assay for acetate (produced during ester hydrolysis, Scheme 2.43) [338], MS analysis of (deuterated) 'pseudo-enantiomeric' products, or time-resolved IR thermogravimetry [339–343]. After all, 34% of random single amino acid replacements yield an inactive protein [344] and you always get what you screen for [345].

Scheme 2.43 shows the use of a surrogate ester substrate bearing a chromogenic (p-nitrophenyl) reporter group for the screening of *Pseudomonas aeruginosa* lipase mutants possessing improved enantioselectivities for the kinetic resolution of an α-chiral long-chain fatty acid [346]. Pure substrate enantiomers were separately tested in 96-well microtiter plates using a plate reader for the readout of enantioselectivity. After four rounds of epPCR at a low mutation rate, mutant A ($E = 11$) was obtained. Sequence analysis of mutant A (and several other positive hits) revealed that position #155 was a 'hot spot' for beneficial variations. Hence, mutant A was improved via saturation mutagenesis through variation of all remaining 19 amino acids at position #155 yielding mutant B ($E = 20$). Another round of epPCR on B gave mutant C ($E = 25$), which could not be further improved. At high mutations rates, epPRC of the wildtype enzyme gave only slightly improved variants D and E ($E = 3.0$ and 6.5, respectively) indicating further 'hot regions'. Combinatorial multiple cassette mutagenesis (CMCM) of the wild-type enzyme in the 'hot region' of amino acids 160–163 gave mutant G ($E = 30$). The latter could be further improved by DNA-shuffling with mutant genes D and E to finally yield mutant J, which exhibited a top value of $E > 51$ among ~40,000 mutants screened.

An example for the successful generation of highly enantioselective esterase mutants capable of hydrolyzing acetate esters of *tert*-alcohols is shown in Scheme 2.44 [347, 348]. In order to improve the modest enantioselectivity of wild-type *Bacillus subtilis* esterase ($E = 5$ and 43), a library of ca. 5000 mutants was constructed, which

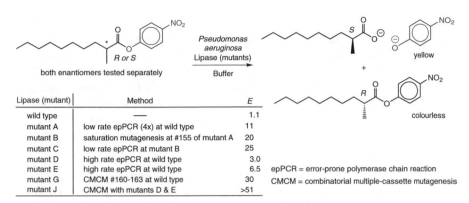

Lipase (mutant)	Method	E
wild type	—	1.1
mutant A	low rate epPCR (4x) at wild type	11
mutant B	saturation mutagenesis at #155 of mutant A	20
mutant C	low rate epPCR at mutant B	25
mutant D	high rate epPCR at wild type	3.0
mutant E	high rate epPCR at wild type	6.5
mutant G	CMCM #160-163 at wild type	30
mutant J	CMCM with mutants D & E	>51

epPCR = error-prone polymerase chain reaction
CMCM = combinatorial multiple-cassette mutagenesis

Scheme 2.43 Screening for *Pseudomonas aeruginosa* lipase mutants showing enhanced enantioselectivities using pure enantiomers of a chromogenic surrogate substrate

encompassed 2800 active variants. Among the latter, the G105A and E188D mutants showed significantly enhanced enantioselectivities for both substrates ($E > 100$). An E188W/M193C double mutant even showed inverted enantiopreference ($E = 64$) for the trifluoromethyl substrate [349]. In order to avoid the undesired modification of the substrate by a chromogenic reporter group, a second-generation screening method was employed based on a commercial test kit: Thus, the acetate formed during ester hydrolysis was activated into acetyl-CoA catalyzed by acetyl-CoA synthase (at the expense of ATP). In a subsequent step, the acetate unit is transferred from acetyl-CoA onto oxaloacetate yielding citrate (catalyzed by citrate synthase). The oxaloacetate required for this reaction is formed by oxidation of L-malate (catalyzed by L-malate dehydrogenase) under consumption of NAD^+ yielding an equimolar amount of NADH, which can be spectrophotometrically monitored at 340 nm [350].

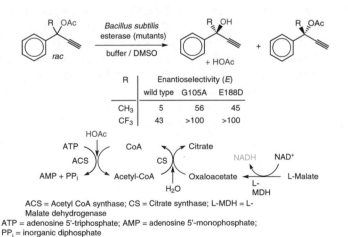

R	Enantioselectivity (E)		
	wild type	G105A	E188D
CH_3	5	56	45
CF_3	43	>100	>100

ACS = Acetyl CoA synthase; CS = Citrate synthase; L-MDH = L-Malate dehydrogenase
ATP = adenosine 5'-triphosphate; AMP = adenosine 5'-monophosphate;
PP_i = inorganic diphosphate

Scheme 2.44 Enantioselectivities of wild-type *Bacillus subtilis* esterase and mutants acting on *tert*-alcohol esters using a multienzyme acetate assay

Model Concepts
In order to avoid extensive enzyme engineering and trial-and-error modifications of substrate structures, several useful 'models' for the more commonly used enzymes have been developed to predict the stereochemical outcome of enzymatic reactions on nonnatural substrates. These models provide a rationale to 'redesign' a substrate or an enzyme, when initial results are not satisfying with respect to reaction rate and/or selectivity. Since the application of such 'models' holds a couple of potential pitfalls, the most important principles underlying their construction and application are discussed here.

Molecular Modeling The structure of an enzyme in crystallized form can be accurately determined by X-ray crystallography [351–354]. Since the tertiary structure of most enzymes is closely related to the preferred form in a dissolved state [355], this method provides the most accurate 3D-description of an active site. However, X-ray data only represent a *static* protein structure, while the chiral recognition process during formation of the enzyme–substrate complex is a complex *dynamic* process. Thus, any attempt of predicting the selectivity of an enzymatic reaction based on X-ray data is comparable to explaining the complex movements in a somersault from a single photographic snapshot.

Although the rapidly increasing number of crystal structures of proteins,[11] which are available through the Protein Data Bank (PDB), encompass widely used enzymes, such as α-chymotrypsin [121], subtilisin [196], and lipases from *Mucor* spp. [9], *Geotrichum candidum* [356], *Candida rugosa* (formerly *cylindracea*) [357], *Candida antarctica* B [358], and *Pseudomonas glumae* [359], for a large number of synthetically useful enzymes, such as pig liver esterase, relevant structural data are not available.

If the amino acid sequence of an enzyme is known either entirely or even in part, computer-assisted calculations can provide a model for its three-dimensional structure [360]. This is done by comparing the amino acid sequence of the enzyme in question with that of other enzymes with known sequence and three-dimensional structure, which serve as blueprint. Depending on the percentage of the homology, i.e., 'overlap', of the amino acid sequences, the results are more or less accurate. In general, an overlap of about ~50–60% is sufficient for good results; less is considered too inaccurate. Tools for the construction of enzyme models, such as Modeller[12] or Phyre²[13] are available via Internet. The utility of various protein structure prediction methods was recently reviewed by Zhang [361].

---

[11]To date (2017), approx. 111.000 protein crystal structures are available.

[12]https://salilab.org/modeller/

[13]http://www.sbg.bio.ic.ac.uk/phyre2/html/page.cgi?id=index

Provided that the three-dimensional structure of an enzyme is available, several methods for predicting the selectivity and its stereochemical preference are possible with various degrees of sophistication and effort [362–364]:

- The enzyme–substrate complex is constructed in its transition state for both enantiomers and the energy value for both (diastereomeric) conformations within the active site of the enzyme are calculated via molecular dynamics (MD). The difference in free energy ($\Delta\Delta G^{\neq}$) – obtained via force field calculations – yields semiquantitative results for the expected selectivity [365].
- The difference in steric interactions during a computer-generated approach of two substrate enantiomers towards an acyl-enzyme intermediate can be used instead [366].
- If the transition state is not known with some certainty, the substrate can be electronically fitted into the active site of the enzyme ('docking'). The orientation of substrate enantiomers with respect to the chemical operator of the enzyme as well as possible substrate movements can be analyzed via MD [367]. This is achieved via (computer-generated) 'heating' of the substrate within the enzyme, followed by a slow electronic 'cooling process', which allows the substrate enantiomers to settle in their position representing the lowest energy minimum. Because selectivities are determined by differences in free energy of transition states, the first approach leads to the most accurate results. A simple free shareware program is AutoDock Vina [368]. However, it should be kept in mind, that errors of up to ±2.5 kcal/M in scoring energies are not uncommon, which renders the estimation of stereoselectivities an educated guess at best (compare Table 1.4).

A representative example for the prediction of the stereochemical outcome of the enzymatic hydrolysis of *rac*-1-phenylethyl acetate catalyzed by *Candida antarctica* lipase B based on its crystal structure (PDB: 5A71) is depicted in Fig. 2.10. The bottom of the active site shows the catalytic triad consisting of Asp-187 and His-224, which activate Ser-105 to perform a nucleophilic attack onto the carbonyl group of the ester moiety, forming a tetrahedral oxy-anion intermediate (see also Scheme 2.1). With the (*R*)-enantiomer (left), the bulky phenyl group is nicely accommodated in the large lipophilic binding site (pink), while the small methyl substituent is pointing upwards into the small pocket (yellow). In contrast, with the (*S*)-enantiomer (right), the phenyl group would clash into the α-helix to the right, while the methyl substituent would be inefficiently bound in the large pocket. Hence, the prediction for the preferred enantiomer is (*R*), also denoted as 'Kazlauskas-rule' (Scheme 2.45).

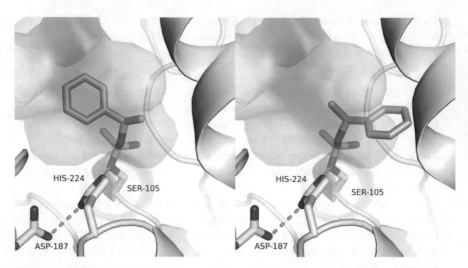

Fig. 2.10 Tetrahedral oxyanion-intermediates during hydrolysis of (*R*)- and (*S*)-1-phenylethyl acetate by *Candida antarctica* lipase B (Graphics prepared by PyMol v. 1.701, courtesy of Georg Steinkellner)

Substrate Model If neither X-ray data nor the amino acid sequence are available for an enzyme – which is not uncommon for synthetically useful enzymes – one can proceed as follows: A set of artificial substrates having a broad variety of structures is subjected to an enzymatic reaction. The results, i.e., the reaction rates and enantioselectivities, then allow to create a general structure of an imagined 'ideal' substrate, which an actual substrate structure should simulate as closely as possible to ensure rapid acceptance by the enzyme and a high enantioselectivity. This idealized substrate structure is then called a 'substrate model' (Fig. 2.11, left). Such models have been developed for PLE [239] and *Candida rugosa* lipase [369, 370]. Of course these crude models only yield reliable predictions if they are based on a substantial number of test substrates.

To ensure optimal selectivity of PLE with methyl carboxylates, the α- and β-substituents should be assigned according to their size (L = large, M = medium and S = small) with the preferably-accepted enantiomer being shown in Fig. 2.11 (left).

Active Site Model Instead of developing an ideal *substrate structure* one also can delineate the structure of the (unknown) *active site* of the enzyme by the method described above. Thus, substrates of varying size and polarity are used as probes to measure the dimensions of the active site in an approach denoted as 'substrate mapping' [371, 372]. Such *active site models* are frequently employed and they usually resemble an arrangement of assumed 'sites' or 'pockets' which are usually box- or cave-shaped. A relatively reliable active-site model for PLE [373] using cubic-space descriptors was based on the evaluation of the results obtained from over 100 substrates (Fig. 2.11, right).

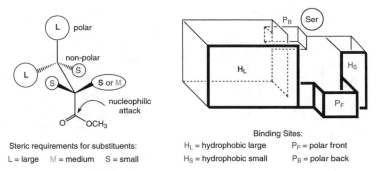

Steric requirements for substituents:
L = large   M = medium   S = small

Binding Sites:
H$_L$ = hydrophobic large   P$_F$ = polar front
H$_S$ = hydrophobic small   P$_B$ = polar back

Fig. 2.11 Substrate model and active site model for porcine liver esterase

The boundaries of the model represent the space available for the accommodation of the substrate. The important binding regions which determine the selectivity of the reaction are two hydrophobic pockets (H$_L$ and H$_S$, with L = large and S = small) and two pockets of more polar character (P$_F$ and P$_B$, with F = front and B = back). The best fit of a substrate is determined by positioning the ester group to be hydrolyzed close to the hydrolytically active serine residue and then arranging the remaining moieties in the H and P pockets.

2.1.3.2 Lipases

Lipases are enzymes which hydrolyze triglycerides into fatty acids and glycerol [374, 375]. Apart from their biological significance, they play an important role in biotechnology, not only for food and oil processing [376–378] but also for the preparation of chiral intermediates [379, 380]. In fact, about 30% of all biotransformations reported to date have been performed with lipases, which presumably constitute the most thoroughly investigated group of enzymes for biotransformations. To date, numerous lipases have been cloned and ~150 crystal structures are available. Although they can hydrolyze and form carboxylic ester bonds like proteases and esterases, their kinetic behaviour and substrate preference are different, which gives rise to some unique properties [381, 382].

The most important difference between lipases and esterases is the physicochemical interaction with their substrates. In contrast to esterases, which show a 'normal' Michaelis-Menten activity depending on the substrate concentration [S] (i.e., a higher [S] leads to an increase in activity), lipases display almost no activity as long as the substrate is in a dissolved monomeric state (Fig. 2.12). However, when the substrate concentration is gradually enhanced beyond its solubility limit by forming a second (lipophilic) phase, a sharp increase in lipase activity takes place [383, 384]. The fact that lipases do not hydrolyze substrates efficiently below a critical concentration (the 'critical micellar concentration', CMC), but display a high activity beyond it, has been called the 'interfacial activation' [385].

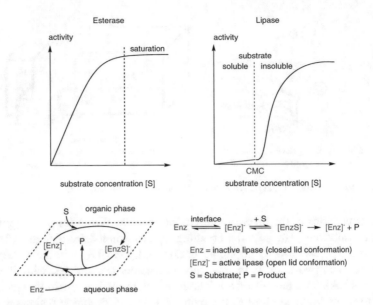

Fig. 2.12 Esterase and lipase kinetics

The molecular reason for this phenomenon is a conformational rearrangement within the enzyme [352]. A freely dissolved lipase in the absence of an aqueous/lipid interface resides in its inactive state [Enz], because a part of the enzyme molecule – the 'lid' – covers the active site. When the enzyme contacts the interface of a biphasic water-oil system, a short α-helix is folded back. Thus, by opening its active site the lipase is rearranged into its active state [Enz]$^{\neq}$.

To ensure optimal activity, lipase-catalyzed hydrolyses thus should be conducted in a biphasic medium. It is sufficient to employ the substrate alone at elevated concentrations, such that it constitutes the second organic phase, or, alternatively, it may be dissolved in a water-immiscible organic solvent such as hexane, a dialkyl ether, or an aromatic solvent. Due to the presence of an interface, physical parameters influencing the mass-transfer of substrate and product between the aqueous and organic phase (such as stirring or shaking speed) have a marked influence on the reaction rate of lipases. Triacylglycerols such as triolein or -butyrin are used as standard substrates for the determination of lipase activity, whereas for esterases p-nitrophenyl acetate is the classic standard.

The fact that many lipases have the ability to hydrolyze esters other than glycerides makes them particularly useful for organic synthesis [386, 387]. Furthermore, some lipases are also able to accept thioesters [388, 389]. In contrast to esterases, lipases have been used for the resolution of racemates more often than for the desymmetrization of *meso*-compounds. Since the natural substrates are esters of a chiral alcohol, glycerol, with an achiral acid, it may be expected that lipases are most useful for hydrolyzing esters of chiral alcohols rather than esters of chiral acids. Although this expectation is true for the majority of substrates (see substrate type III, Scheme 2.45), some lipases also display high selectivity through recognizing the chirality of an acid moiety (substrate type IV).

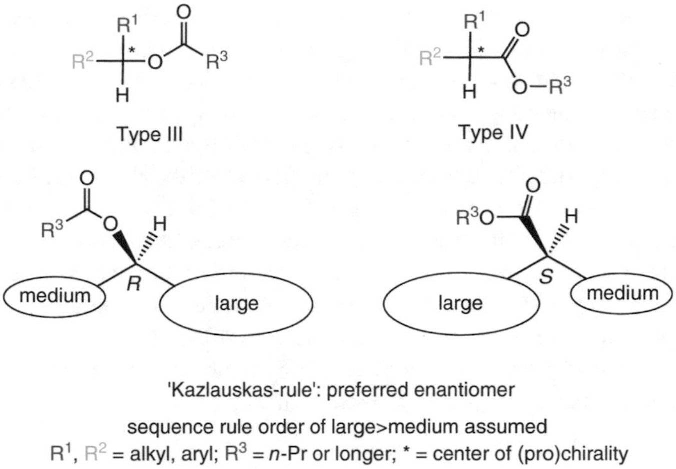

Type III

Type IV

'Kazlauskas-rule': preferred enantiomer

sequence rule order of large>medium assumed

R^1, R^2 = alkyl, aryl; R^3 = n-Pr or longer; * = center of (pro)chirality

Scheme 2.45 Substrate types for lipases

Some of the general rules for substrate-construction are the same as those for esterase-substrates (Scheme 2.20), such as the preferred close location of the chirality center and the necessity of having a hydrogen atom on the carbon atom bearing the chiral or prochiral center. However, other features are different:

- The acid moiety R^3 of lipase-substrate of type III should be of a straight-chain nature possessing at least three to four carbon units to ensure a high lipophilicity of the substrate. Although long-chain fatty acids such as oleates would be advantageous for a fast reaction rate, they do cause operational problems such as a high boiling point of the substrate and they tend to form foams and emulsions during extractive work-up. As a compromise between the two extremes – short chains for ease of handling and long ones for a high reaction rate – n-butanoates or n-butyl esters – are often the first choice.
- Furthermore, the majority of lipases show the same stereochemical preference for esters of secondary alcohols (Scheme 2.45), which is known as the 'Kazlauskas' rule' [370]. Assuming that the Sequence Rule order of substituents R^1 and R^2 is large > medium, the preferably accepted enantiomer lipase-substrate of type III possesses an (R)-configuration at the alcoholic center. The rule for secondary alcohols (Type III) has an accuracy of ≥90%, whereas the predictability for the corresponding α-chiral acids (Type IV) is less reliable.
- Several proteases (such as α-chymotrypsin and subtilisin) and pig liver esterase exhibit a stereochemical preference opposite to that of lipases. This is because the catalytic triad of lipases and proteases – as elucidated by their crystal structures – has been found to be arranged in a mirror-image orientation [390]. Thus, the stereochemical outcome of an asymmetric hydrolysis can often be directed by choosing a hydrolase from a different class [391–394]. Scheme 2.46 depicts the quasi-enantiomeric oxy-anion transition-state intermediates during hydrolysis of a sec-alcohol ester catalyzed by *Candida rugosa* lipase (PDB: 1crl) and the protease subtilisin (PDB: 1sbn). While the

nucleophilic Ser-residues approach from the back, both His are located at the inside, with the oxy-anions pointing outside. Both active sites have suitable space for the large and medium-sized substituents of the *sec*-alcohol moiety (red). The mirror-image orientation of the catalytic center favors opposite enantiomers, which is exemplified by the hydrolytic kinetic resolution of an α-chiral indolyl propionic ester using the (*R*)-selective *Mucor* sp. lipase and the (*S*)-selective protease α-chymotrypsin (Scheme 2.46, bottom) [393]. The activated 2-chloroethyl ester was used to ensure enhanced reaction rates.

- Substrate-type IV represents the general structure of a lesser used ester type for lipases. For type-IV substrates, the alcohol moiety R^3 should preferentially consist of a long straight-chain alcohol such as *n*-butanol. For esters of type IV the stereochemical preference is often (*S*) (Scheme 2.45) but the predictability is less accurate than with type-III substrates [372].

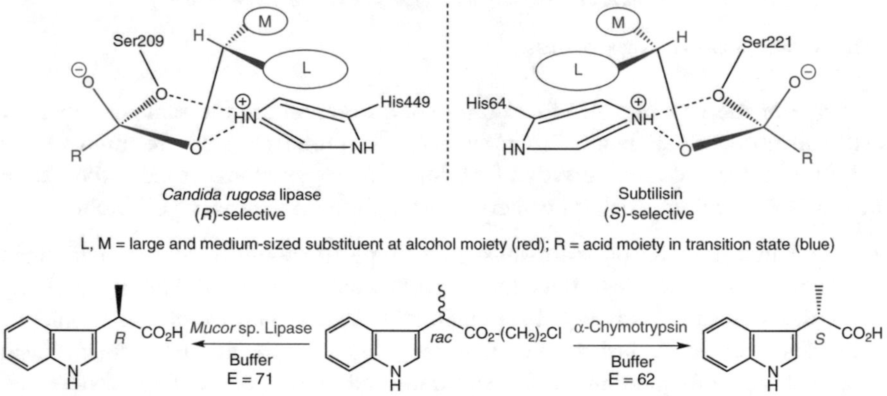

Candida rugosa lipase
(*R*)-selective

Subtilisin
(*S*)-selective

L, M = large and medium-sized substituent at alcohol moiety (red); R = acid moiety in transition state (blue)

Scheme 2.46 Mirror-image orientation of the catalytic machinery of *Candida rugosa* lipase and the protease subtilisin and enantiocomplementary ester hydrolysis using *Mucor* sp. lipase and α-chymotrypsin

A large variety of different lipases are produced by bacteria or fungi and are excreted as extracellular enzymes, which makes their large-scale production particularly easy. The majority of these enzymes are created by the organisms in two isoforms (isoenzymes), usually denoted as type A and B. Both are closely related and usually show the same enantiopreference, but slight structural differences do exist, leading to certain differences in enantioselectivity. Crude technical-grade lipase preparations usually contain both isoforms; the only notable exception is *Candida antarctica* lipase, for which both pure isoforms A and B have been made available through genetic engineering. In contrast to esterases, only a minor fraction of lipases are isolated from mammalian sources such as porcine pancreas. Since some lipases from the same genus (for instance, from *Candida* or *Pseudomonas* sp.) are supplied by different commercial sources, one should be aware of differences in selectivity and activity among the different preparations, while these are usually not in the range of orders of magnitude [395].[14] The actual enzyme content of commercial lipase preparations may vary significantly, from less than 1% up to ~70% – and the selectivity of a lipase preparation from the same microbial source does not necessarily increase with its price! Among the ever increasing number of commercially available lipases only those which have been shown to be of a general applicability are discussed below.

As a rule of thumb, the most widely used lipases may be characterized according to the steric requirements of their preferred substrate esters (Fig. 2.13). Whereas *Aspergillus* sp. lipases are capable of accepting relatively bulky substrates and therefore exhibit low selectivities on 'narrow' ones, *Candida* sp. lipases are more versatile in this regard. Both the *Pseudomonas* and *Mucor* sp. lipases have been found to be highly selective on substrates with limited steric requirements and hence are often unable to accept bulky compounds. Thus, substrates which are recognized with moderate selectivities by a *Candida* lipase, are usually more selectively hydrolyzed by a *Pseudomonas* type. Porcine pancreatic lipase represents a crude mixture of different hydrolytic enzymes and is therefore difficult to predict. However, pure PPL prefers slim substrates.

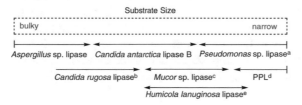

Fig. 2.13 Steric requirements of lipases. (**a**) Lipases from *Pseudomonas cepacia* (syn. *Burkholderia cepacia*), *P. fluorescens*, *P. fragi*, *Chromobacterium viscosum* (syn. *Pseudomonas glumae* or *Burkholderia glumae*); (**b**) syn. *C. cylindracea*; (**c**) *Mucor miehei* (syn. *Rhizomucor miehei*), *M. javanicus* (syn. *Rh. javanicus*); (**d**) pure porcine pancreatic lipase; (**e**) identical to *Thermomyces lanuginosus*.

Porcine Pancreatic Lipase
The cheapest and hence one of the most widely used lipases is isolated from porcine pancreas (PPL) [396–398]. The crude preparations mostly used for biotransformations are called 'pancreatin' or 'steapsin' and contains less than 5% protein. Besides 'true PPL', which is available at a high price in partially purified form, they contain a significant number of other hydrolases. Interestingly enough, in some cases these

---

[14]It seems to be a common phenomenon that microbiologists keep reclassifying microbial species every once in a while. Whether this is to confuse organic chemists, or for other reasons, is often not clear. However, neither the microorganism nor the lipase are changed by a new name.

hydrolase impurities have been shown to be responsible for the highly selective transformation of substrates which were not accepted by purified 'true PPL'. The main hydrolase impurities are α-chymotrypsin, cholesterol esterase, carboxypeptidase B, phospholipases, and other unknown hydrolases. Phospholipases can usually be neglected as undesired hydrolase impurities, because they prefer negatively charged substrate esters which mimic their natural substrates – phospholipids [399, 400]. On the other hand, α-chymotrypsin and cholesterol esterase can be serious competitors in ester hydrolysis. Both of the latter proteins can impair the selectivity of a desired PPL-catalyzed ester hydrolysis by exhibiting a lower selectivity or even opposite stereopreference. Cholesterol esterase and α-chymotrypsin prefer esters of primary and secondary alcohols, whereas 'true PPL' is a highly selective catalyst for esters of primary alcohols only. Thus, any models for PPL should be applied with great caution [401, 402]. Despite the possible interference of different competing hydrolytic enzymes, numerous highly selective applications have been reported with crude PPL [403–405]. Unless otherwise stated, all of the examples shown below have been performed with steapsin.

Regioselective reactions are particularly important in the synthesis of biologically interesting carbohydrates, where selective protection and deprotection of hydroxyl groups is a central problem. Selective removal of acyl groups of peracylated carbohydrates from the anomeric center [406] or from primary hydroxyl groups [407, 408], leaving the secondary acyl groups intact, can be achieved with hydrolytic enzymes or chemical methods, but the regioselective discrimination between secondary acyl groups is a complicated task [409]. PPL can selectively hydrolyze the butanoate ester on position 2 of the 1,6-anhydro-2,3,4-tri-O-butanoyl-galactopyranose derivative shown in Scheme 2.47 [410]. Only a minor fraction of the 2,4-deacylated product was formed.

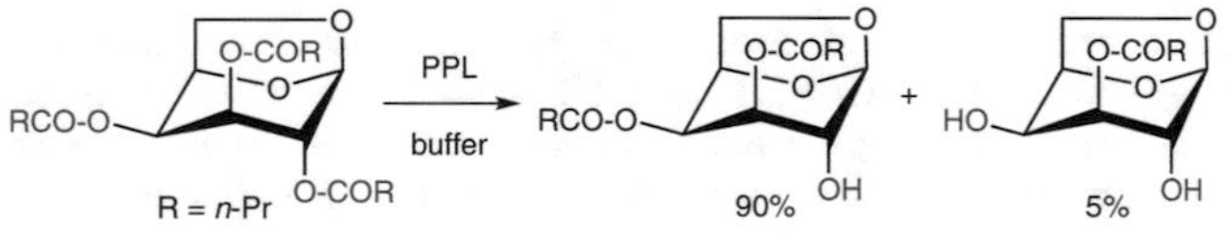

Scheme 2.47 Regioselective hydrolysis of carbohydrate esters by porcine panceatic lipase

A simultaneous regio- and enantioselective hydrolysis of dimethyl 2-methylsuccinate has been reported with PPL [411] with a preference for the (S)-ester and with the hydrolysis taking place at position 4 (Scheme 2.48). The residual unhydrolyzed ester was obtained with >95% e.e. but the monoacid formed (73% e.e.) had to be re-esterified and subjected to a second hydrolytic step in order to be obtained in an optically pure form. It is interesting to note that α-chymotrypsin exhibited the same enantio- but the opposite regioselectivity on this substrate, preferably hydrolyzing the ester at position 1 [412].

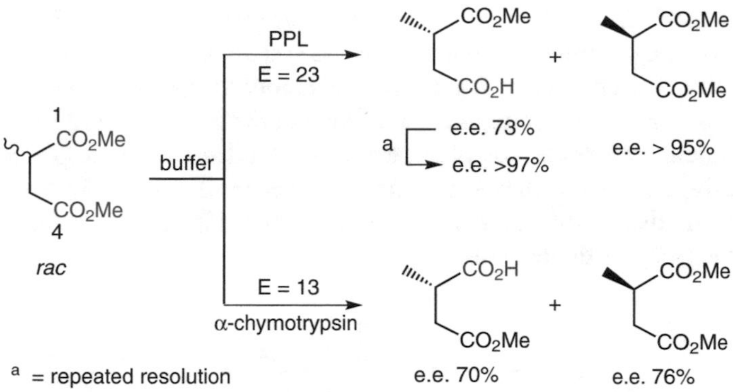

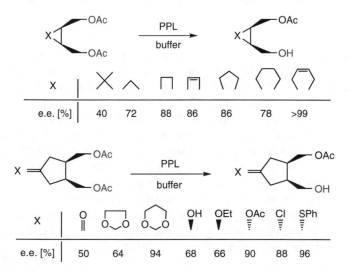

Scheme 2.48 Regio- and enantioselective hydrolysis of dimethyl α-methylsuccinate

The asymmetric hydrolysis of cyclic *meso*-diacetates by PPL proved to be complementary to the PLE-catalyzed hydrolysis of the corresponding *meso*-1,2-dicarboxylates (compare Schemes 2.27 and 2.49). The cyclopentane derivative, which gave low e.e. using the PLE method, was now obtained with 86% e.e. [31, 413]. This selectivity was later improved by substrate modification of the cyclopentane moiety [414], giving access to a number of chiral cyclopentanoid building blocks for the synthesis of carbacyclic prostaglandin I$_2$ derivatives, which are therapeutic agents for the treatment of thrombotic diseases.

X	✕	∧	⊓	⊓	◇	◇	◇
e.e. [%]	40	72	88	86	86	78	>99

X	O‖	O⌣O	O⌣O	OH	OEt	OAc	Cl	SPh
e.e. [%]	50	64	94	68	66	90	88	96

Scheme 2.49 Asymmetric hydrolysis of cyclic *meso*-diacetates by porcine pancreatic lipase

Chiral glycerols, optically active C$_3$-synthons, were obtained by asymmetric hydrolysis of prochiral 1,3-propanediol diesters using PPL (Scheme 2.50) [415]. A remarkable influence of a π-system located on substituents at position 2 on the optical purity of the products indicate that the selectivity of an enzyme does not depend on steric factors alone, but also on electronic issues [293, 296, 416]. Note that a rigid (E)-C=C bond or a bulky aromatic system [417] on the 2-substituent led to an enhanced selectivity of the enzyme as compared to the corresponding saturated analogs. When the configuration of the double bond was Z, a reversal in the stereochemical preference took place, associated with an overall drop of selectivity. Additionally, this study shows a positive influence of a biphasic system (using di-*iso*-propyl ether or toluene [418] as water-immiscible organic cosolvent) on the enantioselectivity of the enzyme.

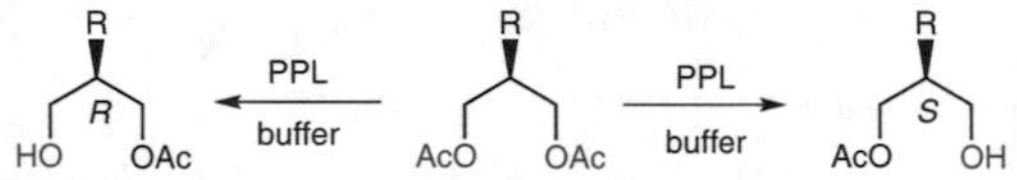

R	Cosolvent	Configuration	e.e. [%]
n-C$_7$H$_{15}$-	i-Pr$_2$O	S	70
(CH$_3$)$_2$CH-(CH$_2$)$_2$-	i-Pr$_2$O	S	72
(E)-n-C$_5$H$_{11}$-CH=CH-	none	S	84
(E)-n-C$_5$H$_{11}$-CH=CH-	i-Pr$_2$O	S	95
(Z)-n-C$_5$H$_{11}$-CH=CH-	i-Pr$_2$O	R	53
(E)-(CH$_3$)$_2$CH-CH=CH-	none	S	90
(E)-(CH$_3$)$_2$CH-CH=CH-	i-Pr$_2$O	S	97
(Z)-(CH$_3$)$_2$CH-CH=CH-	i-Pr$_2$O	R	15

Scheme 2.50 Desymmetrization of prochiral 1,3-propanediol diesters by procine pancreatic lipase

Chiral epoxy alcohols, which are not easily available via the Sharpless procedure due to lack of a directing hydroxy moiety [419], were successfully resolved with PPL (Scheme 2.51). Interestingly, the lipase is not deactivated by a possible reaction with the epoxide moiety [420, 421]. The significant influence of the nature of the acyl moiety on the selectivity of the resolution – again, long-chain fatty acid esters gave better results than the corresponding acetate – may be attributed to the presence of different hydrolytic enzymes present in the crude PPL preparation [422, 423]. In particular α-chymotrypsin and cholesterol esterase are known to hydrolyze acetates of alcohols but not their long-chain counterparts. Thus, they are more likely to be competitors of PPL on short-chain acetates. In order to improve the modest enantioselectivity of crude PPL for industrial applications, a pure hydrolase was isolated from crude pancreatin, which gave perfect enantioselectivity ($E > 100$) in the presence of dioxane as cosolvent [424].

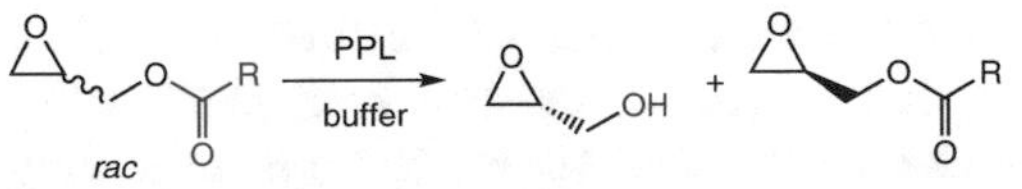

R	e.e. [%]	Selectivity (E)
CH3	53	4
C2H5	88	11
n-C3H7	92	13
n-C4H9	96	16
n-C3H7	>99	>100 *

* Pure immobilised enzyme in presence of 10% dioxane.

Scheme 2.51 Resolution of epoxy esters by porcine pancreatic lipase

During a study on the resolution of the sterically demanding bicyclic acetate shown in Scheme 2.52 [425], which represents an important chiral building block for the synthesis of leukotrienes [426], it was found that crude steapsin is a highly selective catalyst for its resolution. In contrast, pure PPL and α-chymotrypsin were unable to hydrolyze the substrate and cholesterol esterase was able to hydrolyze the ester but with low selectivity. Finally, a novel hydrolase which was isolated from crude PPL proved to be the enzyme responsible for the highly selective transformation.

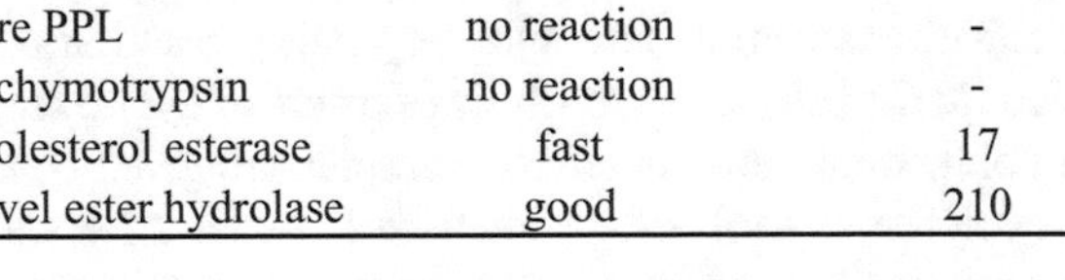

Enzyme	Reaction Rate	Selectivity (E)
crude PPL	good	>200
pure PPL	no reaction	-
α-chymotrypsin	no reaction	-
cholesterol esterase	fast	17
novel ester hydrolase	good	210

Scheme 2.52 Resolution of bicyclic acetate by hydrolases present in crude porcine pancreatic lipase

Certain azlactones, such as oxazolin-5-ones, represent derivatives of activated esters and thus can be hydrolyzed by proteases, esterases, and lipases (Scheme 2.53) [427] to yield N-acyl α-amino acids. When proteases are employed, only products of modest optical purity were obtained due to the fact that the enzymatic reaction rate is in the same order of magnitude as the spontaneous ring opening in the aqueous medium ($k_{spont} \approx k_R$ or k_S).

On the other hand, lipases were found to be more efficient catalysts [428]. Thus, N-benzoyl amino acids of moderate to excellent optical purities were obtained depending on the substituent on C-4. Whereas PPL led to the formation of L-amino acids, the D-counterparts were obtained with a lipase from *Aspergillus niger*. Furthermore, the racemization rate of the two configurationally unstable substrate antipodes under weakly basic conditions at pH 7.6 is sufficiently rapid to provide a dynamic resolution with a theoretical yield of 100% ($k_{rac}^{Sub} \geq k_R$ or k_S, respectively), whereas the products are configurationally stable ($k_{rac}^{Prod} \approx 0$, compare Fig. 2.9).

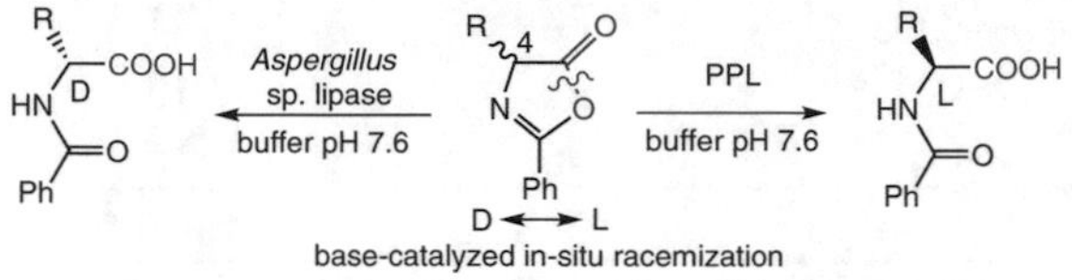

base-catalyzed in-situ racemization

R	Lipase	Configuration	e.e. [%]
Ph-	PPL	L	76
CH3-S-CH2-CH2-	PPL	L	80
Ph-CH2-	PPL	L	>99
Ph-	*Aspergillus* sp.	D	80
CH3-S-CH2-CH2-	*Aspergillus* sp.	D	83
Ph-CH2-	*Aspergillus* sp.	D	>99

Scheme 2.53 Lipase-catalyzed dynamic resolution of oxazolin-5-ones

Candida sp. Lipases

Several crude lipase preparations are available from the yeasts *Candida lipolytica*, *C. antarctica* (CAL), and *C. rugosa* (CRL, syn. *C. cylindracea*). The latter enzyme, the three-dimentional structure of which has been resolved by X-ray analysis [357], has been frequently used for the resolution of esters of secondary alcohols [429–434] and, to a lesser extent, for the resolution of α-substituted carboxylates [435, 436]. The CRL preparations from several commercial sources which contain up to 16% of protein [437] differ to some extent in their activity but their selectivity is very similar [438]. As CRL is able to accommodate relatively bulky esters in its active site, it is the lipase of choice for the selective hydrolysis of esters of cyclic secondary alcohols. To illustrate this point, some representative examples are given below.

The racemic cyclohexyl enol ester shown in Scheme 2.54 was enzymatically resolved by CRL to give a ketoester with an (*S*)-stereocenter on the α-position (77% e.e.) coupled with a diastereoselective protonation of the liberated enol, which led to an (*R*)-configuration on the newly generated center on the γ-carbon atom, only a trace of the (*S,S*)-diastereomer was formed. The remaining (*R*)-enol ester was obtained in optically pure form [439]. In accordance with the substrate model Type IV (Scheme 2.45), no significant hydrolysis on the fully substituted ethyl carboxylate was observed.

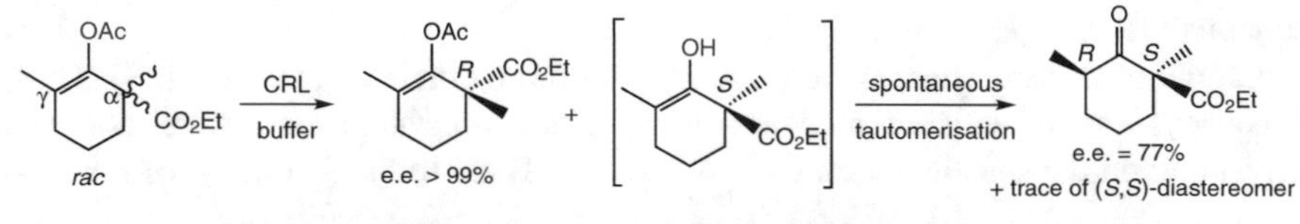

Scheme 2.54 Enzymatic resolution of a cyclic enol ester by *Candida rugosa* lipase

Racemic 2,3-dihydroxy carboxylates, protected as their respective acetonides, were resolved by CRL [440] by using their lipophilic n-butyl esters (Scheme 2.55). It is particularly noteworthy that the bulky α-methyl derivatives could also be transformed, although compounds of this type are usually not accepted by hydrolases.

A number of cyclohexane 1,2,3-triols were obtained in optically active form via resolution of their esters using CRL as shown in Scheme 2.55 [441]. To prevent acyl migration which would lead to racemization of the product, two of the hydroxyl groups in the substrate molecule were protected as the corresponding acetal. In this case, a variation of the acyl chain from acetate to butanoate increased the reaction rate, but had no significant effect on the selectivity of the enzyme.

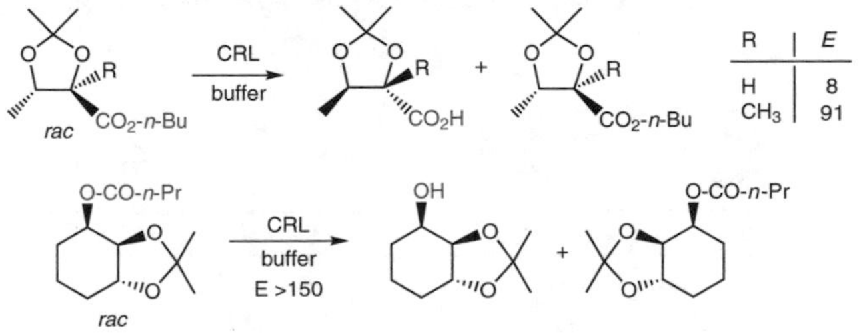

Scheme 2.55 Enzymatic resolution of cyclic esters by *Candida rugosa* lipase

The ideal substrates for CRL are esters of cyclic *sec*-alcohols, which usually give excellent enantioselectivities [425, 442–445]. In contrast, straight-chain substrates are only well resolved when sterically demanding substituents are present (Scheme 2.56) [446]. Esters of *prim*-alcohols usually yield modest stereoselectivities.

In order to provide a general tool which allows to predict the stereochemical outcome of CRL-catalyzed reactions, a substrate model for bicyclic *sec*-alcohols [52, 369, 447, 448] and an active site model [449] have been developed.

The fact that crude *Candida rugosa* lipase occasionally exhibits a moderate selectivity particularly on α-substituted carboxylic esters could be attributed to the presence of two isomeric forms of the enzyme present in the crude preparation [450, 451]. Both forms—denoted as A and B—could be chromatographically separated and were shown to possess identical enantiopreference, but different enantioselectivity (Scheme 2.57). Thus, racemic α-phenyl propionate was resolved with low selectivity ($E = 10$) using crude CRL, whereas isoenzyme A was highly selective ($E > 100$). The isomeric lipase B showed almost the same moderate selectivity as the crude enzyme.

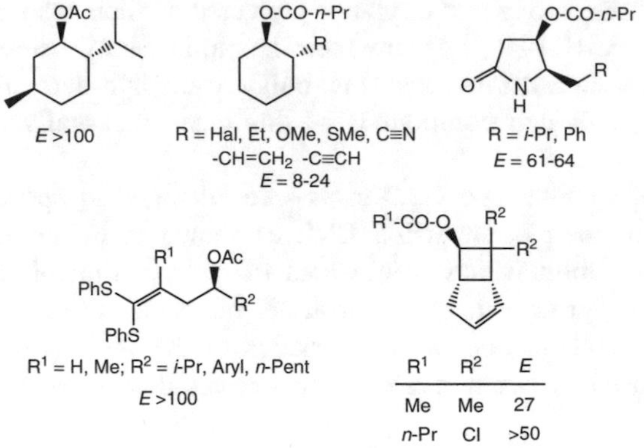

Scheme 2.56 Typical ester substrates for *Candida rugosa* lipase (reacting enantiomer shown)

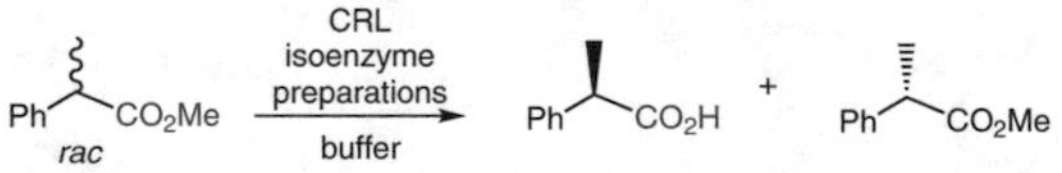

Lipase	Selectivity (*E*)
crude CRL	10
CRL form A	>100
CRL form B	21

Scheme 2.57 Enantioselectivities of isoenzyme preparations of *Candida rugosa* lipase

The most versatile 'champion' lipase for preparative biotransformations is obtained from the basidomycetous yeast *Candida antarctica* (CAL) [452]. As indicated by its name, this yeast was isolated in Antarctica with the aim of finding enzymes with extreme properties to be used in detergent formulations. Like others, the organism produces two isoenzymes A and B, which differ to a significant extent [453]: whereas lipase A (CALA) is Ca^{2+}-dependent and more thermostable, the B-component is less thermotolerant and metal-independent. More important for preparative applications, the substrate-specificity varies a great deal, as the A-lipase is highly active in a nonspecific manner on triglycerides, showing a preference for the *sn*-2 ester group [454] and is not very useful for simple nonnatural esters. On the contrary, the B-component (CALB) is very active on a broad range of nonnatural esters. Both isoenzymes have been made available in pure form through cloning and overexpression in *Aspergillus oryzae* as the host organism [455] and various preparations of this enzyme are produced by Novozymes (DK) in bulk quantities [456]. For the preparative applications discussed below, the B-component has been used more often.

CALB is an exceptionally robust protein which is deactivated only at 50–60 °C,[15] and thus also shows increased resistance towards organic solvents. In contrast to many other lipases, the enzyme shows only weak interfacial activation

---

[15]In immobilized form, the upper operational limit increases to 60–80 °C.

[457], which makes it an intermediate between an esterase and a lipase. Its selectivity could be predicted through computer modeling to a fair extent [458], and for the majority of substrates the Kazlauskas' rule (Scheme 2.45) can be applied. In line with these properties of CALB, selectivity-enhancement by addition of water-miscible organic cosolvents such as *t*-butanol or acetone is possible – a technique which is rather common for esterases. All of these properties make CALB the most widely used lipase both in the hydrolysis [459–464] and synthesis of esters (Sect. 3.1.1).

A representative selection of ester substrates, which have been hydrolyzed in a highly selective fashion is depicted in Scheme 2.58 [249, 465–468]. The wide substrate tolerance of this enzyme is demonstrated by a variety of carboxyl esters bearing a chiral center in the alcohol- or the acid-moiety. In addition, desymmetrization of *meso*-forms was also achieved. In general, good substrates for CALB are somewhat smaller than those for *Candida rugosa* lipase and typically comprise acetate or butyrate esters of *sec*-alcohols in the (ω-1)- or (ω-2)-position with a straight-chain or monocyclic framework.

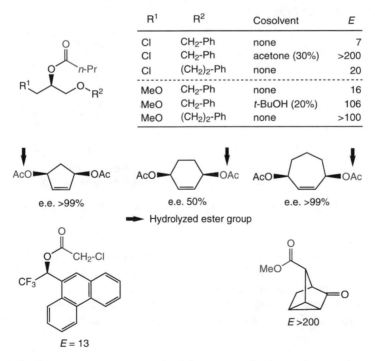

R^1	R^2	Cosolvent	E
Cl	CH$_2$-Ph	none	7
Cl	CH$_2$-Ph	acetone (30%)	>200
Cl	(CH$_2$)$_2$-Ph	none	20
MeO	CH$_2$-Ph	none	16
MeO	CH$_2$-Ph	*t*-BuOH (20%)	106
MeO	(CH$_2$)$_2$-Ph	none	>100

e.e. >99%   e.e. 50%   e.e. >99%

➤ Hydrolyzed ester group

E = 13   E >200

Scheme 2.58 Typical ester substrates for *Candida antarctica* lipase (reacting enantiomer shown)

Pseudomonas sp. Lipases

Bacterial lipases isolated from *Pseudomonas fluorescens*, *P. aeruginosa*, *P. cepacia*, and *P. glumae* are highly selective catalysts [469].[16] They seem to possess a 'narrower' active site than CRL, since they are often unable to accommodate bulky substrates, but they can be extremely selective on 'slim' counterparts [470–474]. Like the majority of the microbial lipases, the commercially available crude *Pseudomonas* sp. lipase preparations (PSL) all possess a stereochemical preference for the hydrolysis of the (*R*)-esters of secondary alcohols, but the selectivity among the different preparations may differ to some extent [475]. Various active-site models for PSL have been proposed [179, 476–478] and the crystal structures of *P. cepacia* and *P. glumae* lipases were elucidated by X-ray analysis [479, 480].

The exceptionally high selectivity of PSL on 'narrow' open-chain esters is demonstrated by the following examples (Scheme 2.59).

The desymmetrization of some prochiral dithioacetal esters possessing up to five bonds between the prochiral center and the ester carbonyl – the site of reaction – proceeded with high selectivity using PSL [481]. This example of a highly selective chiral recognition of a 'remote' chiral/prochiral center is not unusual amongst hydrolytic enzymes [482–484].

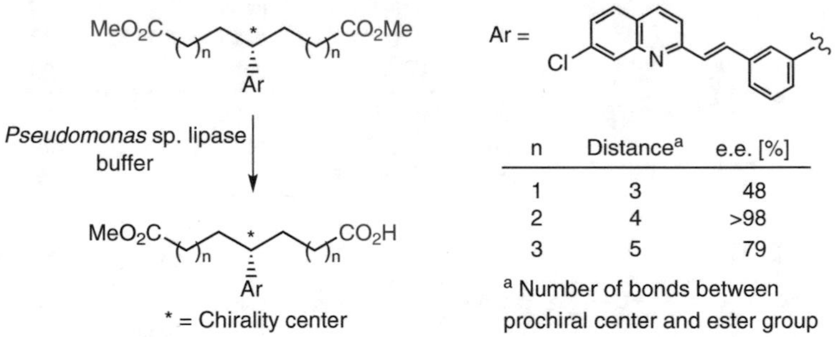

n	Distance[a]	e.e. [%]
1	3	48
2	4	>98
3	5	79

[a] Number of bonds between prochiral center and ester group

* = Chirality center

Scheme 2.59 Desymmetrization of esters having a remote prochiral center by *Pseudomonas* sp. lipase

Chirality need not reside on a sp^3 carbon atom to be recognized by PSL but can be located on a sulfur atom (Scheme 2.60). Thus, optically pure aryl sulfoxides were obtained by lipase-catalyzed resolution of methyl sulfinyl acetates [485] in a biphasic medium containing toluene. The latter compounds are important starting materials for the synthesis of chiral allylic alcohols via the 'SPAC' reaction.

---

[16]Several *Pseudomonas* spp. were reclassified as *Burkholderia* spp.

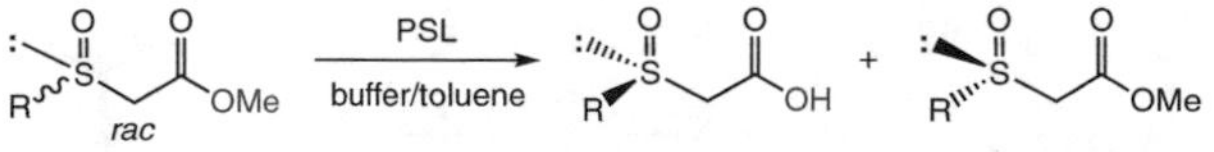

R	e.e. Acid [%]	e.e. Ester [%]	Selectivity (E)
p-Cl-C6H4-	91	>98	>97
Ph-	92	>98	>110
p-NO2-C6H4-	97	>98	>200
c-C6H11-	>98	>98	>200

Scheme 2.60 Resolution of sulfoxide esters by *Pseudomonas* sp. lipase

The selectivity of PSL-catalyzed hydrolyses may be significantly improved by substrate-modification through variation of the nonchiral acyl moiety (Scheme 2.61) [486]. Whereas alkyl- and chloroalkyl esters gave poor selectivities, the introduction of a sulfur atom to furnish the 2-thioacetates proved to be advantageous. Thus, optically active β-hydroxynitriles, precursors of β-hydroxy acids and β-aminoalcohols, were conveniently resolved via the methyl- or phenyl-2-thioacetate derivatives.

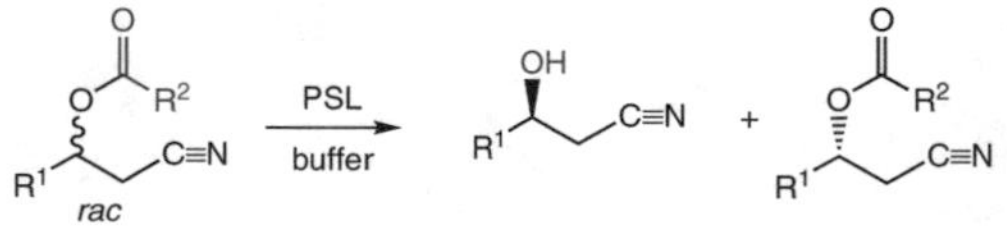

R^1	R^2	Selectivity (E)
Me-	Me	7
Me-	Cl-CH2-	6
Me-	n-C3H7-	2
Me-	Me-O-CH2-	14
Me-	Me-S-CH2-	29
Ph-CH=CH-	Ph-S-CH2-	55
Ph-	Ph-S-CH2-	74

Scheme 2.61 Resolution of β-acyloxynitriles by *Pseudomonas* sp. lipase

An elegant example for a dynamic resolution of an allylic alcohol via enantioselective ester hydrolysis is depicted in Scheme 2.62 [487]. It is based on the combination of an enzyme with a (transition) metal catalyst in the same reactor, which has been termed 'enzyme-metal-combo-catalysis' [488], a technique which became very popular during recent years [489–491]. Thus, *Pseudomonas* sp. lipase hydrolyzed the acetate ester with high specificity, while the in-situ racemization of the substrate enantiomers was effected by a catalytic amount of Pd^{II} leading to the product alcohol in 96% e.e. and 81% yield. However, the lipase has to be chosen with great care, since other hydrolytic enzymes such as acetylcholine esterase and lipases from *Penicillium roqueforti*, *Rhizopus niveus*, and *Chromobacterium viscosum* were incompatible with the metal catalyst.

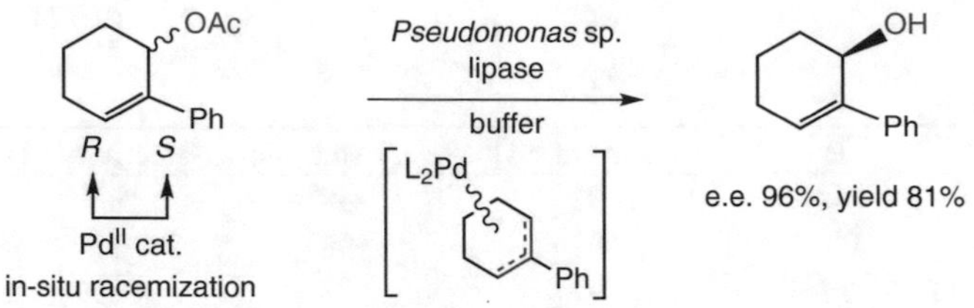

Scheme 2.62 Dynamic resolution of an allylic alcohol ester using *Pseudomonas* sp. lipase and Pd[II] catalysis

The typical substrate for *Pseudomonas* sp. lipases is an (ω-1)-acetate ester bearing a rather small group on one side, whereas remarkable space is available for the large group on the opposite side (Scheme 2.63) [395, 492–495]. Esters of cyclic *sec*-alcohols are well accepted as long as the steric requirements are not too demanding [496–498]. A special feature of PSL is its high selectivity for racemic or prochiral *prim*-alcohols [499, 500], where other lipases often show insufficient stereorecognition.

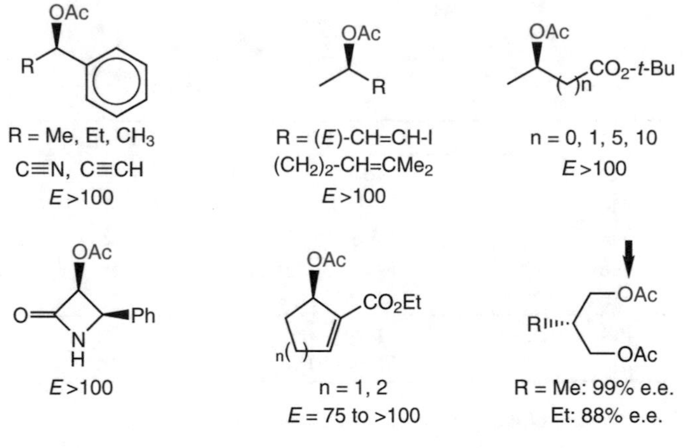

Scheme 2.63 Typical ester substrates for *Pseudomonas* sp. lipase (reacting enantiomer shown)

Mucor sp. Lipases

Lipases from *Mucor* species (MSL) [9, 501] such as *M. miehei* and *M. javanicus* (also denoted as *Rhizomucor*) have frequently been used for biotransformations [429, 502]. With respect to the steric requirements of substrates they seem to be related to the *Pseudomonas* sp. lipases. Like *Candida* and *Pseudomonas* sp. lipases, the different MSL preparations are related in their hydrolytic specificity [38].

A case where only MSL showed good selectivity is shown in Scheme 2.64 [503]. The desymmetrization of *meso*-dibutanoates of a tetrahydrofuran-2,5-dimethanol, which constitutes the central subunit of several naturally occurring polyether antibiotics [504] and platelet-activating-factor (PAF) antagonists, was investigated using different lipases. Whereas crude PPL and CRL showed low selectivity, PSL – as may be expected from its more narrow active site – was significantly better. *Mucor* sp. lipase, however, was completely selective leading to optically pure monoester products. It should be noted that the analogous reaction of the 2,5-unsubstituted acetate (R = H) with PLE at low temperature resulted in the formation of the opposite enantiomer [505].

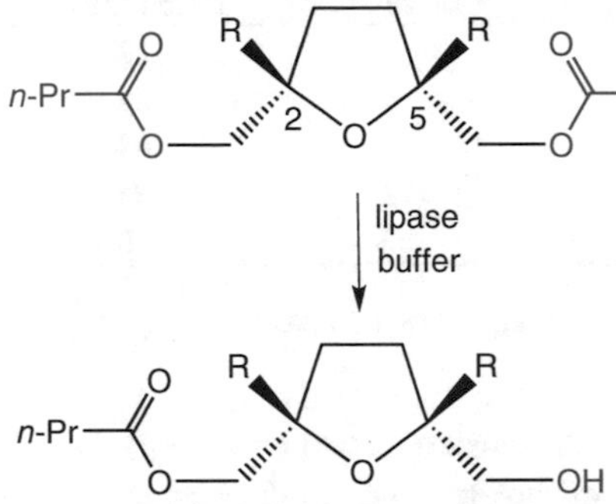

Lipase	R	e.e. [%]
CRL	H	12
PPL	Me	20
PSL	H	81
MSL	H	>99
MSL	Me	>99

Scheme 2.64 Desymmetrization of bis(acyloxy-methyl)tetrahydrofurans by lipases

The majority of lipase-catalyzed transformations have been performed using PPL, CRL, CAL, PSL, and MSL – the 'champion lipases' – and it may be expected that most of the typical lipase substrates may be resolved by choosing one of this group. However, there is a broad potential of other 'niche' lipases which is illustrated by the following examples.

Optically pure cyanohydrins are required for the preparation of synthetic pyrethroids, which are used as more environmentally acceptable insect pestcontrol agents in contrast to the classic highly chlorinated phenol derivatives, such as DDT. Cyanohydrins also constitute important intermediates for the synthesis of chiral α-hydroxy acids, α-hydroxyaldehydes [506] and aminoalcohols [507, 508]. They may be obtained via asymmetric hydrolysis of their respective acetates by microbial lipases (Scheme 2.65) [509]. In the ester hydrolysis mode, only the remaining unaccepted substrate enantiomer can be obtained in high optical purity, because the formed cyanohydrin is spontaneously racemized since via its equilibrium with the corresponding aldehyde, liberating hydrocyanic acid at neutral pH values. However, it has recently been shown that the racemization of the cyanohydrin can be avoided when the hydrolysis is carried out at pH 4.5 [510] or in special nonaqueous solvent systems (see Sect. 3.1.1).

The resolution of the commercially important esters of (*S*)-α-cyano-3-phenoxybenzyl alcohol was only moderately efficient using lipases from *Candida rugosa*, *Pseudomonas*, and *Alcaligenes* sp. (Scheme 2.65). The best selectivities were obtained with lipases from *Chromobacterium* and *Arthrobacter* sp. [511], respectively.

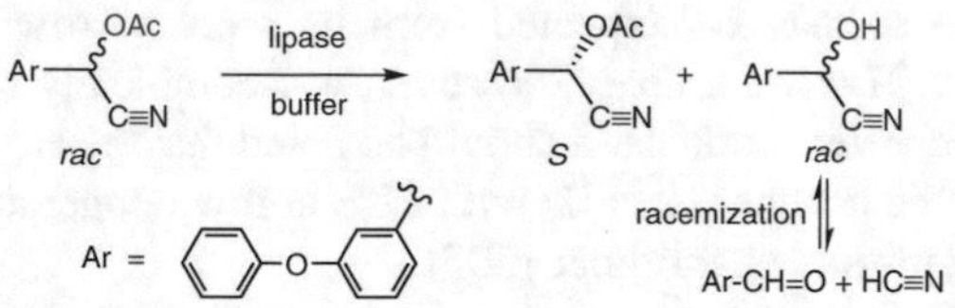

$$Ar = \text{[diphenyl ether aryl group]}$$

Lipase	e.e. Ester [%]	Configuration	Selectivity (E)
CRL	70	R	12
PSL	93	S	88
Alcaligenes sp.	93	S	88
Chromobacterium sp.	96	S	160
Arthrobacter sp.	>99	S	>200

Scheme 2.65 Hydrolysis of cyanohydrin esters using microbial lipases

The epoxy-ester shown in Scheme 2.66 is an important chiral building block for the synthesis of the Ca-channel blocker diltiazem, a potent drug for the treatment of angina pectoris, which is produced at >100 t/year worldwide. Resolution on industrial scale is performed via enantioselective hydrolysis of the corresponding methyl ester using lipases from *Rhizomucor miehei* (E > 100) or an extracellular lipase from *Serratia marcescens* (E = 135) in a membrane reactor [512, 513]. The (undesired) carboxylic acid enantiomer undergoes spontaneous decarboxylation yielding p-methoxyphenyl acetaldehyde, which is removed via extraction of the corresponding bisulfite adduct.

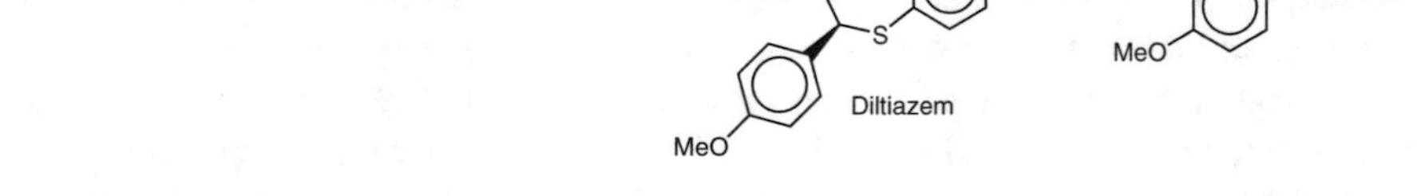

Scheme 2.66 Lipase-catalyzed resolution of an epoxy-ester on industrial scale

The power of lipases for the asymmetric synthesis of pharmaceuticals on industrial scale was demonstrated by the development of an improved process for the manufacture of pregabalin (Lyrica™), which is a widely employed γ-aminobutyric acid (GABA) analog used for the treatment of nervous disorders including epilepsy, anxiety and social phobia (Scheme 2.67). The initial process relied on tedious resolution of *rac*-pregabalin via diastereomer crystallization with (S)-mandelate. The second generation process introduces stereochemistry at a very early stage employing a hydrolytic kinetic resolution of a readily accessible β-cyano-malonate diester as starting material [514]. Screening of several carboxyl ester hydrolases revealed that none of the obvious suspects (*Streptomyces griseus* protease, pig liver esterase), nor the 'champion-lipases' (from *Candida antarctica, Mucor* or *Pseudomonas* sp.) showed sufficient enantioselectivity for the chiral center at the remote β-position. Finally, two (*S*)-selective lipases were identified, of which *Thermomyces lanuginosus* lipase was selected for its perfect enantioselectivity and superior reaction rate. This enzyme was the first lipase produced by GMOs which is employed in detergent formulations since 1990 under the trademark Lipolase™ [515]. Enzymatic hydrolysis stops at the monoester stage and furnishes the hemiester with desired (*S*)-configuration at the β-center. Extractive separation of the non-reacted (*R*)-diester followed by base-catalyzed racemization allows its ex-situ recycling. Thermal decarboxylation of the (*S*)-hemiester yields the corresponding β-cyanoester without racemization. Alkaline ester hydrolysis and catalytic reduction of the nitrile in a one-pot procedure gave (*S*)-pregabalin in perfect e.e. After careful optimization, the lipase-mediated resolution process could be performed at 3M substrate concentration at a batch-size of 3.5 t in an 8 m^3 reactor with a TTN of ~10^5. The enzyme-based process allowed to reduce the usage of organic solvents (by 92%), Raney Ni (by 87%), and starting material (by 39%) and eliminated the need of mandelic acid. Overall, the E factor (i.e. the ratio of the mass of waste per mass of product [516]) was cut from 87 to a mere 17.

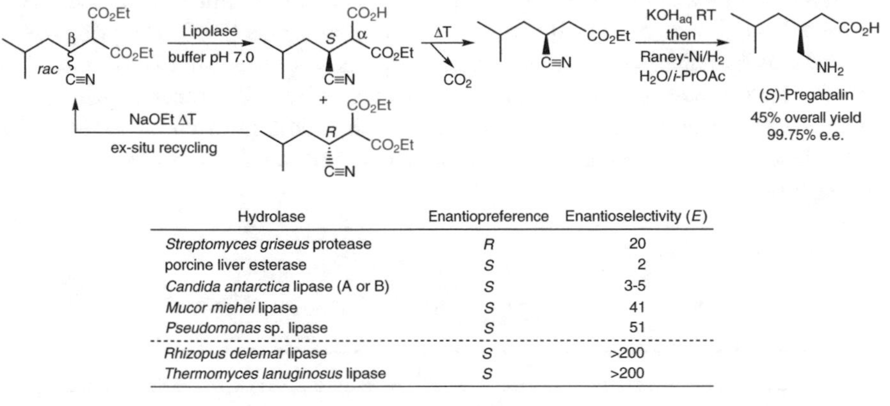

Hydrolase	Enantiopreference	Enantioselectivity (*E*)
Streptomyces griseus protease	*R*	20
porcine liver esterase	*S*	2
Candida antarctica lipase (A or B)	*S*	3-5
Mucor miehei lipase	*S*	41
Pseudomonas sp. lipase	*S*	51
Rhizopus delemar lipase	*S*	>200
Thermomyces lanuginosus lipase	*S*	>200

Scheme 2.67 Lipase-assisted chemoenzymatic synthesis of pregabalin on industrial scale

A lesser known lipase is obtained from the mold *Geotrichum candidum* [517, 518]. The three-dimensional structure of this enzyme has been elucidated by X-ray crystallography [356] showing it to be a serine hydrolase (like MSL), with a catalytic triad consisting of an *Glu*-His-Ser sequence, in contrast to the more usual *Asp*-His-Ser counterpart. It has a high sequence homology to CRL (~40%) and shows a similar preference for more bulky substrates like *Candida rugosa* lipase.

Another extracellular lipase, called 'cutinase', [519] is produced by the plant-pathogenic microorganism *Fusarium solani pisi* for the hydrolysis of cutin—a wax ester which is excreted by plants in order to protect their leaves against microbial attack [520]. The enzyme has been purified to homogeneity [521] and has been made readily available by genetic engineering [522]. Due to its modest stereoselectivities it has not been used widely for asymmetric ester hydrolyses [523], but its unique ability to act on macroscopic polymer esters made it a prime candidate for the enzymatic modification of polyesters [524] and their biodegradation [525].

Optimization of Selectivity
Most of the general techniques for an enzymatic selectivity enhancement such as adjustment of temperature [526], buffer type and pH [307], and the kinetic param-eters of the reaction which were described for the hydrolysis of esters using esterases and proteases, are applicable to lipase-catalyzed reactions as well. Fur-thermore, the switch to another enzyme to obtain a better selectivity is relatively easy due to the large number of available lipases. Substrate modification involving not only the chiral alcohol moiety of an ester but also its acyl group [527], as described above, is a valuable technique for the selectivity improvement of lipase-catalyzed transformations. Bearing in mind that lipases are subject to a strong induced-fit and pronounced interfacial activation (Fig. 2.12), medium engineering with lipases is generally more effective by applying biphasic systems (aqueous buffer plus a water-*immiscible* organic solvent) instead of monophasic solvents (buffer plus a water-*miscible* organic cosolvent).

Enantioselective Inhibition of Lipases The addition of weak chiral bases such as amines or aminoalcohols has been found to have a strong influence on the selec-tivity of *Candida rugosa* [303] and *Pseudomonas* sp. lipase [528]. The principle of this selectivity enhancement was elaborated as early as 1930! [529]. As shown in Scheme 2.68, the resolution of 2-aryloxypropionates by CRL proceeds with low to moderate selectivity in aqueous buffer alone. The addition of chiral bases of the morphinan-type to the medium led to a significant improvement of about one order of magnitude.

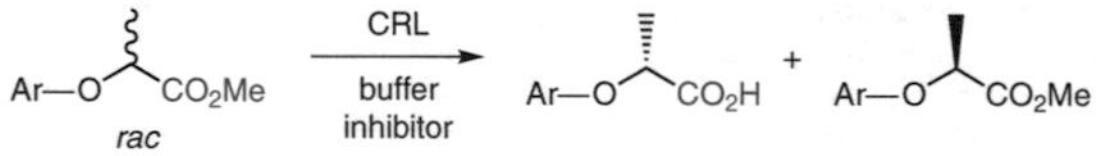

Ar-	Inhibitor	Selectivity (E)
2,4-dichlorophenyl-	none	1
2,4-dichlorophenyl-	dextro- or levomethorphan[a]	20
2,4-dichlorophenyl-	DMPA[b]	23
4-chlorophenyl-	none	17
4-chlorophenyl-	dextro- or levomethorphan[a]	>100

[a] Dextro- or levo-methorphan = D- or L-3-methoxy-N-methylmorphinane.

[b] N,N-Dimethyl-4-methoxyphenethylamine.

Scheme 2.68 Selectivity enhancement of *Candida rugosa* lipase by enantioselective inhibition

Kinetic inhibition experiments revealed that the molecular action of the base on the lipase is a noncompetitive inhibition – i.e., the base attaches itself to the lipase at a site other than the active site, also denoted as 'allosteric effect' – which inhibits the transformation of one enantiomer but not that of its mirror image. Surprisingly, the chirality of the base has only a marginal impact on the selectivity enhancement effect. The applicability of this method – impeded by the high cost of morphinan alkaloids and their questionable use for large-scale synthesis – has been extended by the use of more simple amines such as N, N-dimethyl-4-methoxyphenethylamine (DMPA) [250].

2.1.3.3 Hydrolysis of Lactones

Owing to their cyclic structure, lactones are more stable than open-chain esters and thus are generally not hydrolyzed by 'standard' ester hydrolases. They can be hydrolyzed by lactonases [530], which are involved in the metabolism of aldoses [531] and the deactivation of bioactive lactones, such as N-acyl homoserine lactone [532]. In the hydrolytic kinetic resolution of lactones, the separation of the formed (water-soluble) hydroxycarboxylic acid from unreacted (lipophilic) lactone is particularly easy via extraction using an aqueous-organic system.

Crude PPL has been shown to hydrolyse γ-substituted α-amino lactones with moderate to good enantioselectivity, however, the identity of the enzyme responsible remains unknown (Scheme 2.69) [533].

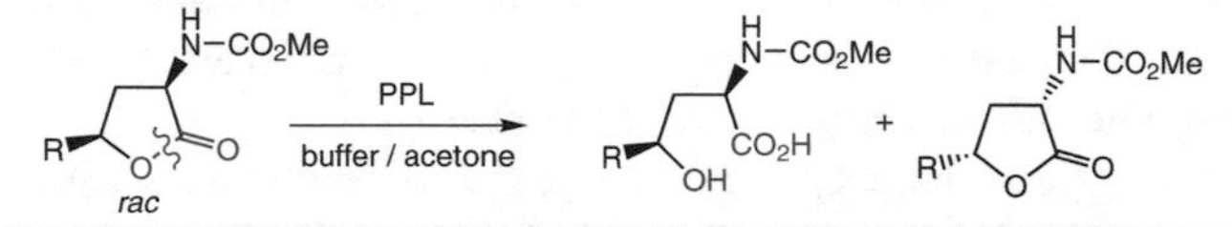

R	e.e. Acid [%]	e.e. Lactone [%]	Selectivity (E)
H-	71	62	11
Ph-	86	32	18
CH$_2$=CH-	90	95	70

Scheme 2.69 Enantioselective hydrolysis of γ-lactones by porcine pancreatic lipase

Well-defined lactonases were identified in bacteria [534–536] and fungi. The most prominent example for the use of a lactonase comprises the resolution of DL-pantolactone, which is required for the synthesis of calcium pantothenate (vitamin B$_5$, Scheme 2.70). The latter is used as vitamin supplement, feed additive, and in cosmetics. An lactonase from *Fusarium oxysporum* cleaves the D-enantiomer from racemic pantolactone forming D-pantoate in 96% e.e. by leaving the L-enantiomer behind. After simple extractive separation, the unwanted L-lactone is thermally racemized and resubjected to the resolution process. In order to optimize the industrial-scale process, which is performed at ca. 3500 t/year, the lactonase has been cloned and overexpressed into *Aspergillus oryzae* [537]. A corresponding enantiocomplementary L-specific lactonase was identified in *Agrobacterium tumefaciens* [538].

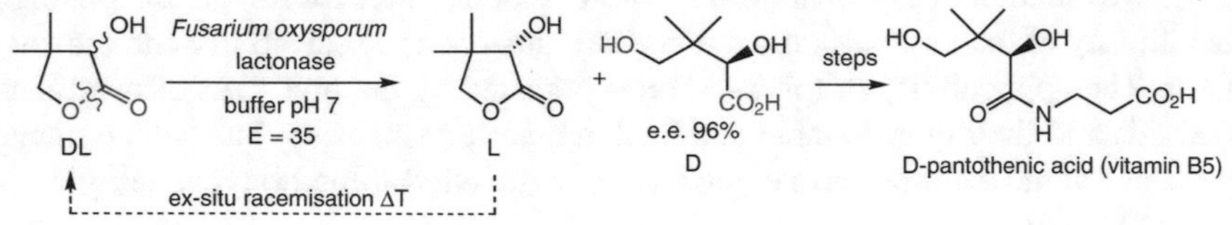

Scheme 2.70 Resolution of pantolactone using a lactonase

More recently, lactonases were employed in the biodegradation of mycotoxins, e.g. zearalenone [539]. Several lactonases were shown to possess also promiscuous phosphotriesterase-activities, which makes them valuable in the bioremediation of neurotoxic organophosphorus agents, which were widely employed as chemical warfare agents and as pesticides, such as parathion [540, 541].

2.1.4   Hydrolysis and Formation of Phosphate Esters

The hydrolysis of phosphate esters can be equally achieved by chemical methods and by phosphate ester hydrolases (phosphatases) and the application of enzymes for this reaction is only advantageous if the substrate is susceptible to decomposition. Thus, the enzymatic hydrolysis of phosphates has found only a limited number of applications. The same is true concerning the enantioselective hydrolyses of racemic phosphates affording a kinetic resolution.

In contrast, the *formation* of phosphate esters is of importance, particularly when regio- or enantioselective phosphorylation is required. On the one hand, numerous bioactive agents display their highest activity only when phosphorylated, for instance the hallucinogen psilocibin found in 'magic mushrooms'. On the other hand, phosphorylation is a common strategy to increase the solubility of orally administered drugs [542, 543]. For instance, prednisolone phosphate is a water-soluble prodrug used as immunosuppressant and the non-natural nucleoside phosphate fludarabin is employed as cytostatic for the treatment of leukemia (Scheme 2.71).

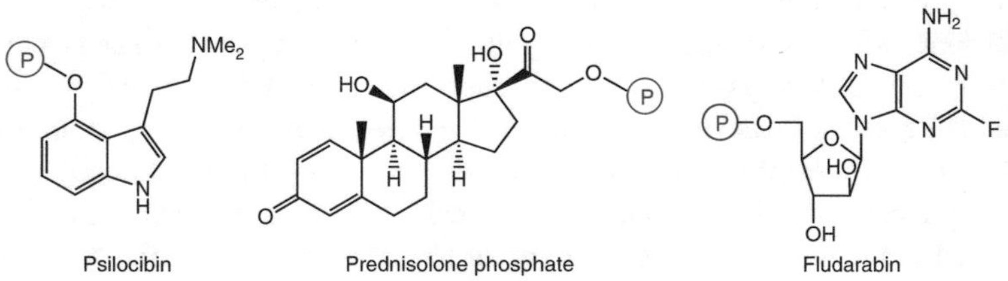

Scheme 2.71 Phosphorylated bioactive compounds

Furthermore, a significant proportion of essential cofactors (or cosubstrates) for other enzyme-catalyzed reactions involve phosphate esters. Adenosine triphosphate (ATP) represents the phosphate donor for most biological phosphorylation reactions and hence constitutes the universal 'energy-currency' in biological systems. For many redox-reactions, nicotinamide adenine dinucleotide phosphate (NADP$^+$) or glucose-6-phosphate (G6P) are an essential cofactor or cosubstrate (Sect. 2.2.1). Dihydroxyacetone phosphate (DHAP) is an important activated cosubstrate for enzymatic aldol reactions (Sect. 2.4.1), thiamine diphosphate (TDP) is an essential cofactor for enzymatic acyloin and benzoin condensations (Sect. 2.4.2) and pyridoxal-5'-phosphate (PLP) serves as molecular shuttle for transamination reactions (Sect. 2.6.2). Glycosyl phosphates are essential for glycosyl transfer reactions catalyzed by carbohydrate phosphorylases (Sect. 2.6.1). In addition, the emerging field of metabolic engineering creates a substantial market for phosphorylated metabolites [544].

Hydrolysis of Phosphate Esters
Chemoselective Hydrolysis of Phosphate Esters Chemical hydrolysis of polyprenyl pyrophosphates is hampered by side reactions due to the lability of the molecule. Hydrolysis catalyzed by acid phosphatase – an enzyme named because it displays its pH-optimum in the acidic range – readily afforded the corresponding dephosphorylated products in acceptable yields [545].

The product from a DHAP-depending aldolase-reaction is a sensitive 2-oxo-1,3,4-triol, which is phosphorylated at position 1 (Scheme 2.72). Mild dephosphorylation by using acid phosphatase without cumbersome isolation of the polar phosphorylated intermediate is a standard method to obtain the chiral polyol product [546–549] in good yield. In the latter example, it was transformed into the sex pheromone of the pine bark beetle (+)-*exo*-brevicomin.

Enantioselective Hydrolysis of Phosphate Esters In comparison with the hydrolysis of carboxyl esters, enantioselective hydrolyses of phosphate esters have been seldom reported due to problems to handle charged species. Acid phosphatases were applied to the kinetic resolution of serine and threonine via hydrolysis of the corresponding O-phosphate esters (Scheme 2.73) [550]. As for the resolutions of amino acid derivatives using proteases, the natural L-enantiomer was hydrolyzed in the case of threonine O-phosphate, leaving the D-counterpart behind ($E > 200$). After separation of the D-phosphate from L-threonine, the D-enantiomer could be dephosphorylated using an unspecific alkaline phosphatase – an enzyme with the name derived from having its pH-optimum in the alkaline region. Interestingly, the N151D mutant exhibited an opposite enantiopreference for the D-enantiomer in case of DL-serine-O-phosphate ($E = 18$) [551].

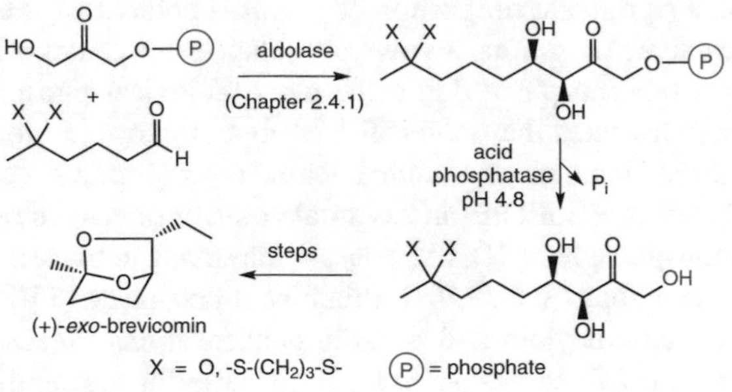

Scheme 2.72 Chemoselective enzymatic hydrolysis of phosphate esters

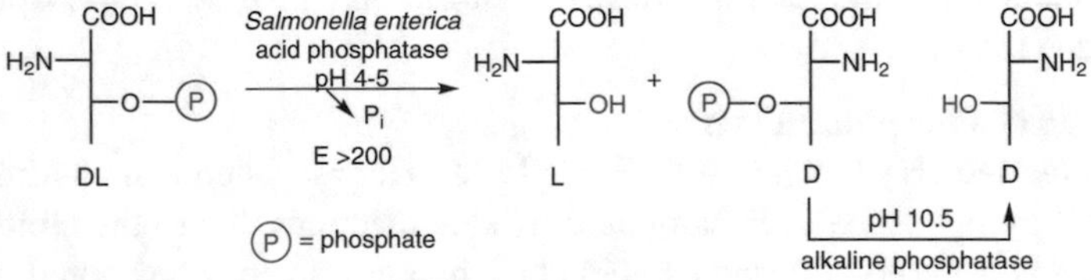

Scheme 2.73 Resolution of *rac*-threonine *O*-phosphate using acid phosphatase

Carbocyclic nucleoside analogs with potential antiviral activity, such as aristeromycin [552] and fluorinated analogs of guanosine [553], were resolved via their 5′-phosphates using a 5′-ribonucleotide phosphohydrolase from snake venom (see Scheme 2.74). After separation, the nonaccepted enantiomer, possessing a configuration opposite to that of the natural ribose moiety, was dephosphorylated by unspecific alkaline phosphatase.

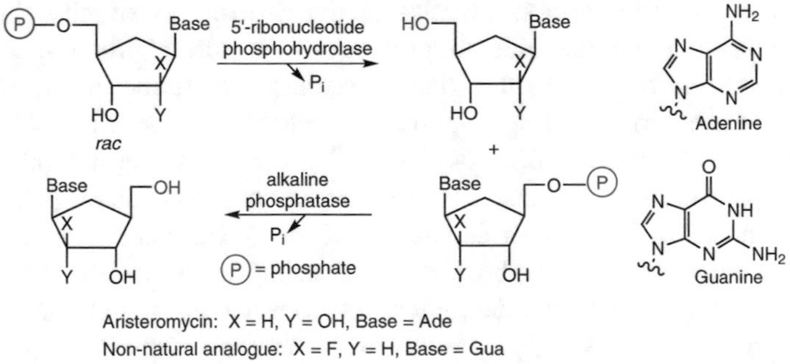

Aristeromycin: X = H, Y = OH, Base = Ade
Non-natural analogue: X = F, Y = H, Base = Gua

Scheme 2.74 Resolution of carbocyclic nucleoside analogs

ATP-Dependent Phosphorylation Employing Kinases

The selective phosphorylation of a polyhydroxy compound by classic chemical methods using POCl$_3$ or phosphorochloridates is tedious since it usually requires a number of protection and deprotection steps. Furthermore, over-phosphorylation leading to undesired oligophosphate esters as byproducts is a common problem.

Employing enzymes for the regioselective formation of phosphate esters can eliminate many of these disadvantages thus making these syntheses more efficient. Additionally, enantioselective transformations via desymmetrization of prochiral or *meso*-diols or through racemate resolution is also possible.

In biological systems, phosphate esters are usually synthesized by means of phosphorylating transferases called kinases, which catalyze the transfer of a phosphate moiety (more rarely a di-[17] or triphosphate moiety) from an energy-rich phosphate donor, such as ATP, onto a nucleophile alcohol.[18] Due to the high price of these phosphate donors, they cannot be employed in stoichiometric amounts.[19] As with all cofactors in general, ATP cannot be replaced by less expensive man-made chemical equivalents, which requires efficient in-situ regeneration to render enzymatic phosphorylations more economic. Fortunately, ATP recycling has become feasible on a molar scale [554–557].

ATP Recycling In living organisms, ATP is regenerated by metabolic processes, but for biocatalytic transformations performed in vitro using purified enzymes, this does not occur. The (hypothetical) addition of stoichiometric amounts of these cofactors would not only be undesirable from a commercial standpoint but also for thermodynamic reasons, because accumulation of the consumed cofactor (most commonly the corresponding diphosphate, ADP) can tip the equilibrium of the reaction in the reverse direction. Thus, nucleoside triphosphate cofactors, such as ATP, are used only in catalytic amounts and are continuously regenerated during the course of the reaction by an auxiliary system which usually consists of a second kinase enzyme and a stoichiometric quantity of a cheap high-energy phosphate donor (Scheme 2.75, Table 2.1). As nucleoside triphosphates are intrinsically unstable in solution, the triphosphate species is typically recycled a few 100 times [558, 559]. Sophisticated reaction engineering using a macromolecular ATP-PEG-construct in a membrane reactor has raised the ATP-cycle number to a solitary record number of ~20,000 mol product/mol ATP [560]. The total turnover numbers (TTN) concerning the enzyme performance are the range of ~10^6–10^8 mol of product per mol of enzyme.

The pro's and con's of the commonly used ATP-regenerating systems are as follows:

- The use of the phosphoenol pyruvate (PEP)/pyruvate kinase system is probably the most useful method for the regeneration of nucleoside triphosphates [561]. PEP is not only very stable towards spontaneous hydrolysis but it is also a very strong phosphorylating agent (Table 2.1). Furthermore, nucleosides other than adenosine phosphates are also accepted by pyruvate kinase. The drawbacks of this system is the considerable cost of PEP due to its more complex synthesis [562, 563] and the fact that pyruvate kinase is inhibited by pyruvate at higher concentrations.
- Acetyl phosphate can be easily synthesized from acetic anhydride and phosphoric acid and is therefore much cheaper than PEP [564]; together with acetate

---

[17] Also termed 'pyro-phosphates'

[18] Phosphorylations involving C, N or S are very rare.

[19] The retail price for one mole of ATP is about US $4500, bulk prices are about one tenth of that.

kinase it is a commonly used regeneration system [565]. However, acetyl phosphate is modestly stable in aqueous solution ($t_{1/2}$ at pH 4.2–7.2 ca. 7–21 h) and its phosphoryl donor potential is lower than that of PEP. As for pyruvate kinase, acetate kinase also can accept nucleoside phosphates other than adenosine, and it is inhibited by acetate. Regeneration of other nucleoside triphosphates (GTP, UTP, and CTP) or the corresponding 2′-deoxynucleoside triphosphates – which are important substrates for enzyme-catalyzed glycosyl transfer reactions (Sect. 2.6.1) [561, 566, 567] – can be accomplished in the same manner using the PEP- or acetate kinase systems.

- As an alternative to expensive phosphenol pyruvate and hydrolytically unstable acetyl phosphate, the use of creatine kinase together with creatine phosphate has been proposed for ATP-recycling [568]. Due to the (controversially debated) role of creatine as food supplement for bodybuilding, creatine phosphate is relatively inexpensive and it is an equally strong phosphate donor as acetyl phosphate.

- Promising ATP-recycling methods for large-scale applications use cheap inorganic polyphosphate as phosphate donor and polyphosphate kinase, respectively [569, 570]. Polyphosphate is a ubiquitous natural polymer of tens to hundreds of orthophosphate residues linked by a high-energy phosphoanhydride bond, which is believed to be an ancient energy carrier preceding ATP in the prebiotic age [571]. It is widely used as an acidulant additive in soft drinks. Polyphosphate kinase from *E. coli* accepts also other nucleoside diphosphates and yields up to 40 regeneration cycles [572], but owing to the limited phosphate donor strength of poly/pyrophosphate, equilibrium yields for ATP are ≥85% [573].

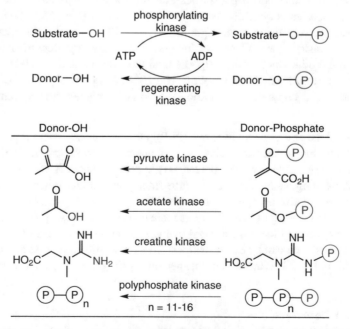

Scheme 2.75 Use of kinases for the enzymatic phosphorylation of alcohols and ATP recycling

Table 2.1 Standard free energy of phosphate donors upon hydrolysis

Phosphorylating agent	ΔG° [kcal mol^{-1}]
Phosphoenol pyruvate	-14.8
Acetyl phosphate	-10.3
Creatine phosphate	-10.3
Poly/pyrophosphate	-8.0
ATP $\rightarrow$ ADP + P$_i$	-7.3

Two further ATP-recycling systems use carbamoyl phosphate (NH_2-CO-O-P) and methoxycarbonyl phosphate (MeO-CO-O-P) as nonnatural phosphate donors together with carbamate kinase and acetate kinase, respectively [574, 575]. Both systems lead to the formation of carbamic acid and methyl carbonate as unstable by-products, which readily decompose forming NH_3 + CO_2 or MeOH + CO_2, thereby driving the equilibrium towards completion. Unfortunately, both phosphate donors undergo spontaneous hydrolysis in aqueous media, which severely limits their applicability, hence these systems have not been widely employed.

A number of reactions which consume ATP generate AMP rather than ADP as a product, only few produce adenosine [576]. ATP may be recycled from AMP using polyphosphate-AMP phosphotransferase and polyphosphate kinase in a tandem-process at the expense of inorganic polyphosphate as phosphate donor for both steps (Scheme 2.76). Alternatively, the combination of adenosine kinase and adenylate kinase were used (Scheme 2.76) [577].

Scheme 2.76 Step-wise enzymatic recycling of ATP from AMP via ADP

Regioselective Phosphorylation The selective phosphorylation of hexoses and a few pentoses (e.g. D-arabinose) on the primary alcohol moiety can be achieved by hexokinase (Scheme 2.77) [578, 579]. The other (secondary) hydroxyl groups can be either removed or they can be exchanged for a fluorine atom, amino groups are tolerated on C2 [580], even thia- or aza-analogs or glucals are accepted. Such modified hexose analogs represent potent enzyme inhibitors and are therefore of interest as potential pharmaceuticals or pharmacological probes. The most important compound in Scheme 2.77 is glucose-6-phosphate (R_{ax} = H; R_{eq} = OH; X = O), which serves as a hydride source during the recycling of NAD(P)H when using glucose-6-phosphate dehydrogenase [581, 582] (Sect. 2.2.1).

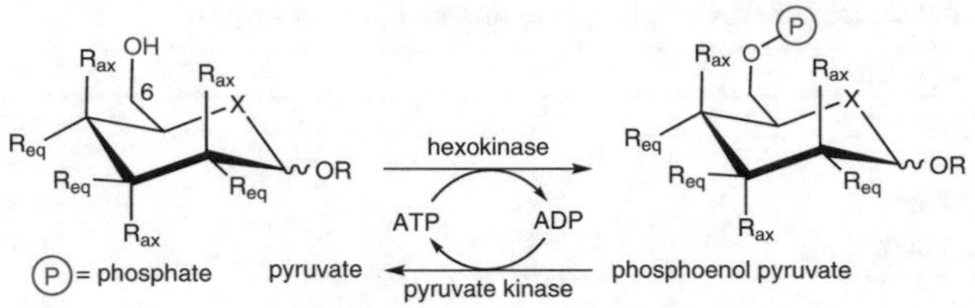

Substituents	Variations tolerated
R	H, Me
R_{eq}	H, F, OH, NH_2, AcNH
R_{ax}	H, F
X	O, S, NH

Scheme 2.77 Regioselective phosphorylation of hexose derivatives by hexokinase

Another important phosphate species, which is needed as a cosubstrate for DHAP-dependent aldolase reactions, is dihydroxyacetone phosphate (Scheme 2.78, also see Sect. 2.4.1). Its chemical synthesis using phosphorus oxychloride is hampered by moderate yields and the generation of side products. Enzymatic phosphorylation, however, gives significantly enhanced yields of a product which is sufficiently pure so that it can be used without isolation for the subsequent carboligation step [583, 584].

5-Phospho-D-ribosyl-α-1-pyrophosphate (PRPP) serves as a key intermediate in the biosynthesis of purine, pyrimidine, and pyridine nucleotides, such as nucleotide cofactors [ATP, UTP, GTP, CTP, and NAD(P)H]. It was synthesized on a large scale from D-ribose using two consecutive phosphorylating steps [585] (Scheme 2.79). First, D-ribose was phosphorylated at the *prim*-hydroxy group using ribokinase. Subsequently a pyrophosphate moiety was transferred from ATP onto the anomeric center in the α-position by PRPP synthase. In this latter step, AMP (rather than ADP) was generated, which required adenylate kinase and pyruvate kinase for step-wise regeneration of ATP. Phosphoenol pyruvate (PEP) served as phosphate donor in all phosphorylation steps. The PRPP thus obtained was subsequently transformed into orotidine monophosphate (O-5-P) via enzymatic linkage of the nucleobase by orotidine-5′-pyrophosphorylase (a transferase), followed by decarboxylation of O-5-P by orotidine-5′-phosphate decarboxylase (a lyase) yielding UMP in 73% overall yield.

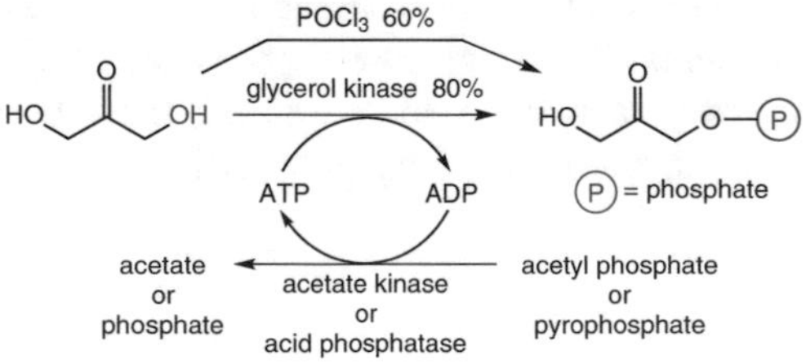

Scheme 2.78 Phosphorylation of dihydroxyacetone by glycerol kinase

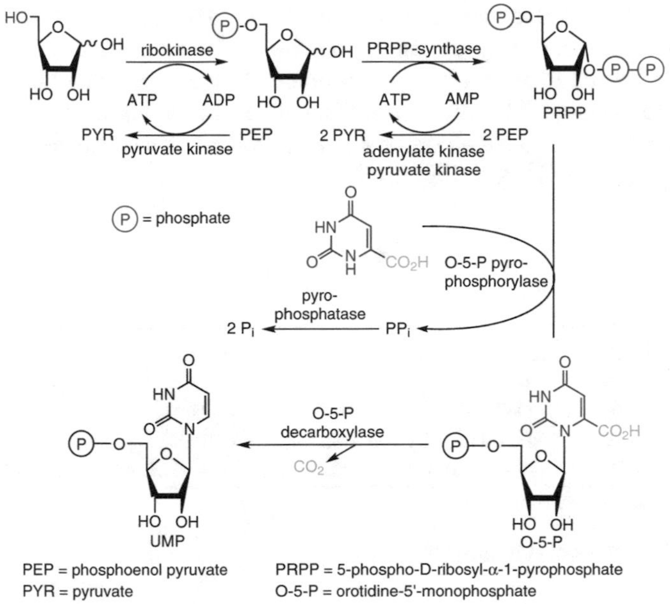

Scheme 2.79 Phosphorylation of D-ribose and enzymatic synthesis of UMP

Enantioselective Phosphorylation Glycerol kinase [586] is not only able to accept its natural substrate, glycerol, to form *sn*-glycerol-3-phosphate [587], or close analogs of it such as dihydroxyacetone (see Scheme 2.78), but it is also able to transform a large variety of prochiral or racemic primary alcohols into chiral phosphates (Scheme 2.80) [588–590]. The latter compounds represent synthetic precursors to phospholipids [591] and their analogs [592].

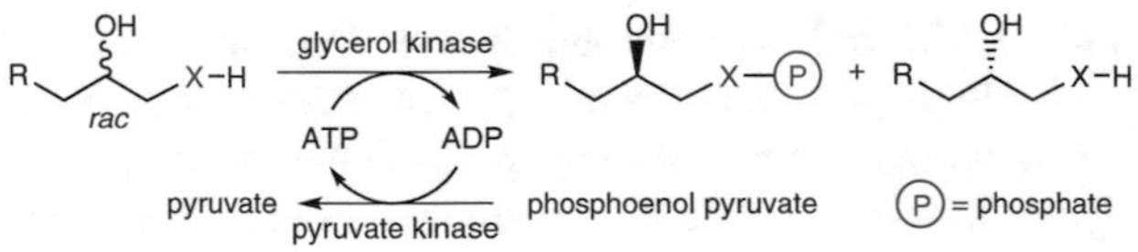

R	X	e.e. Phosphate [%]	e.e. Alcohol [%]
Cl	O	>94	88
SH	O	94	n.d.
CH$_3$O	O	90	n.d.
Br	O	90	n.d.
OH	NH	>94	94

Scheme 2.80 Enantioselective phosphorylation of glycerol derivatives

As depicted in Scheme 2.80, the glycerol backbone of the substrates may be varied quite widely without affecting the high specificity of glycerol kinase. In resolutions of racemic substrates, both the phosphorylated species produced and the remaining substrate alcohols were obtained with moderate to good optical purities (88 to >94%). Interestingly, the phosphorylation of the aminoalcohol shown in the last entry occurred in an enantio- and chemoselective manner on the more nucleophilic nitrogen atom. The evaluation of the data obtained from more than 50 substrates permitted the construction of a general model of a substrate that would be accepted by glycerol kinase (Fig. 2.14).

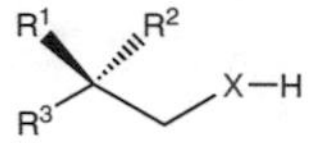

$$R^1 \diagdown\hspace{-0.3em}\diagup R^2$$
$$R^3 \diagup\hspace{-2em}\diagdown X{-}H$$

Position	Requirements
X	O, NH
R^1	preferably OH, also H or F, but not NH_2
R^2	H, OH (as hydrated ketone), small alkyl groups[a]
R^3	small groups, preferably polar, e.g. $-CH_2-OH$, $-CH_2-Cl$

[a] Depending on enzyme source.

Fig. 2.14 Substrate model for glycerol kinase

ATP-Independent Phosphorylation Employing Phosphatases

Enzymatic phosphorylation at the expense of ATP catalysed by kinases is predominantly involved in biological activation and messaging processes required for bio-*synthesis*. Like most enzymes from primary metabolism, kinases posess a limited substrate spectrum, which – together with the requirement for ATP recycling – severely limits their applicability for the phosphorylation of non-natural substrates. In contrast, phosphate ester hydrolases (phosphatases) usually display a much broader substrate spectrum because they are found in bio*degradation* pathways. Although the ability of phosphatases to catalyse phosphate-transfer reactions yielding phosphate esters was already recognized in 1948 [593, 594], it was only recently, that the potential of ATP-independent phosphorylation was recognized [595, 596].

In order to enable phosphatases to catalyse phosphate transfer reactions, their mechanism of action must proceed through a covalent enzyme-phosphate intermediate in analogy to the acyl-enzyme intermediate in ester hydrolysis (Scheme 2.81, compare Scheme 2.1) [597]. In the hydrolysis mode, the phosphate ester is attacked by a nucleophilic His-residue[20] releasing ROH and forming a covalent enzyme-phosphate intermediate. The latter is attacked by water – through assistance of another His-residue – yielding phosphate and liberating His. In the trans-phosphorylation mode, the His-phosphate intermediate is preferably formed at the expense of an energy-rich di-, tri- or polyphosphate. Attack of the substrate alcohol R-OH yields the phosphate ester.

In practice, hydrolysis and trans-phosphorylation are taking place simultaneously and their relative rates depend on the reaction conditions, the type of

---

[20]In acid phosphatases, the active site nucleophile is usually a His (in AphA-St it is a carboxylate, Asp), in alkaline phosphatases it is a Ser or Thr residue.

substrates and the relative reactivity of the enzyme-phosphate intermediate with water or the alcohol substrate. In order to boost trans-phosphorylation over hydrolysis, the following strategies have been developed:

- In presence if a high-energy phosphate donor and elevated concentrations of the substrate alcohol, trans-phosphorylation is usually a fast process, which gives acceptable yields of the desired phosphate ester, which has to be recovered to prevent hydrolysis upon extended reaction times (kinetic control) [598–600].
- Kinetic control is facilitated by using a flow-system, where the phosphate donor and the substrate are continuously pumped through a column containing the immobilized phosphatase. By adjusting the flow rate, the residence (contact) time between enzyme and reactants can be tuned such that the maximum conversion is reached when the product leaves the column. Since the immobilized enzyme stays behind, no undesired hydrolysis can occur [601].
- Phosphatase mutants have been designed which show a greatly diminished hydrolysis rates but maintain good trans-phosphorylation activities [602].
- Reversal of hydrolysis is possible for selected substrates (e.g. glycerol), which are tolerated at very high concentrations (~70–95% v/v) at reduced water activities (thermodynamic control), where inorganic phosphate serves as phosphate donor [603, 604]. In practice, however, workup is problematic due to highly viscous reaction systems. A schematic representation of the time course of kinetic versus thermodynamic control is given in Fig. 2.19.

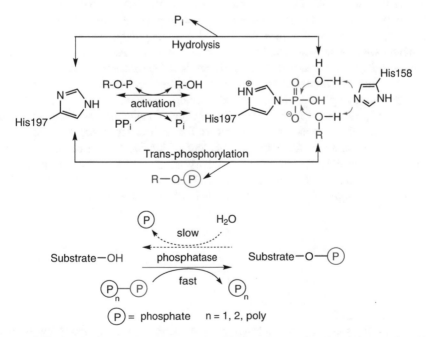

Scheme 2.81 Mechanism of phosphate ester hydrolysis and trans-phosphorylation catalyzed by phosphatase PhoN-Se via a covalent enzyme-phosphate intermediate

Acid phosphatases from *Shigella flexneri* (PhoN-Sf) [605] and *Salmonella enterica* (PhoN-Se) [606] phosphorylate not only simple alcohols, polyols and cyclic and aromatic alcohols, but also many simple carbohydrates using pyrophosphate as phosphate donor with a predominant regioselectivitity in favor of *prim*-hydroxy groups, only PhoN-Se converts *sec*-alcohols at low rates [607]. On the contrary, stereoselectivities observed so far were modest.

Regioselective monophosphorylation of diols using PhoN-Sf, PhoN-Se and PiACP (from *Prevotella intermedia*) using PP_i, PPP_i or polyP as phosphate donor revealed that on short-chain diols, only mono-phosphorylation occurred, bis-phosphates were formed to a small extent with long-chain analogs. Exclusive *O*-phosphorylation was observed with an aminoalcohol [608].

Table 2.2. Regioselective phosphorylation of diols and aminoalcohols using phosphatases

Substrate	Conc. [mM]	Enzyme	Product	Conc. [mM]
HO⁀OH	500	PhoN-Sf	mono/bis-P >99/<1	176
HO⁀=⁀OH	300	PhoN-Se	mono/bis-P 90/10	120
HO⁀OH	500	PhoN-Sf	mono/bis-P 90/10	305
HO⁀OH	500	PhoN-Sf	mono/bis-P 85/15	187
H₂N⁀OH	500	PhoN-Se	O-P versus N-P >99/<1	85

Nucleotides are not only important intermediates for the synthesis of pharmaceuticals, but they also are widely used as flavor-enhancers and are thus produced on industrial scale (~16,000 t/a). Among them are inosine- (5-IMP) and guanosine 5′-monophosphate (5-GMP), whose biological activity depends on the position of the phosphate group: Whereas the 2′- and 3′-monophosphates are tasteless, the 5′-regioisomer is responsible for the 'umami'-taste.[21] As nucleosides, such as inosine and guanosine can be efficiently produced by fermentation, access to the corresponding 5′-nucleotides depends on regioselective phosphorylation. Non-specific acid phosphatases from *Morganella morganii* [609] and *Escherichia blattae* [610] were employed in trans-phosphorylation using PP_i as cheap phosphate donor. In order to meet the requirements for an industrial process, the native enzymes were mutated for enhanced binding of inosine by the phospho-enzyme intermediate, which favors trans-phosphorylation, while undesired hydrolysis is largely suppressed. The S72F/G74D/I153T-triple mutant was able to produce $140\ g\ L^{-1}$ of inosine 5′-phosphate with a yield of 71% from inosine (Scheme 2.82).

---

[21]Together with sweetness, saltiness, bitterness and sourness, 'umami' or 'savory'-taste is one of the five basic tastes.

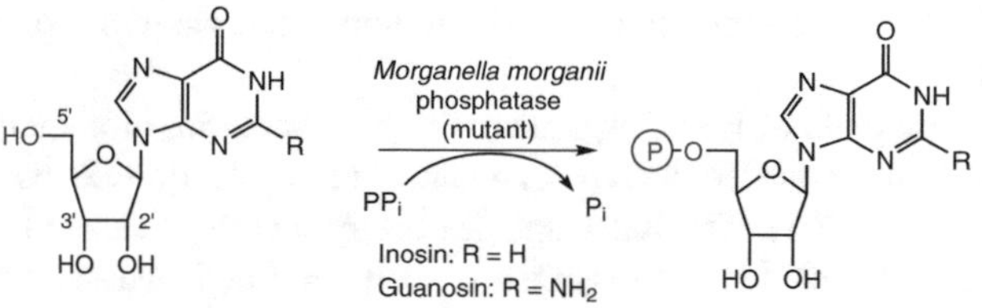

Scheme 2.82 Regioselective phosphorylation of nucleosides

The merits of enzymatic phosphorylation over chemical methods is the lack of side reactions owing to the mild reaction conditions. On lab-scale, the use of ATP-dependent phosphorylation is advantageous due to higher conversions and better selectivities of kinases despite the increased complexity connected with ATP recycling. On industrial scale, direct phosphate transfer from pyrophosphate mediated by phosphatases is preferable.

2.1.5 Hydrolysis of Epoxides

Chiral epoxides and vicinal diols (employed as their corresponding cyclic sulfate or sulfite esters as reactive intermediates) are extensively employed high-value intermediates for the synthesis of enantiomerically pure bioactive compounds due to their ability to react with a broad variety of nucleophiles [611, 612]. As a consequence, extensive efforts have been devoted to the development of catalytic methods for their production. Although several chemical strategies are known for preparing them from optically active precursors, or via asymmetric syntheses involving desymmetrization or resolution methods [613], none of them is of general applicability and each of them has its merits and limits. Thus, the Sharpless epoxidation gives excellent stereoselectivities and predictable configurations of epoxides, but it is limited to allylic alcohols [614]. On the other hand, the Jacobsen epoxidation is applicable to nonfunctionalized alkenes [615]. The latter gives high selectivities for *cis*-alkenes, whereas the results obtained with *trans*- and terminal olefins were less satisfactory. As an alternative, a number of biocatalytic processes for the preparation of enantiopure epoxides via direct or indirect methods are available [616–619]. Among them, microbial epoxidation of alkenes would be particularly attractive by providing a direct access to optically pure epoxides, but this technique requires sophisticated fermentation and process engineering (Sect. 2.3.3.3) [620]. In contrast, the use of hydrolase enzymes for this purpose would be clearly advantageous. An analogous metal-based chemocatalyst for the asymmetric hydrolysis of epoxides is available [621, 622].

Enzymes catalyzing the regio- and enantiospecific hydrolysis of epoxides – epoxide hydrolases (EH)[22] [623] – play a key role in the metabolism of xenobiotics.

---

[22]Epoxide hydrolases have been also called 'epoxide hydratases' or 'epoxide hydrases'.

In living cells, aromatics and olefins can be metabolized via two different pathways (Scheme 2.83).

In prokaryotic cells of lower organisms such as bacteria, dioxygenases catalyze the cycloaddition of molecular oxygen onto the C=C double bond forming a dioxetane (Sect. 2.3.3.7). The latter species are reductively cleaved into *cis*-diols. In eukaryotic cells of higher organisms such as fungi, yeasts and mammals, enzymatic epoxidation mediated by monooxygenases (Sect. 2.3.3.3) is the major degradation pathway. Due to the electrophilic character of epoxides, they represent powerful alkylating agents which makes them incompatible with living cells: they are toxic, cancerogenic, and teratogenic agents. In order to eliminate them from the cell, epoxide hydrolases catalyze their degradation into biologically more innocuous *trans*-1,2-diols, which can be further metabolized or excreted due to their enhanced water solubility. As a consequence, most of the epoxide hydrolase activity found in higher organisms is located in organs, such as the liver, which are responsible for the detoxification of xenobiotics [624, 625].

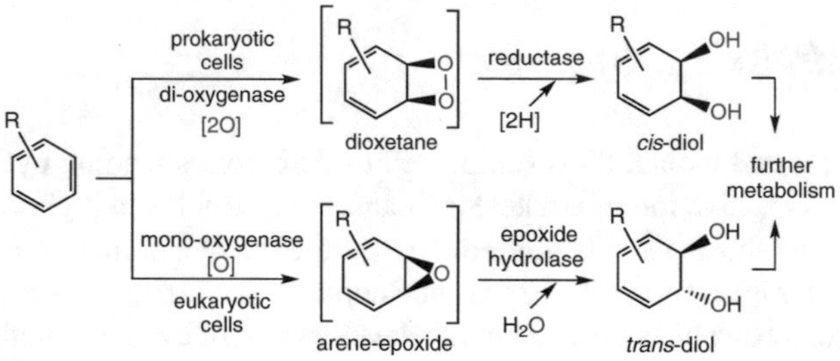

Scheme 2.83 Oxidative biodegradation of aromatics

Enzyme Mechanism and Stereochemical Implications
The mechanism of epoxide hydrolase-catalyzed hydrolysis has been elucidated from microsomal epoxide hydrolase (MEH) and bacterial enzymes and involves the *trans*-antiperiplanar addition of water to epoxides to give vicinal diol products. In general, the reaction occurs with *inversion* of configuration at the oxirane carbon atom to which the addition takes place and involves neither cofactors nor metal ions [626]. Two types of mechanism are known (Scheme 2.84).

S_N2-**Type Mechanism** A carboxylate residue – aspartate – performs a nucleophilic attack on the (usually less hindered) epoxide carbon atom by forming a covalent glycol-monoester intermediate [627–629]. The latter species can be regarded as a 'chemically inverted' acyl-enzyme intermediate in serine hydrolase reactions (Scheme 2.1). In order to avoid the occurrence of a charged oxy-anion, a proton from an adjacent Tyr-residue is simultaneously transferred. In a second step, the ester bond of the glycol monoester intermediate is hydrolyzed by a hydroxyl ion which is provided from water with the aid of a base – histidine [630] – thereby liberating the glycol. Finally, proton-migration from His to Tyr closes the catalytic cycle. This mechanism shows striking similarities to that of haloalkane dehalogenases, where a halide is displaced by an aspartate residue in a similar manner (Scheme 2.230)

[631, 632]. In addition, a mechanistic relationship with β-glycosidases which act via formation of a covalent glycosyl-enzyme intermediate by retaining the configuration at the anomeric center is obvious (Scheme 2.217) [633].

Borderline-S_N2-Type Mechanism Some enzymes, such as limonene-1,2-epoxide hydrolase, have been shown to operate via a single-step push-pull mechanism [634]. General acid catalysis by a protonated aspartic acid weakens the oxirane to facilitate a simultaneous nucleophilic attack of hydroxyl ion, which is provided by deprotonation of H_2O via an aspartate anion. Due to the borderline-S_N2-character of this mechanism, the nucleophile preferentially attacks the higher substituted carbon atom bearing the more stabilized δ^+-charge. After liberation of the glycol, proton-transfer between both Asp-residues closes the cycle.

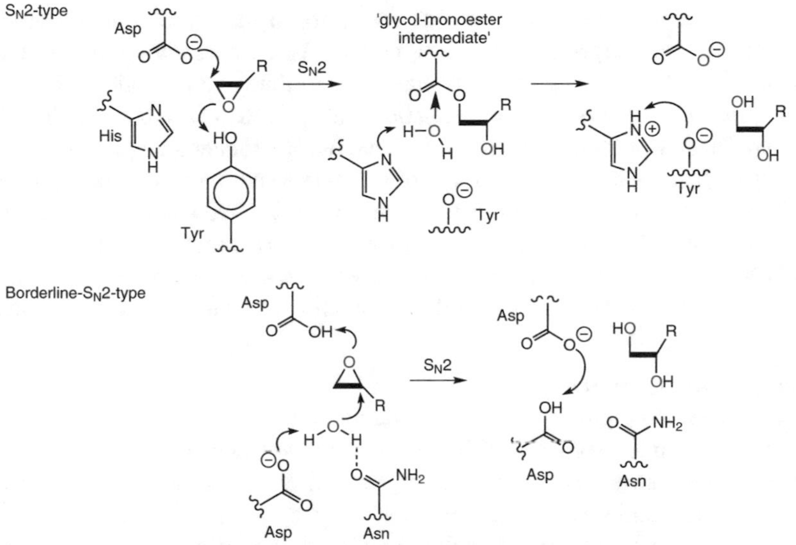

Scheme 2.84 S_N2- and borderline-S_N2-type mechanism of epoxide hydrolases

The above-mentioned facts have important consequences on the stereochemical course of the kinetic resolution of nonsymmetrically substituted epoxides. In contrast to the majority of kinetic resolutions of esters (e.g., by ester hydrolysis using proteases, esterases, and lipases) where the absolute configuration of the stereogenic center always remains the same throughout the reaction, the enzymatic hydrolysis of epoxides may take place via two different pathways (Scheme 2.85).

- Attack of the (formal) hydroxide ion on the less hindered (unsubstituted) oxirane carbon atom causes *retention* of configuration and leads to a hetero-chiral product mixture of enantiomeric diol and nonreacted epoxide.
- Attack on the stereogenic center leads to *inversion* and furnishes homochiral products possessing the same sense of chirality.

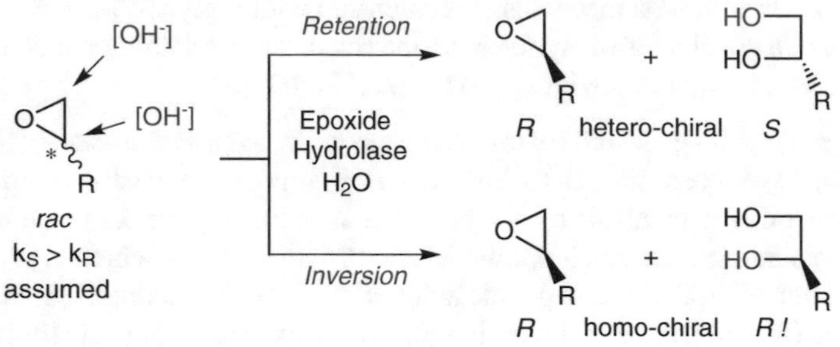

Scheme 2.85 Enzymatic hydrolysis of epoxides proceeding with retention or inversion of configuration

Although retention of configuration seems to be the more common pathway, inversion has been reported depending on the substrate structure and the type of enzyme [635, 636]. As a consequence, the absolute configuration of *both the product and the substrate* from a kinetic resolution of a racemic epoxide has to be determined separately in order to elucidate the stereochemical pathway. As may be deduced from Scheme 2.85, the use of the enantiomeric ratio is only appropriate to describe the enantioselectivity of an epoxide hydrolase as long as its regioselectivity is uniform, i.e., *only* inversion *or* retention is taking place, but *E*-values are inapplicable where mixed pathways, i.e., retention *and* inversion, are detected [637]. For the solution to this stereochemical problem, various methods were proposed [638].

Hepatic Epoxide Hydrolases
To date, two main types of epoxide hydrolases from liver tissue have been characterized, i.e., a microsomal (MEH) and a cytosolic enzyme (CEH), which are different in their substrate specificities. In general, MEH has been shown to possess higher activities and selectivities compared to its cytosolic counterpart.

Although pure MEH can be isolated from the liver of pigs, rabbits, mice, guinea pigs [639], or rats [640], a crude preparation of liver microsomes or even the $9000 \times g$ supernatant of homogenized liver was employed as a source for EH activity with little difference from that of the purified enzyme being observed [641]. However, other enzyme-catalyzed side-reactions such as ester hydrolysis may occur with crude preparations.

Cyclic *cis-meso*-epoxides can be asymmetrically hydrolyzed using hepatic epoxide hydrolases to give *trans*-diols. In this case, the (*S*)-configured oxirane carbon atom is preferentially attacked and inverted to yield an (*R,R*)-diol (Scheme 2.86) [642, 643]. Among hepatic epoxide hydrolases, the microsomal enzyme was more selective than the cytosolic counterpart. In comparison, microbial epoxide hydrolases were considerably more stereoselective and showed stereo-complementary preferences [644–646].

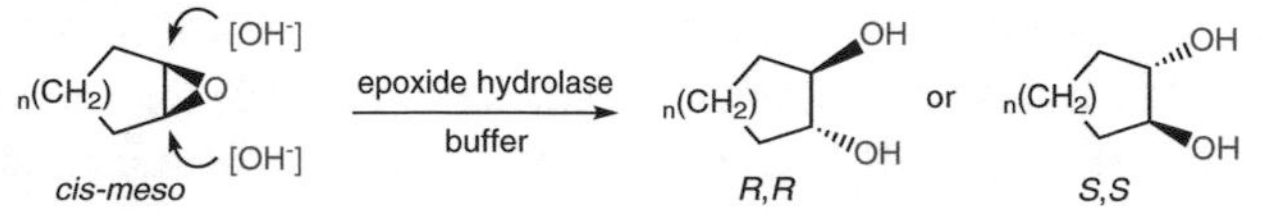

Scheme 2.86 Desymmetrization of cyclic *cis-meso*-epoxides by hepatic epoxide hydrolases

n	Enzyme	Diol	
		Config.	E.e. [%]
1	microsomal EH	*R,R*	90
1	cytosolic EH	*R,R*	60
2	microsomal EH	*R,R*	94
2	cytosolic EH	*R,R*	22
1	*Rhodotorula glutinis* CIMW147	*R,R*	>98
2	*Rhodococcus erythropolis* DCL14 (mutant)	*S,S*	97
2	*Sphingomonas* sp. HXN-200	*R,R*	99
3	*Rhodococcus erythropolis* DCL14 (mutant)	*S,S*	98

Utilizing steroid substrates, MEH was able to hydrolyze not only epoxides, but also the corresponding heteroatom derivatives such as aziridines to form *trans*-1,2-aminoalcohols albeit at slower rates (Scheme 2.87) [647]. The thiirane, however, was inert towards enzymatic hydrolysis. The enzyme responsible for this activity was assumed to be the same microsomal epoxide hydrolase.

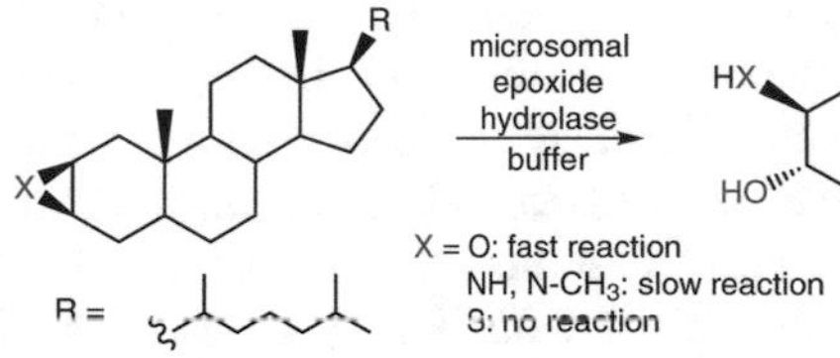

Scheme 2.87 Enzymatic hydrolysis of steroid epoxides and aziridines by microsomal epoxide hydrolase

Although many studies have been undertaken with hepatic epoxide hydrolases due to their importance in detoxification mechanisms [648], enzymes from these sources are unsuitable for preparative-scale transformations, since they cannot be obtained in reasonable amounts. In contrast, epoxide hydrolases from microbial sources are easy to produce by overexpression.

Microbial Epoxide Hydrolases

Although it was known for several years that microorganisms possess epoxide hydrolases, they were only scarcely applied to preparative organic transformations [649–652]. Thus, the hydrolysis of epoxides, which was occasionally observed during the microbial epoxidation of alkenes as an undesired side reaction causing product degradation, was usually neglected, and systematic studies were undertaken later on. It should be emphasized, that for practical reasons, many preparative-scale reactions were performed by using whole-cell preparations or crude cell-free extracts containing an unknown number of epoxide hydrolases. Some microbial epoxide hydrolases have been purified and characterized [653–657].

As a result, an impressive amount of knowledge on microbial epoxide hydrolases from various sources – such as bacteria, filamentous fungi, and yeasts – has been gathered and featured in several reviews [658–666]. The data available to date indicate that the enantioselectivities of enzymes from certain microbial sources can be correlated to the substitutional pattern of various types of substrates [667]:

- Red yeasts (e.g., *Rhodotorula* or *Rhodosporidium* sp.) give best enantioselectivities with monosubstituted oxiranes.
- Fungal cells (e.g., *Aspergillus* and *Beauveria* sp.) are best suited for styrene-oxide-type substrates.
- Bacterial enzymes (in particular derived from *Actinomycetes* such as *Rhodococcus, Nocardia* and *Sphingomonas* sp.) are the catalysts of choice for more highly substituted 2,2- and 2,3-disubstituted epoxides.

Monosubstituted oxiranes represent highly flexible and rather 'slim' molecules, which make chiral recognition a difficult task [668–671]. Thus, the majority of attempts to achieve highly selective transformations using epoxide hydrolases from bacterial and fungal origin failed for this class of substrates. The only notable exceptions were found among red yeasts, such as *Rhodotorula araucarae* CBS 6031, *Rhodosporidium toruloides* CBS 349, *Trichosporon* sp. UOFS Y-1118, and *Rhodotorula glutinis* CIMW 147. Regardless of the enzyme source, the enantiopreference for the (R)-enantiomer was predominant and the regioselectivity prevailed for the sterically less hindered carbon atom (Scheme 2.88).

Styrene oxide-type epoxides have to be regarded as a special group of substrates, as they possess a benzylic carbon atom, which facilitates the formation of a carbenium ion through resonance stabilization by the adjacent aromatic moiety (Scheme 2.89). Thus, attack at this position is electronically facilitated, although it is sterically hindered, and mixed regiochemical pathways (proceeding via retention *and* inversion) are particularly common within this group of substrates. As a consequence, E-values can only be applied to cases of single stereochemical pathways. The biocatalysts of choice were found among the fungal epoxide hydrolases, such as *Aspergillus niger* LCP 521 [672], *Beauveria densa* CMC 3240 and *Beauveria bassiana* ATCC 7159. Under certain circumstances, *Rhodotorula glutinis* CIMW 147 might serve as well [673–675].

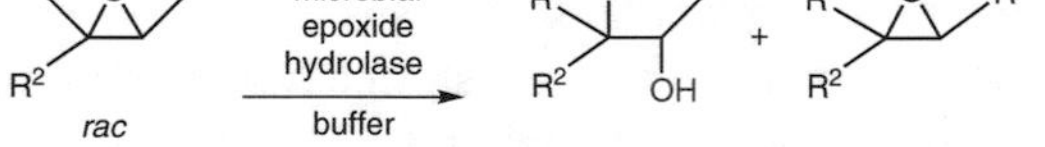

R^1	Enzyme Source	Selectivity[a]
CH$_2$Cl, C(CH$_3$)$_2$O(CO)C(CH$_3$)$_3$, CH$_2$OCH$_2$Ph, t-C$_4$H$_9$	bacterial	-
n-C$_3$H$_7$, n-C$_4$H$_9$, n-C$_5$H$_{11}$, n-C$_6$H$_{13}$, n-C$_8$H$_{18}$, n-C$_{10}$H$_{21}$	bacterial	±
n-C$_6$H$_{13}$	fungal	-
CH$_2$OH, C$_2$H$_5$, CH$_2$Cl, CH$_2$OCH$_2$Ph	yeast	- to ±
CH$_3$, n-C$_2$H$_5$	yeast	+
n-C$_3$H$_7$, n-C$_4$H$_9$, n-C$_5$H$_{11}$, n-C$_6$H$_{13}$	yeast	++

[a] Enantioselectivities are denoted as (-) = low (E <4), (±) = moderate (E = 4 - 12), (+) = good (E = 13 - 50), (++) excellent (E >50).

Scheme 2.88 Microbial resolution of monosubstituted epoxides (R^2, R^3 = H)

Among the sterically more demanding substrates, 2,2-disubstituted oxiranes were hydrolyzed in virtually complete enantioselectivities using enzymes from bacterial sources ($E > 200$), in particular *Mycobacterium* NCIMB 10420, *Rhodococcus* (NCIMB 1216, DSM 43338, IFO 3730) and closely related *Nocardia* spp. (Scheme 2.90) [676, 677]. All bacterial epoxide hydrolases exhibited a preference for the (S)-enantiomer. In those cases where the regioselectivity was determined, attack was found to exclusively occur at the unsubstituted oxirane carbon atom.

In contrast to 2,2-disubstituted epoxides, mixed regioselectivities are common for 2,3-disubstituted analogs and, as a consequence, E-values are not applicable (Table 2.3, Scheme 2.88) [678]. This is understandable, bearing in mind that the steric requirements at both oxirane positions are similar. Whereas fungal enzymes were less useful, yeast and bacterial epoxide hydrolases proved to be highly selective.

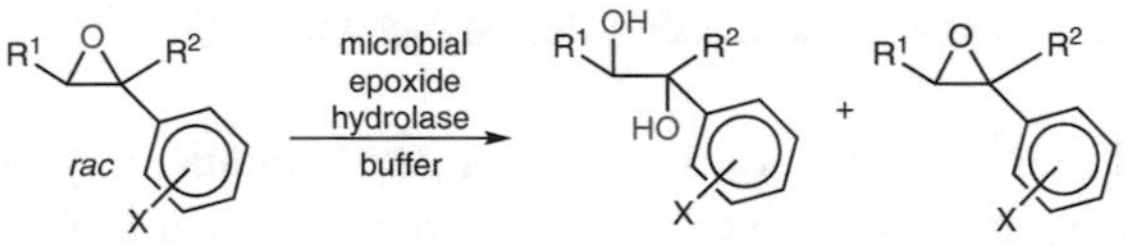

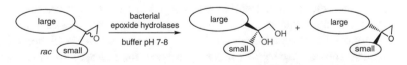

R^1	R^2	X	Enzyme Source	Selectivity[a]
H	H	p-CH$_3$, o-Cl, p-Cl	bacterial	$\pm$
H	CH$_3$	H	bacterial	$\pm$
H	H	o-CH$_3$, o-Hal	yeast	-
H	H	H	yeast	$\pm$
H	H	p-F, p-Cl, p-Br, p-CH$_3$	yeast	+
CH$_3$	H	H	yeast	++
H	CH$_3$	H	fungal	-
		indene oxide	fungal	+
CH$_3$	H	H	fungal	++
H	H	H	fungal	++
H	H	p-NO$_2$	fungal	++

[a] Enantioselectivities are denoted as (-) = low ($E < 4$), ($\pm$) = moderate ($E = 4$ - 12), (+) = good ($E = 13$ - 50), (++) excellent ($E > 50$).

Scheme 2.89 Microbial resolution of styrene oxide-type oxiranes

Small	Large	Enzyme Source	Selectivity[a]
CH$_3$	n-C$_5$H$_{11}$	fungal	$\pm$
CH$_3$	(CH$_2$)$_2$Ph, CH$_2$Ph	bacterial	$\pm$
C$_2$H$_5$	n-C$_5$H$_{11}$	bacterial	+
CH$_3$	n-C$_4$H$_9$, n-C$_5$H$_{11}$, n-C$_7$H$_{15}$, n-C$_9$H$_{19}$, (CH$_2$)$_4$Br, (CH$_2$)$_3$CH=CH$_2$	bacterial	++

[a] Enantioselectivities are denoted as (-) = low ($E < 4$), ($\pm$) = moderate ($E = 4$ - 12), (+) = good ($E = 13$ - 50), (++) excellent ($E > 50$).

Scheme 2.90 Enzymatic resolution of 2,2-disubstituted epoxides using microbial epoxide hydrolases

Table 2.3 Microbial resolution of 2,3-disubstituted epoxides (for substrate structures see Scheme 2.88)

R^1	R^2	R^3	Enzyme source	Selectivity[a]
CH_3	H	n-C_5H_{11}	Fungal	$\pm$
H	CH_3	n-C_5H_{11}	Fungal	$\pm$
CH_3	H	CH_3	Yeast	++
H	CH_3	CH_3	Yeast	++
H	C_2H_5	n-C_3H_7	Bacterial	$\pm$
C_2H_5	H	n-C_4H_9	bacterial	$\pm$
H	CH_3	n-C_4H_9, n-C_5H_{11}, n-C_9H_{19}	Bacterial	++
CH_3	H	n-C_4H_9	Bacterial[b]	++

[a]Enantioselectivities are denoted as $(-)$ = low $(E < 4)$, $(\pm)$ = moderate $(E = 4-12)$, $(+)$ = good $(E = 13-50)$, $(++)$ excellent $(E > 50)$
[b]Enantioconvergent pathway, i.e., a sole stereoisomeric diol was formed

To date, only limited data are available on the enzymatic hydrolysis of trisub-stituted epoxides [679–684]. For example, a racemic allylic terpene alcohol containing a trisubstituted epoxide moiety was hydrolyzed by whole cells of *Helminthosporium sativum* to yield the (*S,S*)-diol with concomitant oxidation of the terminal alcoholic group (Scheme 2.91). The mirror image (*R,S*)-epoxide was not transformed. Both optically pure enantiomers were then chemically converted into a juvenile hormone [685].

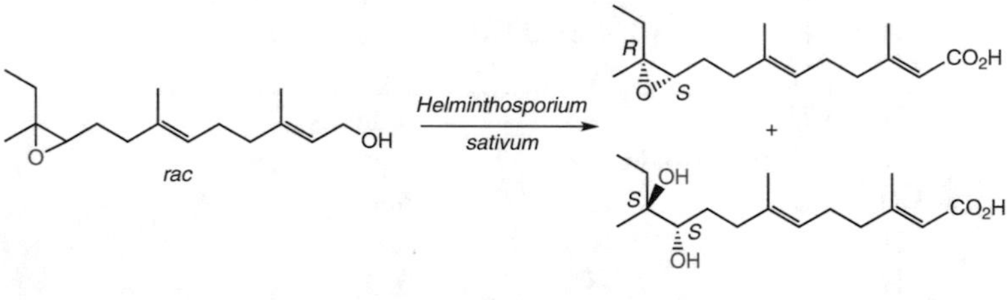

Scheme 2.91 Microbial resolution of a trisubstituted epoxide

In order to circumvent the disadvantages of kinetic resolution, several protocols were developed towards the *enantioconvergent* hydrolysis of epoxides, which lead to a single enantiomeric vicinal diol as the sole product from the racemate.

The first technique made use of two fungal epoxide hydrolases possessing matching opposite regio- and enantioselectivity for styrene oxide (Scheme 2.92) [686]. Resting cells of *Aspergillus niger* hydrolyzed the (*R*)-epoxide via attack at the less hindered carbon atom to yield the (*R*)-diol of moderate optical purity. The (*S*)-epoxide remained unchanged and was recovered in 96% e.e. In contrast, *Beauveria bassiana* exhibited the *opposite* enantio- and regioselectivity. It hydro-lyzed the (*S*)-enantiomer but with an unusual *inversion of configuration* via attack at the more hindered benzylic position. As a result, the (*R*)-diol was obtained from the (*S*)-epoxide leaving the (*R*)-epoxide behind. By combining both microbes in a single reactor, an elegant deracemization technique was accomplished making use of both stereo-complementary pathways. Whereas *Aspergillus* hydrolyzed the (*R*)-

epoxide with retention, *Beauveria* converted the (*S*)-counterpart with inversion. As a result, (*R*)-phenylethane-1,2-diol was obtained in 89% e.e. and 92% yield.

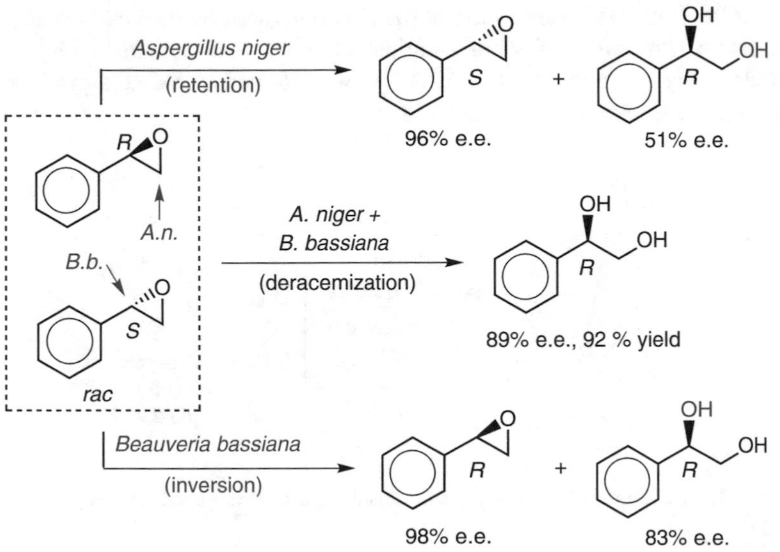

Scheme 2.92 Microbial resolution and deracemization of styrene oxide

For 2,2-disubstituted oxiranes, this technique was not applicable because it would require an enzyme performing a highly unfavored nucleophilic attack on a fully substituted carbon atom. In this case, a one-pot two-step sequence consisting of combined bio- and chemocatalysis was successful (Scheme 2.93) [687]. In the first step, 2,2-disubstituted oxiranes were kinetically resolved by using bacterial epoxide hydrolases in excellent selectivity. The biohydrolysis proceeds exclusively via attack at the unsubstituted carbon atom with complete *retention* at the stereogenic center. By contrast, acid-catalyzed hydrolysis of the remaining nonconverted enantiomer under carefully controlled conditions proceeds via an S_N2-borderline mechanism with *inversion* of configuration. Thus, combination of both steps in a one-pot resolution-inversion sequence yields the corresponding (*S*)-1,2-diols in virtually enantiopure form and in excellent yields (>90%).

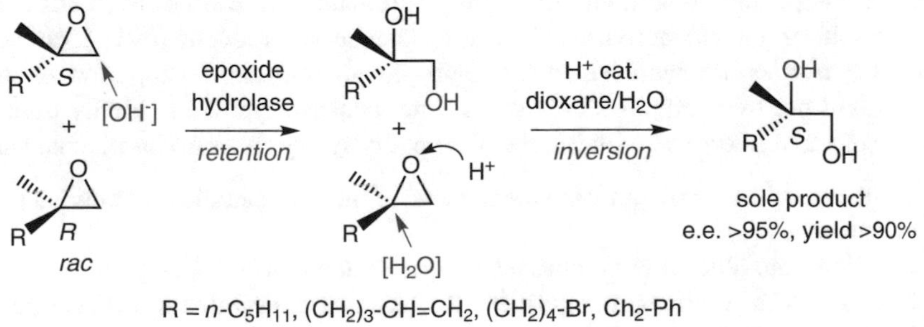

R = *n*-C_5H_{11}, $(CH_2)_3$-CH=CH_2, $(CH_2)_4$-Br, Ch_2-Ph

Scheme 2.93 Deracemization of 2,2-disubstituted oxiranes using combined bio- and chemo-catalysis

An exceptional case for an enantioconvergent biocatalytic hydrolysis of a (±)-cis-2,3-epoxyalkane is shown in Scheme 2.94 [688]. Based on ^{18}O-labeling experiments, the stereochemical pathway of this reaction was elucidated to proceed via attack of the (formal) hydroxyl ion at the (S)-configured oxirane carbon atom with concomitant *inversion* of configuration at both enantiomers with *opposite* regioselectivity. As a result, the (R,R)-diol was formed as the sole product in up to 97% e.e. in almost quantitative yield.

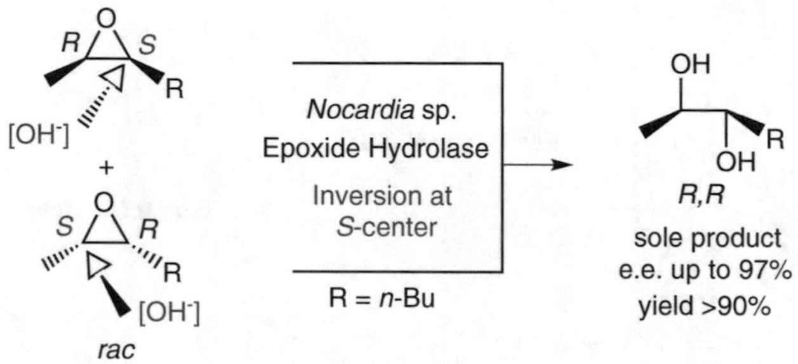

Scheme 2.94 Deracemization of 2,3-disubstituted oxiranes via enantioconvergent enzymatic hydrolysis

Enzymatic epoxide hydrolysis has been successfully upscaled to multigram batches using resting microbial cells containing (overexpressed) epoxide hydrolases. In order to avoid enzyme deactivation by the toxic substrate and to overcome solubility problems, aqueous-organic two-phase systems consisting of an alkane (hexane, i-octane) or an ether (MeOtBu, iPr$_2$O) were employed [689, 690].

As an alternative to the enzymatic hydrolysis of epoxides, nonracemic vicinal diols may be obtained from epoxides via the nucleophilic ring-opening by nitrite catalyzed by halohydrin dehalogenase (a lyase). The corresponding nitrite-monoesters are spontaneously hydrolyzed to yield diols. For the application of this technique see Sect. 2.7.2.

2.1.6 Hydrolysis of Nitriles

Organic compounds containing nitrile groups are found in the environment not only as a result of human activities, but also as natural products [691]. Naturally occurring nitriles are synthesized by plants, fungi, bacteria, algae, sponges, and insects, but not by mammals. This is puzzling, because cyanide is highly toxic to living cells and interferes with biochemical pathways by three major mechanisms:

- Tight chelation to di- and trivalent metal atoms in metalloenzymes such as cytochromes
- Addition onto aldehydes or ketones to form cyanohydrin derivatives
- Reaction with Schiff-base intermediates (e.g., in transamination reactions) to form stable nitrile derivatives [692]

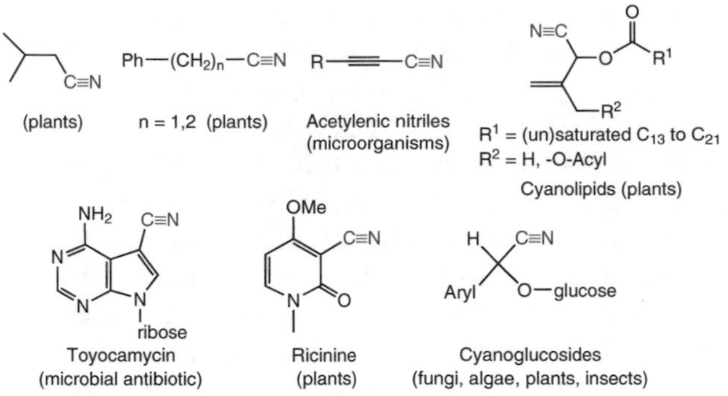

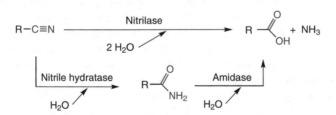

Scheme 2.95 Naturally occurring organic nitriles

As shown in Scheme 2.95, natural nitriles include cyanogenic glucosides which are produced by a wide range of plants including major crops such as almond, cassava [693] and sorghum (millet). Plants and microorganisms are also able of producing aliphatic or aromatic nitriles, such as cyanolipids, ricinine, and phenylacetonitrile [694]. These compounds can serve not only as a nitrogen storage, but also as protecting agents against attack by hungry predators. However, if one species has developed a defence mechanism, an invader will try to undermine it with a counterstrategy. As a consequence, it is not unexpected that there are several biochemical pathways for nitrile degradation, such as oxidation and – more important – by hydrolysis. Enzyme-catalyzed hydrolysis of nitriles may occur via two different pathways depending on steric and electronic factors of the substrate structure [695–698] (Scheme 2.96).

$$R-C{\equiv}N \xrightarrow[2\,H_2O]{\text{Nitrilase}} R\!\!\begin{array}{c}O\\ \|\\ \end{array}\!\!OH + NH_3$$

Scheme 2.96 General pathways of the enzymatic hydrolysis of nitriles

- Aliphatic nitriles are often metabolized in two stages. First they are converted to the corresponding carboxamide by a *nitrile hydratase* [699–701] and then to the carboxylic acid by an *amidase* enzyme (a protease) [702].
- Aromatic, heterocyclic, and certain unsaturated aliphatic nitriles are often directly hydrolyzed to the corresponding acids without formation of the intermediate free amide by a so-called *nitrilase* enzyme. The nitrile hydratase and nitrilase enzyme use distinctively different mechanisms of action.

Nitrile hydratases are generally induced by amides and are known to possess a tightly bound metal atom (Co^{2+} or Fe^{3+} [703]) which is required for catalysis [704–

709]. In nitrile hydratase from *Bevibacterium* sp., the central metal is octahedrally coordinated to two NH-amide groups from the backbone and three Cys–SH residues, two of which are post-translationally modified into a Cys-sulfenic (–SOH) and a Cys-sulfinic (–SO$_2$H) moiety. This claw-like setting is required to firmly bind the non-heme iron or the non-corrinoid cobalt in a pseudo-porphyrin arrangement [710–712] (Scheme 2.97. The remaining axial ligand (X) is either a water molecule (Co^{2+}) [713] or nitric oxide (NO) which binds to Fe^{3+} [714, 715]. Quite remarkably, the activity of the latter protein is regulated by light: in the dark, the enzyme is inactive, because NO occupies the binding site for the substrate. Upon irradiation with visible light, NO dissociates and activity is switched on.

Three proposals for the mechanism of metal-depending nitrile hydratases have been suggested, the most plausible assumes direct coordination of the nitrile to the metal, which (by acting as Lewis-acid) increases the electrophilicity of the carbon atom to allow attack of a water-molecule. The hydroxy-imino-species thus formed tautomerizes to form the carboxamide [716–718].

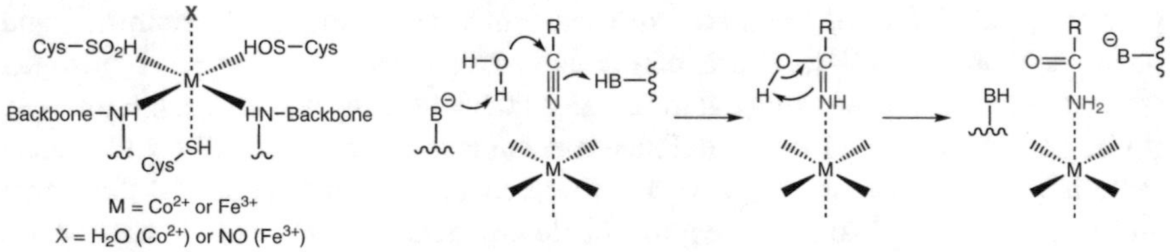

Scheme 2.97 Coordination sphere of Fe^{3+} and mechanism of *Brevibacterium* sp. nitrile hydratase

Nitrile hydratases from different sources are very similar to each other in terms of their substrate spectrum and accept a broad range of aliphatic, aromatic and arylaliphatic nitriles, generally with low or marginal stereoselectivities [719].

On the other hand, nitrilases operate by a completely different mechanism (Scheme 2.98). They possess neither coordinated metal atoms, nor cofactors, but act through an essential nucleophilic sulfhydryl residue of a cysteine [720, 721], which is encoded in the nitrilase-sequence motif Glu–Lys–Cys [722]. The mechanism of nitrilases is similar to general base-catalyzed nitrile hydrolysis: Nucleophilic attack by the sulfhydryl residue on the nitrile carbon atom forms an enzyme-bound thioimidate intermediate, which is hydrated to give a tetrahedral intermediate. After the elimination of ammonia, an acyl-enzyme intermediate is formed, which (like in serine hydrolases) is hydrolyzed to yield a carboxylic acid [723]. According to their substrate specificities, nitrilases have been classified into three subtypes, aliphatic nitrilases, aromatic nitrilases and arylacetonitrilases, of which the latter are often enantioselective, which gives them the greatest potential for biotransformations [724].

Enzymatic hydrolysis of nitriles is not only interesting from an academic standpoint, but also from a biotechnological point of view [725–732]. Cyanide represents a widely applicable C$_1$-synthon – a 'water-stable carbanion' – but the conditions usually required for the chemical hydrolysis of nitriles present several disadvantages. The reactions usually require either extreme pH, which is incompatible with other hydrolyzable groups that may be present. Alternative methods rely on metal catalysts, e.g. Raney-copper or manganese dioxide. Overall, energy consumption is high and unwanted side-products arising from over-hydrolysis or decomposition are common. Considerable amounts of salts are formed during neutralization. Using enzymatic methods, conducted at physiological pH, most of these drawbacks can be avoided. Additionally, these transformations can often be achieved in a chemo-, regio-, and enantioselective manner. Due to the fact that isolated nitrile-hydrolyzing enzymes are often very sensitive [696], the majority of transformations have been performed using sturdy whole-cell systems.

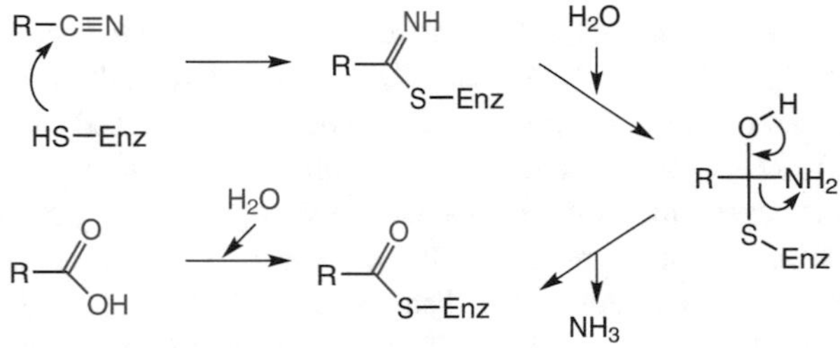

Scheme 2.98 Mechanism of nitrilases

Another important aspect is the enzymatic hydrolysis of cyanide and nitriles for the detoxification of industrial effluents [733–736].

Chemoselective Hydrolysis of Nitriles
The microorganisms used as sources of nitrile-hydrolyzing enzymes usually belong to the genera *Bacillus, Brevibacterium, Micrococcus, Rhodococcus, Pseudomonas*, and *Bacteridium* and they generally show a broad metabolic diversity. Depending on the source of carbon and nitrogen – acting as 'inducer' – added to the culture medium, either nitrilases or nitrile hydratases are predominantly produced by the cell. Thus, the desired hydrolytic pathway leading to an amide or a carboxylic acid can often be biologically 'switched on' during the growth of the culture by using aliphatic or aromatic nitriles as inducers. In order to avoid substrate inhibition (which is a more common phenomenon with nitrile-hydrolyzing enzymes than product inhibition [737]) the substrates are fed continuously to the culture.

Acrylamide is one of the most important commodity chemicals for the synthesis of various polymers and is produced in an amount of about 2 Mt/year worldwide. In its conventional synthesis, the hydration of acrylonitrile is performed with copper catalysts. However, the preparative procedure for the catalyst, difficulties in its regeneration, problems associated with separation and purification of the formed acrylamide, undesired polymerization and over-hydrolysis are serious drawbacks. Using whole cells of *Brevibacterium* sp. [738, 739], *Pseudomonas chlororapis* [740, 741] or *Rhodococcus rhodochrous* [742] acrylonitrile can be converted into acrylamide in yields of >99%; the formation of byproducts such as acrylic acid is circumvented by blocking of the amidase activity. The scale of this biotransformation exceeds 600,000 t/year [743] (Scheme 2.99).

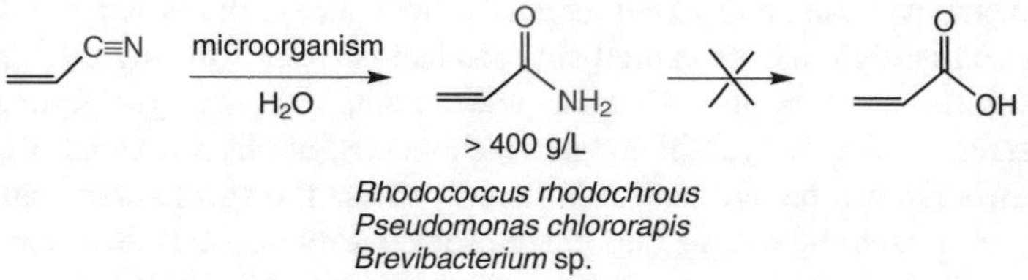

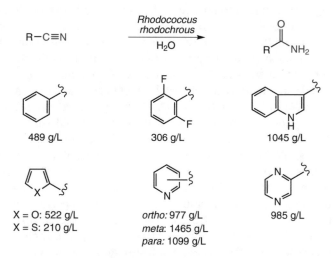

Scheme 2.99 Chemoselective microbial hydrolysis of acrylonitrile

Aromatic and heteroaromatic nitriles were selectively transformed into the corresponding amides by a *Rhodococcus rhodochrous* strain [744]; the products accumulated in the culture medium in significant amounts (Scheme 2.100). In contrast to the hydrolysis performed by chemical means, the biochemical transformations were highly selective and occurred without the formation of the corresponding carboxylic acids.

Important from a commercial standpoint was that *o*-, *m*-, and *p*-substituted cyanopyridines were accepted as substrates [745, 746] to give picolinamide (a pharmaceutical), nicotinamide (a vitamin), and isonicotinamide (a precursor for isonicotinic acid hydrazide, a tuberculostatic) (Scheme 2.100). Extremely high productivities were obtained due to the fact that the less soluble carboxamide product readily crystallized from the reaction medium in 100% purity. Nicotinamide – enzymatically produced on a scale of 6000 t/year – is an important nutritional factor and is therefore widely used as a vitamin additive for food and feed supplies [747]. Pyrazinamide is used as a tuberculostatic.

Scheme 2.100 Chemoselective microbial hydrolysis of aromatic and heteroaromatic nitriles yielding carboxamides (product concentrations)

By changing the biochemical pathway through using modified culture conditions, the enzymatic pathways of nitrile hydrolysis are switched and the corresponding carboxylic acids can be obtained (see Scheme 2.101). For instance, p-aminobenzoic acid, a member of the vitamin B group, was obtained from p-aminobenzonitrile using whole cells of *Rhodococcus rhodochrous* [748]. Similarly, the antimycobacterial agent pyrazinoic acid was prepared in excellent purity from cyanopyrazine [749]. Like nicotinamide, nicotinic acid is a vitamin used as an animal feed supplement, in medicine, and also as a biostimulator for the formation of activated sludge. Microbial hydrolysis of 3-cyanopyridine using *Rhodococcus rhodochrous* [750] or *Nocardia rhodochrous* [751] proceeds quantitatively, whereas chemical hydrolysis is hampered by moderate yields.

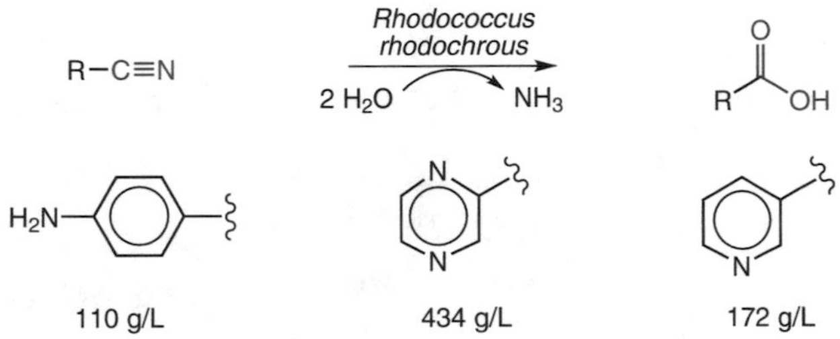

Scheme 2.101 Chemoselective microbial hydrolysis of aromatic and heteroaromatic nitriles yielding carboxylic acids (product concentrations)

Regioselective Hydrolysis of Dinitriles

The selective hydrolysis of one nitrile group out of several in a molecule is generally impossible using traditional chemical catalysis and the reactions usually result in the formation of complex product mixtures. In contrast, whole microbial cells can be very efficient for this purpose [752] (Scheme 2.102).

For instance, 1,3- and 1,4-dicyanobenzenes were selectively hydrolyzed by *Rhodococcus rhodochrous* to give the corresponding monoacids [753, 754]. In the aliphatic series, tranexamic acid (*trans*-4-aminomethyl-cyclohexane-1-carboxylic acid), which is a hemostatic agent, is synthesized from *trans*-1,4-dicyanocyclohexane. Complete mono-hydrolysis was achieved by using an *Acremonium* sp. [755]. The outcome of regioselective nitrile hydrolysis is believed to depend on the distance of the nitrile moieties and the presence of other polar groups within the substrate [756, 757].

5-Cyanovaleramide is required for the synthesis of the herbicide azafenidin. The chemical hydration of adiponitrile results in significant formation of the undesired di-amide and generates large amounts of waste products. In contrast, selective mono-hydration was achieved employing *Rhodococcus ruber* CGMCC3090 [758] or *Pseudomonas chlororaphis* B23. The latter process was upscaled using whole cells immobilized in Ca alginate beads to convert 12.7 t of adiponitrile with a selectivity of 96% [759].

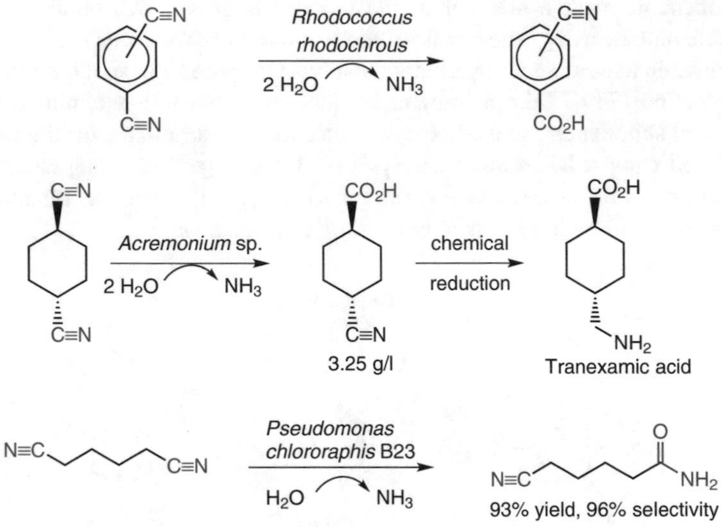

Scheme 2.102 Regioselective microbial hydrolysis of dinitriles

Enantioselective Hydrolysis of Nitriles
While most biocatalytic hydrolyses of nitriles make use of the mild reactions conditions and the chemo- and regioselectivity of nitrile-hydrolyzing enzymes, their stereoselectivity has been investigated more scarcely. It seems to be a common trend that both nitrilases and nitrile hydratases are often less specific with respect to the chirality of the substrate and that in nitrile hydratase-amidase pathways enantiodiscrimination often occurs during the hydrolysis of the intermediate carboxamide by the amidase [760] (Scheme 2.96). As a rule, the 'natural' L-configured enantiomer is usually converted into the acid leaving the D-counterpart behind. This is not unexpected bearing in mind the high specificities of proteases on α-substituted carboxamides (see Sect. 2.1.2).

Desymmetrization of Prochiral Dinitriles Prochiral α,α-disubstituted malono-nitriles can be hydrolyzed in an asymmetric manner by the aid of *Rhodococcus rhodochrous* [761] (Scheme 2.103). In accordance with the above-mentioned trend, the dinitrile was nonselectively hydrolyzed by the nitrile hydratase in the cells to give the dicarboxamide. In a second consecutive step, the latter was subsequently transformed by the amidase with high selectivity for the pro-(R) amide group to yield the (R)-amide-acid in 96% e.e. and 92% yield. This pathway was confirmed by the fact that identical results were obtained when the dicarboxamide was used as substrate. The nonracemic amide-acid product thus obtained serves as a starting material for the synthesis of nonnatural α-methyl-α-amino acids [762].

In contrast, prochiral glutarodinitriles were stereoselectively hydrolyzed via two steps using whole microbial cells: in a first step, a stereoselective nitrile hydratase furnished the (R)-monoamide, which was further hydrolyzed to the corresponding carboxylic acid by an amidase [726, 763]. The cyano-acids thus obtained served as building blocks for the synthesis of cholesterol-lowering drugs from the statin family. An impressive example for the development of stereoselective enzymes derived from the metagenome is the discovery of >130 novel nitrilases from biotope-specific environmental DNA libraries [764, 765]. Among these enzymes, 22 nitrilases showed (S)-selectivity for the desymmetrization of the unprotcted 3-hydroxyglutarodinitrile (Scheme 2.103, R = H), while one produced the mirror-image (R)-enantiomer in 95–98% e.e. [766].

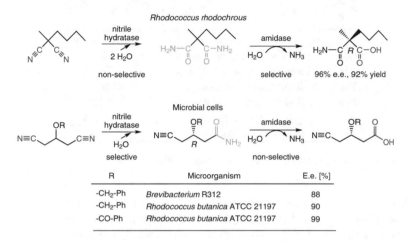

R	Microorganism	E.e. [%]
-CH₂-Ph	Brevibacterium R312	88
-CH₂-Ph	Rhodococcus butanica ATCC 21197	90
-CO-Ph	Rhodococcus butanica ATCC 21197	99

Scheme 2.103 Asymmetric microbial hydrolysis of a prochiral dinitrile

Kinetic and Dynamic Resolution of *rac*-Nitriles α-Hydroxy and α-amino acids can be obtained from the corresponding α-hydroxynitriles (cyanohydrins) and α-aminonitriles [767], which are easily synthesized in racemic form from the corresponding aldehyde precursors by addition of hydrogen cyanide or a Strecker synthesis, respectively (Schemes 2.104 and 2.105). In aqueous systems, cyanohydrins are stereochemically labile and undergo spontaneous racemization via HCN elimination, which furnishes a dynamic resolution process. From aliphatic *rac*-cyanohydrins, whole cells of *Torulopsis candida* yielded the corresponding (S)-α-hydroxy acids [768], while (R)-mandelic acid is produced from *rac*-mandelonitrile on an industrial scale by employing resting cells of *Alcaligenes faecalis* [769] in >90 % yield [770, 771].

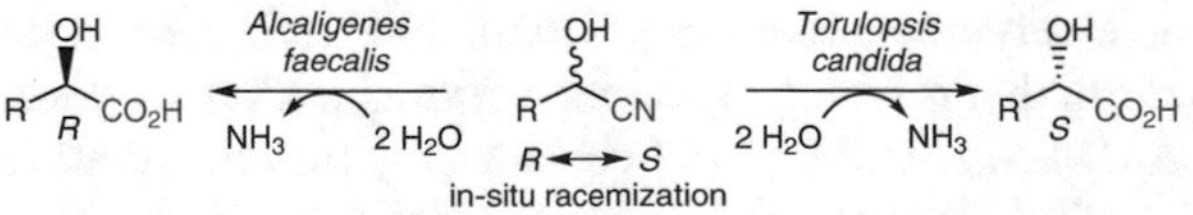

R	Microorganism	e.e. α-Hydroxyacid [%]
(CH₃)₂CH-	*Torulopsis candida*	>90
(CH₃)₂CH-CH₂-	*Torulopsis candida*	>95
Ph-	*Alcaligenes faecalis*	~100

Scheme 2.104 Stereocomplementary enantioselective hydrolysis of α-hydroxynitriles

In a related fashion, α-aminonitriles are enzymatically hydrolyzed to yield α-amino acids (Scheme 2.105). Whereas the enantiorecognition in *Brevibacterium imperiale* or *Pseudomonas putida* occurs through an amidase [772, 773], *Rhodococcus rhodochrous* PA-43, *Acinetobacter* sp. APN, and *Aspergillus fumigatus* possess enantiocomplementary nitrilases [772, 774, 775].

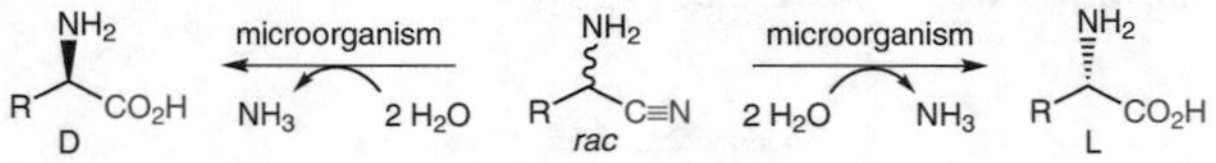

R	Microorganism	e.e. α-Aminoacid [%]
Leu	*Rhodococcus rhodochrous*	90 (L)
Ala	*Rhodococcus rhodochrous*	57 (D)
Val	*Rhodococcus rhodochrous*	100 (L)
Met	*Rhodococcus rhodochrous*	96 (L)
PhGly	*Aspergillus fumigatus*	80 (L)
Ala	*Acinetobacter* sp.	74 (L)

Scheme 2.105 Enantioselective hydrolysis of α-aminonitriles

Many kinetic resolutions of *rac*-nitriles were performed in search of a method to produce (*S*)-configurated α-arylpropionic acids, such as ketoprofen, ibuprofen, or naproxen, which are widely used as nonsteroidal antiinflammatory agents. Overall, enantioselectivities depended on the strain used, and whether a nitrilase- or nitrile hydratase-amidase pathway was dominant, which determines the nature of (enantiomeric) products consisting of a mixture of nitrile/carboxylic acid or amide/carboxylic acid, respectively [770, 776–779].

For organisms which express both pathways for nitrile hydrolysis, the stereochemical pathways can be very complex. The latter is illustrated by the microbial resolution of α-aryl-substituted propionitriles using a *Rhodococcus butanica* strain (Scheme 2.106) [780]. Formation of the 'natural' L-acid and the D-amide indicates the presence of an L-specific amidase and a nonspecific nitrile hydratase. However, the occurrence of the (*S*)-nitrile in case of Ibuprofen (R = *i*-Bu, e.e. 73%) proves the enantioselectivity of the nitrile hydratase [777]. In a related approach, *Brevibacterium imperiale* was used for the resolution of structurally related α-aryloxypropionic nitriles [781].

As a substitute for (expensive) commercial enzyme preparations for nitrile-hydrolysis, whole-cell preparations are recommended: *Rhodococcus* R312 [782][23] contains both nitrile-hydrolyzing metabolic pathways, whereas *Rhodococcus* DSM 11397 and *Pseudomonas* DSM 11387 contain only nitrile hydratase (no nitrilase) and nitrilase (no nitrile hydratase) activity, respectively [783].

---

[23]The strain was formerly denoted as *Brevibacterium* and is available as CBS 717.73.

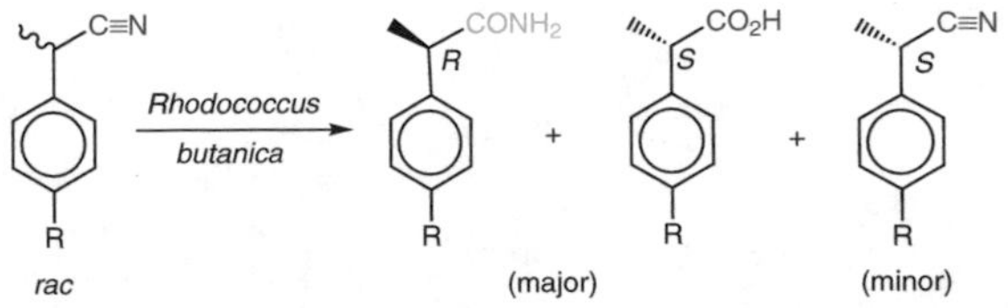

R	e.e. Amide [%]	e.e. Acid [%]	e.e. Nitrile [%]
$(CH_3)_2CH\text{-}CH_2\text{-}$	99	87	73
Cl	76	>99	-
OCH_3	99	99	-

Scheme 2.106 Enantioselective hydrolysis of α-aryl propionitriles

Enzymatic nitrile hydrolysis is a simple and convenient method to selectively obtain the corresponding carboxamides or carboxylic acids, depending on the type of enzyme(s) employed. Due to the sensitivity of nitrile-hydrolysing enzymes, whole microbial (resting) cells are used, in particular on industrial scale. Although excellent chemo- and regioselectivities are common, stereoselectivities may vary and are often incomplete.

2.2 Reduction Reactions

The enzymes employed for the majority of redox reactions are classified into three categories: dehydrogenases, oxygenases and oxidases (Scheme 2.144) [784–786]. Among them, alcohol dehydrogenases – also termed carbonyl reductases – have been widely used for the reduction of carbonyl groups (aldehydes, ketones) and ene-reductases are employed for the bioreduction of (electronically activated) carbon-carbon double bonds. In contrast, the asymmetric bioreduction of C=N-bonds is only feasible for special types of substrates, such as (cyclic) Schiff-base type imines, or in the reductive amination of α-keto acids yielding α-amino acids.

Since reduction usually implies the transformation of a planar sp^2-hybridized carbon into a tetrahedral sp^3-atom, it goes in hand with the generation of a stereogenic center and represents a desymmetrization reaction (Scheme 2.107). In contrast, the corresponding reverse process (e.g., alcohol oxidation or dehydrogenation) leads to the destruction of a chiral center, which is generally of limited use.

In contrast, oxygenases – named for using molecular oxygen as cosubstrate – have been shown to be particularly useful for oxidation reactions since they catalyze the functionalization of nonactivated C–H or C=C bonds, as well as electron-rich heteroatoms, affording C–H hydroxylation, C=C epoxidation, and thioether-oxidation, respectively (Sect. 2.3.3). Oxidases, which are responsible for the transfer of electrons, have gained increasing importance for the oxidation of alcohols (Sect. 2.3.1) and amines (Sect. 2.3.2) more recently.

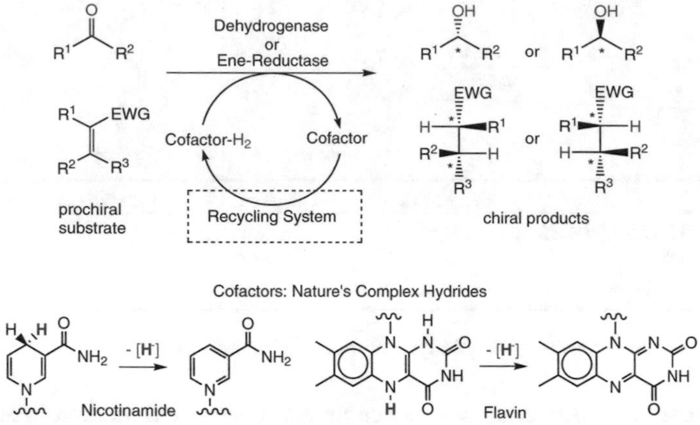

Scheme 2.107 Reduction reactions catalyzed by dehydrogenases (*EWG* electron withdrawing group)

2.2.1 Recycling of Cofactors

The major and crucial distinction between redox enzymes and hydrolases described in the previous chapter, is that the former require redox cofactors, which donate or accept the chemical equivalents for reduction (or oxidation). For the majority of redox enzymes, nicotinamide adenine dinucleotide [NAD(H)] and its respective phosphate [NADP(H)] are required by about 80% and 10% of redox enzymes, respectively. Flavines (FMN, FAD) and pyrroloquinoline quinone (PQQ) are encountered more rarely. The nicotinamide cofactors – resembling 'Nature's complex hydrides' – have two features in common, i.e., they are relatively unstable molecules and they are prohibitively expensive if used in stoichiometric amounts.[24] In addition, they cannot be replaced by more economical man-made substitutes. Since it is only the *oxidation state* of the cofactor which changes during the reaction, while the remainder of the complex structure stays intact, the cofactor may be regenerated in situ by using a second concurrent redox-reaction to allow it to re-enter the reaction cycle. Thus, the expensive cofactor is needed only in catalytic amounts, which leads to a drastic reduction in cost. The efficiency of such a recycling process is measured by the number of cycles which can be achieved before a cofactor molecule is finally destroyed. It is expressed as the 'total turnover number' (TTN, Sect. 1.4.3) – which is the total number of moles of product formed per mole of cofactor during its entire lifetime.[25] As a rule of thumb, a few thousand cycles (10^3–10^4) are sufficient for redox reactions on a laboratory scale, whereas for technical purposes, total turnover numbers of at least 10^5 are highly desirable. The economic barrier to large-scale reactions posed by cofactor

---

[24]The current prices for 1 mole are: NAD$^+$ US \$1400, NADH US \$2600, NADP$^+$ US \$18,000, NADPH US \$70,000.

[25]In cofactor recycling the TTN is sometimes called 'cycle number'.

costs has been recognized for many years and a large part of the research effort concerning dehydrogenases has been expended in order to solve the problem of cofactor recycling [555, 787–790].

Cofactor recycling is no problem when whole microbial cells are used as biocatalysts for redox reactions. In this case, inexpensive sources of redox equivalents such as carbohydrates can be used since the microorganism possesses all the enzymes and cofactors which are required for metabolism. The advantages and disadvantages of using whole-cell systems are discussed in Sect. 2.2.3.

Recycling of Reduced Nicotinamide Cofactors
The easiest but least efficient method of regenerating NADH from NAD^+ is the nonenzymic reduction using a reducing agent such as sodium dithionite ($Na_2S_2O_4$) [791]. Since the corresponding turnover numbers of this process are very low (TTN $\leq$ 100), this method has only historical interest. Similarly, electrochemical [792–794] and photochemical regeneration methods [795–798] suffer from insufficient electron transport causing side-reactions and show low to moderate turnover numbers (TTN $\leq$ 1000).[26] On the other hand, enzymic methods for NADH or NADPH recycling have been shown to be much more efficient and nowadays these represent the methods of choice. They may be conveniently subdivided into coupled-substrate and coupled-enzyme types.

Coupled-Substrate Process Aiming at keeping things as simple as possible, the cofactor required for the transformation of the main substrate is constantly regenerated by addition of a second auxiliary substrate (H-donor) which is transformed by the *same* enzyme, but into the *opposite* direction (Scheme 2.108) [799–801]. To shift the equilibrium of the reaction in the desired direction, the donor must be applied in excess [802]. In principle, this approach is applicable to both directions of redox reactions [803] and it constitutes a biological variant of a transfer-hydrogenation. Although the use of a single enzyme simultaneously catalyzing two reactions appears elegant, some significant disadvantages are often encountered in coupled-substrate cofactor recycling:

- The overall efficiency of the process is limited since the enzyme's activity is distributed between both the substrate (hydrogen acceptor) and the auxiliary hydrogen donor.
- Enzyme inhibition caused by the high concentrations of the auxiliary substrate – cosubstrate inhibition – is common, in particular when highly reactive carbonyl species such as acetaldehyde or cyclohexenone are generated in the recycling process.
- The product has to be purified from large amounts of auxiliary substrate used in excess.

---

[26]For example, if the reduction of $NAD(P)^+$ to NAD(P)H is 95% selective for hydride transfer onto the *p*-position of the nicotinamide ring, after 100 turnovers the residual activity of the cofactor would be 0.95^{100} being equivalent to only ~0.6%.

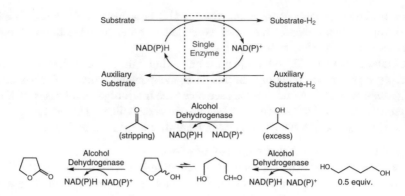

Scheme 2.108 Cofactor recycling by the coupled-substrate method

As a consequence, the coupled-substrate cofactor recycling only works with exceptionally sturdy dehydrogenases, which can tolerate high concentrations of a sacrificial *sec*-alcohol, such as 2-propanol, as hydride donor, but due to its simplicity it is the method of choice for industrial scale bioreductions. An elegant example to overcome some drawbacks of coupled-substrate nicotinamide recycling makes use of 1,4-butanediol as auxiliary substrate. Hydride abstraction yields 4-hydroxybutanal, which spontaneously cyclises to yield a lactol, which is irreversibly oxidised (by delivering a second hydride) to the corresponding butyrolactone. This drives the equilibrium without requirement for an excess of auxiliary substrate [804, 805]. A special technique avoiding some of these drawbacks makes use of gas-membranes and is discussed in Sect. 3.3.

Coupled-Enzyme Approach   The use of two independent enzymes is more advantageous (Scheme 2.109). In this case, the two parallel redox reactions – i.e., conversion of the main substrate plus cofactor recycling – are catalyzed by *two different* enzymes [806]. To achieve optimal results, both of the enzymes should have sufficiently different specificities for their respective substrates whereupon the two enzymatic reactions can proceed independently from each other and, as a consequence, both the substrate and the auxiliary substrate do not have to compete for the active site of a single enzyme, but are independently converted by the two biocatalysts.

Several excellent methods, each having its own particular pros and cons, have been developed to regenerate NADH. On the other hand, NADPH may be regenerated sufficiently on a lab scale but a really inexpensive and reliable method is still needed for industrial-scale applications.

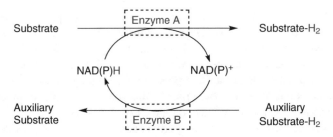

Scheme 2.109   Cofactor recycling by the coupled-enzyme method

The best and most widely used method for recycling NADH uses formate dehydrogenase (FDH), which is obtained from methanol-utilizing microorganisms, to catalyze the oxidation of formate to CO_2 (Scheme 2.110) [807, 808]. This method has the advantage that both the auxiliary substrate and the coproduct are innocuous to enzymes and CO_2 is easily removed from the reaction, which drives the reaction out of equilibrium. FDH is commercially available, readily immobilized and reasonably stable, if protected from autooxidation [809] and trace metals. The only disadvantage of this system is the high cost of FDH and its low specific activity (3 U/mg). However, both drawbacks can be readily circumvented by using an immobilized [810] or membrane-retained FDH system [811]. Overall, the formate/FDH system is the most convenient and most economical method for regenerating NADH, particularly for large-scale and repetitious applications, with TTNs (mol product/mol cofactor) approaching 600,000. The regeneration system based on FDH from *Candida boidinii* used as a technical-grade biocatalyst is limited by being specific for NADH [812]. This drawback has been circumvented by application of a genetically engineered formate dehydrogenase from *Pseudomonas* sp., which also accepts NADPH [813–815].

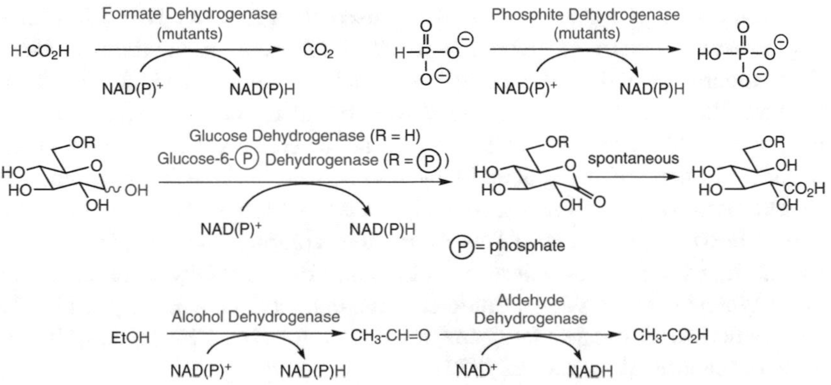

Scheme 2.110   Enzymatic regeneration of reduced nicotinamide cofactors

Another widely used method for recycling NAD(P)H makes use of the oxidation of glucose, catalyzed by glucose dehydrogenase (GDH, Scheme 2.110)

[816, 817]. The equilibrium is shifted towards the product because the gluconolactone formed is spontaneously hydrolyzed to give gluconic acid. The glucose dehydrogenase from *Bacillus cereus* is highly stable [818] and accepts either NAD$^+$ or NADP$^+$ with high specific activity. Like FDH, however, GDH is expensive and product isolation from polar gluconate may complicate the workup. In the absence of purification problems, this method is attractive for laboratory use, and it is certainly a convenient way to regenerate NADPH.

Similarly, glucose-6-phosphate dehydrogenase (G6PDH) catalyzes the oxidation of glucose-6-phosphate (G6P) to 6-phosphogluconolactone, which spontaneously hydrolyzes to the corresponding phosphogluconate (Scheme 2.110). The enzyme from *Leuconostoc mesenteroides* is inexpensive, stable and accepts both NAD$^+$ and NADP$^+$ [582, 819], whereas yeast-G6PDH accepts only NADP$^+$. A major disadvantage of this system is the high cost of G6P. Thus, if used on a large scale, it may be enzymatically prepared from glucose using hexokinase and this involves the regeneration of ATP using kinases (see pp. 107–109). Alternatively, glucose-6-sulfate and G6PDH from *Saccharomyces cerevisiae* may be used to regenerate NADPH [820]. The sulfate does not act as an acid catalyst for the hydrolysis of NADPH and is more easily prepared than the corresponding phosphate [821]. Overall, the G6P/G6PDH system complements glucose/GDH as an excellent method for regenerating NADPH and is a good method for regenerating NADH.

More recently, phosphite dehydrogenase has been shown to offer a promising alternative [822, 823]: The equilibrium is extremely favorable, both phosphite and phosphate are inoccuous to enzymes and act as buffer. The wild-type enzyme from *Pseudomonas stutzeri* accepts only NAD$^+$ [824], but thermostable mutants were generated which are also able to reduce NADP$^+$ [825–827].

Ethanol and alcohol dehydrogenase (ADH) have been used in the past to regenerate NADH and NADPH [828, 829]. The low to moderate cost of ADH and the volatility of both ethanol and acetaldehyde make this system attractive for lab-scale reactions. An alcohol dehydrogenase from yeast reduces NAD$^+$, while an ADH from *Leuconostoc mesenteroides* is used to regenerate NADPH (Scheme 2.110). However, due to the low redox potential, only activated carbonyl substrates such as aldehydes and cyclic ketones are reduced in good yields. With other substrates, the equilibrium must be driven by using ethanol in excess or by removing acetaldehyde. The latter may be achieved by sweeping with nitrogen [830] or by further oxidizing acetaldehyde to acetate [831], using aldehyde dehydrogenase thereby generating a second equivalent of reduced cofactor. All of these methods, however, give low TTNs or involve complex multi-enzyme systems. Furthermore, even low concentrations of ethanol or acetaldehyde inhibit or deactivate enzymes. Alternatively, a crude cell-free extract from baker's yeast has been recommended as an (unspecified) enzyme source for NADPH recycling by using glucose as the ultimate reductant [832].

A particularly attractive alternative for the regeneration of NADH makes use of hydrogenase enzymes, so called because they are able to accept molecular hydrogen directly as the hydrogen donor [833, 834]. The latter is strongly reducing, innocuous to enzymes and nicotinamide cofactors, and its consumption leaves no byproduct. For organic chemists, however, this method is of limited use because hydrogenase is usually isolated from strict anaerobic organisms. Thus, the enzyme is sensitive to oxidation, is not commercially available and requires sophisticated fermentation procedures for its production.[27] Furthermore, some of the organic dyes, which serve as mediators for the transport of redox equivalents from the donor onto the cofactor are relatively toxic.

Recycling of Oxidized Nicotinamide Cofactors

For oxidation, reduction reactions can be run in reverse, although the equilibrium is strongly disfavoured. The best and most widely applied method for the regeneration of nicotinamide cofactors in their oxidized form involves the use of glutamate dehydrogenase (GluDH) which catalyzes the reductive amination of α-ketoglutarate to give L-glutamate (Scheme 2.111) [837, 838]. Both NADH and NADPH are accepted as cofactors. In addition, α-keto-adipate can be used instead of the corresponding glutarate [839], leading to the formation of a high-value byproduct, L-α-aminoadipate.

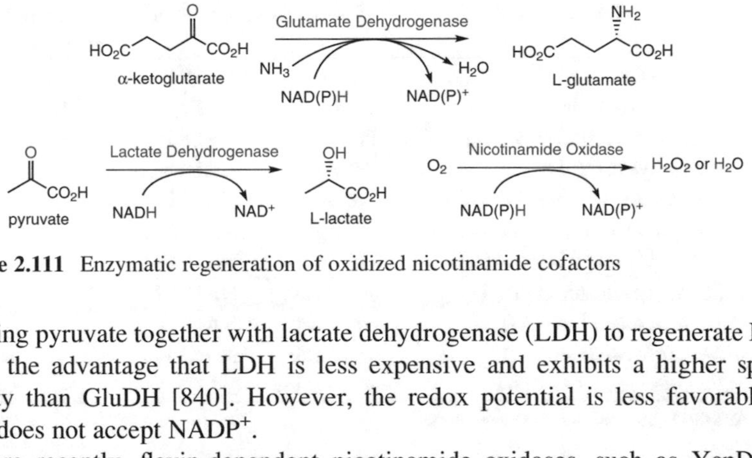

Scheme 2.111 Enzymatic regeneration of oxidized nicotinamide cofactors

Using pyruvate together with lactate dehydrogenase (LDH) to regenerate NAD^+ offers the advantage that LDH is less expensive and exhibits a higher specific activity than GluDH [840]. However, the redox potential is less favorable and LDH does not accept $NADP^+$.

More recently, flavin-dependent nicotinamide oxidases, such as YcnD from *Bacillus subtilis* [841] or an enzyme from *Lactobacillus sanfranciscensis* [842] were employed for the (irreversible) oxidation of nicotinamide cofactors at the expense of molecular oxygen producing H_2O_2 or (more advantageous) H_2O via a two- or four-electron transfer reaction, respectively [843–845]. Hydrogen peroxide can be destroyed by addition of catalase and in general, both NADH and NADPH are accepted about equally well.

Acetaldehyde and yeast-ADH have also been used to regenerate NAD^+ from NADH [846]. Although reasonable total turnover numbers were achieved (10^3–10^4), the above-mentioned disadvantages of enzyme deactivation and self-condensation of acetaldehyde outweigh the merits of the low cost of yeast-ADH and the volatility of the reagents involved.

---

[27]For an O_2-tolerant hydrogenase from *Ralstonia eutropha* see [835, 836].

2.2.2 Reduction of Aldehydes and Ketones Using Isolated Enzymes

A broad range of ketones can be reduced stereoselectively using dehydrogenases to furnish chiral secondary alcohols [847–850]. During the course of the reaction, the enzyme delivers the hydride preferentially either from the *si*- or the *re*-side of the ketone to give (*R*)- or (*S*)-alcohols, respectively. The stereochemical course of the reaction, which is mainly dependent on the steric requirements of the substrate, may be predicted for most dehydrogenases from a simple model which is generally referred to as 'Prelog's rule' (Scheme 2.112) [851].

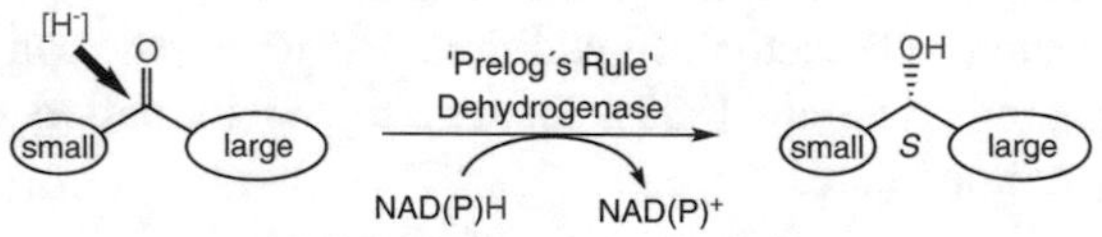

Dehydrogenase	Specificity	Cofactor	Commercially available
yeast-ADH	Prelog	NADH	+
horse liver-ADH	Prelog	NADH	+
Thermoanaerobium brockii-ADH	Prelog[a]	NADPH	+
Hydroxysteroid-DH	Prelog	NADH	+
Rhodococcus ruber ADH-A	Prelog	NADH	+
Rhodococcus erythropolis ADH	Prelog	NADH	+
Candida parapsilosis-ADH	Prelog	NADH	+
Lactobacillus brevis ADH	anti-Prelog	NADPH	+
Lactobacillus kefir-ADH	anti-Prelog	NADPH	+
Mucor javanicus-ADH	anti-Prelog	NADPH	-
Pseudomonas sp.-ADH	anti-Prelog	NADH	-

[a] Anti-Prelog specificity on small ketones.

Scheme 2.112 Prelog's rule for the asymmetric reduction of ketones

It is based on the stereochemistry of microbial reductions using *Curvularia falcata* cells and it states that the dehydrogenase delivers the hydride from the *re*-face of a prochiral ketone to furnish the corresponding (*S*)-configured alcohol. The majority of the commercially available dehydrogenases used for the stereospecific reduction of ketones [such as yeast alcohol dehydrogenase (YADH), horse liver alcohol dehydrogenase (HLADH)] and the majority of microorganisms (for instance, baker's yeast) follow Prelog's rule [852]. *Thermoanaerobium brockii* alcohol dehydrogenase (TBADH) also obeys this rule when large ketones are used as substrates, but the stereopreference is reversed with small substrates. Microbial dehydrogenases which lead to the formation of anti-Prelog configurated (*R*)-alcohols are known to a lesser extent, and even fewer are commercially available, e.g., from *Lactobacillus* sp. [853–855]. Other ADHs from *Curvularia falcata* [856], *Mucor javanicus* and *Pseudomonas* sp. [857] are of limited use as long as they are not commercially available.

The substrate range of commercially available alcohol dehydrogenases has been mapped including aldehydes, (acyclic, aromatic, and unsaturated) ketones, diketones and various oxo-esters [858]. The most commonly used dehydrogenases are shown in Fig. 2.15, with reference to their preferred size of their substrates [859].

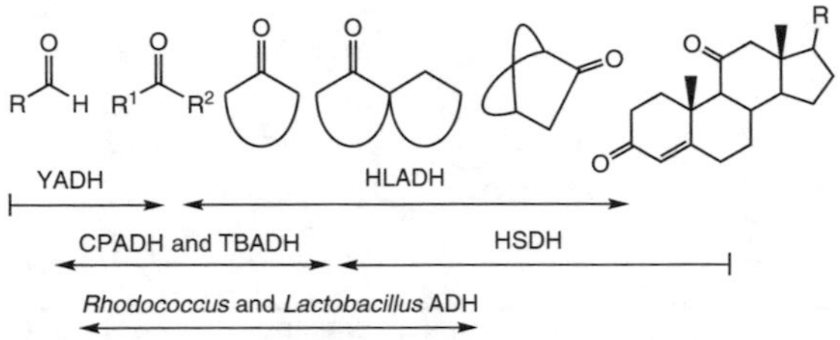

Fig. 2.15 Preferred substrate size for dehydrogenases. *YADH* yeast alcohol dehydrogenase, *HLADH* horse liver alcohol dehydrogenase, *CPADH Candida parapsilosis* alcohol dehydrogenase, *TBADH Thermoanaerobium brockii* alcohol dehydrogenase, *HSDH* hydroxysteroid dehydrogenase

Yeast ADH has a very narrow substrate specificity and, in general, only accepts aldehydes and methyl ketones [860, 861]. Therefore, cyclic ketones and those bearing carbon chains larger than a methyl group are not accepted as substrates. Thus, YADH is only of limited use for the preparation of small chiral secondary alcohols.

Horse liver ADH is a very universal enzyme with a broad substrate specificity and excellent stereoselectivity. Historically, it is the most widely used dehydrogenase in biotransformations [862, 863] and its mechanism was elucidated [863] on the basis of its crystal structure [864]. Although the primary sequence is quite different, the tertiary structure of HLADH is similar to that of YADH [865]. The most useful applications of HLADH are found in the reduction of medium-ring monocyclic ketones (four- to nine-membered ring systems) and bicyclic ketones [866–868]. Sterically demanding molecules which are larger than decalines are not readily accepted and acyclic ketones are usually reduced with modest enantioselectivities [869, 870]. HLADH consists of two isoenzymes (HLADH-E and HLADH-S[28]), which differ in their substrate preference [871].

A considerable number of monocyclic and bicyclic racemic ketones have been resolved using HLADH with fair to excellent specificities [872–874]. Even sterically demanding cage-shaped polycyclic ketones were readily accepted [875, 876] (Scheme 2.113). For instance, *rac*-2-twistanone was reduced to give the *exo*-alcohol and the enantiomeric ketone in 90% and 68% e.e., respectively [877]. Also *O*- and *S*-heterocyclic ketones were shown to be good substrates (Scheme 2.113) [878–880]. Thus, (±)-bicyclo[4.3.0]nonan-3-ones bearing either an O or S atom in position 8 were resolved with excellent selectivities

----
[28]The prefix 'S' stands for steroids, 'E' stands for ethanol.

[870]. Attempted reduction of the corresponding N-heterocyclic ketones led to deactivation of the enzyme via complexation of the essential Zn^{2+} ion in the active site [881].

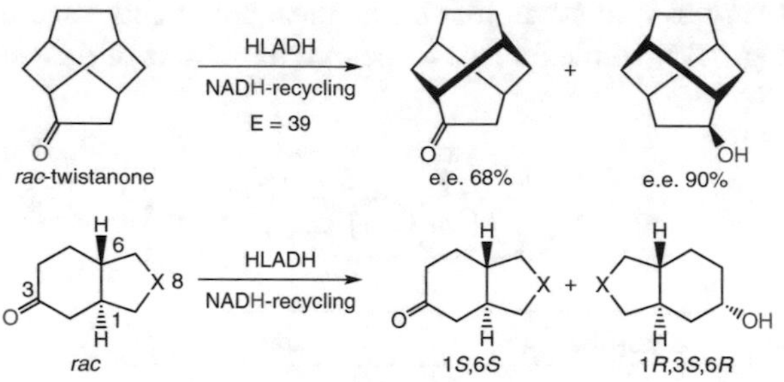

X	Selectivity (E)	e.e. Ketone [%]	e.e. Alcohol [%]
O	>120	60	>97
S	>110	53	>97

Scheme 2.113 Kinetic resolution of bi- and polycyclic ketones using horse liver alcohol dehydrogenase (HLADH)

Every kinetic resolution of bi- and polycyclic ketones suffers from one particular drawback because the bridgehead carbon atoms make it impossible to recycle the undesired 'wrong' enantiomer via racemization. Hence the desymmetrization of prochiral diketones, making use of the enantioface- or enantiotopos-specificity of HLADH, is of advantage. For instance, both the *cis-* and *trans-*forms of the decalinediones shown in Scheme 2.114 were reduced to give (S)-alcohols with excellent optical purity. Similar results were obtained with unsaturated derivatives [828, 882].

The wide substrate tolerance of HLADH encompassing nonnatural compounds is demonstrated by the resolution of organometallic derivatives possessing axial chirality [883]. For instance, the racemic tricarbonyl cyclopentadienyl manganese aldehyde shown in Scheme 2.115 was enantioselectively reduced to give the (R)-alcohol and the residual (S)-aldehyde with excellent optical purities [884].

In order to predict the stereochemical outcome of HLADH-catalyzed reductions, a number of models have been developed, each of which having its own merits. The most useful substrate model based on a flattened cyclohexanone ring is shown in Fig. 2.16 [885]. It shows the Zn^{2+} in the catalytic site which coordinates to the carbonyl oxygen atom and the nucleophilic attack of the hydride occurring from the bottom face. The preferred orientation of the substrate relative to the hydride delivered from NADH can be estimated by placing the substituents into the 'allowed' and 'forbidden' zones.

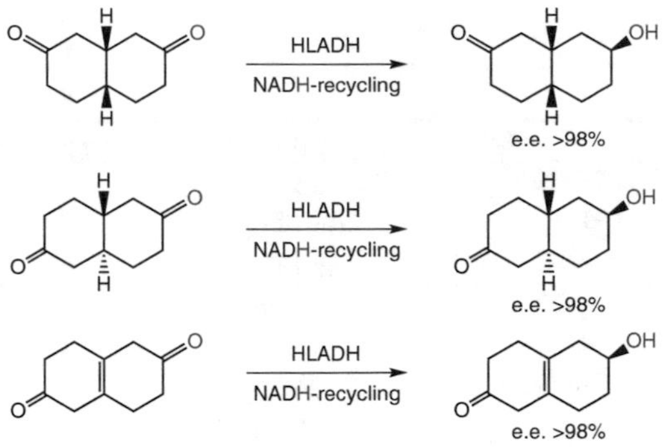

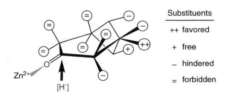

Scheme 2.114 Desymmetrization of prochiral diketones using HLADH

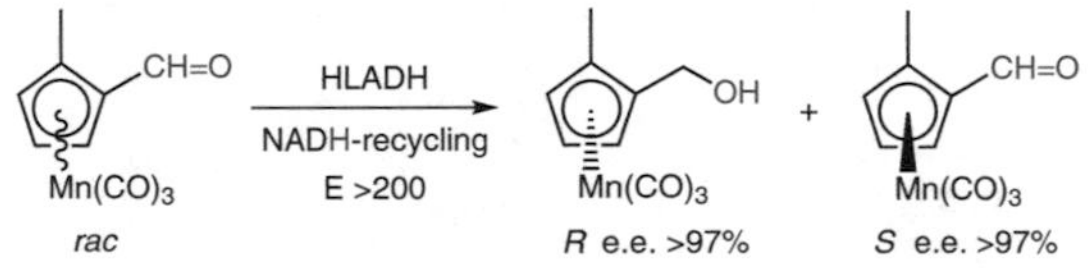

Scheme 2.115 Enantioselective reduction of an organometallic aldehyde using HLADH

Fig. 2.16 Substrate model for HLADH for cyclic ketones

    YADH and HLADH are less useful for the asymmetric reduction of open-chain ketones, but this gap is efficiently covered by a range of alcohol dehydrogenases from mesophilic bacteria, such as *Rhodococcus* (ADH-A) and *Lactobacillus* (LBADH, LKADH), and thermophilic *Thermoanaerobacter* [886] and *Thermoanaerobium* (TBADH) strains (Scheme 2.116) [311, 887–890]. Some of these enzymes are remarkably thermostable (up to 85 °C) and can tolerate the presence of organic solvents such as *iso*propanol, which serves as hydrogen-donor for NADP-recycling in a coupled-substrate approach [891–893].

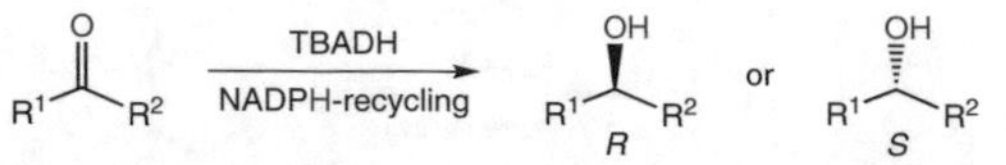

R^1	R^2	Specificity	Configuration	e.e. [%]
CH_3	$CH(CH_3)_2$	Anti-Prelog	R	86
CH_3	C_2H_5	Anti-Prelog	R	48
CH_3	$cyclo$-C_3H_5	Anti-Prelog	R	44
CH_3	n-C_3H_7	Prelog	S	79
CH_3	$C{\equiv}CH$	Prelog	S	86
Cl-CH_2-	CH_2-CO_2Et	Prelog	R[a]	90
CF_3	Ph	Prelog	R[a]	94
CH_3	CH_2-$CH(CH_3)_2$	Prelog	S	95
C_2H_5	n-C_3H_7	Prelog	S	97
C_2H_5	$(CH_2)_2$-CO_2Me	Prelog	S	98
CH_3	$(CH_2)_3$-Cl	Prelog	S	98
CH_3	n-C_5H_{11}	Prelog	S	99
CH_3	$(CH_2)_5$-Cl	Prelog	S	>99
C_2H_5	$(CH_2)_3$-Cl	Prelog	S	>99
n-C_3H_7	n-C_3H_7		no reaction	

[a] Switch in CIP-sequence order.

Scheme 2.116 Asymmetric reduction of ketones using *Thermoanaerobium brockii* alcohol dehydrogenase (TBADH)

Open-chain methyl- and ethyl-ketones are readily reduced by TBADH to furnish the corresponding secondary alcohols, generally with excellent specificities [894]. Similarly, ω-haloalkyl- [817, 895] and methyl- or trifluoromethyl ketones possessing heterocyclic substituents were converted into the corresponding secondary alcohols with excellent optical purities [896, 897]. However, α,β-unsaturated ketones and ketones where both substituents are larger than ethyl are not accepted. In general TBADH obeys Prelog's rule with 'normal-sized' ketones leading to (*S*)-alcohols, but the stereoselectivity was found to be reversed with small substrates. In order to predict the stereochemical outcome of TBADH reductions, an active site model based on a quadrant rule was proposed [898].

The key to access both stereoisomers of a *sec*-alcohol via asymmetric carbonyl reduction is the availability of stereocomplementary dehydrogenases. For open-chain ketones bearing a small and large substituent at each side, this is feasible by using an appropriate enzyme showing Prelog or anti-Prelog specificity. Whereas dehydrogenases from *Rhodococcus ruber*, *R. erythropolis*, and *Candida parapsilosis* produce the Prelog enantiomer, *Lactobacillus* ADHs furnish the corresponding mirror-image product, usually with high stereoselectivity (Scheme 2.117) [899]. In an analogous fashion, α-ketocarboxylic acids were reduced to the corresponding enantiomeric α-hydroxyacids using stereocomplementary lactate dehydrogenases (LDH) [900–903], or hydroxyisocaproate dehydrogenases (HicDHs) [904, 905].

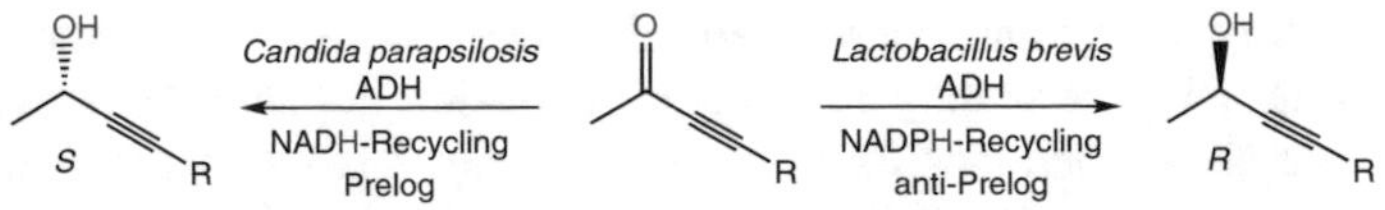

R	*Candida parapsilosis* ADH		*Lactobacillus* brevis ADH	
	Configuration	E.e. [%]	Configuration	E.e. [%]
H	(*S*)	49	(*R*)	60
SiMe$_3$	(*S*)	57	(*R*)	>99
SiMe$_2$Ph	(*S*)	>99	(*R*)	>99
Ph	(*S*)	>99	(*R*)	>99
2-Pyridyl	(*S*)	>99	(*R*)	>99

Scheme 2.117 Stereocomplementary bioreduction using a Prelog and anti-Prelog dehydrogenase

Hydroxysteroid dehydrogenases (HSDH) are ideally suited enzymes for the reduction of bulky mono- [906] and bicyclic ketones (Scheme 2.118) [907]. This is not surprising if one thinks of the steric requirements of their natural substrates: steroids [908, 909]. For instance, bicyclo[3.2.0]heptan-6-one systems were reduced with HSDH with very low selectivity when substituents in the adjacent 7-position were small (R^1, R^2 = H), but TBADH showed an excellent enantioselectivity with this 'slim' ketone. When the steric requirements of the substrate were increased by additional methyl- or chloro-substituents adjacent to the carbonyl group, the situation changed. Then, HSDH became a very specific catalyst and TBADH (or HLADH) proved to be unable to accept the bulky substrates [910, 911]. The switch in the stereochemical preference is not surprising and can be explained by Prelog's rule: with the unsubstituted ketone, the position 5 is 'larger' than position 7. However, when the hydrogen atoms on carbon atom 7 are replaced by sterically demanding chlorine or methyl groups, the situation is reversed.

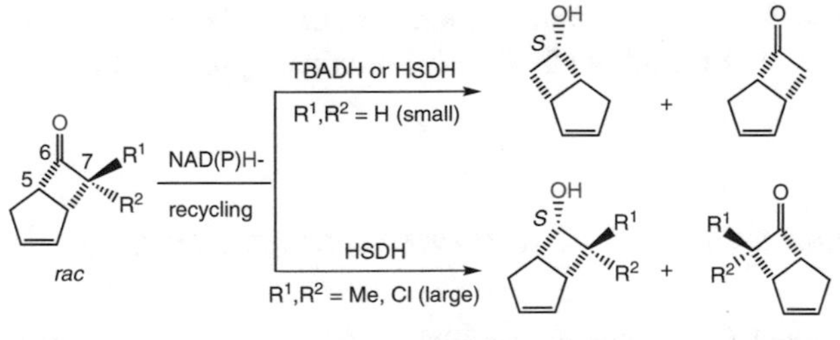

R^1	R^2	Enzyme	e.e. Alcohol [%]
H	H	HSDH	≤10
H	H	TBADH	>95
Cl	Cl	HSDH	>95
Me	Me[a]	HSDH	>95

[a] No reaction was observed with HLADH or TBADH.

Scheme 2.118 Kinetic resolution of sterically demanding ketones using hydroxysteroid dehydrogenase (HSDH)

The majority of synthetically useful ketones can be transformed into the corresponding chiral secondary alcohols by choosing the appropriate dehydrogenase from the above-mentioned set of enzymes (Fig. 2.15). Other enzymes, which have been shown to be useful for specific types of carbonyl substrates, are mentioned below.

One general limitation of alcohol dehydrogenases is their inability to convert sterically demanding ketones bearing bulky groups on both sides. This limitation was overcome by identification of two special ADHs from *Ralstonia* sp. DSM 6428 and *Sphingobium yanoikuyae* DSM 6900 [912]. The former enzyme reduced aryl-alkyl ketones bearing *n*-propyl- to *n*-pentyl chains with excellent Prelog-specificity [913].

The natural role of glycerol dehydrogenase is the interconversion of glycerol and dihydroxyacetone. The enzyme is commercially available from different sources and has been used for the stereoselective reduction of α-hydroxyketones [837]. Glycerol DH has been found to tolerate some structural variation of its natural substrate – dihydroxyacetone – including cyclic derivatives. An enzyme from *Geotrichum candidum* was shown to reduce not only α- but also β-ketoesters with high selectivity [914].

Enzymes from thermophilic organisms (which grow in the hostile environment of hot springs with temperatures ranging from 70 to 100 °C) have recently received much attention [915–918]. Thermostable enzymes are not only stable to heat but, in general, also show enhanced stability in the presence of common protein denaturants and organic solvents. Since they are not restricted to working in the narrow temperature range which is set for mesophilic, 'normal' enzymes (20–40 °C), an influence of the temperature on the selectivity can be studied over a wider range. For instance, the diastereoselectivity of the HLADH-catalyzed reduction of 3-cyano-4,4-dimethyl-cyclohexanone is diminished at 45 °C (the upper operational limit for HLADH) when compared with that observed at 5 °C [919]. On the other hand, a temperature-dependent *reversal* of the enantiospecificity of an alcohol dehydrogenase from *Thermoanaerobacter ethanolicus* could be achieved when the temperature was raised to 65 °C [920] (compare pp. 75–76).

2.2.3   Reduction of Aldehydes and Ketones Using Whole Cells

Instead of isolated dehydrogenases, which require sophisticated cofactor recycling, whole microbial cells can be employed. They contain multiple dehydrogenases which are able to accept nonnatural substrates, all the necessary cofactors and the metabolic pathways for their regeneration. Thus, cofactor recycling can be omitted since it is automatically done by the living cell. Therefore, cheap carbon sources such as saccharose or glucose can be used as auxiliary substrates for asymmetric reduction reactions. Furthermore, all the enzymes and cofactors are well protected within their natural cellular environment.

However, these distinct advantages have to be taken into consideration alongside some significant drawbacks:

- The productivity of microbial conversions is usually low since the majority of nonnatural substrates are toxic to living organisms and are therefore only tolerated at low concentrations (~0.1–0.3% per volume).
- The large amount of biomass present in the reaction medium causes low recovery, particularly when the product is stored inside the cells and not excreted into the medium. Since only a small fraction (typically 0.5–2%) of the auxiliary cosubstrate is used for coenzyme recycling, while the bulk is metabolized forming polar byproducts, product purification is troublesome and monitoring of the reaction becomes difficult.
- Finally, different strains of a microorganism most likely possess different specificities; thus it is important to use exactly the same culture to obtain comparable results with the literature [921].
- Stereoselectivities may vary to a great extent due to the presence of multiple enzymes. If *two* enzymes, each with high but *opposite* stereochemical preference, compete for the same substrate, the optical purity of the product is determined by the relative rates of the individual reactions. The latter, in turn, depend on the substrate concentration. At concentrations below saturation, the relative rates are determined by the ratio V_{max}/K_M for each enzyme. On the other hand, when saturation is reached using elevated substrate concentrations, the relative rates mainly depend on the ratio of k_{cat} of the two reactions. Consequently, when two (or more) enzymes are involved in the transformation of enantiomeric substrates, the optical purity of the product becomes a function of the substrate concentration, because the values of K_M and k_{cat} for the substrate enantiomers are different for both competing enzymes. With yeasts, it is a well-known phenomenon that lower substrate concentrations often give higher e.e.$_p$s [922].

The following general techniques can be applied to enhance the selectivity of microbial reduction reactions:

- Substrate modification, e.g., by variation of protecting groups which can be removed after the transformation [923–925]
- Variation of the metabolic parameters by immobilization [926–928]
- Using cells of different age [929]
- Variation of the fermentation conditions [930–932]
- Screening of microorganisms to obtain strains with the optimum properties (a hard task for nonmicrobiologists) [933, 934]
- Selective inhibition of one of the competing enzymes (see below)

Reduction of Aldehydes and Ketones by Baker's Yeast

Asymmetric Reduction of Ketones Baker's yeast (*Saccharomyces cerevisiae*) is by far the most widely used microorganism for the asymmetric reduction of ketones [935–939]. It is ideal for nonmicrobiologists, since it is readily available at a very reasonable price and its use does not require sterile fermenters but can be handled using standard laboratory equipment. Thus, it is not surprising that yeast-catalyzed transformations of nonnatural compounds leading to chiral products have been reported from the beginning of the twentieth century [940] and the first comprehensive review which covers almost all the different strategies of yeast-reductions dates back to 1949! [941].

A wide range of functional groups within the ketone are tolerated, including heterocyclic- [942, 943], fluoro- [944–947], chloro- [948], bromo- [949], perfluoro-alkyl- [950], cyano-, azido-, nitro- [951–953], hydroxyl- [954, 955], sulfur- [956–958], and dithianyl groups [959]. Even organometallic derivatives [960, 961], such as silyl- [962] and germyl groups [963] are accepted.

Simple aliphatic and aromatic ketones are reduced by fermenting yeast according to Prelog's rule to give the corresponding (S)-alcohols in good optical purities (Scheme 2.119) [861]. Long-chain ketones such as n-propyl-n-butylketone and several bulky phenyl ketones are not accepted; however, one long alkyl chain is tolerated if the other moiety is the methyl group [964, 965]. As might be expected, best stereoselectivities were achieved with groups of greatly different size.

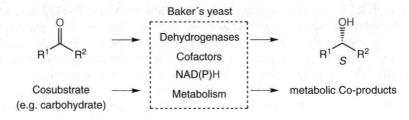

R^1	R^2	e.e. [%]
Me	Et	67
Me	CF$_3$	>80
CF$_3$	CH$_2$-Br	>80
Me	n-Bu	82
Me	Ph	89
Me	CH$_2$-OH	91
Me	(CH$_2$)$_2$-CH=C(CH$_3$)$_2$	94
Me	c-C$_6$H$_{11}$	>95
Me	C(CH$_3$)$_2$-NO$_2$	>96

Scheme 2.119 Reduction of aliphatic and aromatic ketones using baker's yeast

Acyclic β-ketoesters (Scheme 2.120) are readily reduced by yeast to yield β-hydroxyesters [966, 967], which serve as chiral starting materials for the synthesis of β-lactams [968], insect pheromones [969], and carotenoids [970]. It is obvious that the enantioselectivity and the stereochemical preference for the re- or the si-side of the β-ketoester depends on the relative size of the alkoxy moiety and the ω-substituent of the ketone, which directs the nucleophilic attack of the hydride occurring according to Prelog's rule (Scheme 2.120). Therefore, the absolute configuration of the newly generated *sec*-alcoholic center may be directed by substrate modification using either the corresponding short- or long-chain alkyl ester, which switches the relative size of the substituents flanking the carbonyl group [971].

In baker's yeast, the reason for this divergent behavior is not due to an alternative fit of the substrates in a single enzyme, but rather due to the presence of a number of different dehydrogenases, possessing opposite stereochemical preferences, which compete for the substrate [922, 972]. A D-specific enzyme – belonging to the fatty acid synthetase complex – shows a higher activity towards β-ketoesters having a short-chain alcohol moiety, such as methyl esters. By contrast, an L-enzyme is more active on long-chain counterparts, e.g., octyl esters. Therefore, the stereochemical direction of the reduction may be controlled by careful design of the substrate, or by selective inhibition of one of the competing dehydrogenases.

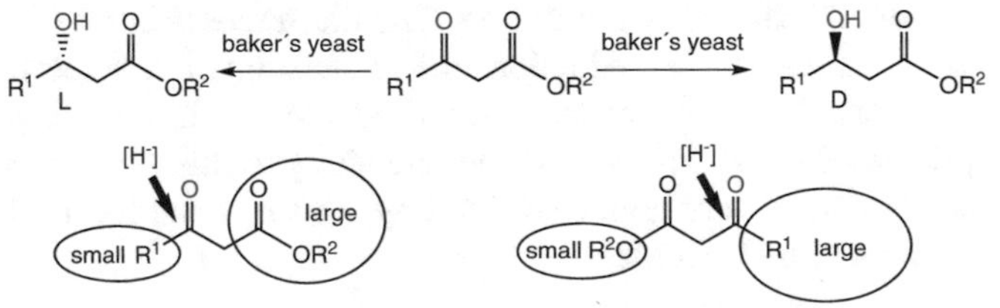

R^1	R^2	Configuration	e.e. [%]
Cl-CH$_2$-	CH$_3$	D	64
Cl-CH$_2$-	C$_2$H$_5$	D	54
Cl-CH$_2$-	n-C$_3$H$_7$	D	27
Cl-CH$_2$-	n-C$_5$H$_{11}$	L	77
Cl-CH$_2$	n-C$_8$H$_{17}$	L	97
(CH$_3$)$_2$C=CH-(CH$_2$)$_2$-	CH$_3$	D	92
CCl$_3$	C$_2$H$_5$	D	85
CH$_3$	C$_2$H$_5$	L	>96
N$_3$-CH$_2$-	C$_2$H$_5$	L	80
Br-CH$_2$-	n-C$_8$H$_{17}$	L	100
C$_2$H$_5$-	n-C$_8$H$_{17}$	L	95

Scheme 2.120 Reduction of acyclic β-ketoesters using baker's yeast

Inhibition of the L-enzyme (which leads to the increased formation of D-β-hydroxyesters) was accomplished by addition of unsaturated compounds such as allyl alcohol [973] or methyl vinyl ketone [974]. The same effect was observed when the yeast cells were immobilized by entrapment into a polyurethane gel [975, 976]. As expected, L-enzyme inhibitors led to a considerable increase in the optical purity of D-β-hydroxyesters.

On the contrary, various haloacetates [977], thioethers [978], and allyl bromide [979] are inhibitors for the D-enzyme, which leads to an increased formation of the L-enantiomer.

Diasteroselective Reduction of Ketones by Baker's Yeast Asymmetric microbial reduction of α-substituted ketones leads to the formation of diastereomeric *syn-* and *anti*-products. Because the chiral center on the α-position of the ketone is stereochemically labile, rapid in-situ racemization of the substrate enantiomers occurs via enolization – leading to dynamic resolution [64, 980, 981]. Thus, the ratio between the diastereomeric *syn-* and *anti*-products is not 1:1, but is determined by the selectivities of the enzymes involved in the reduction process [982]. Under optimal conditions it can even be as high as 100:0 [983]. When the chiral center is moved to the β- or γ-position, in situ racemization is impossible and, as a consequence, *syn/anti*-diastereomers are always obtained in a 1:1 ratio.

Diastereoselective yeast-reduction of ketones has been mainly applied to α-monosubstituted β-ketoesters leading to the formation of diastereomeric *syn-* and *anti*-β-hydroxyesters (Scheme 2.121) [984–987]. With small α-substituents, the formation of *syn*-diastereomers predominates, but the diaselectivity is reversed when the substituents are increased in size. The diastereoselectivity (i.e., the *syn/anti*-ratio) of yeast-catalyzed reductions of α-substituted β-ketoesters can be predicted from the relative size of the α-substituent versus the carboxylate moiety using a simple model [988]. In any case, the selectivity for the newly generated *sec*-alcohol center is always very high (indicated by the e.e.s) and its absolute configuration is determined by Prelog's rule.

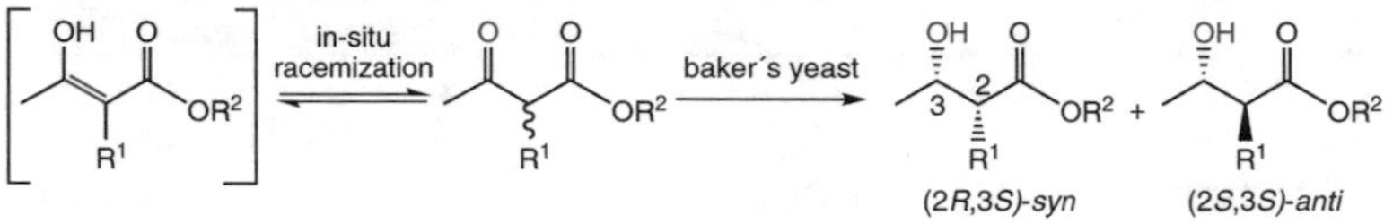

R^1	R^2	e.e. [%]		Ratio
rac		*syn*	*anti*	*syn/anti*
CH$_3$	C$_2$H$_5$	100	100	83:17
CH$_3$	Ph-CH$_2$-	100	80	67:33
CH$_2$=CH-CH$_2$-	C$_2$H$_5$	100	100	25:75
Ph-CH$_2$-	C$_2$H$_5$	100	100	33:67
Ph-S-	CH$_3$	>96	>96	17:83

Scheme 2.121 Diastereoselective reduction of α-substituted β-ketoesters using baker's yeast

The yeast-reduction of cyclic β-ketoesters exclusively leads to the corresponding *syn*-β-hydroxy-esters (Scheme 2.122) [989–991]. The corresponding *anti*-diastereomers cannot be formed because rotation around the α,β-carbon–carbon bond is impossible with such cyclic structures. Furthermore, the reductions are generally more stereoselective than the corresponding acyclic substrate due to the enhanced rigidity of the system. Thus, it can be worthwhile to create a sulfur-containing ring in the substrate temporarily and to remove the heteroatom after the biotransformation to obtain the desired open-chain product (e.g., by Raney-Ni reduction) in order to benefit from enhanced selectivities (compare Scheme 2.41).

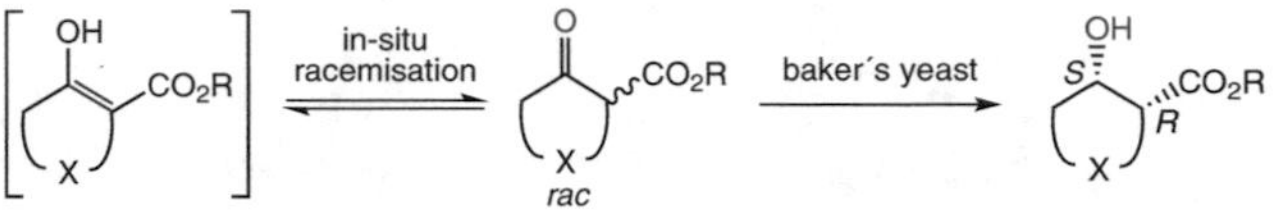

X	R	e.e. [%]
-(CH$_2$)$_2$-	C$_2$H$_5$	>98
-(CH$_2$)$_3$-	C$_2$H$_5$	86-99[a]
-S-CH$_2$-	CH$_3$	85
-CH$_2$-S-	CH$_3$	>95

a Depending on the yeast strain used.

Scheme 2.122 Yeast-reduction of cyclic β-ketoesters

The biocatalytic reduction of α-substituted β-ketoesters with concomitant dynamic resolution has been proven to be extremely flexible (Scheme 2.123) [992–997]. Thus, by choosing the appropriate microorganism possessing the desired enantio- and diastereoselectivity, each of the four possible diastereomeric products were obtained in excellent enantiomeric and diastereomeric purity. As expected, the corresponding Prelog-configurated products with respect to the newly generated *sec*-alcohol center (pathways A, B) were obtained by using baker's yeast and the mold *Geotrichum candidum*, respectively. For the diastereomers possessing the opposite configuration at the alcoholic center (anti-Prelog pathways C, D) other microorganisms had to be employed.

As long as the α-substituent consists of an alkyl- or aryl-group, dynamic resolution is readily achieved, leading to chemical yields far beyond the 50% which would be the maximum for a classic kinetic resolution. However, in-situ racemization is not possible due to electronic reasons for α-hydroxy- [998], α-alkylthio- [984], α-azido- [999], or α-acetylamino derivatives [1000]. Consequently, they are subject to kinetic resolution. The same holds for substrates which are fully substituted at the α-position, due to the impossibility of form the corresponding enolate.

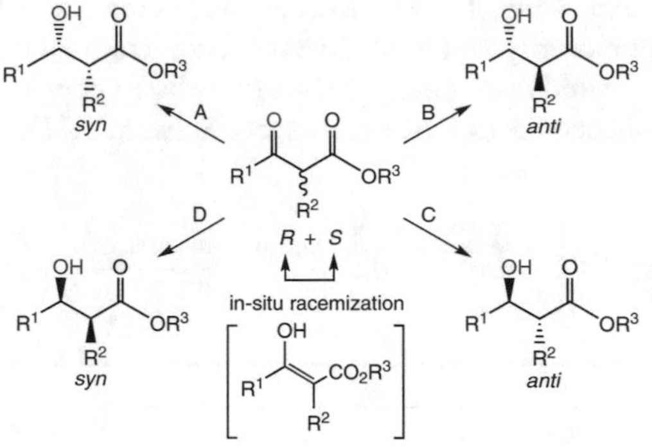

Scheme 2.123 Stereocomplementary microbial reduction of α-substituted β-ketoesters

Path- way	R^1	R^2	R^3	Biocatalyst	yield [%]	d.e. [%]	e.e. [%]	Refer ences
A	Me	allyl	Et	baker's yeast	94	92	>99	[988]
A	Me	Me	n-Octyl	baker's yeast	82	90	>98	[989]
B	Me	Me	Et	Geotrichum candidum	80	>98	>98	[990]
B	Et	Me	Et	Geotrichum candidum	80	96	91	[991]
C	4-MeOC$_6$H$_4$-	Cl	Et	Sporotrichum exile	52	96	98	[992]
D	4-MeOC$_6$H$_4$-	Cl	Me	Mucor ambiguus	58	>98	>99	[993]

α-Ketoesters and α-ketoamides can be asymmetrically reduced to furnish the corresponding α-hydroxy derivatives. Thus, following Prelog's rule, (*S*)-lactate [1001] and (*R*)-mandelate esters [982] were obtained from pyruvate and α-keto-phenylacetic esters by fermenting baker's yeast in excellent optical purity (e.e. 91–100%).

Cyclic β-diketones are selectively reduced to give β-hydroxyketones without the formation of dihydroxy products (Scheme 2.124) [1002–1005]. It is important, however, that the highly acidic protons on the α-carbon atom are fully replaced by substituents in order to avoid the (spontaneous) chemical condensation of the substrate with acetaldehyde, which is always present in yeast fermentations and to avoid racemization of the α-monosubstituted β-hydroxyketone formed as product. Again, with small-size rings, the corresponding *syn*-products are formed predominantly, usually with excellent optical purity. However, the diastereoselectivity becomes less predictable and the yields drop when the rings are enlarged. Again, the stereochemistry at the newly formed secondary alcohol center can be predicted by Prelog's rule.

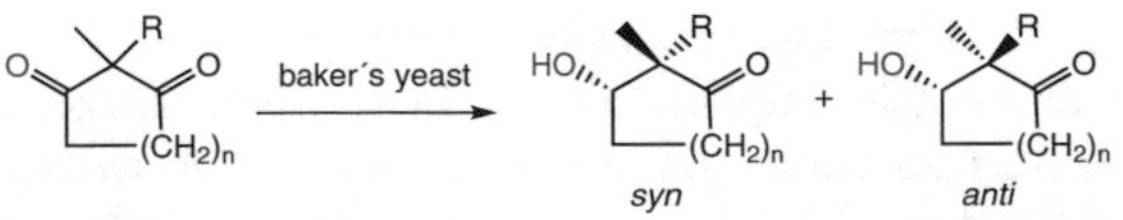

R	n	e.e. *syn* [%]	*syn/anti*
CH$_2$=CH-CH$_2$-	1	>98	90:10
HC≡C-CH$_2$-	1	>90	100:0
N≡C-(CH$_2$)$_2$-	1	>98	96:4
CH$_2$=CH-CH$_2$-	2	>98	45:55
HC≡C-CH$_2$-	2	>98	27:73
N≡C-(CH$_2$)$_2$-	2	>98	30:70
CH$_2$=CH-CH$_2$-	3	>98	100:0
CH$_2$=CH-CH$_2$-	4	>98	82:18
CH$_2$=CH-CH$_2$-	5	>98	no reaction

Scheme 2.124 Yeast reduction of cyclic β-diketones

In contrast, the reduction of α-diketones does not stop at the α-hydroxyketone (acyloin) stage but leads to the formation of vicinal diols (Scheme 2.125). In general, the less hindered carbonyl group is quickly reduced in a first step to give the (S)-α-hydroxyketone according to Prelog's rule, but further reduction of the (usually more sterically hindered) remaining carbonyl group yields the corresponding diols predominantly in the *anti*-configuration as the final product [1006, 1007].

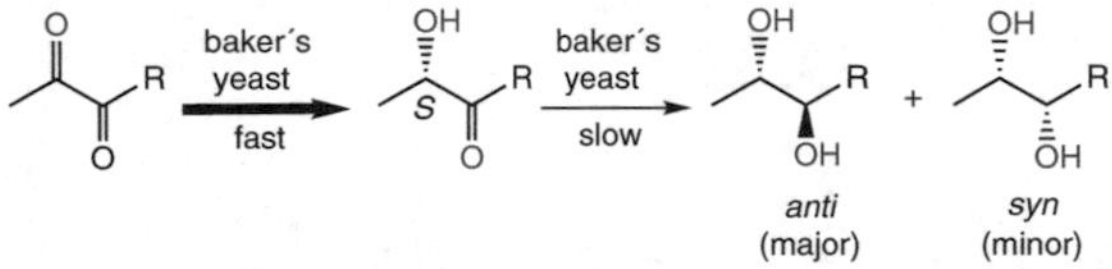

R	e.e. *anti*-diol [%]	*anti/syn*
Ph-	94	>95:<5
1,3-dithian-2-yl-	97	95:5
Ph-S-CH$_2$-	>97	86:14

Scheme 2.125 Yeast reduction of α-diketones

Secondary alcohols possessing the *anti-Prelog configuration* can be obtained from yeast reductions via substrate modification (Scheme 2.120) or through enzyme inhibition. If these techniques are unsuccessful, the use of microorganisms other than yeast [854, 857, 1008–1012], such as *Pichia farinosa* [1013], *Geotrichum candidum* [1014, 1015], and *Yarrowia lipolytica* [1016] may be of an advantage. Even plant cell cultures such as *Gardenia* may be employed for this purpose [1017, 1018]. However, in this case the help of a microbiologist is recommended for organic chemists.

The high level of technology available for the biocatalytic reduction of carbonyl compounds has allowed its implementation for the industrial-scale production of chiral building blocks containing *sec*-alcohol moieties. The system applied – either whole microbial (wild-type) cells, a designer bug containing a dehydrogenase plus cofactor-recycling enzyme, or the use of isolated enzymes – depends on the case and is mainly dependent on the economic and the patent situation. Representative examples are depicted in Scheme 2.126 [1019–1025].

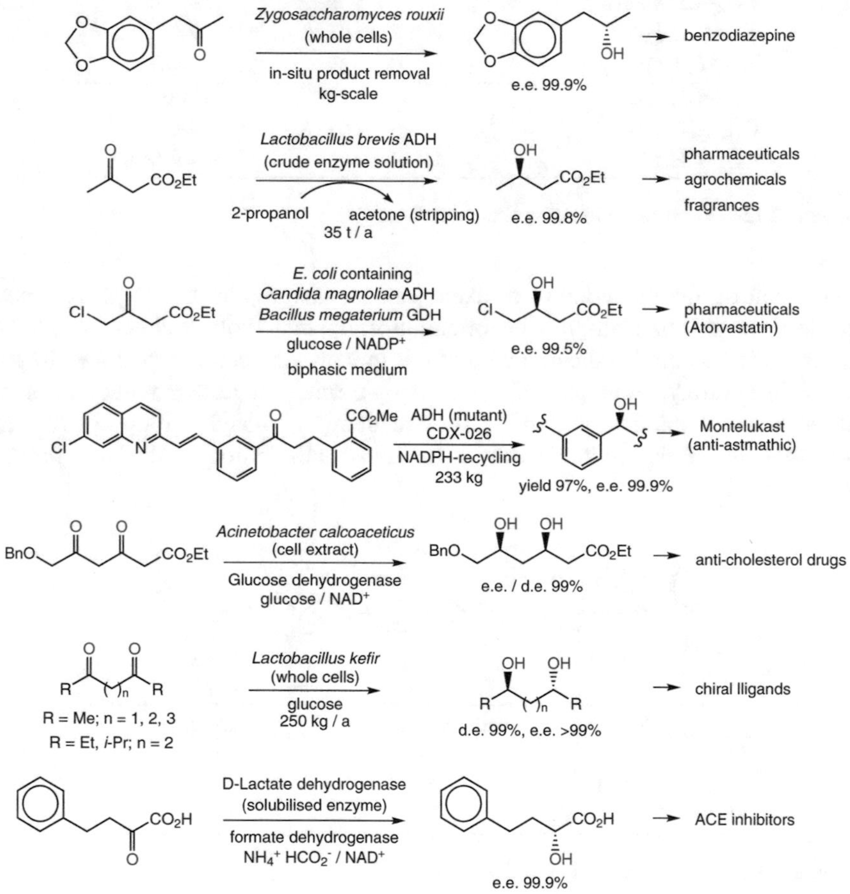

Scheme 2.126 Industrial-scale bioreduction of carbonyl compounds

Deracemization via Biocatalytic Stereoinversion Racemic secondary alcohols may be converted into a single enantiomer via stereoinversion which proceeds through a two-step redox sequence (Scheme 2.127) [37, 1031, 1032]: In a first step, one enantiomer from the racemic mixture is selectively oxidized to the corresponding ketone while the other enantiomer remains unaffected. Then, the ketone is reduced in a second subsequent step by another redox-enzyme displaying *opposite stereochemical preference.* Overall, this process constitutes a deracemization technique, which leads to the formation of a single enantiomer in 100% theoretical yield from the racemate [1028]. Due to the presence of two consecutive oxidation–reduction reactions, the net redox balance of the process is zero and no external cofactor recycling is required since the redox equivalents are exchanged between both steps in a closed loop. In order to achieve a high optical purity of the product, at least one of the steps has to be irreversible for entropic reasons [1026, 1033].

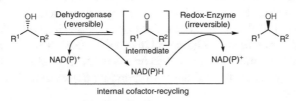

R^1	R^2	Microorganism(s)	Yield [%]	e.e. [%]	References
Me	CH$_2$CO$_2$Et	*Geotrichum candidum*	67	96	[1023]
Me	*p*-Cl-C$_6$H$_4$	*Geotrichum candidum*	97	96	[1024]
CH$_2$OH	Ph	*Candida parapsilosis*	~100	~100	[1025]
Me	(CH$_2$)$_2$CH=CMe$_2$	*Bacillus stearothermophilus* + *Yarrowia lipolytica*	91	~100	[1026]
Ph	CO$_2$H	*Pseudomonas polycolor* + *Micrococcus freudenreichii*	70	>99	[1027]

Scheme 2.127 Deracemization of *sec*-alcohols via microbial stereoinversion

The origin of the irreversibility of microbial/enzymatic deracemization of *sec*-alcohols proceeding through an oxidation–reduction sequence depends on the type of microorganism and is unknown in most cases. For instance, deracemization of various terminal (±)-1,2-diols by the yeast *Candida parapsilosis* has been claimed to operate via an (*R*)-specific NAD$^+$-linked dehydrogenase and an irreversible (*S*)-specific NADPH-dependent reductase [1033]. Along these lines, the enzymatic stereoinversion of (biologically inactive) D-carnitine to furnish the desired bioactive L-enantiomer[29] was accomplished by using two stereocomplementary carnitine dehydrogenases. Due to the fact that both dehydrogenase enzymes are NAD(H)-dependent, the end point of the process was close to equilibrium (64%) [1034]. By contrast, the stereoinversion of β-hydroxyesters using the fungus *Geotrichum candidum* required molecular oxygen, which would suggest the involvement of an alcohol oxidase rather than an alcohol dehydrogenase [1035]. Recently, the mechanism of enzymatic stereoinversion catalyzed by stereocomplementary

---

[29]L-Carnitine is an essential factor for the transport of long-chain fatty acids across mitochondrial membranes and is used in the treatment of certain dysfunctions of skeletal muscles, acute hypoglycemia, and heart disorders

dehydrogenases was elucidated to depend on the opposite cofactor-dependence for NADH and NADPH of the dehydrogenases involved [1036, 1037].

Microbial stereoinversion of *sec*-alcohols has become quite popular [1038]. For instance, the deracemization of simple secondary alcohols proceeds with excellent results using the fungi *Geotrichum candidum* or *Candida parapsilosis*. In case the oxidation and reduction cannot be performed by a single species, two microorganisms may be used instead. For instance, *Bacillus stearothermophilus* and *Yarrowia lipolytica* or *Pseudomonas polycolor* and *Micrococcus freudenreichii* were coupled for the deracemization of the pheromone sulcatol and mandelic acid, respectively. In a similar fashion, (±)-pantoyl lactone – a key intermediate for the synthesis of pantothenic acid [1039] – was deracemized by using resting cells of *Rhodococcus erythropolis* or *Candida* sp. (Scheme 2.128) [1040, 1041]. Thus, L-pantoyl lactone is oxidized to the α-ketolactone, which in turn is reduced by another dehydrogenase present in the organisms to yield the corresponding (*R*)-D-pantoyl lactone in 100% theoretical yield.

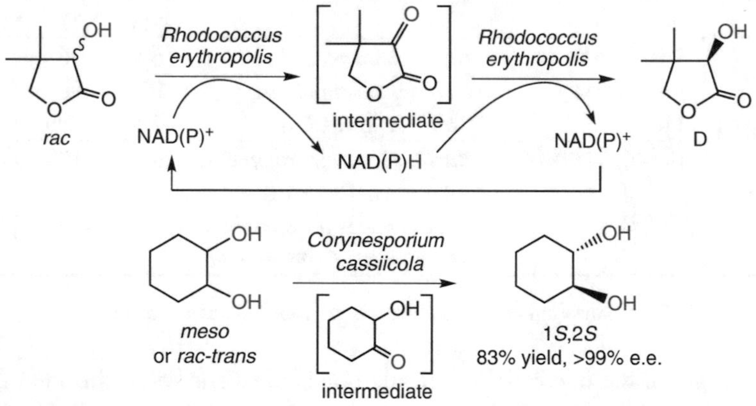

Scheme 2.128 Microbial deracemization of pantoyl lactone and 1,2-cyclohexanediol

Microbial stereoinversion has been shown to be extremely flexible, as it is also applicable to *sec*-diols possessing *two* stereocenters [1042–1044]. Thus, *meso*- or *rac-trans*-cyclohexane-1,2-diol was deracemized by *Corynesporium cassiicola* DSM 62475 to give the (1*S*,2*S*)-enantiomer as the sole product in >99% e.e. and 83% yield. The process was shown to proceed in a stepwise fashion via the corresponding hydroxyketone as intermediate, which was detected in small amounts. More important is the deracemization of *rac-trans*-indane-1,2-diol, which was accomplished with excellent results in a similar fashion. The (1*S*,2*S*)-isomer is a central building block for the anti-HIV-agent indinavir [1045].

2.2.4   Reduction of C=N Bonds

According to recent estimates, chiral amine moieties are present in ~40% of active pharmaceutical ingredients and ~20% of agrochemicals [1046], and hence are

attractive targets for asymmetric synthesis. Established approaches for their preparation are the asymmetric addition of nucleophiles across imines (including the
famous Mannich and Strecker reactions), asymmetric C-H amination and
hydroamination and the asymmetric reduction of enamines and imines, which are
either pre-formed or occur as intermediates in reductive amination [1047, 1048].

In this context, the asymmetric reduction of imines using NAD(P)H-dependent
enzymes – imine reductases – represents an attractive option for biocatalysis
[1049–1051]. This reaction can, in theory, give access to almost any *prim-*, *sec-*
and *tert-*amine while the imine (or iminium) substrate is either pre-formed in a
separate condensation step or is generated in-situ.

In Nature, C=N bond reduction is widespread and occurs in the biosynthesis of
cofactors, most prominent 5,6,7,8-tetrahydrofolate and 5,6,7,8-dihydropterine
(Scheme 2.129). The same holds for the reduction of a 2-thiazoline moiety in the
biosynthesis of bacterial iron binding proteins (siderophores), for example
Yersiniabactin. Furthermore, several alkaloids, such as coniine (from poisonous
hemlock) and reticuline (from opium poppy) are derived via reduction of their
imine precursors, as the cyclic amino acids L-pipecolate and L-proline are obtained
from the corresponding Δ^1-imino-precursors. Although these NADPH-dependent
C=N reductases are highly efficient, their substrate scope is very narrow and hence
their importance lies in their physiological significance, rather than in their biocatalytic potential.

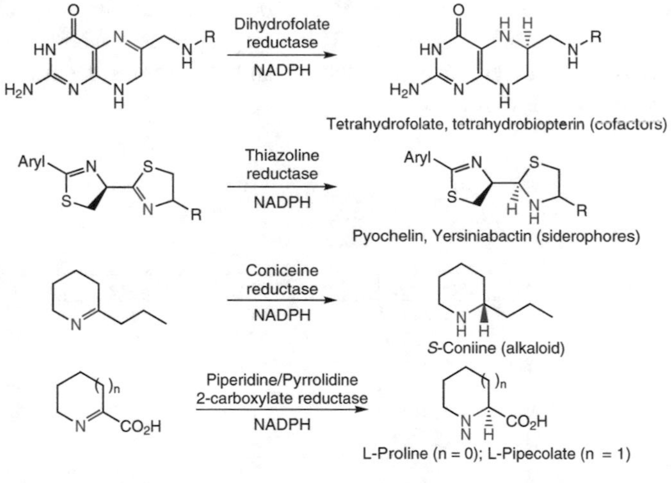

Scheme 2.129   Natural products derived via enzymatic C=N bond reduction

Consequently, the search for imine reductases of general applicability resorted to
the use of whole microbial cells. In particular, yeasts [1052] (which have proven
useful for carbonyl reduction) and bacteria [1053] were chosen in the early studies
directed to the bioreduction of imines. Unfortunately, none of these proof-of-
principle studies were investigated further, the responsible enzyme(s) were not
identified, and some reports were not reproducible [1054].

Imines – and even more so iminium species – are rather reactive compounds due to their electrophilic character, which makes them susceptible to attack by a wide range of nucleophiles, including water. As a consequence, imines are notoriously unstable in aqueous systems at physiological conditions. The only notable exception are cyclic five- and six-membered imines and those bearing a carboxylic acid moiety on the imine carbon, which provides a stabilizing H-bond.

Imine reductases possessing a broad substrate tolerance were first identified from a large-scale screening encompassing 688 microbial strains including yeasts, bacteria, actinomycetes and fungi for their activity on 2-methyl-1-pyrroline as hydrolytically inert test substrate (Scheme 2.130, $n = 1$, R=CH_3) [1055]. Among all cultures tested, only five *Streptomyces* sp. turned out to be active, two of which showed satisfactory stereoselectivities, but with stereo-complementary behaviour: *Streptomyces* sp. GF3587 afforded (*R*)-2-methylpyrrolidine in 99% e.e. and strain GF3546 gave the (*S*)-enantiomer in 81% e.e. The responsible enzymes were purified, cloned and heterologously expressed [1056] and they constitute the first members of the now rapidly growing family of imine reductases. The substrate scope of both enzymes encompasses 5-, 6- and 7-membered cyclic imines, 3,4-dihydroisoquinolones and 3,4-dihydro-β-carbolines with excellent levels of stereoselectivity (Scheme 2.130). An α,β-unsaturated imine was chemoselectively reduced at the C=N bond and the alkene remained intact [1057, 1058]. In general, biotransformations using imine reductases have so far been carried out using resting cells of *E. coli* in which the respective imine reductase is heterologously expressed. This setup ensures NADPH recycling by the host cell through metabolism of glucose and overcomes the limited thermal stability of many imine reductases.

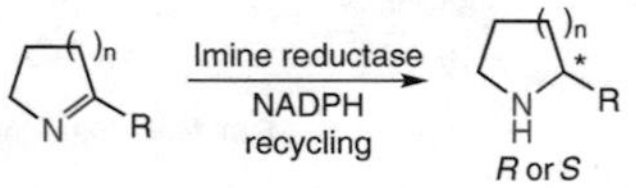

R	n	Imine reductase[a]	Conv. [%]	Config.	E.e. [%]
Me	1	GF3587	>98	R	>98
Me	1	GF3546	57	S	>95
Me	2	GF3587	>98	R	>98
Me	2	GF3546	>98	S	>98
Me	3	GF3587	>98	R	>98
Me	3	GF3546	>98	S	>98
n-Pr	2	GF3587	>98	R	>98
p-F-C_6H_4	2	GF3546	42	R*	98
p-MeO-C_6H_4	2	GF3587	50	S*	>98

[a] Whole cells of *E. coli* BL21 (DE3) containing (*R*)- or (*S*)-imine reductase from *Streptomyces* sp. GF3587 or GF3546, resp; * switch in CIP sequence priority.

Scheme 2.130 Asymmetric reduction of cyclic imines using stereocomplementary imine reductases

The discovery of the first imine reductases with a broad substrate scope resulted in identification of a broad range of additional enzymes via a sequence homology search. The hits have been used to generate an electronic library of several hundred putative imine reductases [1059]. Most imine reductases known to date originate from *Streptomyces*, but are also found in *Mycobacteria*, *Bacillus* and *Pseudonocardia* sp., and their physiological role and hence their natural substrate(s) are unknown. Although positively charged N-alkylated iminium species are reduced, electronically related carbonyl compounds are unreactive [1060]. Mechanistically, imine reductases do not possess a metal ion (such as Zn^{2+}) found in many alcohol dehydrogenases, but act through general acid-base catalysis. Imine reduction is assumed to proceed through hydride delivery from NAD(P)H onto the electrophilic imine carbon with concomitant protonation at N involving an Asp, His or Tyr residue acting as Brønsted acid to overcome the formation of an (energetically unfavourable) amide intermediate (R_2N^-). The large majority of imine reductases prefers NADPH as cofactor, only a few enzymes can use both nicotinamide species about equally well.

Reductive Amination of Ketones
In contrast to the reduction of hydrolytically stable cyclic imines which constitutes a viable protocol for the formation of cyclic *sec*-amines, open-chain imines derived from ketones and ammonia or short-chain *prim*-amines (e.g. methyl- or *n*-butylamine) are converted by imine reductases at low rates, which results in low to modest conversions (typically 50–70%) and requires high enzyme loadings [1061, 1062]. In addition, whole-cell preparations and crude cell lysates containing imine reductases are plagued with competing ketone reduction by alcohol dehydrogenases and the amine donor has to be employed in excess (typically ~50:1) to drive the equilibrium towards imine formation. Recently, imine reductases were identified which catalysed the reductive amination of cyclic, aliphatic and aromatic ketones (e.g. 2-hexanone or cyclohexanone) using ammonia and small aliphatic *prim*-amines (preferably methylamine) with encouraging results [1063].

Reductive Amination of α-Ketocarboxylic Acids
The (reversible) transformation of an α-ketocarboxylic acid in presence of ammonia and one equivalent of NAD(P)H furnishes the corresponding α-amino acid and is catalyzed by amino acid dehydrogenases [EC 1.4.1.X] [1064]. This reaction bears a strong resemblance to imine reduction and it formally represents a reductive amination (Scheme 2.131). A vast number of L-amino acid dehydrogenases from diverse organisms as well as variants engineered for industrial application has been described [1065–1068]. Stereo-complementary D-selective enzymes have been developed via protein engineering of *meso*-diaminopimelic acid D-dehydrogenases [1069, 1070] and they have been applied to the pilot-plant scale production of the non-natural amino acid D-5,5,5-trifluoromethylnorvaline [1071].

As deduced for L-Leu-dehydrogenase [1072], the α-ketoacid substrate is positioned in the active site between two Lys-residues (Scheme 2.131). Nucleophilic attack by NH_3 leads to a hemiaminal intermediate, which eliminates H_2O to form an iminium species. The latter is reduced by a hydride from nicotinamide forming the

L-amino acid. In contrast to imine reductases, amino acid dehydrogenases catalyse *both* imine formation and C=N bond reduction. Since this mechanism is highly tuned for α-keto/α-amino acids, it is clear that only ammonia is accepted as amine donor and a neutral imine (Schiff base) lacking the carboxylate moiety on the imine carbon atom cannot be accepted as substrate. Due to the importance of α-amino acids, both D- and L-amino acid dehydrogenases are important enzymes in industrial processes.

Among the various amino acid dehydrogenases, Leu-DH has captured an important role for the synthesis of nonproteinogenic L-α-amino acids via asymmetric reductive amination of the corresponding α-ketoacids [1073, 1074]. A range of protease inhibitors used for the treatment of tumors and viral infections contain sterically hindered amino acids as key element for their biological action. The latter cannot be synthetized via the conventional (protease-dependent) methods (Sect. 2.1.2), but they are produced on industrial-scale making use of the relaxed substrate specificity of LeuDH in combination with NADH recycling using the formate dehydrogenase/formate system [1075]. In particular, L-*t*-leucine is a key intermediate for the synthesis of the HIV protease inhibitor Atazanavir.

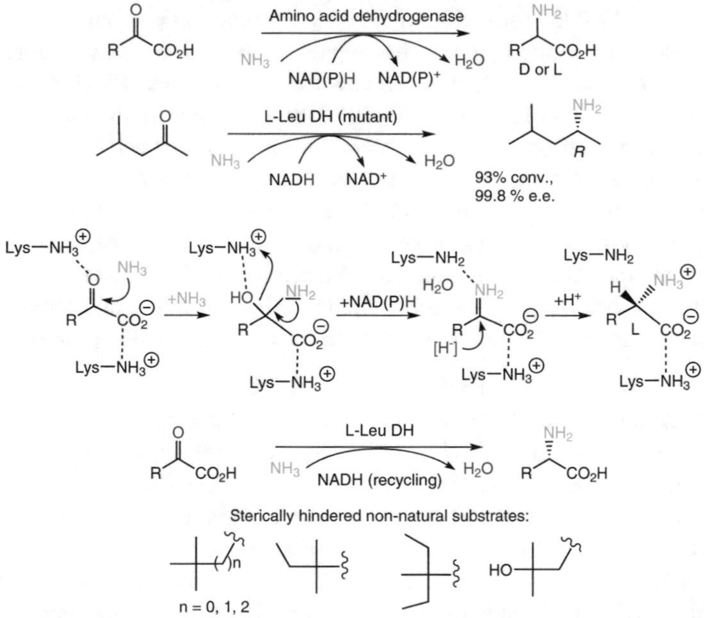

Scheme 2.131 Reductive amination of α-ketocarboxylic acids using D- and L-amino acid dehydrogenases (*top*); reductive amination of a methyl ketone using a L-leucine dehydrogenase mutant (*center*); mechanism of L-leucine dehydrogenase

In analogy to α-amino acid dehydrogenases, nicotinamide-depending enzymes for the reductive amination of ketones lacking the carboxylate moiety would be termed 'amine dehydrogenases'. Since naturally occurring enzymes of this type are unknown to date,[30] artificial amine dehydrogenases have been created by semi-rational protein design using existing α-amino acid dehydrogenase scaffolds. In a pioneering study, L-LeuDH from *Bacillus stearothermophilus* was chosen as starting point for directed evolution [1077]. A total of 19 amino acid residues was selected for mutagenesis including combinatorial active-site saturation (CAST). After several rounds of directed evolution, a quadruple variant (K68S, E114V, N261L, V291C) was obtained, which showed a reasonable specific activity of 0.69 U/mg in the reductive amination of 4-methyl-2-pentanone forming the corresponding (*R*)-amine in 99.8% e.e. Not surprisingly, two of the mutations introduced (K68S, N261L) involved those in binding of the natural substrate's carboxylate group. Their replacement by more unpolar amino acid residues completely abolished the restriction to α-amino/α-ketocarboxylic acids. This strategy was later successfully extended to L-phenylalanine dehydrogenase [1078].

2.2.5   Reduction of C=C-Bonds

The asymmetric (bio)catalytic reduction of C=C-bonds goes in hand with the creation of (up to) two chiral centers and is thus one of the most widely employed strategies for the production of chiral compounds. Whereas *cis*-hydrogenation using transition-metal based homogeneous catalysts has been developed to an impressive standard [1079], stereocomplementary asymmetric *trans*-hydrogenation is less sophisticated [1080].

The biocatalytic counterpart for the stereoselective reduction of alkenes is catalyzed by flavin-dependent ene-reductases [EC 1.3.1.31], which are members of the 'old yellow enzyme' family (OYE, Scheme 2.132) [1081, 1082], first described in the 1930s by O. Warburg, who first demonstrated the requirement of a low molecular weight 'cofactor' for enzymatic catalysis [1083]. These enzymes are widely distributed in microorganisms and in plants. Some of them occur in well-defined pathways, e.g., in the biosynthesis of secondary metabolites, such as morphine [1084] and jasmonic acid [1085]. Others are involved in the detoxification of xenobiotics [1086], such as nitro esters [1087] and nitro-aromatics [1088] like trinitrotoluene (TNT) [1089]. *Ene*-reductases should not be confused with '*enoate* reductases',[31] which contain an Fe_4S_4-cluster in addition to flavin. These enzymes are found in strict anaerobic organisms and are very sensitive towards molecular oxygen, which makes them of limited use for preparative biotransformations [1090, 1091].

The catalytic mechanism of the asymmetric reduction of alkenes catalyzed by ene-reductases from the old yellow enzyme family has been studied in great detail

---

[30]For a rare exception see [1076].

[31]More precisely 2-enoate reductase. Since these enzymes belong to the same EC class 1.3.1.X, ene-reductases and enoate reductases are often confused.

[1092] and it has been shown that a hydride (derived from a reduced flavin cofactor) is stereoselectively transferred onto Cβ, while a Tyr-residue adds a proton (which is ultimately derived from the solvent) onto Cα from the opposite side (Scheme 2.132). As a consequence of the stereochemistry of this mechanism, the overall addition of [H₂] proceeds in a *trans*-fashion with absolute stereospecificity [1093]. This reaction is generally denoted as the 'oxidative half reaction'. The catalytic cycle is completed by the so-called 'reductive half reaction' via reduction of the oxidized flavin cofactor at the expense of NAD(P)H, which is ultimately derived from an external H-source via another redox reaction, which is employed for cofactor-recycling (Scheme 2.110). In contrast to alcohol dehydrogenases (carbonyl reductases), which show a rather pronounced preference for either NADH or NADPH [1094], ene-reductases are more flexible in this respect: some enzymes are very specific [1095], others are able to accept both cofactors equally well [1096, 1097]. Overall, the reaction resembles an asymmetric Michael-type addition of a chiral hydride onto an enone and, as a consequence of the mechanism, nonactivated C=C bonds are therefore completely unreactive [1098]. Although the overall hydride pathway appears rather complex, practical problems are minimal since flavin cofactors are usually tightly bound to the enzyme and are thereby protected from the environment.

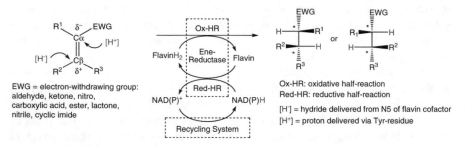

Scheme 2.132 Asymmetric bioreduction of activated alkenes using flavin-dependent ene-reductases

Although the remarkable synthetic potential of ene-reductases has been recognized long ago, preparative-scale applications were severely impeded by two major problems: Simple to use whole-cell systems, such as baker's yeasts [1099], and fungi, such as *Geotrichum candidum, Rhodotorula rubra, Beauveria bassiana* [1100] and *Aspergillus niger,* are plagued by undesired side reactions, particularly carbonyl reduction (catalyzed by alcohol dehydrogenases/carbonyl reductases) or ester hydrolysis (mediated by carboxyl ester hydrolases) [794]. On the other hand, the first generation of isolated (cloned) C=C bond reducing enzymes (enoate reductases) were obtained from (strict or facultative) anaerobes, such as *Clostridia* [1101] or methanogenic *Proteus* sp. [1102], which were inapplicable to preparative-scale transformations due to their sensitivity towards traces of molecular oxygen. It was only recently, that this bottleneck was resolved by providing oxygen-stable OYEs from bacteria, plants, and yeasts [1103–1111].

The following crude guidelines for the asymmetric bioreduction of activated alkenes using ene-reductases can be delineated:

- Only C=C-bonds which are 'activated' by electron-withdrawing substituents (EWG) are reduced (Scheme 2.133) [1112], electronically 'isolated' double bonds are not accepted [1113]. In a rough approximation, the activating capability of an EWG goes in line with its electron-withdrawing strength. With activated, conjugated 1,3-dienes only the α,β-bond is selectively reduced, leaving the nonactivated γ,δ-bond behind (Scheme 2.133). In a similar manner, cumulated 1,2-dienes (allenes) mainly give the corresponding 2-alkenes. A rearrangement of the allene to give an acetylene may be observed occasionally [1114].
- Acetylenic triple bonds yield the corresponding (E)-alkenes [1115]. The latter may be subject to further (slow) reduction.

The following functional groups may serve as 'activating' groups:

- α,β-Unsaturated carboxaldehydes (enals) are quickly reduced in a clean fashion yielding saturated aldehydes when pure ene-reductases are used. In contrast, whole-cell reductions are heavily plagued by competing carbonyl reduction, which often outcompetes the ene-reductase to furnish the corresponding allylic alcohol (thereby depleting the substrate) and/or the saturated *prim*-alcohol (via over-reduction of the desired product) [1116, 1117]. These undesired side-reactions sometimes allow to use an allylic alcohol as substrate, which is transformed via the corresponding enal by whole cells [1118] (Scheme 2.133).
- α,β-Unsaturated ketones (enones) are good substrates for ene-reductases. With whole cells, competing carbonyl-reduction is slower as compared to enals and the product distribution depends on the relative rates of competing carbonyl- and ene-reductases [1119, 1120] (Scheme 2.134).
- α,β-Unsaturated nitro compounds can be readily transformed into chiral nitro-alkanes. Depending on the type of OYE, reductive biodegradation may occur via the Nef-pathway [1121]. Due to the high acidity of nitroalkanes, any chiral center at Cα is prone to racemization, whereas Cβ-analogs are perfectly stable [1122] (Scheme 2.135).
- Cyclic imides, such as maleimide, are readily reduced without competing side reactions.
- α,β-Unsaturated carboxylic acids or esters have to be regarded as 'borderline'-substrates:

  Simple α,β-unsaturated *mono*-carboxylic acids or *mono*-esters are not readily reduced by OYEs (they are substrates for 'enoate-reductases'). However, the presence of an additional electron-withdrawing group (which alone would not be sufficient to act as activator), such as halogen, helps to boost the degree of activation [1123] (Scheme 2.136). Consequently, *di*-carboxylic acids and *di*-esters are accepted by OYEs, although ester hydrolysis is a common side-reaction when using whole cells. Due to their reduced carbonyl activity, carboxylic acids are less activated than the corresponding esters.

- Only few reports are available regarding α,β-unsaturated lactones, their degree of activation parallels that of esters.
- α,β-Unsaturated nitriles may be used as substrates [1124].
- Sometimes the absolute (R/S)-configuration of the product can be controlled by starting with (E)- or (Z)-alkenes (Scheme 2.136) [1125].
- Steric hindrance at Cβ (where the hydride has to be delivered) seems to play an important role for OYEs. Consequently, sterically demanding substituents at the C=C-bond are more easily tolerated in the α-position.

The bioreduction of citral using the ene-reductase OPR3 (12-oxophytodienoic acid reductase) proceeds in a clean fashion yielding the fragrance compound (R)-citronellal in excellent chemical and optical yields (Scheme 2.133). The latter is a central intermediate for various terpenoid odorants, such as citronellyl nitrile and menthol.

In contrast, baker's yeast reduction of a closely related enal bearing a carboxylic ester group yielded the saturated *prim*-alcohol as the major product due to over-reduction of the aldehyde moiety. The less activated C=C bond adjacent to the ester remained unchanged [1126]. Instead of starting with an enal (whose aldehyde moiety would be quickly reduced by baker's yeast) the corresponding allylic alcohol may serve as substrate in whole-cell bioreductions. Thus, geraniol gave (R)-citronellol in >97% e.e. [1127] and in a similar fashion only the α,β-bond was reduced in a conjugated 2,4-diene-1-ol [1128]. It is important to note that in whole-cell transformations the C=C-reduction of allylic alcohols always occurs at the aldehyde stage, which is reversibly formed as intermediate (Scheme 2.133).

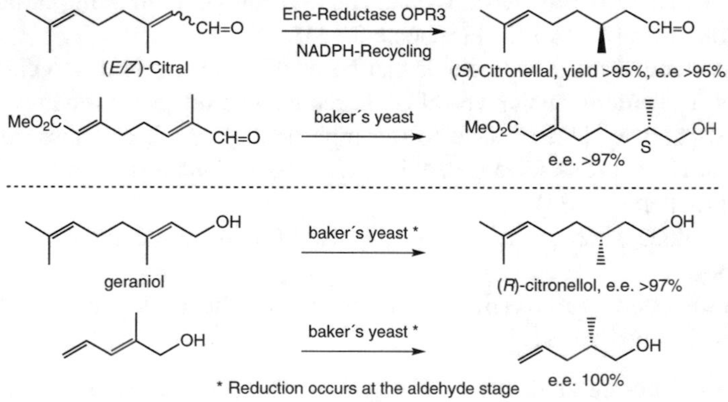

Scheme 2.133 Asymmetric bioreduction of enals and allylic alcohols using isolated ene-reductase and baker's yeast

In contrast to aldehydes, over-reduction is less pronounced on α,β-unsaturated ketones (Scheme 2.134). Nonracemic levodione, which is a precursor for the synthesis of carotenoids, such as astaxanthin and zeaxanthin, was obtained in 80% yield and >95% e.e. via yeast-mediated reduction of ketoisophorone. Two other products arising from reduction of the carbonyl moieties were formed in minor amounts [1129]. In contrast, no trace of carbonyl reduction was observed using ene-reductase OPR3 (Scheme 2.134). (3R,3′R)-Zeaxanthin is used in human nutrition and healthcare products and is considered for the treatment of age-related macula degeneration [1130].

Nitro-olefins are readily reduced by ene-reductases to form chiral nitro-alkanes (Scheme 2.135) [1131]. Using ene-reductase OPR1 or baker's yeast, the corresponding (R)-nitroalkanes were obtained in high e.e. Surprisingly, the mirror-image product was formed by using isoenzyme OPR2, which is highly homologous to OPR3 (53%).

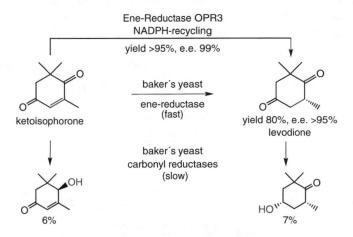

Scheme 2.134 Asymmetric bioreduction of α,β-unsaturated diketone

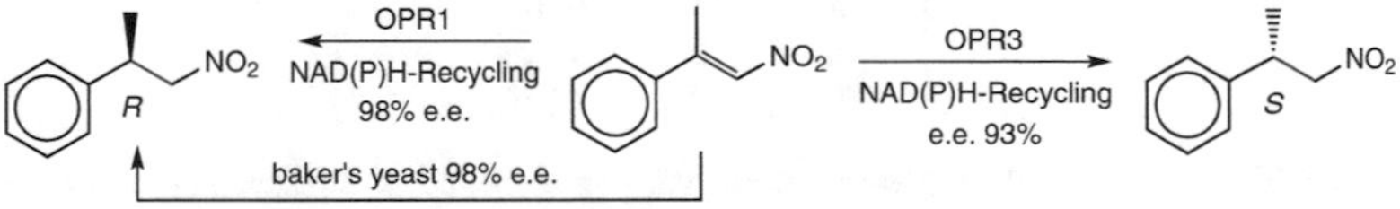

Scheme 2.135 Stereocomplementary bioreduction of nitro-olefins

Simple *mono*carboxylic esters require the presence of an additional activating group, such as a halogen atom or a second ester, to be accepted by OYEs (Scheme 2.136). α-Substituted butenedioic esters were readily reduced by ene-reductases YqjM and OPR1 with excellent specificities, while the stereochemical outcome could be controlled by choice of the ene-reductase or by using an (E)- or (Z)-configurated substrate: (R)-2-Methylsuccinate was obtained by using OPR1, regardless of the (E/Z)-configuration of the substrate. In contrast, with YqjM, the configuration of the product switched when an (E)-fumarate ester was used instead of a (Z)-maleate [1132]. This bioreduction was upscaled to 70g batch size with subsequent in situ ester hydrolysis [1133].

The suitability of β-substituted α-haloacrylate esters as substrates for ene-reductases was first proven using baker's yeast and the absolute configuration of the product was shown to depend on the (E/Z)-configuration of the substrate [1134]. While the chiral recognition of the (Z)-alkenes was perfect, the (E)-isomers gave products with lower e.e. and it was shown that the microbial reduction took place on the carboxylic acid stage, which were formed enzymatically by hydrolysis of the starting esters prior to the reduction step [1135].

Undesired ester hydrolysis can be avoided when isolated ene-reductases are used. In this case, excellent results were achieved with methyl α-haloacrylates, which were reduced to (R)- or (S)-α-halopropionates in a stereo-divergent fashion depending on the ene-reductase employed [1136]. In contrast, halogen atoms in the β-position proved to be unsuitable. Although α,β-dihalo derivatives were rapidly reduced, the saturated 2,3-dihalo esters thus formed spontaneously underwent HX-elimination yielding an α-haloacrylate ester. Overall, this sequence consists of a (formal) reductive β-dehalogenation (Scheme 2.136) [1137].

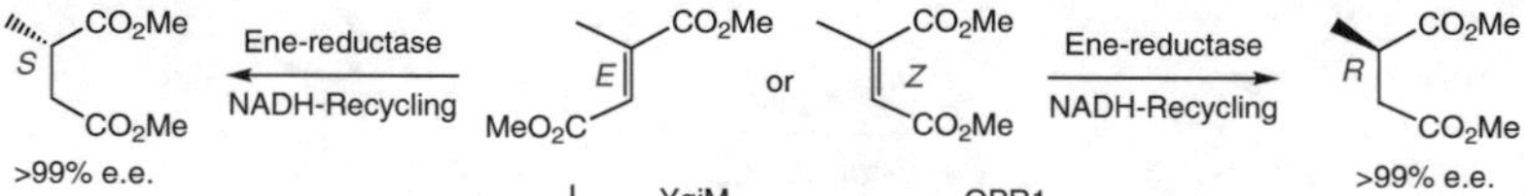

Configuration	YqjM		OPR1	
	c [%]	e.e. [%]	c [%]	e.e. [%]
Z	93	>99 (R)	91	>99 (R)
E	70	>99 (S!)	99	>99 (R)

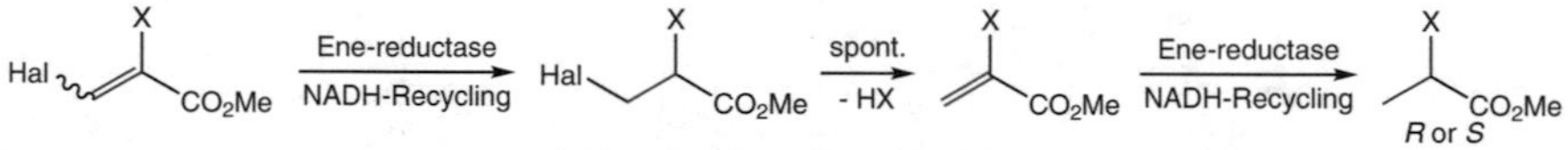

X	Enzyme	Conv. [%]	E.e. [%]
Cl	OYE3	>99	89 (S)
	NCR	50	>99 (R)
Br	OYE3	>99	95 (S)
	NCR	>99	99 (R)
I	OYE1	54	90 (S)
	OPR1	5	40 (R)

Scheme 2.136 Stereocontrol of ene-reduction via enzyme-type or substrate-configuration and reduction of haloacrylate esters

Only few reports are available on the asymmetric bioreduction of α,β-unsaturated lactones. For instance, β-substituted five-membered ring lactones were readily reduced by baker's yeast to give the (R)-configured saturated analogs [1138]. More recent investigations using isolated ene-reductases revealed that butyrolactones bearing an α-substituent were reduced with excellent stereoselectivity, albeit at a slow rate. In contrast, β-substituted analogs were converted considerably faster. A chiral center in γ-substituted butyrolactones was nicely recognized and led to kinetic resolution with E-values of up to E = 49 (Scheme 2.137) [1139].

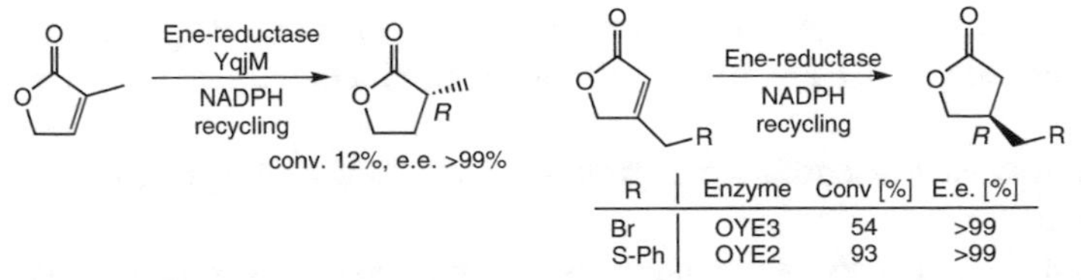

Scheme 2.137 Asymmetric bioreduction of α,β-unsaturated butyrolactones

The reverse reaction catalyzed by ene-reductases – α,β-desaturation of carbonyl compounds – is thermodynamically strongly disfavored because it involves the breakage of two C–H σ-bonds, which are not energetically compensated for by the newly formed C=C π-bond. Hence it is not feasible using $NAD(P)^+$ as driving force in a coupled-enzyme system (Scheme 2.132). However, it was shown to occur in the disproportionation of enones, which has been described as 'dismutase' or 'aromatase' activity of OYEs (Scheme 2.138) [1140]. While one enone molecule is reduced, the other is desaturated yielding the corresponding dienone, which quickly tautomerizes to form a phenol, thereby providing a large driving force of about −30 kcal/M, which drives the reaction to completion. Overall, this redox-neutral process is independent on nicotinamide cofactors, requires only a single ene-reductase and represents a coupled-substrate variant for C=C bioreduction. This method has been exploited using sacrificial hydrogen donors, such as cyclohex-2-enones, 2-tetralone and five-membered heterocyclic 3-ones as sacrificial hydrogen donors [1141].

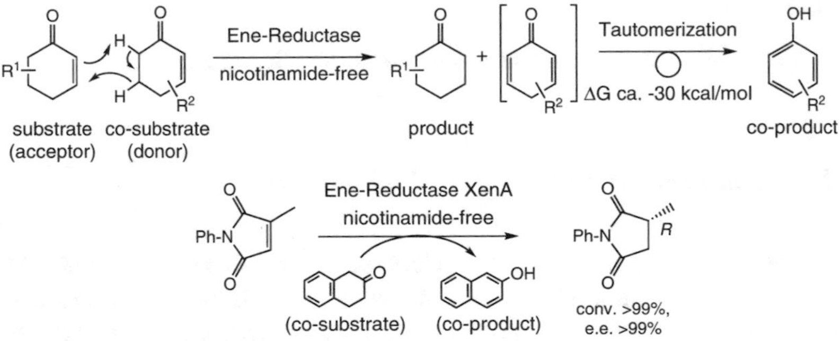

Scheme 2.138 Disproportionation of enones and nicotinamide-independent hydrogen transfer

2.3 Oxidation Reactions

Oxidation constitutes one of the key steps for the introduction of functional groups into the raw materials of organic synthesis which are almost invariably an alkane, an alkene, or an aromatic molecule.[32] Traditional methodology is plagued by several drawbacks, that is

---

[32]It is an alarming fact that ≥90% of the hydrocarbons derived from crude oil are 'wasted' for energy-production (forming CO_2), the small remainder of ≤10% is used as raw material for the chemical industry to produce (long-lasting) products [1142].

- Many oxidants are based on metal ions such as copper, manganese, iron, nickel, or chromium, which are often environmentally incompatible when used on large scale.
- Undesired side reactions are common due to a lack of chemoselective oxidation methods.
- The most inexpensive and innocuous oxidant, molecular oxygen, cannot be used efficiently.
- It is extremely difficult to perform oxidations in a regio- and stereoselective fashion.

Therefore, organohalogens have been widely used as intermediates for the synthesis of oxygenated compounds, which has led to severe environmental problems due to recalcitrant halogenated organic compounds.

Many of the drawbacks mentioned above can be circumvented by using biological oxidation, in particular for those cases where stereoselectivity is required [1143–1145].

The biooxidation reactions discussed in this chapter are grouped according to their requirement for the oxidant, i.e.:

- Dehydrogenation depending on a nicotinamide cofactor [NAD(P)H] (Sect. 2.3.1)
- Oxidation and oxygenation at the expense of molecular oxygen (Sect. 2.3.1, 2.3.2 and 2.3.3)
- Peroxidation reactions requiring hydrogen peroxide or a derivative thereof (Sect. 2.3.4)

For a classification of biooxidation reactions see Scheme 2.144.

2.3.1 Oxidation of Alcohols and Aldehydes

Oxidations of primary and secondary alcohols to furnish aldehydes and ketones, respectively, are common chemical reactions that rarely present insurmountable problems to the synthetic organic chemist. These reactions can be catalysed by alcohol dehydrogenases together with $NAD(P)^+$-recycling. However, in contrast to the corresponding (carbonyl) reduction reactions, alcohol oxidation using dehydrogenases have been reported to a lesser extent for the following reasons [1146]:

- Oxidations of alcohols using $NAD(P)^+$-dependent dehydrogenases are thermodynamically unfavorable. Thus, the recycling of the oxidized nicotinamide cofactor becomes a complicated issue (see Scheme 2.111).
- Enzymatic oxidations usually work best at elevated pH (8–9) where nicotinamide cofactors and (particularly aldehydic) products are unstable.
- Lipophilic aldehydes or ketones are often more tightly bound onto the hydrophobic active site of dehydrogenases than the more hydrophilic substrate alcohol. Hence, product inhibition is a common phenomenon, in particular when reactive aldehydes are involved [846].

- Oxidation of a secondary alcohol involves the *destruction* of an asymmetric center ($sp^3 \rightarrow sp^2$ hybrid) and is therefore of limited synthetic use.

Regioselective Oxidation of Polyols
The enzyme-catalyzed oxidation of alcohols is only of practical interest to the synthetic organic chemist if complex molecules such as polyols are involved (Scheme 2.139) [1147–1151]. Such compounds present selectivity-problems with conventional chemical oxidants, which requires protection–deprotection steps. In contrast, numerous sugars and related polyhydroxy compounds obtained from renewable resources have been selectively oxidized in a single step into the corresponding keto-ols or ketoacids using a variety of microorganisms, for example the vinegar-producing bacterium *Acetobacter* (Scheme 2.139). The regioselective microbial oxidation of D-sorbitol (obtained by catalytic hydrogenation of D-glucose) by *Acetobacter suboxydans* yields L-sorbose, which represents the key step in the famous Reichstein–Grüssner process for the production of L-ascorbic acid (vitamin C).

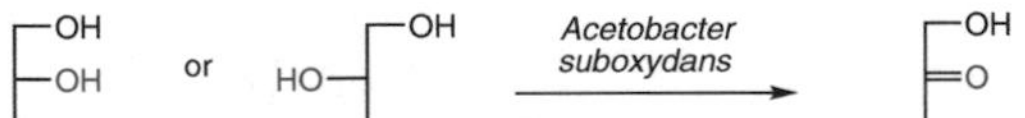

Substrate Polyol	Product Keto-alcohol	References
adonitol	L-adonulose	[1144]
D-sorbitol	L-sorbose	[1145]
L-fucitol	4-keto-L-fucose	[1146]
D-gluconic acid	5-keto-D-gluconic acid	[1147]
1-deoxy-D-sorbitol	6-deoxy-L-sorbose	[1148]

Scheme 2.139 Regioselective oxidation of polyols by *Acetobacter suboxydans*

Kinetic Resolution and Desymmetrization of Alcohols by Oxidation
Among the enzymatic systems for NAD(P)⁺-recycling described in Scheme 2.111, the use of a flavin mononucleotide (FMN) dependent nicotinamide oxidase is preferable, because it requires only molecular oxygen and is virtually irreversible [1152]. To avoid enzyme deactivation, the hydrogen peroxide produced during this two-electron transfer process is removed using catalase [1153]. In conjunction with HLADH, this system was employed for the kinetic resolution of mono-, bi-, and polycyclic secondary alcohols [873, 875, 1154, 1155]. Alcohols bearing an electron-withdrawing group (e.h. halogen, MeO, etc.) in the α-position form a strong internal H-bond and are difficult to oxidize by this method [1156].

Terminal glycols were regio- and enantioselectively oxidized at their *prim*-hydroxy group to yield L-α-hydroxyacids using a co-immobilized alcohol and aldehyde dehydrogenase system (Scheme 2.140). In the first step, kinetic resolution of the diol furnished a mixture of L-hydroxyaldehyde and the remaining D-diol. In order to avoid enzyme deactivation by the aldehyde species, it was oxidized in-situ by an aldehyde dehydrogenase to yield the more innocuous L-hydroxyacid in high optical purity [878, 1157]. An analogous double-step oxidation of *prim*-alcohols yielding carboxylic acids using a single ADH variant was recently reported [1158].

In contrast to the resolution of secondary alcohols, where the more simple lipase technology is recommended instead of a redox reaction, desymmetrization of prochiral or *meso-prim*-diols is a valuable method for the synthesis of chiral lactones (Scheme 2.140) [1159].

As a rule of thumb, oxidation of the (*S*)- or pro-(*S*)-hydroxyl group occurs selectively with HLADH. In the case of 1,4- and 1,5-diols, the intermediate γ- and δ-hydroxyaldehydes spontaneously cyclize to form the more stable five- and six-membered hemiacetals (lactols). The latter are further oxidized in a subsequent step by HLADH to form γ- or δ-lactones by maintaining (*S*)- or pro-(*S*) specificity [1160]. Both steps – desymmetrization of the prochiral or *meso*-diol and kinetic resolution of the intermediate lactol – are often highly stereoselective. Enantiopure lactones were derived from *cis-meso*-2,3-dimethylbutane-1,4-diol and the cyclic thia-analog [1161] and similar results were obtained with sterically demanding bicyclic *meso*-diols [1162].

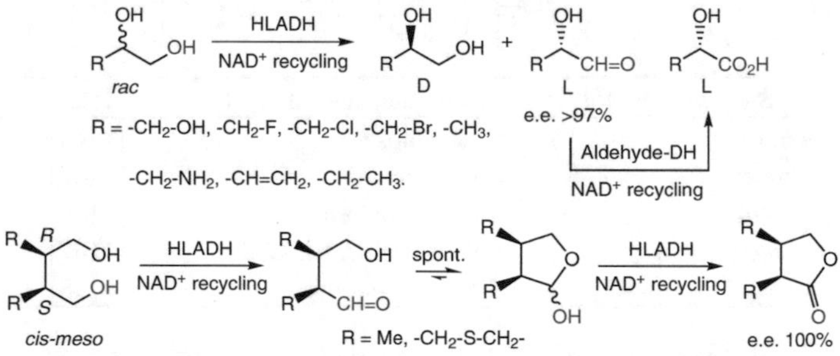

Scheme 2.140 Kinetic resolution of 1,2-diols and desymmetrization of *meso*-diols by a HLADH/ aldehyde DH system

The issues of NAD(P)$^+$-recycling can be circumvented by using nicotinamide-independent alcohol oxidases [1163]. These enzymes are either metal- (Cu, Fe) or flavin-dependent and catalyse the oxidation of *prim*- and *sec*-alcohols at the expense of O_2 as oxidant by producing H_2O_2 as by-product, which is usually destroyed by catalase.[33] With *sec*-alcohols, the reaction stops at the ketone stage, but 'over-oxidation' of aldehydes obtained from *prim*-alcohols yields carboxylic acids (Scheme 2.141, top). This activity is often observed with flavin-depending alcohol oxidases. For synthetic applications, alcohol oxidation is a crucial step in the functionalization of carbohydrates and polyols, because it yields aldehydes or ketones, which are excellent acceptors for *C*-, *N*-, *O*- and *S*-nucleophiles. Thereby, a given carbon backbone derived from natural resources can be conveniently extended.

---

[33]Many *E. coli* strains, which are used as host for the heterologous overexpression of alcohol oxidases, contain a strong native catalase activity.

Fortunately, Nature provides a broad arsenal of alcohol oxidases acting on hexoses (Scheme 2.141 bottom) [1164]: Oxidation of the most reactive anomeric hydroxyl group by glucose or hexose oxidase yields the corresponding lactone, while galactose oxidase selectively oxidizes the *prim*-OH to the aldehyde moiety, pyranose oxidase predominantly forms 2-ketoses. Of particular interest is a recently discovered flavin-dependent 5-hydroxymethyl furfural oxidase (HMF oxidase), which converts HMF via a three-step sequence to furan-2,5-dioic acid (FDC) [1165]. HMF is obtained on multi-ton scale via acid-catalyzed thermal triple dehydration of hexoses and FDC is a promising replacement for (fossil-derived) terephthalic acid, ~50 mio t of which is annualy converted into PET-polymers.

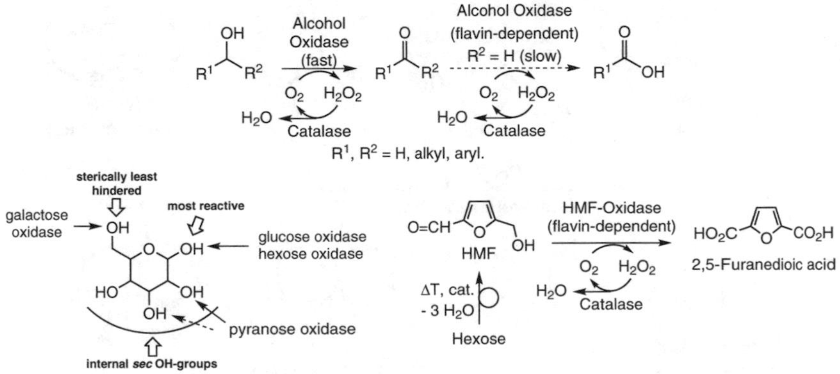

Scheme 2.141 Oxidation of alcohols and aldehydes using alcohol oxidases

2.3.2   Oxidation of Amines

In close analogy to the oxidation of alcohols using alcohol oxidases, amines can be oxidized by imine oxidases at the expense of O_2 with concomitant production of H_2O_2. The main biochemical role of enzymes from microbial origin is the oxidative degradation of amines, which provides essential NH_3 for assimilation and growth. In particular, the flavin-dependent monoamine oxidase from *Aspergillus niger* (MAO-N) has served as platform for the development of numerous mutants, which oxidize *prim*-, *sec*- and *tert*-amines with high stereoselectivities [1166–1169].

In contrast to alcohol oxidation, which furnishes stable carbonyl compounds, amine oxidation yields unstable imines, which cannot be easily isolated and stored. However, two ingenious strategies have been deployed to render amine oxidation as a synthetically useful tool.

Cyclic Deracemization of Amines
Enantioselective oxidation of an amine bearing an adjacent chiral C atom by an amine oxidase yields the corresponding achiral imine, while the non-converted enantiomer remains untouched. In order to overcome the 50% limit of kinetic resolution, the imine can be (non-stereoselectively) reduced in situ using a mild reducing agent, such as amine-borane complex, which yields an equimolar amount

of 25% amine enantiomers. Hence, after a single oxidation-reduction cycle, the enantiomeric composition of the starting amine is 75:25, another cycle renders 87.5:12.5 and so forth (Scheme 2.142). Overall, the reacting enantiomer is gradually depleted while the non-reacting counterpart accumulates. Although this process – termed cyclic deracemization [1170] – lacks elegance at a quick glance, it furnishes an e.e. of 93.4 e.e. after only four cycles, whereas after seven cycles the e.e. is >99%. Starting from wild-type MAO-N, which is able to oxidize only simple primary amines, a panel of variants was developed through several rounds of random mutagenesis and directed evolution combined with rational design, which accept also *sec*- and *tert*-amines [1171].

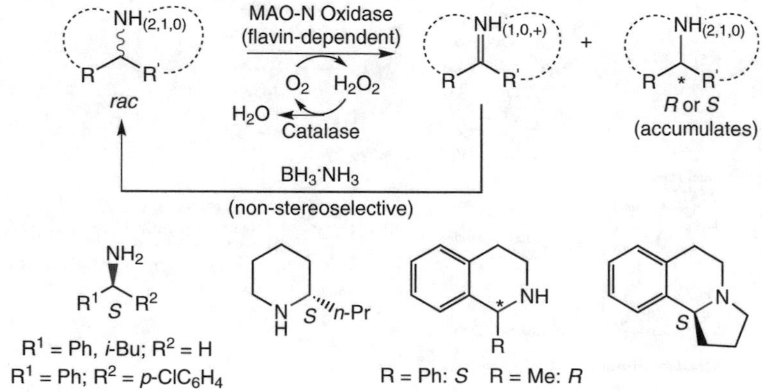

Scheme 2.142 Cyclic deracemization of amines using monoamine oxidase (variants) combined with non-stereoselective imine reduction; accumulated product enantiomers are shown

Desymmetrization of cyclic *sec*-amines.

Asymmetric oxidation of (bi)cyclic *sec*-amines by MAO-N breaks the symmetry of these *meso*-structures and yields chiral imines. Being good electrophiles, the latter can be trapped in-situ as bisulfite adducts, which are directly converted into the corresponding α-aminonitriles with high *trans*-diastereoselectivity (30:1 to 100:1) by treatment with NaCN. Acid catalyzed methanolysis in a Strecker-like protocol yields the corresponding methyl carboxylate in 88% overall yield and >99% e.e. Non-natural mono-, bi- and tricyclic L-proline analogs of this type are key building blocks for the synthesis of peptidomimetic protease inhibitors, such as Boceprevir and Telaprevir, which are used for the treatment of chronic Hepatitis C (Scheme 2.143) [1172].

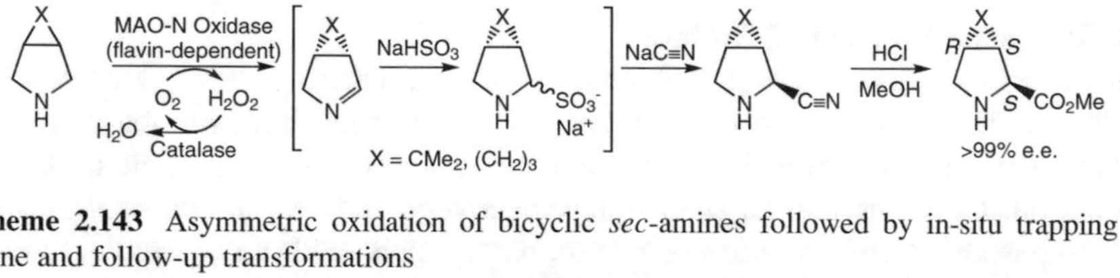

Scheme 2.143 Asymmetric oxidation of bicyclic *sec*-amines followed by in-situ trapping of imine and follow-up transformations

2.3.3   Oxygenation Reactions

Enzymes which catalyze the direct incorporation of molecular oxygen into an organic molecule are called '*oxygen*ases' [1173–1176]. Enzymatic oxygenation reactions are particularly intriguing since direct oxyfunctionalization of nonactivated organic compounds remains a largely unresolved challenge to synthetic chemistry. On the one hand, there are numerous (catalytic) oxidation processes developed by industry to convert simple raw materials, such as alkanes, alkenes and aromatics at the expense of O_2 into more valuable intermediate products, such as alcohols, aldehydes, ketones and carboxylic acids.[34] However, the catalysts employed are highly sophisticated and thus show a very narrow substrate range, which limits their applicability to a single (or few) substrate(s) and they cannot be used on lab-scale for a wider range of organic compounds, except for the Wacker-oxidation of alkenes. Unsurmountable problems persist where regio- or enantiospecificity is desired.

The appealing use of O_2 as zero-priced oxidant comes with some drawbacks: Its triplet ground state makes it kinetically unreactive with organic molecules, which are in the singlet state, and it tends to form radical species, which are difficult to control and tend to cause side reactions. Furthermore, it is a four-electron oxidant, which makes it difficult to terminate an oxidation process at an intermediate stage, leading to 'over-oxidation'. However, nature has managed to tame O_2 to a remarkable extent and several highly selective oxygenation reactions may be achieved by means of biocatalysts.

Oxygen-transfer from molecular oxygen into organic acceptor molecules may proceed through three different mechanisms (Scheme 2.144).

- Monooxygenases incorporate *one* oxygen atom from molecular oxygen into the substrate, the other is reduced at the expense of a donor (usually NADH or NADPH) to form water [1177–1179]. Overall this is a four-electron transfer, comprising two electrons each from the substrate and the cofactor.
- Dioxygenases simultaneously incorporate *both* oxygen atoms of O_2 into the substrate by forming a peroxy-species in a two-electron transfer.[35]
- Oxidases catalyze electron-transfer onto molecular oxygen, which proceeds via a two- or (more rarely) a four-electron transfer yielding either hydrogen peroxide or water, respectively, as byproduct. Incorporation of O into the substrate does

---

[34]The most important processes with respect to scale are: *p*-xylene → terephthalic acid, ethylene → ethylene oxide, ethylene → acetaldehyde, ethylene/HOAc → vinyl acetate, methanol → formaldehyde, acetaldehyde → acetic acid.

[35]Although they are redox enzymes belonging to EC class 1, occasionally they have been misleadingly called 'oxygen transferases'.

not occur. Oxidases include flavoprotein oxidases (such as alcohol, amine and nicotinamide oxidases), metallo-flavin oxidases (aldehyde oxidase) and heme-protein oxidases (catalase, H_2O_2-specific peroxidases [1180]).

- In a related fashion, per*oxidases* catalyze two-electron oxidations at the expense of hydrogen peroxide forming water.
- Per*oxygenases* incorporate O into a substrate from H_2O_2.

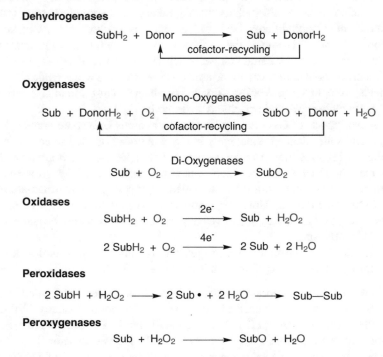

Dehydrogenases

$$SubH_2 + Donor \longrightarrow Sub + DonorH_2$$
cofactor-recycling

Oxygenases

Mono-Oxygenases
$$Sub + DonorH_2 + O_2 \longrightarrow SubO + Donor + H_2O$$
cofactor-recycling

Di-Oxygenases
$$Sub + O_2 \longrightarrow SubO_2$$

Oxidases

$$SubH_2 + O_2 \xrightarrow{2e^-} Sub + H_2O_2$$

$$2\,SubH_2 + O_2 \xrightarrow{4e^-} 2\,Sub + 2\,H_2O$$

Peroxidases

$$2\,SubH + H_2O_2 \longrightarrow 2\,Sub\bullet + 2\,H_2O \longrightarrow Sub-Sub$$

Peroxygenases

$$Sub + H_2O_2 \longrightarrow SubO + H_2O$$

Scheme 2.144 Systematics of enzymatic oxidation reactions (donor = nicotinamide)

Monooxygenases

Although the reaction mechanisms of various monooxygenases differ greatly depending on the subtype of enzyme, their mode of oxygen-transfer is the same: Whereas one of the oxygen atoms from O_2 is transferred onto the substrate, the other is reduced to form a water molecule. The latter requires two electrons, which are derived from a cofactor, usually NADH or NADPH, serving as 'donor' (Scheme 2.144).

The net reaction and a number of synthetically useful monooxygenation reactions are shown in Scheme 2.145.

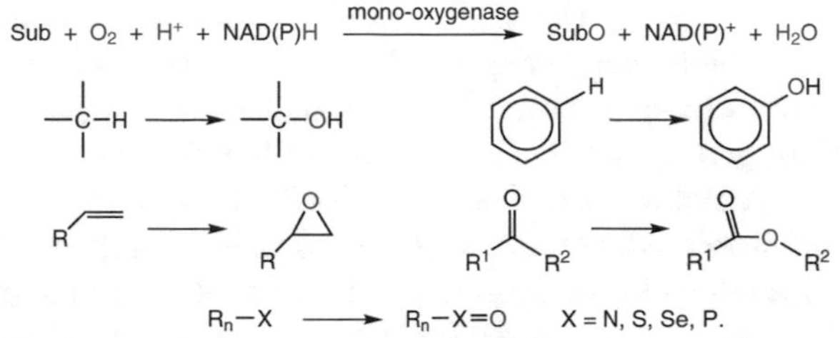

$$Sub + O_2 + H^+ + NAD(P)H \xrightarrow{\text{mono-oxygenase}} SubO + NAD(P)^+ + H_2O$$

$$-\overset{|}{\underset{|}{C}}-H \longrightarrow -\overset{|}{\underset{|}{C}}-OH$$

$$R_n-X \longrightarrow R_n-X=O \qquad X = N, S, Se, P.$$

Substrate	Product	Type of Reaction	Type of Cofactor
alkane	alcohol	hydroxylation	metal-dependent
aromatic	phenol	hydroxylation	metal-dependent
alkene	epoxide	epoxidation	metal-dependent
heteroatom[a]	heteroatom-oxide	heteroatom oxidation	flavin-dependent
ketone	ester/lactone	Baeyer-Villiger	flavin-dependent

[a] N, S, Se or P.

Scheme 2.145 Monooxygenase catalyzed reactions and their typical (but not exclusive) cofactor-dependence

The generation of the activated oxygen-transferring species is mediated either by cofactors containing a transition metal (Fe or Cu) or by a heteroaromatic system (a pteridin [1181] or flavin [1182–1184]). The catalytic cycle of the iron-depending monooxygenases, the majority of which belong to the cytochrome P-450 type (Cyt P-450) [1185–1189], has been deduced largely from studies on the camphor hydroxylase of *Pseudomonas putida* [1190, 1191]. A summary of the catalytic cycle is depicted in Scheme 2.146.

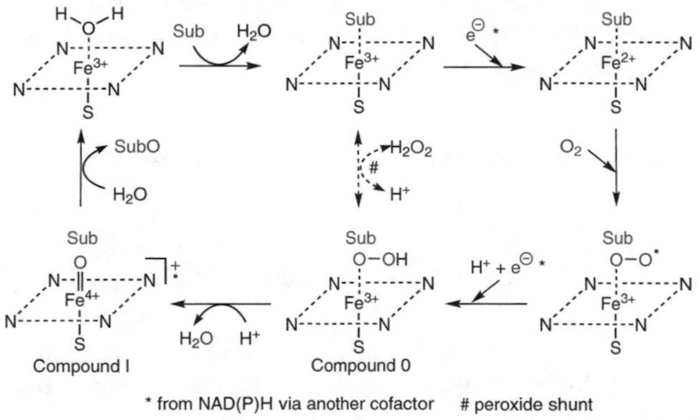

* from NAD(P)H via another cofactor   # peroxide shunt

Scheme 2.146 Catalytic cycle of cytochrome P-450-dependent monooxygenases

The iron species is coordinated equatorially by a heme moiety and axially by the sulfur atom of a cysteine residue. Catalysis occurs in the remaining sixth coordination site. After binding of the substrate (Sub) by replacing a water molecule in a hydrophobic pocket adjacent to the porphine [1192], the iron is reduced to the ferrous state (Fe^{3+} → Fe^{2+}). The single electron is delivered from NAD(P)H via another cofactor, which (depending on the enzyme) is a flavin, an iron-sulfur protein (ferredoxin) or a cytochrome b$_5$. Next, molecular oxygen is bound to give a Cyt P-450 dioxygen complex. Delivery of a second electron and protonation forms Compound 0. Protonation cleaves the O–OH bond with expulsion of water and forms the ultimate oxidizing Fe^{4+}=O species called Compound I, which – as a strong electrophile – attacks the substrate [1193]. Expulsion of the product (SubO) reforms the Fe^{3+} species and closes the catalytic cycle. Put simply, Cyt P-450 resembles an oxidation by a hypervalent transition metal oxidant (nature's permanganate).

Despite the fact that the mechanism of Cyt P-450 enzymes has been intensively investigated over half a century [1194], many mechanistic details are still poorly understood and it was only recently, that the existence of an Fe^{5+} species was ruled out [1195].

Aside from the productive cycle, Compound 0 can be formed directly by H$_2$O$_2$ through the so-called 'peroxide-shunt'. This obviates the necessity for additional electron-transport components described above, because no single-electron transfer occurs. However, so far, the use of P-450 enzymes in the peroxygenase-mode is impeded by limited enzyme stabilities in presence of H$_2$O$_2$ [1196]. Under certain conditions, Compound 0 may liberate H$_2$O$_2$, or Compound I may decompose forming H$_2$O, which wastes NAD(P)H in futile cycles, processes which are called 'uncoupling'. Hence, it is not surprising, that P-450 enzymes are comparatively slow catalysts with typical TOFs of ~1 s^{-1}.

Cyt P-450 enzymes got their name from their hemoprotein character: P stands for 'pigment' and 450 reflects the absorption of the CO-complex at 450 nm. To date, more than 200,000 distinct Cyt P-450 enzymes are known and these proteins are classified into four major groups (bacterial, mitochondrial, microsomal and self-sufficient Cyt) according to the mode of the electron-transport and the interaction between the subunits [1197]. A simplified schematic organization of Cyt P-450 systems is depicted in Fig. 2.17.

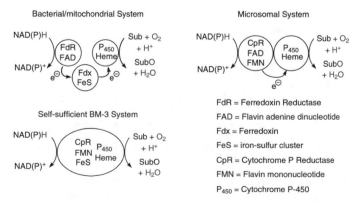

Fig. 2.17 Schematic organization and electron-transport of cytochrome P-450 monooxygenases

Bacterial and mitochondrial Cyt P-450 systems depend on three proteins: the P-450 monooxygenase with its heme unit, which performs the actual oxygenation of the substrate, a ferredoxin reductase, which accepts hydride equivalents from nicotinamide via an FAD cofactor and ferredoxin, which acts as electron shuttle between them using an iron-sulfur cluster as electron carrier [1198]. The microsomal system is somewhat simpler, as electron transfer occurs directly between the cytochrome P reductase (possessing an FMN and FAD cofactor) and the Cyt P-450 enzyme and thus does not require the ferredoxin. The minimal Cyt P-450 system BM-3 is derived from *Bacillus megaterium* and it consists of a single (fusion) protein, which is made up of two domains, a cytochrome P reductase (containing FMN and the FeS cluster) and the P-450 enzyme. It is evident, that for its simplicity the latter system has been the prime target of studies directed towards the development of enzymatic oxygenation systems for preparative-scale applications [1199–1201].

Due to their inherent complexity of the electron-transport chain [1202], Cyt P450 monooxygenase-catalyzed systems are generally employed as whole microbial host cells ('designer bugs'), which co-express all the required proteins, including those required for NAD(P)H-recycling, in particular when applied to large-scale reactions [1203–1205].

In contrast, flavin-dependent monooxygenases (see Scheme 2.147 and Table of Scheme 2.145) use a different mechanism which involves a flavin cofactor [1206–1208]. First, NADPH reduces the Enz-FAD complex thereby breaking its aromaticity. The $FADH_2$ so formed is oxidized by molecular oxygen via Michael-type addition yielding a hydroperoxide (FAD-4a-OOH) or peroxyflavin-species (FAD-4a-OO$^-$), depending on its protonation state, which is determined by the molecular environment of the enzyme's active site. The latter can either perform an electrophilic or nucleophilic oxidation, such as alkene epoxidation or Baeyer-Villiger oxidations, respectively [1209]. In contrast to P-450 enzymes, flavin-dependent monooxygenases show negligible uncoupling.

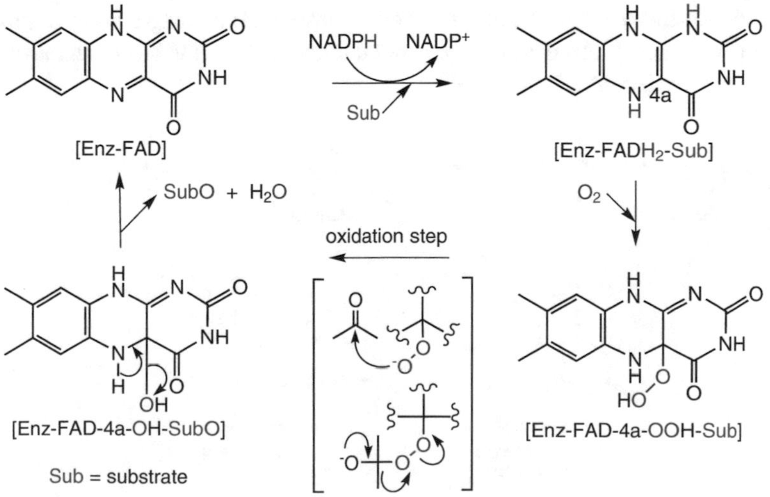

Scheme 2.147 Catalytic cycle of flavin-dependent monooxygenases

In Baeyer-Villiger oxidations, the peroxy-anion performs a nucleophilic attack on the carbonyl group of the aldehyde or ketone substrate. The tetrahedral species thus formed (corresponding to a Criegee-intermediate) collapses via rearrangement of the carbon-framework forming the product ester or lactone, respectively. Finally, water is eliminated from the FAD-4a-OH species to reform FAD. In addition to the Baeyer-Villiger oxidation, the flavin-4a-hydroperoxy species can also mediate the hydroxylation of aromatics [1210–1214], alkene epoxidation [1215, 1216] and heteroatom-oxidation, desaturation and oxidative C–C coupling [1217–1219]. Put simply, the FAD-OOH species resembles nature's m-chloroperbenzoic acid and the mechanism parallels the oxidation of organic compounds by peroxides or peracids [1220]. Flavin-dependent monooxygenases, such as Baeyer-Villigerases, can be used in isolated form together with a suitable nicotinamide-cofactor recycling system.

2.3.3.1 Hydroxylation of Alkanes

The hydroxylation of nonactivated centers in hydrocarbons is one of the most useful biotransformations [1175, 1221–1226] due to the fact that this process has only very few counterparts in traditional organic synthesis [1227–1229]. In general, the relative reactivity of carbon atoms in bio-hydroxylation reactions declines in the order of secondary > tertiary > primary [1230], which is in contrast to radical reactions (tertiary > secondary > primary) [1231].

Straight-chain hydrocarbons are preferably hydroxylated at the (ω-1)-position, and with whole cells, the corresponding 2-alkanol is quickly oxidized to the corresponding methyl ketone, which further undergoes Baeyer-Villiger oxidation forming the corresponding *prim*-acetate ester. The latter is hydrolysed to acetate and a *prim*-alcohol, both of which are channeled into the β-oxidation pathway yielding CO_2 and water (Scheme 2.148, top). Due to the inability to stop this metabolic pathway for n-alkanes at a certain intermediate stage, is of no practical value for synthetic biotransformations, but it takes care for the bioremediation of oil spills in the environment.

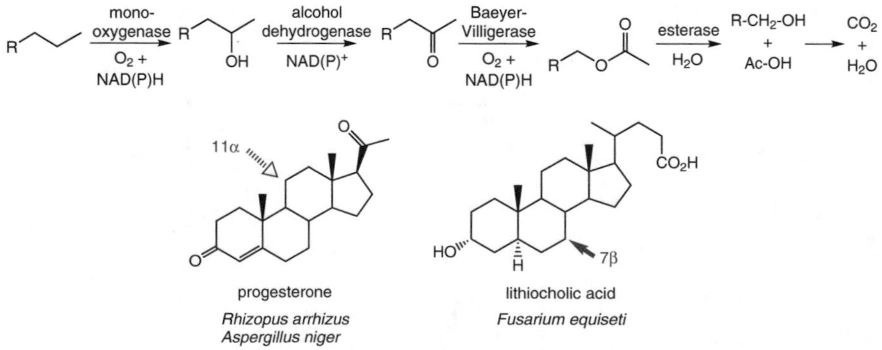

Scheme 2.148 Regio- and stereoselective microbial hydroxylation of steroids

In contrast, there are two main groups of hydrocarbon molecules, which can be efficiently transformed by microbial hydroxylation – *steroids* and *terpenoids*. Their common property is that they possess a large main framework, which impedes the metabolic degradation of their hydroxylated products.

Intense research on the stereoselective hydroxylation of alkanes started in the late 1940s in the steroid field, driven by the demand for pharmaceuticals [1232–1237]. In the meantime, some of the hydroxylation processes, e.g., 9α- and 16α-hydroxylation of the steroid framework [1238, 1239], have been developed to the scale of industrial production. Nowadays, virtually any center in a steroid can be selectively hydroxylated by choosing the appropriate microorganism, for a comprehensive list see [1239]. For example, hydroxylation of progesterone in the 11α-position by *Rhizopus arrhizus* [1240] or *Aspergillus niger* [1241] made roughly half of the 37 steps of the conventional chemical synthesis redundant and made 11α-hydroxyprogesterone available for hormone therapy at a reasonable cost (Scheme 2.148, bottom). A highly selective hydroxylation of lithiocholic acid in position 7β was achieved by using *Fusarium equiseti* [1242]. The product (ursodeoxycholic acid) is capable of dissolving cholesterol and thus can be used in the therapy of gallstones.

In search for new drugs, active pharmaceutical ingredients (APIs) are often subjected to microbial hydroxylation. For instance, the regioselective allylic hydroxylation of the potent cholesterol-lowering drug simvastatin was achieved using *Nocardia autotrophica* to yield 6-β-hydroxy-simvastatin together with some minor side-products [1243]. An impressive amount of 15 kg of product was obtained from a 19 m^3 reactor (Scheme 2.149).

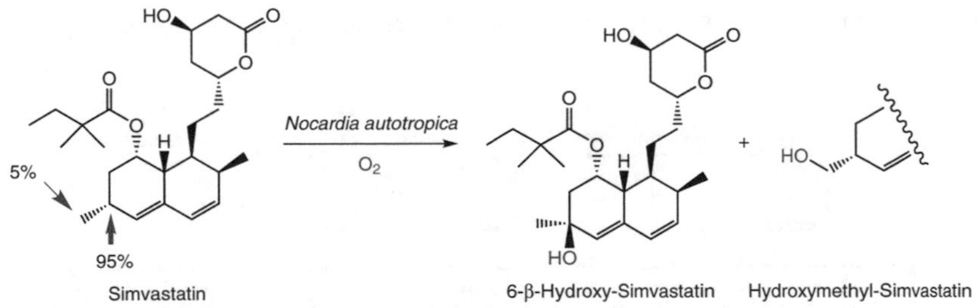

Scheme 2.149 Regioselective microbial hydroxylation of HMG-CoA reductase inhibitor Simvastatin

Optically active β-hydroxy-*iso*butyric acid has been used as a starting material for the synthesis of vitamins (α-tocopherol [1244]), fragrance components (muscone [1245]) and antibiotics (calcimycin [1246]). Both enantiomers may be obtained by asymmetric hydroxylation of *iso*-butyric acid [1247, 1248] (Scheme 2.150). An intensive screening program using 725 strains of molds, yeasts and bacteria revealed that, depending on the microorganism, either the (*R*)- or the (*S*)-β-hydroxy-*iso*-butyric acid was formed in varying optical purity. Best results were obtained using selected *Candida* and *Pseudomonas* strains.

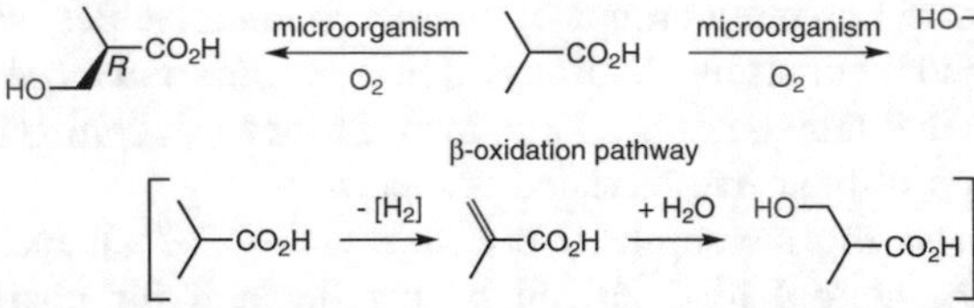

Scheme 2.150 Asymmetric microbial hydroxylation of *iso*butyric acid via the β-oxidation pathway

Microorganism	Configuration	e.e. [%]
Pseudomonas putida ATCC 21244	S	>95
Candida rugosa IFO 750	S	99
Candida rugosa IFO 1542	R	97

Although the mechanism of this reaction was initially assumed to be a 'direct' hydroxylation at position β, detailed studies showed that it proceeds via the conventional β-oxidation pathway involved in fatty acid metabolism [1249]. An analogous sequence is the basis for the transformation of 4-trimethylammonium butanoate to the β-hydroxy derivative carnitine (compare Scheme 2.203) [1250].

The production of new olfactory compounds for the aroma and fragrance industry was the powerful driving force in the research on the hydroxylation of terpenes [1008, 1251–1253]. For instance, 1,4-cineole, a major constituent of eucalyptus oil, was regioselectively hydroxylated by *Streptomyces griseus* to give 8-hydroxycineole as the major product along with minor amounts of *exo*- and *endo*-2-hydroxy derivatives, with low optical purity (Scheme 2.151) [1254]. On the other hand, when *Bacillus cereus* was used, (2R)-*exo*- and (2R)-*endo*-hydroxycineoles were exclusively formed in a ratio of 7:1, both in excellent enantiomeric excess [1255].

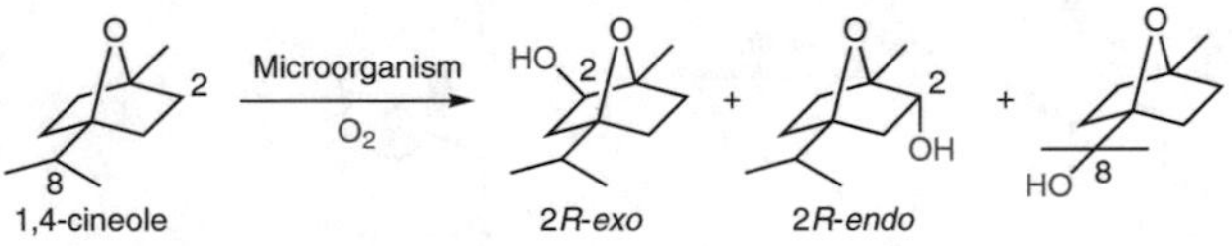

Microorganism	e.e. [%] *exo*	e.e. [%] *endo*	*exo*/*endo* ratio
Streptomyces griseus	46	74	1:1.7[a]
Bacillus cereus	94	94	7:1

[a] 8-Hydroxycineole was the major product.

Scheme 2.151 Microbial hydroxylation of 1,4-cineole

Among the many hundreds of microorganisms tested for their capability to perform hydroxylation of nonnatural aliphatic compounds, fungi have been more often used than bacteria. Among them, the fungus *Beauveria bassiana* ATCC 7159 (formerly denoted as *B. sulfurescens*) has been studied most thoroughly [1256–1260]. In general the presence of a polar group in the substrate such as an acetamide, benzamide or *p*-toluene-sulfonamide moiety proved to be advantageous in order to firmly bind the substrate in the active site [1261]. Hydroxylation occurs at a distance of 3.3–6.2 Å from the polar anchor group. With cycloalkane rings of different size, hydroxylation preferentially occurred in the order cycloheptyl > cyclohexyl > cyclopentyl.

In the majority of cases, hydroxylation by *Beauveria bassiana* occurs in a *regioselective* manner, but high *enantioselectivity* is not always observed. As shown in Scheme 2.152, both enantiomers of the *N*-benzyl-protected bicyclic lactam are hydroxylated with high regioselectivity in position 11, but the reaction showed very low enantioselectivity. On the other hand, when the lactam moiety was replaced by a sterically more accessible polar benzoyl-amide, which functions as polar anchor group, high enantiodifferentiation occurred. The (1*R*)-enantiomer was hydroxylated at carbon 12 and the (1*S*)-counterpart gave the 11-hydroxylated product [1262]. A minor amount of 6-*exo*-alcohol was formed with low enantiomeric excess.

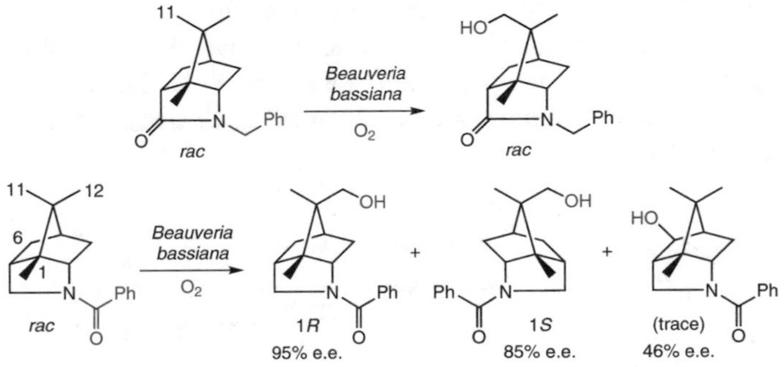

Scheme 2.152 Regio- and enantioselective hydroxylation by *Beauveria bassiana*

In order to provide a tool to predict the stereochemical outcome of hydroxylations using *Beauveria bassiana*, an active site model [1263, 1264] and a substrate model containing a polar anchor group were developed [1265].

In summary, (bio)hydroxylation of sterically demanding hydrocarbon compounds is feasible by using one of the many microorganisms used to date, but it is difficult to predict the likely site of oxidation for any novel substrate using monooxygenases. However, there are three strategies which can be employed to improve regio- and/or stereoselectivity in biocatalytic hydroxylation procedures:

- Broad screening of different strains[36]
- Substrate modification, particularly by introduction of a polar anchor group [1266–1269]
- Variation of the culture by stressing the metabolism of the cells

---

[36]The following strains have been used more frequently: *Aspergillus niger, Cunninghamella blakesleeana, Bacillus megaterium, Bacillus cereus, Mucor plumbeus, Mortierella alpina, Curvularia lunata, Helminthosporium sativum, Pseudomonas putida, Rhizopus arrhizus, Rhizopus nigricans, Beauveria bassiana.*

2.3.3.2 Hydroxylation of Aromatic Compounds

Regiospecific hydroxylation of aromatic compounds by purely chemical methods is notoriously difficult. There are reagents for *o*- and *p*-hydroxylation available [1270, 1271], but some of them are explosive and byproducts are usually obtained [1272]. The selective bio-hydroxylation of aromatics in the *o*- and *p*-position to existing substituents can be achieved by using monooxygenases. In contrast, *m*-hydroxylation is rarely observed for electronic reasons [1273]. Mechanistically, it has been proposed that in eukaryotic cells (fungi, yeasts and higher organisms) the reaction proceeds predominantly via epoxidation of the aromatic species which leads to an unstable arene-oxide (Scheme 2.83) [1274]. Rearrangement of the latter involving the migration of a hydride anion (NIH-shift) forms the phenolic product [1275].

Like steroids and terpenes, aromatic compounds are regioselectively hydroxylated by using whole cells [1276–1280]. For instance, 6-hydroxynicotinic acid is produced from nicotinic acid by *Pseudomonas acidovorans* or *Achromobacter xylosoxidans* on a scale of 20 t/a [1281]. Racemic prenalterol, a compound with important pharmacological activity as a β-blocker, was obtained by regioselective *p*-hydroxylation of a simple aromatic precursor using *Cunninghamella echinulata* (Scheme 2.153) [1282].

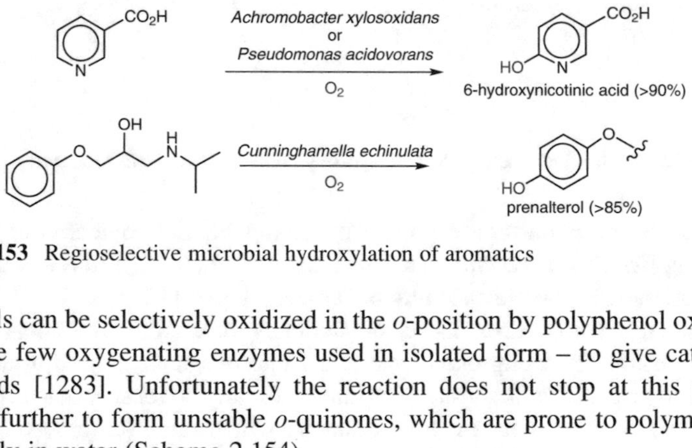

Scheme 2.153 Regioselective microbial hydroxylation of aromatics

Phenols can be selectively oxidized in the *o*-position by polyphenol oxidase[37] – one of the few oxygenating enzymes used in isolated form – to give catechols in high yields [1283]. Unfortunately the reaction does not stop at this point but proceeds further to form unstable *o*-quinones, which are prone to polymerization, particularly in water (Scheme 2.154).

Two techniques have been developed to solve the problem of *o*-quinone instability:

- One way to prevent *o*-quinone formation is by maintaining a reducing environment by addition of ascorbic acid, which prevents the over-oxidation and leads to the accumulation of catechols. Ascorbate, however, like many other reductants can act as an inhibitor of polyphenol oxidase and hence the concentration

---

[37] Also called tyrosinase, catechol oxidase, cresolase.

of the reducing agent must be kept at a minimum. In addition, a borate buffer which leads to the formation of a catechol-borate complex is advantageous [1284].

- Polymerization, which requires the presence of water, can be avoided if the reaction is performed in a lipophilic organic solvent such as chloroform (Sect. 3.1.6) [1285].

The following rules for phenol hydroxylation have been deduced for polyphenol oxidase:

- A remarkable range of simple phenols are accepted, as long as the substituent R is in the *p*-position; *m*- and *o*-derivatives are unreactive, some electron-rich nonphenolic species such as *p*-toluidine are accepted.
- For electronic reasons, the reactivity decreases if the nature of the R group is changed from electron-donating to electron-withdrawing.
- Bulky phenols (*p-tert*-butylphenol and 1- or 2-naphthols) are not substrates.

R = H-, Me-, MeO-, HO_2C-$(CH_2)_2$-, HO-$(CH_2)_{1,2}$-, PhCO-$NHCH_2$-

Scheme 2.154 *o*-Hydroxylation of phenols by polyphenol oxidase

The synthetic utility of this reaction was demonstrated by the oxidation of amino acids and -alcohols containing an electron-rich *p*-hydroxyphenyl moiety (Scheme 2.154). Thus, L-DOPA (3,4-dihydroxyphenyl alanine) used for the treatment of Parkinson's disease, D-3,4-dihydroxy-phenylglycine and L-epinephrine (adrenaline) were synthesized from their *p*-monohydroxy precursors without racemization in good yield.

2.3.3.3   Epoxidation of Alkenes

Chiral epoxides are extensively employed high-value intermediates in the synthesis of chiral compounds due to their ability to react with a broad variety of nucleophiles. In recent years a lot of research has been devoted to the development of catalytic methods for their production [611, 1286]. The Katsuki-Sharpless method for the asymmetric epoxidation of allylic alcohols [1287, 1288] and the Jacobsen-catalysts for the epoxidation of nonfunctionalized olefins are now widely applied and reliable procedures [615, 1289]. Although high selectivities have been achieved for the epoxidation of *cis*-alkenes, the selectivities achieved with *trans*- and terminal olefins were less satisfactory using the latter methods.

In contrast, the strength of enzymatic epoxidation, catalyzed by monooxygenases, is in the preparation of small and nonfunctionalized epoxides, where traditional methods are limited [617, 1290]. Despite the wide distribution of monooxygenases within all types of organisms, their capability to epoxidize alkenes seems to be associated mainly with alkane- and alkene-utilizing bacteria, whereas fungi are applicable to a lesser extent [1175, 1291–1295].

Like C-H hydroxylation, epoxidation of alkenes catalyzed by monooxygenases is preferably performed on a preparative scale with whole microbial cells. Toxic effects of the epoxide formed, which accumulates in the cells, where it reacts with cellular enzymes, and its further (undesired) metabolism catalyzed by epoxide hydrolases in whole cells (Sect. 2.1.5) can be minimized by employing biphasic media. Alternatively, the alkene itself can constitute the organic phase into which the product is removed, away from the cells. However, the bulk apolar phase tends to damage the cell membranes, which reduces and eventually abolishes all enzyme activity [1296]. In any case, these methods require bioengineering skills [1297].

Once the problems of product toxicity were surmounted by sophisticated process engineering, microbial epoxidation of alkenes became also feasible on an industrial scale [1298, 1299]. The latter was achieved by using organic-aqueous two-phase systems or by evaporation of volatile epoxides. For instance, the epoxy-phosphonic acid derivative 'fosfomycin' [1300], whose enantiospecific synthesis by classical methods would have been extremely difficult, was obtained by a microbial epoxidation of the corresponding olefinic substrate using *Penicillium spinulosum*.

The most intensively studied microbial epoxidizing agent is the ω-hydroxylase system of *Pseudomonas oleovorans* [1301, 1302]. It consists of three protein components: the actual nonheme iron ω-hydroxylase and the electron-transport chain consisting of rubredoxin and NADH-dependent rubredoxin reductase. It catalyzes not only the hydroxylation of aliphatic C–H bonds, but also the epoxidation of alkenes [1303, 1304]. The following rules can be formulated for epoxidations using *Pseudomonas oleovorans* (Scheme 2.155).

- Terminal, acyclic alkenes of moderate chain length (e.g. 1-octene) are converted into (*R*)-1,2-epoxides of high enantiomeric excess along with varying amounts of ω-en-1-ols or 1-als [1305], the ratio of which depends on the chain length of the substrate [1306, 1307]. In contrast, alkane hydroxylation predominates over epoxidation for short (propene, 1-butene) and long-chain olefins.
- α,ω-Dienes are transformed into the corresponding terminal (*R,R*)-bis-epoxides.

- Cyclic, branched and internal olefins, aromatic compounds and alkene units which are conjugated to an aromatic system are not epoxidized [1308].
- To avoid problems arising from the toxicity of the epoxide [1309] a water-immiscible organic cosolvent such as hexane can be added [1310, 1311].

Besides *Pseudomonas oleovorans* numerous bacteria have been shown to epoxidize alkenes [1312, 1313]. As shown in Scheme 2.155, the optical purity of epoxides depends on the strain used, although the absolute configuration is usually (*R*) [1314]. This concept has been applied to the synthesis of chiral alkyl and aryl gycidyl ethers [1315, 1316]. The latter are of interest for the preparation of enantiopure 3-substituted 1-alkylamino-2-propanols, which are widely used as β-adrenergic receptor-blocking agents [1317].

The structural restrictions for substrates elaborated for *Pseudomonas oleovorans* (see above) could be overcome by using different microorganisms. As can be seen from Scheme 2.155, nonterminal alkenes can be epoxidized by *Mycobacterium* or *Xanthobacter* spp. [1318]. On the other hand, *Nocardia corallina* converted branched alkenes into the corresponding (*R*)-epoxides in good optical purities (Scheme 2.156). Aiming at the improvement of the efficiency of microbial epoxidation protocols, a styrene monooxygenase (StyA) and reductase StyB required for electron-transport were co-expressed into *E. coli* to furnish a designer-bug for the asymmetric epoxidation of styrene-type substrates [1319, 1320].

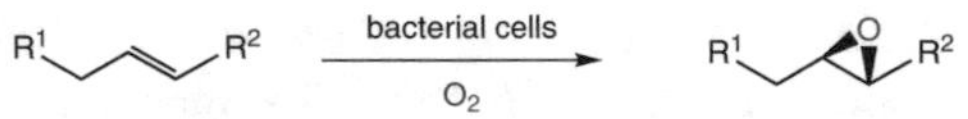

$$R^1 \diagdown \diagup R^2 \xrightarrow[\text{O}_2]{\text{bacterial cells}} R^1 \diagdown \triangleleft^{\text{O}} R^2$$

Microorganism	R^1	R^2	Configuration	e.e. [%]
Pseudomonas	$n\text{-C}_5\text{H}_{11}$	H	*R*	70-80
oleovorans	H	H	*R*	86
	$\text{NH}_2\text{CO-CH}_2\text{-C}_6\text{H}_4\text{-O}$	H	S^a	97
	$\text{CH}_3\text{O(CH}_2)_2\text{-C}_6\text{H}_4\text{-O}$	H	S^a	98
Corynebacterium	CH_3	H	*R*	70
equi	$n\text{-C}_{13}\text{H}_{27}$	H	*R*	~100
Mycobacterium	H	H	*R*	98
sp.	Ph-O	H	S^a	80
Xanthobacter	Cl	H	S^a	98
Py2	CH_3	CH_3	*R,R*	78
Nocardia sp. IP1	Cl	H	S^a	98
	CH_3	H	*R*	98

[a] Switch in CIP-sequence priority.

Scheme 2.155 Microbial epoxidation of alkenes

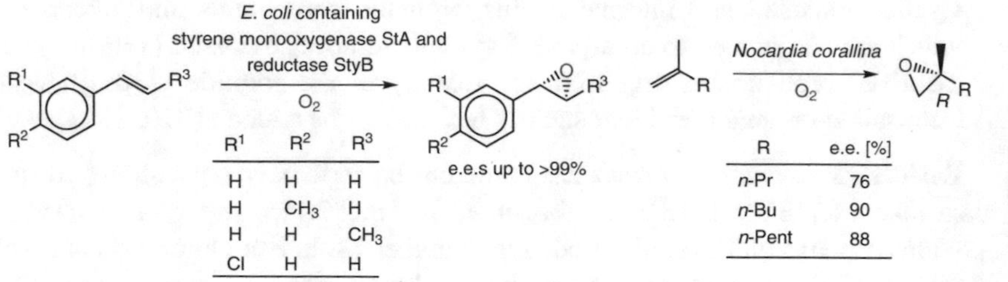

Scheme 2.156 Epoxidation of styrene derivatives and branched alkenes using cloned mono-oxygenase and *Nocardia corallina*

2.3.3.4 Sulfoxidation Reactions

Chiral sulfoxides are not only common pharmacophores in active pharmaceutical ingredients, but they have also been extensively employed as asymmetric auxiliary group that assist stereoselective reactions. The sulfoxide functional group activates adjacent carbon–hydrogen bonds to allow proton abstraction by bases, and the corresponding anions can be alkylated [1321] or acylated [1322] with high diastereoselectivity. Similarly, thermal elimination [1323] and reduction of α-keto sulfoxides [1324] can proceed with transfer of chirality from sulfur to carbon. In spite of this great potential as valuable chiral relay reagents, with rare exceptions [1325], no general method is available for the synthesis of sulfoxides possessing high enantiomeric purities.

An alternative approach involves the use of enzymatic sulfur-oxygenation reactions catalyzed by monooxygenases [1326, 1327]. The main types of enzymatic sulfur oxygenation are shown in Scheme 2.157. The direct oxidation of a thioether by means of a dioxygenase, which directly affords the corresponding sulfone, is of no synthetic use since no generation of chirality is involved. On the other hand, the stepwise oxidation involving a chiral sulfoxide, which is catalyzed by monooxygenases or peroxidases,[38] offers two possible ways of obtaining chiral sulfoxides.

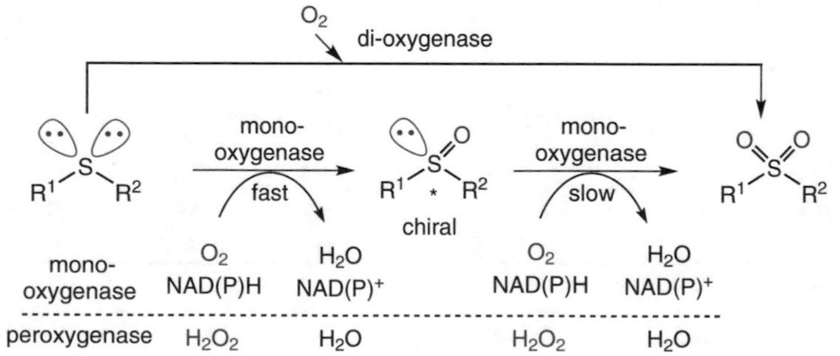

Scheme 2.157 Enzymatic sulfur oxygenation reactions

---

[38]Since O is incorporated into the substrate, peroxidases performing thioether oxidation should be correctly termed as 'peroxygenases', however, this distinction is often not made.

- The asymmetric monooxidation of a thioether leading to a chiral sulfoxide resembles a desymmetrization of a prochiral substrate and is therefore of high synthetic value.
- The kinetic resolution of a racemic sulfoxide during which one enantiomer is oxidized to yield an achiral sulfone is feasible but it has been shown to proceed with low selectivities.

The first asymmetric sulfur oxygenation using cells of *Aspergillus niger* was reported in the early 1960s [1328]. Since this time it was shown that the enantiomeric excess and the absolute configuration of the sulfoxide not only depend on the species but also on the strain of microorganism used [1329]. In general, the formation of (*R*)-sulfoxides predominates.

Thioethers can be asymmetrically oxidized both by bacteria (e.g., *Corynebacterium equi* [1330], *Rhodococcus equi* [1331]) and fungi (e.g., *Helminthosporium* sp. [1332] and *Mortierella isabellina* [1333]). Even baker's yeast has this capacity [1334, 1335]. As shown in Scheme 2.158, a large variety of aryl-alkyl thioethers were oxidized to yield sulfoxides with good to excellent optical purities [1336–1338]. The second oxidation step was usually negligible, but with certain substrates the undesired formation of the corresponding sulfone was observed.

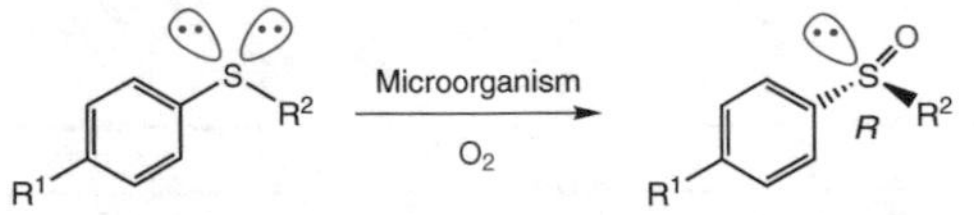

Microorganism	R^1	R^2	e.e. [%]
Mortierella	$(CH_3)_2CH$	CH_3	82
isabellina	H	$(CH_3)_2CH$	83
	H	C_2H_5	85
	C_2H_5	CH_3	90
	H	$n\text{-}C_3H_7$	~100
	Br	CH_3	~100[a]
Corynebacterium	H	CH_3	92
equi	CH_3	CH_3	97
	H	$n\text{-}C_4H_9$	~100
	H	$CH_2\text{-}CH=CH_2$	~100
baker's yeast	CH_3	CH_3	92

[a] Some sulfone was formed in this case.

Scheme 2.158 Microbial oxidation of aryl-alkyl thioethers

The transformation of thioacetals into mono- or bis-sulfoxides presents intriguing stereochemical possibilities. In a symmetric thioacetal of an aldehyde other than formaldehyde, the sulfur atoms are enantiotopic and each of them contains two diastereotopic nonbonded pairs of electrons (Scheme 2.159). Unfortunately, most of the products from asymmetric oxidation of thioacetals are of low to moderate optical purity [1339, 1340]. Two exceptions, however, are worth mentioning. Oxidation of 2-*tert*-butyl-1,3-dithiane by *Helminthosporium* sp. gave the (1*S*,2*R*)-monosulfoxide in 72% optical purity [1341] and formaldehyde thioacetals were oxidized by *Corynebacterium equi* to yield (*R*)-sulfoxide-sulfone products [1342] with excellent e.e.

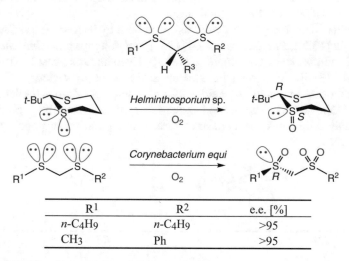

R^1	R^2	e.e. [%]
n-C$_4$H$_9$	*n*-C$_4$H$_9$	>95
CH$_3$	Ph	>95

Scheme 2.159 Microbial oxidation of dithioacetals

In order to avoid poor recoveries of the water-soluble sulfoxide from considerable amounts of biomass when using whole cells, isolated (flavin-dependent) monooxygenases have been successfully employed for thioether-oxidations together with NAD(P)H-recycling (for biocatalytic sulfur oxidation using peroxidases, see Sect. 2.3.4). In particular, cyclohexanone monooxygenase (CHMO) [1343–1345], phenylacetone monooxygenase (PAMO) [1346] and hydroxyacetone monooxygenase (HAPMO) [1347, 1348] were shown to be most useful. Overall, recoveries and stereoselectivities were high, and some enzymes exhibited stereo-complementary properties on selected substrates (Scheme 2.160).

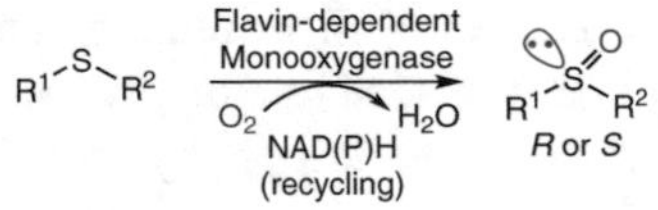

R^1	R^2	Enzyme*	Config.	Conv. [%]	E.e. [%]
Me	c-Hexyl	CHMO	R	80	98
t-Bu	CH=CH$_2$	CHMO	R	78	98
Me	Ph	CHMO	R	88	99
Et	p-F-C$_6$H$_4$	CHMO	S	96	93
CH$_2$-CN	Ph	CHMO	R	90	92
CH$_2$-CN	p-Me-C$_6$H$_4$	CHMO	S	95	98
Me	c-Hexyl	HAPMO	S	>99	>99
Me	Ph	HAPMO	S	>97	>99
Me	2-Pyridyl	HAPMO	S	>99	>99
Me	4-Pyridyl	HAPMO	R	63	>99

* CHMO = Cyclohexanone monooxygenase, HAPMO = hydroxyacetone monooxygenase.

Scheme 2.160 Enzymatic oxidation of thioethers using isolated monooxygenases

2.3.3.5   Baeyer-Villiger Reactions

Oxidation of ketones by peracids – the Baeyer-Villiger reaction [1349, 1350] – is a reliable and useful method for preparing esters or lactones (Scheme 2.161). The mechanism comprises a two-step process, in which the peracid attacks the carbonyl group of the ketone to form the so-called tetrahedral 'Criegee-intermediate' [1351]. The fragmentation of this unstable species, which proceeds via expulsion of a carboxylate ion going in hand with migration of a carbon–carbon bond, leads to the formation of an ester or a lactone. The regiochemistry of oxygen insertion of the chemical and the enzymatic Baeyer-Villiger reaction can usually be predicted by assuming that the carbon atom best able to support a positive charge will migrate preferentially, i. e. *tert*-alkyl > *sec*-alkyl ~ phenyl > *prim*-alkyl > methyl [1352].

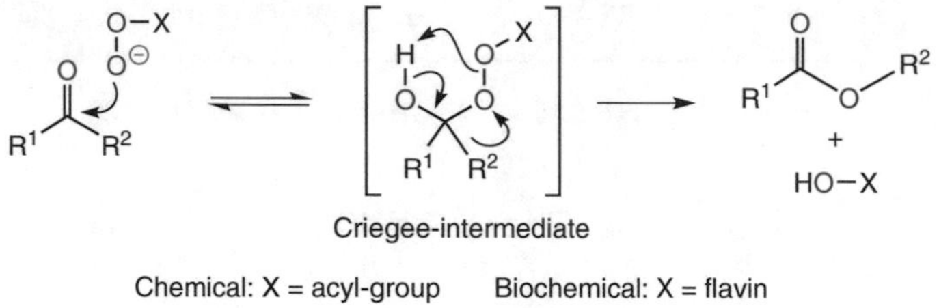

Criegee-intermediate

Chemical: X = acyl-group   Biochemical: X = flavin

Scheme 2.161 Mechanism of the chemical and biochemical Baeyer-Villiger oxidation

All mechanistic studies on enzymatic Baeyer-Villiger reactions support the hypothesis that conventional and enzymatic reactions are closely related [1206, 1353]. The oxidized flavin cofactor peroxy species (FAD-4a-OO$^-$, see Scheme 2.147) plays the role of a nucleophile similar to the peracid. The strength of enzyme-catalyzed Baeyer–Villiger reactions resides in the recognition of chirality [1354–1356], which has been accomplished by conventional means with moderate selectivities [1357].

The enzymatic Baeyer-Villiger oxidation of ketones is catalyzed by flavin-dependent monooxygenases and plays an important role in the breakdown of carbon structures containing a ketone moiety (Scheme 2.148). Early studies were performed by using whole microbial cells, particularly in view of avoiding the necessity for NAD(P)H-recycling [1358]. However, whole-cell Baeyer–Villiger oxidations often suffer from low yields due to side reactions catalyzed by competing hydrolytic enzymes. Furthermore, some of the most potent strains, such as *Acinetobacter calcoaceticus*, are potentially pathogenic and therefore have to be handled with extra care (see the Appendix, Chap. 5).[39]

To avoid further degradation of esters and lactones in microbial Baeyer-Villiger reactions catalyzed by hydrolytic enzymes and to maximize product accumulation, the following approaches are possible:

- Blocking of the hydrolytic enzymes by selective hydrolase-inhibitors such as tetraethyl pyrophosphate (TEPP [1359]) or diethyl *p*-nitrophenylphosphate (paraoxon). However, all of these inhibitors are highly toxic and have to be handled with extreme caution.
- Development of mutant strains lacking lactone-hydrolases.
- Application of nonnatural ketones, whose lactone products are not substrates for the hydrolytic enzymes.
- Use of isolated Baeyer-Villigerases together with NAD(P)H-recycling is nowadays the method of choice.

---

[39]*Acinetobacter calcoaceticus* NCIMB 9871 is a class-II pathogen.

Prochiral (symmetric) ketones can be asymmetrically oxidized by a bacterial cyclohexanone monooxygenase from an *Acinetobacter* sp. to yield the corresponding lactones [1360, 1361]. As depicted in Scheme 2.162, oxygen insertion occurred on both sides of the ketone depending on the substituent in the 4-position. Whereas in the majority of cases products having the (*S*)-configuration were obtained, a switch to the (*R*)-lactone was observed with sterically demanding 4-*n*-butylcyclohexanone. Simple models are available, which allow the prediction of the stereochemical outcome of Baeyer-Villiger oxidations catalyzed by cyclohexanone monooxygenase of *Acinetobacter* and *Pseudomonas* sp. by determination of which group within the Criegee-intermediate is prone to migration [1362, 1363].

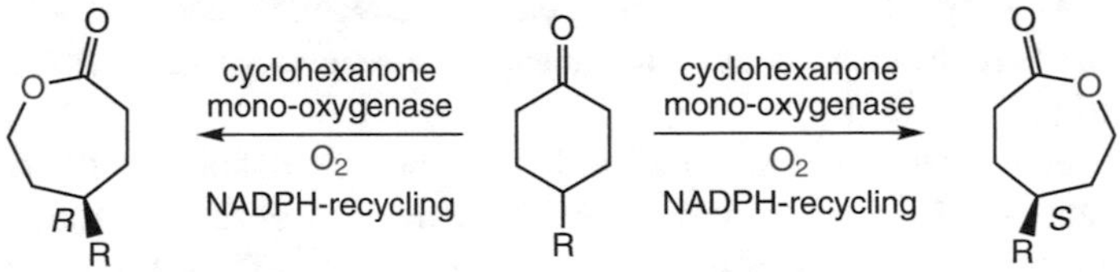

R	Configuration	e.e. [%]
CH₃-O-	S	75
Et-	S	>98
n-Pr-	S	>98
t-Bu-	S	>98
n-Bu-	R	52

Scheme 2.162 Desymmetrization of prochiral ketones via enzymatic Baeyer-Villiger oxidation

Racemic (nonsymmetric) ketones can be resolved via two pathways. The 'classic' form of a kinetic resolution involves a transformation in which one enantiomer reacts and its counterpart remains unchanged [1364]. For example, bicyclic haloketones, which were used for the synthesis of antiviral 6'-fluoro-carbocyclic nucleoside analogs, were resolved by using an *Acinetobacter* sp. [1365] (Scheme 2.163). Both enantiomers were obtained with >95% optical purity. Interestingly, the enantioselectivity of this microbial oxidation depends on the presence of the halogen atoms, since the dehalogenated bicyclo[2.2.1]heptan-2-one was transformed with low selectivity. On the other hand, replacement of the halogens by methoxy- or hydroxy groups gave rise to compounds which were not accepted as substrates.

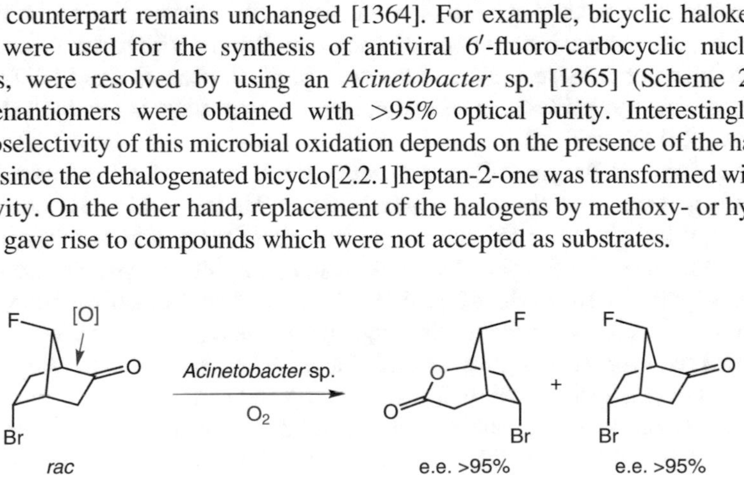

Scheme 2.163 Microbial Baeyer-Villiger oxidation of a bicyclic ketone involving 'classic' resolution

The biological Baeyer-Villiger oxidation of a racemic ketone does not have to follow the 'classic' kinetic resolution format as described above, but can proceed via a 'nonclassic' route involving oxidation of *both* enantiomers with opposite regioselectivity. Thus, oxygen insertion occurs on the *two opposite sides* of the ketone at each of the enantiomers. As shown in Scheme 2.164, *both* enantiomers of the bicyclo[3.2.0]heptenones were microbially oxidized, but in an *enantiodivergent* manner [1366, 1367]. Oxygen insertion on the (5*R*)-ketone occurred as expected, adjacent to C7, forming the 3-oxabicyclic lactone. On the other hand, the (5*S*)-ketone underwent oxygen insertion in the 'wrong sense' towards C5, which led to the 2-oxabicyclic species. The synthetic utility of this system has been proven by the large-scale oxidation using an *E. coli* designer bug harboring cyclohexanone monooxygenase together with a suitable NADPH-recycling enzyme [1368, 1369]. In order to minimize product toxicity, in-situ substrate-feeding product removal (SFPR) was applied [1370, 1371].

It has been shown that the molecular reasons of enantiodivergent Baeyer-Villiger reactions [1372, 1373] can either be the docking of the substrate in a single enzyme in two opposite modes or due to the presence of different monooxygenases present in the microbial cells [1374].

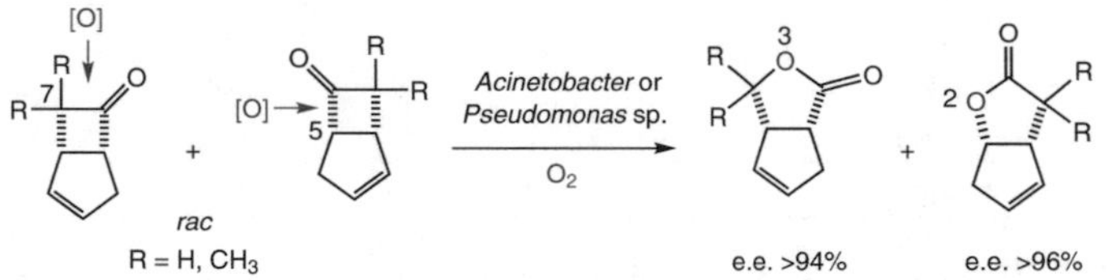

Scheme 2.164 Enantiodivergent microbial Baeyer-Villiger oxidation involving 'nonclassic' resolution

In order to overcome problems associated with whole-cell Baeyer-Villliger oxidations, an impressive number of bacterial 'Baeyer-Villigerases' possessing opposite stereopreference [1375, 1376] were purified, characterized [1377–1380] and cloned into a suitable (nonpathogenic) host, such as baker's yeast [1381–1385] or *E. coli* [1386].

The majority of Baeyer-Villigerases are NADPH-dependent, but several candidates (e. g. from *Pseudomonas putida*) accept NADH, which is more easily recycled [1387]. In order to facilitate cofactor recycling, a selfsufficient fusion protein consisting of a Baeyer-Villigerase and a phosphite dehydrogenase unit for NADH-recycling were designed [1388]. The overall performance of the fusion-protein was comparable to that of the single (non-fused) proteins. A clever concept of internal cofactor recycling for isolated Baeyer-Villigerases was developed using a coupled enzyme system [1389]. Thus, the substrate ketone is not used as such, but is rather produced by enzymatic oxidation of the corresponding alcohol (at the expense of NADP$^+$ or NAD$^+$, resp., Sect. 2.3.1) using a dehydrogenase from *Thermoanaerobium brockii* or from *Pseudomonas* sp. In a second step, the monooxygenase generates the lactone by consuming the reduced cofactor. Therefore, the NAD(P)H is concurrently recycled in a closed loop via 'hydrogen-borrowing' (Scheme 3.42).

Investigation of the regio- and enantioselectivity of Baeyer-Villigerases from bacteria from an industrial wastewater treatment plant cloned into *E. coli* revealed the following trends (Scheme 2.165):

- Prochiral 4-substituted cyclohexanones underwent desymmetrization yielding (*R*)- (e.e.$_{max}$ 60%) or (*S*)-4-alkyl-ε-caprolactones (e.e.$_{max}$ > 99%), depending on the enzyme used and on the size of the substituent.
- Racemic 2-substituted cyclohexanones underwent 'classic' kinetic resolution with absolute regioselectivity for oxygen-insertion at the predicted side to afford enantiomeric pairs of (*S*)-lactone and unreacted (*R*)-ketone with excellent enantioselectivities ($E \geq 200$).
- In contrast, 3-substituted cyclohexanones furnished 'non-classic' kinetic resolution via oxygen-insertion at both sides with different regioselectivities to furnish regio-isomeric lactones. The enantioselectivites depended on the enzyme used and on size of the substituent [1390].

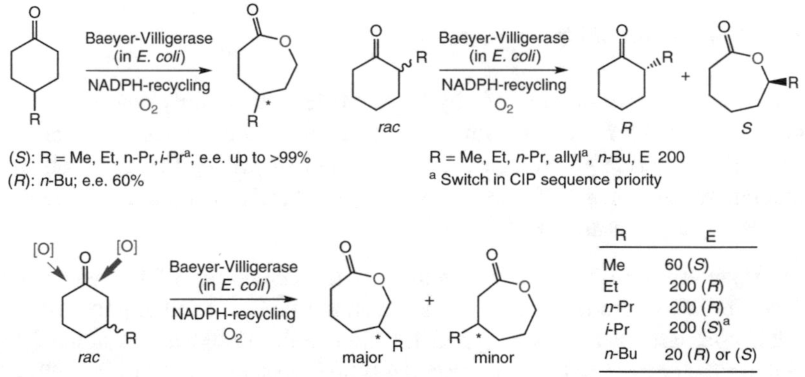
Baeyer-Villigerase
(in E. coli)
NADPH-recycling
O2
(S): R = Me, Et, n-Pr, i-Pr^a; e.e. up to >99%
(R): n-Bu; e.e. 60%
rac
R
S
R = Me, Et, n-Pr, allyl^a, n-Bu, E 200
^a Switch in CIP sequence priority
[O]
major
minor
R E
Me 60 (S)
Et 200 (R)
n-Pr 200 (R)
i-Pr 200 (S)^a
n-Bu 20 (R) or (S)

Scheme 2.165 Regio- and enantioselective Baeyer-Villiger oxidation using cloned Baeyer-Villigerases via desymmetrization, 'classic' and 'nonclassic' kinetic resolution

Dioxygenases

Typical dioxygenase reactions, during which *two* oxygen atoms are simultaneously transferred onto the substrate, are shown in Scheme 2.166. Insertion of O_2 into a C–H or C=C bond yields a highly reactive and unstable hydro- or endo-peroxide species, respectively, which is subject to (enzymatic or nonenzymatic) reduction or rearrangement to yield stable mono- or di-hydroxy products [1391].

- Non-conjugated 1,4-dienes may be oxidized by lipoxygenases at the allylic position to furnish an allyl *hydro*peroxide which, upon chemical reduction (e.g., by sodium borohydride) yields an allylic alcohol. In living systems, the formation of lipid peroxides is considered to be involved in some serious diseases and malfunctions including arteriosclerosis and cancer [1392].

* Alternatively, an *endo*-peroxide may be formed, whose enzymatic reduction leads to a vicinal diol (Scheme 2.166). The latter reaction resembles the cyclo-addition of singlet-oxygen onto an unsaturated system. In mammals, it occurs in the biosynthesis of prostaglandins and leukotrienes, while in prokaryotic (bacterial) cells it constitutes the initial step in the oxidative biodegradation of aromatics (Scheme 2.83) [1393, 1394].

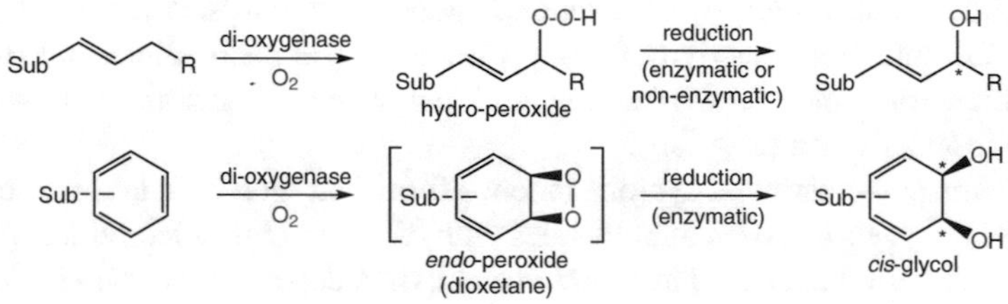

Scheme 2.166 Dioxygenase-catalyzed reactions

2.3.3.6 Formation of Hydroperoxides

The biocatalytic formation of hydroperoxides is mainly associated with dioxygenase activity found in plants, such as peas, peanuts, cucumbers, and potatoes as well as marine green algae. Thus, it is not surprising that the (nonnatural) compounds transformed so far have a strong structural resemblance to the natural substrates – (poly)unsaturated fatty acids.

Allylic Hydroperoxidation Lipoxygenase is a nonheme iron dioxygenase which catalyzes the incorporation of dioxygen into an allylic C-H bond of polyunsaturated fatty acids possessing a nonconjugated 1,4-diene unit through a radical mechanism by forming the corresponding conjugated allylic hydroperoxides [1395–1397]. The enzyme from soybean has received the most attention in terms of a detailed characterization because of its early discovery [1398], ease of isolation and acceptable stability [1399, 1400]. The following characteristics can be given for soybean lipoxygenase-catalyzed oxidations:

* The enzyme has a preference for all-(Z) configurated 1,4-dienes at an appropriate location in the carbon chain of polyunsaturated fatty acids, (E,Z)- and (Z,E)-1,4-dienes are accepted at slower rates [1401].
* (E,E)-1,4-Dienes and conjugated 1,3-dienes are generally not oxidized.
* The configuration at the newly formed oxygenated chiral center is predominantly (S), although not exclusively [1402, 1403].

Oxidation of the natural substrate (Z,Z)-9,12-octadecadienoic acid (linoleic acid) proceeds highly selectively (95% e.e.) and leads to peroxide formation at carbon 13 (the 'distal' region) along with traces of 9-oxygenated product [1404] (the 'proximal' region, Scheme 2.167) [1405].

In addition, it has been shown that soybean lipoxygenase can also be used for the oxidation of nonnatural 1,4-dienes, as long as the substrate is carefully designed to effectively mimic a fatty acid [1406], insufficient reaction rates can be improved by an increased the oxygen pressure (up to 50 bar) [1407]. The (Z,Z)-1,4-diene moiety of several long-chain alcohols could be oxidized by attachment of a prosthetic group (PG), consisting of a polar $(CH_2)_n$–CO_2H or a CH_2–O–$(CH_2)_2$–OH unit, which serves as a surrogate of the carboxylate moiety (Scheme 2.167) [1408]. Oxidation occurred with high regioselectivity at the 'normal' (distal) site and the optical purity of the peroxides was >97%. After chemical reduction of the hydroperoxide (e.g., by Ph_3P [1409]) and removal of the prosthetic group, the corresponding secondary alcohols were obtained with retention of configuration [1410].

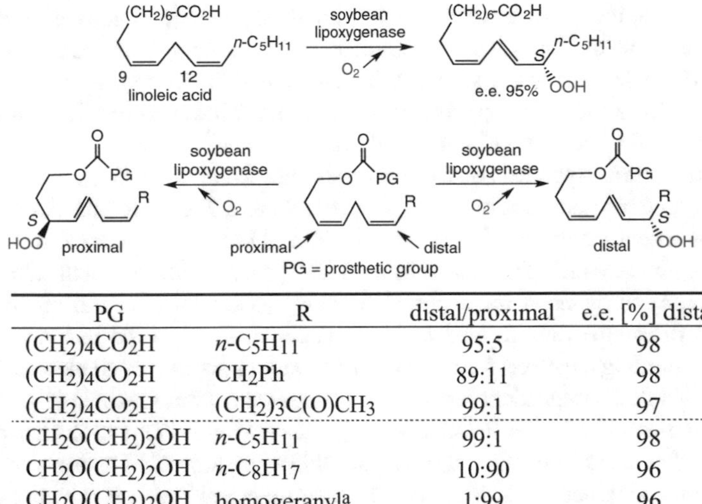

PG	R	distal/proximal	e.e. [%] distal
$(CH_2)_4CO_2H$	n-C_5H_{11}	95:5	98
$(CH_2)_4CO_2H$	CH_2Ph	89:11	98
$(CH_2)_4CO_2H$	$(CH_2)_3C(O)CH_3$	99:1	97
$CH_2O(CH_2)_2OH$	n-C_5H_{11}	99:1	98
$CH_2O(CH_2)_2OH$	n-C_8H_{17}	10:90	96
$CH_2O(CH_2)_2OH$	homogeranyl[a]	1:99	96

[a] = $(CH_2)_2CH=CCH_3(CH_2)_2CH=C(CH_3)_2$.

Scheme 2.167 Natural and non-natural substrates of soybean lipoxygenase

In addition, the regioselectivity of the oxidation could be inverted from 'normal' (distal) to 'abnormal' (proximal) by changing the length of the distal substituent R and the spacer arm linking the prosthetic group PG (Table 2.4 and Scheme 2.167). Enhancing the lipophilicity of R from n-C_5 to n-C_{10} led to an increased reaction at the 'abnormal' site to form predominantly the proximal oxidation product. In contrast, extension of the spacer arm caused a switch to the 'distal' product.

Table 2.4   Variation of prosthetic groups (for formulas see Scheme 2.167)

PG	Variation	R	Distal/proximal
$(CH_2)_4CO_2H$	Distal	$n\text{-}C_5H_{11}$	95:5
$(CH_2)_4CO_2H$	Distal	$n\text{-}C_8H_{17}$	1:1
$(CH_2)_4CO_2H$	Distal	$n\text{-}C_{10}H_{21}$	27:73
$(CH_2)_2CO_2H$	Proximal	$n\text{-}C_8H_{17}$	20:80
$(CH_2)_4CO_2H$	Proximal	$n\text{-}C_8H_{17}$	1:1
$(CH_2)_6CO_2H$	Proximal	$n\text{-}C_8H_{17}$	85:15

2.3.3.7 Dihydroxylation of Aromatic and Conjugated C=C Bonds

cis-Dihydroxylation by bacterial dioxygenases constitutes the initial key step in the oxidative degradation pathway for aromatic compounds (Scheme 2.168) [1411, 1412], which is crucial for the bioremediation of toxic pollutants from contaminated sites. In 'wild-type' microorganisms, the chiral *cis*-glycols initially formed are rapidly further oxidized by dihydrodiol dehydrogenase(s), involving rearomatization of the diol intermediate with concomitant loss of chirality [1413]. The use of mutant strains with blocked dehydrogenase activity [1414], however, allows the chiral glycols to accumulate in the medium, from which they can be isolated in good yield [1415, 1416].

Bacterial Rieske-type iron dioxygenases are multicomponent enzymes, that contain an a non-heme oxygenase component, which contains a 2Fe-2S Rieske cluster together with an adjacent catalytic Fe^{3+}-center (Fig. 2.18) [1417]. The latter forms a side-on complex with O_2, which performs a (formal) [2+2] cycloaddition with the C=C bond in the substrate, as deduced for naphthalene dioxygenase [1418]. The highly reactive (putative) dioxetane thus formed is immediately reduced to the corresponding *cis*-glycol by shuttling electrons from NAD(P)H through a sophisticated electron-transport system, via a flavin-dependent ferredoxin reductase and a ferredoxin [1419, 1420] onto the dioxygenase, like in Cyt P-450 monooxygenases (Fig. 2.17) [1421]. Hence, it is not surprising that Rieske-type dioxygenases are able to perform C-H hydroxylation, C=C epoxidation and thioether oxidation besides their main activity – *cis*-dihydroxylation of alkenes. In contrast, Cyt P-450 enzymes cannot catalyze dihydroxylations due to their Fe^{4+}=O center (Compound I, Scheme 2.146). Given the complexity of this mechanism, it is evident that C=C-dihydroxylations cannot be performed with cell-free systems.

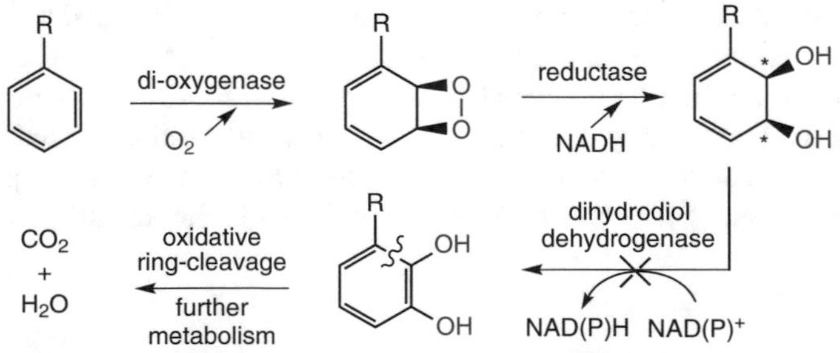

Scheme 2.168 Oxidative degradation of aromatics by bacterial dioxygenases

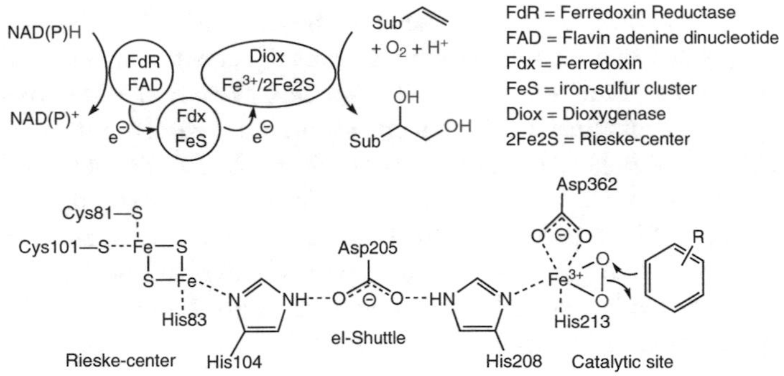

Fig. 2.18 Electron-transport chain and catalytic center of Rieske-type naphthalene dioxygenase

Mutant strains of *Pseudomonas putida* harboring toluene dioxygenase or naphthalene dioxygenase lacking the dihydrodiol dehydrogenase show broad substrate tolerance for ring substituents R^1 and R^2 by maintaining excellent stereospecificity (Scheme 2.169). Consequently, they have been widely employed for the asymmetric dihydroxylation of substituted aromatics [1422] and an impressive number of >300 arenes have been converted into the corresponding chiral *cis*-glycols with excellent optical purities, even on ton-scale with productivities of 20 g L^{-1} h^{-1} and product titers exceeding 80 g L^{-1} [1423–1427]. The stereoselectivity of the oxygen addition can be predicted with some accuracy using a substrate model (Scheme 2.169) [1428–1430].

The substrates need not necessarily be mono- or poly-substituted aromatic compounds such as those shown in Scheme 2.169, but may also be extended monocyclic- [1431], polycyclic- [1432], and heterocyclic derivatives [1433, 1434].

In order to gain access to products showing opposite configuration, a substrate modification approach using *p*-substituted benzene derivatives was developed. Thus, when *p*-iodo derivatives were used instead of the unsubstituted counterparts, the orientation of the oxygen addition was reversed, caused by the switch in relative size of substituents (I > F, I > CH$_3$). Subsequent removal of the iodine (which served as directing group) by catalytic hydrogenation led to mirror-image products [1435].

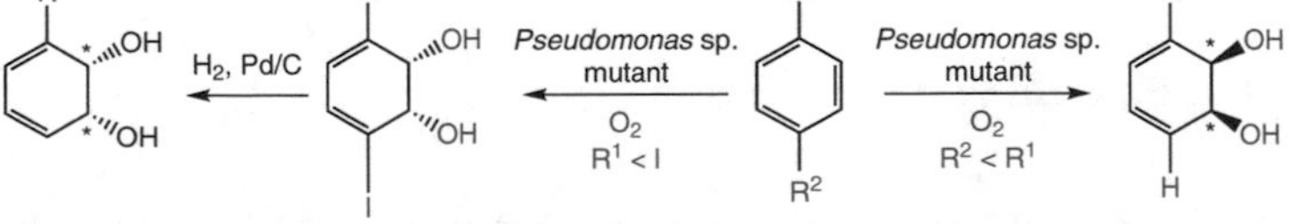

Relative size	R^1	R^2
$R^1 > R^2$	H, Me, Et, *n*-Pr, *i*-Pr, *n*-Bu, *t*-Bu, Et-O,*n*-Pr-O, Halogen, CF$_3$, Ph, Ph-CH$_2$,Ph-CO,CH$_2$=CH, CH$_2$=CH-CH$_2$, HC C	H
$R^2 > R^1$	F, Me	I

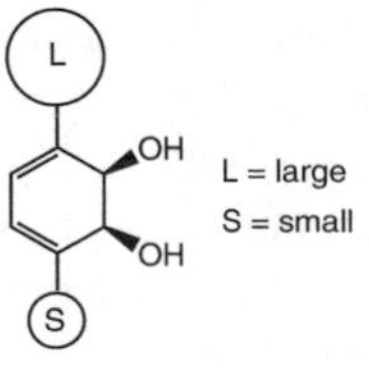

Scheme 2.169 Enantiocomplementary synthesis of *cis*-glycols

The substrate tolerance encompasses also nonaromatic C=C bonds, provided they are conjugated to aromatic systems (such as styrenes) or an additional alkene unit [1436–1438] yielding *ertho*-diols. Thus, *Pseudomonas putida* harboring toluene dioxygenase or naphthalene dioxygenase was able to oxidize a range of styrene-type alkenes and conjugated di- and -trienes (Scheme 2.170). The stereoselectivities were excellent for cyclic substrates but they dropped for open-chain derivatives (e.e.$_{max}$ 88%) [1439]. Depending on the substrate and the type of enzyme, hydroxylation at benzylic or allylic positions were observed as side reactions. In contrast, isolated olefinic bonds react sluggishly and can only be dihydroxylated using dioxygenase mutants with varying success [1440].

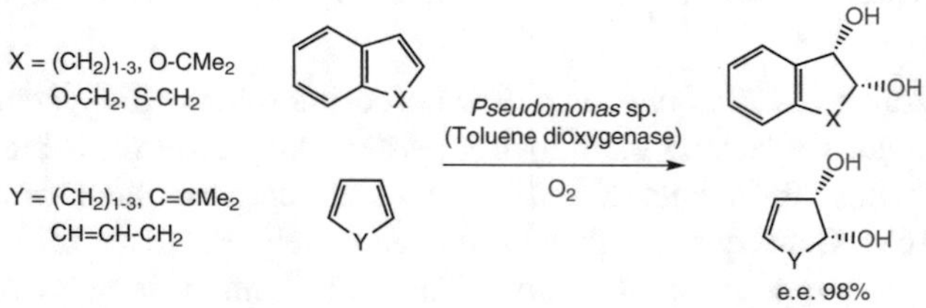

Scheme 2.170 Dihydroxylation of conjugated alkenes using toluene dioxygenase

The synthetic potential of nonracemic *cis*-diols derived via microbial dihydroxylation has been exploited over the years for the synthesis a number of bioactive compounds. Cyclohexanoids have been prepared by making use of the possibility of functionalizing every carbon atom of the glycol in a stereocontrolled way. For instance, (+)-pinitol [1441] and D-*myo*-inositol derivatives [1442] were obtained using this approach. Cyclopentanoid synthons for the synthesis of prostaglandins and terpenes were prepared by a ring-opening/closure sequence [1443]. Rare carbohydrates such as D- and L-erythrose [1444] and L-ribonolactone [1445] were obtained from chlorobenzene as were pyrrolizidine alkaloids [1446]. Furthermore, a bio-inspired synthesis of the blue pigment indigo was developed on a commercial scale using the microbial dihydroxylation of indol [1447], and indene served as starting point for the synthesis of the antiviral agent Indinavir [1448, 1449].

2.3.4 Peroxidation Reactions

Driven by the inability to use molecular oxygen as an oxidant efficiently for the transformation of organic compounds, chemists have used it in a partially reduced form – i.e., hydrogen peroxide [1450] or derivatives thereof, such as *t*-butyl and cumyl hydroperoxide. H_2O_2 offers some significant advantages as it is cheap and environmentally benign – the only byproduct of oxidation being water. However, it is relatively stable and needs to be activated. This is generally accomplished either with organic or inorganic 'promoters' to furnish organic hydro- or endo-peroxides, peroxycarboxylic acids or hypervalent transition metal complexes based on V and Mo. Owing to these drawbacks, the number of industrial-scale oxidation processes using H_2O_2 as the oxidant is very limited.[40] On the other hand, biocatalytic activation of H_2O_2 by peroxidases allow a number of synthetically useful and often highly enantioselective peroxidation reactions, which offer a valuable alternative to traditional chemical methodology.

Peroxidases [EC 1.11.1.X] are a heterogeneous group of redox enzymes found ubiquitously in various sources [1451], such as plants [1452], microorganisms [1453] and animals. They are often named after their sources (e.g., horseradish peroxidase, lacto- and myeloperoxidase) or akin to their substrates (e.g., cytochrome c-, halo- and lignin peroxidase). Although the biological role of these enzymes is quite diverse— ranging from disproportionation of H_2O_2, free radical oligomerization and polymerization of electron-rich aromatics to the oxidation and halogenation of organic substrates—they have in common that they accept hydrogen peroxide or an alkyl hydroperoxide as oxidant. In line with these diverse catalytic activities, the mechanism of action may be quite different and can involve a heme unit, selenium (glutathione peroxidase) [1454], vanadium (bromoperoxidase) [1455, 1456], manganese (manganese peroxidase) [1457] and flavin at the active site (flavoperoxidase) [1458]. The largest group of peroxidases studied so far are heme-enzymes with ferric protoporphyrin IX (protoheme) as the prosthetic group. Their catalytic cycle bears strong similarities to that of heme-dependent monooxygenases (Sect. 2.3.3, Scheme 2.146), but their pathways are more complex (Scheme 2.171). The mechanism of heme-dependent peroxidase catalysis has been largely deduced from horseradish peroxidase [1178, 1459–1461] and its key features are described as follows:

In its native state, the Fe^{3+} species is coordinated equatorially by a heme unit and axially by a histidine residue and is therefore very similar to cytochrome P 450 [1462]. Activation occurs via a two-electron oxidation at the expense of H_2O_2 via the peroxide-shunt to form Compound I. The latter contains a $Fe^{+4}=O$ π-radical moiety and is two oxidation steps above the Fe^{+3} ground state; (in the monooxygenase pathway, the Fe^{3+}-ground state is oxidized by O_2, which requires two additional electrons from a nicotinamide cofactor to cover the net redox balance). Compound I represents the central hypervalent oxidizing species, which can react along two major pathways:

- **Peroxidase-path:** Abstraction of a single electron from an electron-rich substrate such as a phenol, an enol or halide forms a substrate radical and yields an $Fe^{+4}=O$ species denoted Compound II. Since the latter is still one oxidation step above the Fe^{+3}-ground state, this process can occur a second time forming another substrate radical, to finally re-form the enzyme in its native state.
- **Peroxygenase-path:** Alternatively, Compound I can incorporate an O-atom onto a substrate via a two-electron transfer in a single step. Although formally this reaction should be denoted as 'peroxygenase'-activity, this distinction is not always made.

---

[40] A well known industrial-scale process is the oxidation of propene to propene oxide using *tert*-Bu–OOH.

Due to the fact that – in contrast to monooxygenases – no external nicotinamide cofactor is required in the peroxidase cycles, peroxidases are highly attractive for preparative biotransformations. A number of synthetically useful reactions can be achieved (Scheme 2.172) [1463–1465]. Depending on the enzyme–substrate combination, the replacement of hydrogen peroxide by *tert*-butyl hydroperoxide may be beneficial.

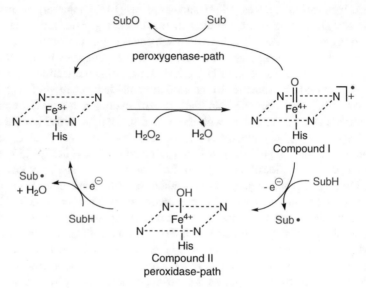

Scheme 2.171 Catalytic cycles of heme-dependent peroxidases

Oxidative coupling (peroxidase-path)

$$2\ SubH + H_2O_2 \longrightarrow 2\ Sub \bullet + 2\ H_2O \longrightarrow Sub\text{—}Sub$$

Oxidative halogenation (peroxidase-path)

$$SubH + H_2O_2 + Hal^- + H^+ \longrightarrow Sub\text{-}Hal + 2\ H_2O$$

Oxygen transfer (peroxygenase-path)

$$Sub + H_2O_2 \longrightarrow SubO + H_2O$$

Scheme 2.172 Synthetically useful peroxidase and peroxygenase reactions

Oxidative Coupling

This reaction is commonly denoted as the 'classical' peroxidase activity, since it was the first type of peroxidase-reaction discovered.

It is mainly restricted to heme peroxidases and it involves the one-electron oxidation of phenols (e.g., guaiacol, resorcinol) and anilines (e.g., aniline, *o*-

dianisidine), forming resonance-stabilized radicals. The latter undergo spontaneous inter- or intramolecular C–C, C–O and/ or C–N coupling to yield dimers or polymers [1466–1468]. In certain cases, dimers, such as biaryls and aryl-ethers, have been obtained (Scheme 2.173) [1469, 1470].

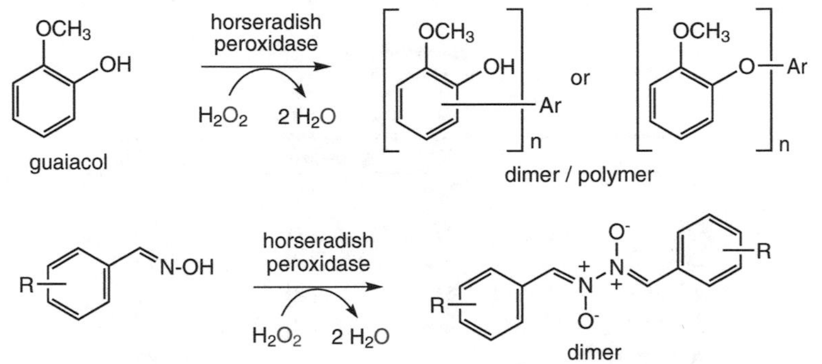

Scheme 2.173 Peroxidase-catalyzed oxidative coupling of aromatics and arylether formation

Oxidative Halogenation
A class of peroxidases – *halo*peroxidases – specializes in the peroxidation of halides (Cl⁻, Br⁻, I⁻ but not F⁻), thereby creating reactive halogenating species (such as hypohalite), which in turn form haloorganic compounds [1471, 1472]. These reactions are described in Sect. 2.7.1.

Oxygen Transfer
From a synthetic viewpoint, selective oxygen transfer via the peroxygenase-path is particularly intriguing, because it is comparable to those catalyzed by monooxygenases with one significant advantage – it is independent of redox cofactors, such as NAD(P)H. Among the various types of reactions – C–H bond oxidation, alkene epoxidation and heteroatom oxidation – the most useful transformations are described below.

Hydroxylation of C–H Bonds Heme-dependent chloroperoxidase (CPO) from the marine fungus *Caldariomyces fumago* has been found to effect the hydroxylation of C–H bonds. The large-scale production of CPO is facilitated by the fact that it is an extracellular enzyme, which is excreted into the fermentation medium [1473–1475] and its crystal structure has been solved [1474]. In order to become susceptible towards hydroxylation by CPO, the C–H bonds have to be activated by a π-electron system. In the allylic position, hydroxylation is not very efficient [1476], but benzylic or propargylic hydroxylation is readily effected to furnish the corresponding *sec*-alcohols in high e.e. (Scheme 2.174) [1477, 1478]. CPO is very sensitive with respect to the substrate structure as the stereochemistry of products was reversed from (*R*) to (*S*) when the alkyl chain was extended from methyl to an ethyl analog, albeit at a slow reaction rate. The selectivity of CPO-catalyzed propargylic hydroxylation was found to be sensitive with respect to the polarity and the alkyne chain length [1479]. In addition, hydroxylation of aromatic C–H bonds seems to be possible, as long as electron-rich (hetero)aromatics, such as indol are used [1480, 1481].

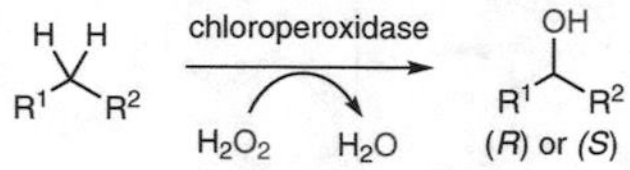

R^1	R^2	Configuration	e.e. [%]
Ph	Me	(R)	97
Ph	Et	(S)	88
Et-C≡C-	Me	(R)	91
n-Pr-C≡C-	Me	(R)	87
AcO-CH$_2$-C≡C-	Me	(R)	95
AcO-(CH$_2$)$_2$-C≡C-	Me	(R)	83
Br-CH$_2$-C≡C-	Me	(R)	94
Br-(CH$_2$)$_2$-C≡C-	Me	(R)	94

Scheme 2.174 Benzylic and propargylic C–H hydroxylations

Epoxidation of Alkenes Due to the fact that the asymmetric epoxidation of alkenes using monooxygenase systems is impeded by the toxicity of epoxides to microbial cells, the use of H_2O_2-depending peroxidases represents a valuable alternative.

Chloroperoxidase-catalyzed epoxidation of alkenes proceeds with excellent enantioselectivites (Scheme 2.175) [1482, 1483]. For styrene oxide it was demonstrated that all the oxygen in the product is derived from hydrogen peroxide, which proves the validity of direct oxygen-transfer via the peroxygenase-path (Scheme 2.171) [1484]. Unfunctionalized *cis*-alkenes [1485] and 1,1-disubstituted olefins [1486, 1487] were epoxidized with excellent selectivities. On the other hand, aliphatic terminal and *trans*-1,2-disubstituted alkenes were epoxidized in low yields and moderate enantioselectivities [1488].

Sulfoxidation Heteroatom oxidation catalyzed by (halo)peroxidases has been observed in a variety of organic compounds. *N*-Oxidation in amines, for instance, can lead to the formation of the corresponding aliphatic *N*-oxides or aromatic nitroso or nitro compounds. From a preparative standpoint, however, sulfoxidation of thioethers is of greater importance since it was shown to proceed in a highly stereo- and enantioselective fashion. Moreover, depending on the source of the haloperoxidase, chiral sulfoxides of opposite configuration could be obtained (Scheme 2.176).

Chloroperoxidase from *Caldariomyces fumago* is a selective catalyst for the oxidation of methylthioethers to furnish (R)-sulfoxides. Initial results were

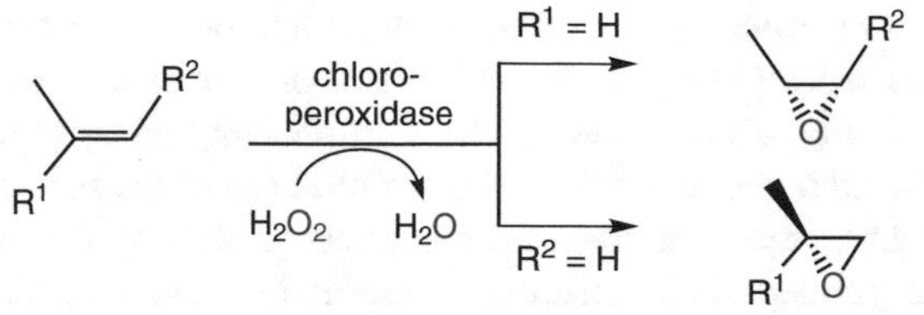

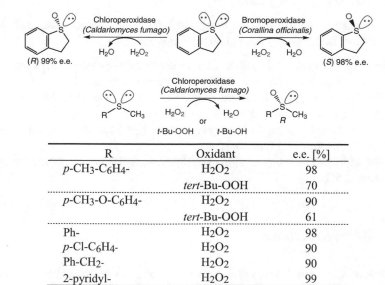

R^1	R^2	e.e. [%]
H	n-C$_4$H$_9$-	96
H	(CH$_3$)$_2$CH-CH$_2$-	94
H	Ph-	96
Ph	H	89
CH$_2$-CO$_2$Et	H	94
(CH$_2$)$_2$-Br	H	85
n-C$_5$H$_{11}$	H	95

Scheme 2.175 Asymmetric epoxidation of alkenes using chloroperoxidase

disappointing, as low e.e.'s were reported [1489]. The latter were caused by substantial nonenzymatic oxidation by hydrogen peroxide, which could be suppressed by optimization of the reaction conditions: whereas the use of *tert*-butylhydroperoxide was unsuccessful, the best results were obtained by maintaining the concentration of H$_2$O$_2$ at a constant low level [1490, 1491].

A vanadium-dependent haloperoxidase from the marine alga *Corallina officinalis* was shown to possess a matching opposite enantiopreference by forming (*S*)-sulfoxides [1492, 1493]. Although simple open-chain thioethers were not well transformed, cyclic analogs were ideal candidates [1494].

R	Oxidant	e.e. [%]
p-CH$_3$-C$_6$H$_4$-	H$_2$O$_2$	98
	tert-Bu-OOH	70
p-CH$_3$-O-C$_6$H$_4$-	H$_2$O$_2$	90
	tert-Bu-OOH	61
Ph-	H$_2$O$_2$	98
p-Cl-C$_6$H$_4$-	H$_2$O$_2$	90
Ph-CH$_2$-	H$_2$O$_2$	90
2-pyridyl-	H$_2$O$_2$	99

Scheme 2.176 Stereocomplementary oxidation of thioethers by haloperoxidases

Although on a superficial glimpse, peroxidases appear to be more easy to use than monooxygenases, several points limit their practical application considerably: Given the tendency of H_2O_2 to deactivate proteins in general, the operational stability of peroxidases in the presence of substantial concentrations of the oxidant is very limited and in order to achieve reasonable turnover numbers (usually only a few hundred) it has to be continuously added via autotitration using a H_2O_2-sensitive electrode ('peroxystat' [1495]) by maintaining a constant low level (~10 mM). Although this also minimizes spontaneous (non-selective) oxidation, reaction rates are modest too. Consequently, at this point, peroxygenases are no serious competitors for monooxygenases.

2.4 Formation of Carbon–Carbon Bonds

The majority of enzymatic reactions exploited for biotransformations involve functional group manipulations via *bond-breaking* reactions. The following enzymatic systems, which are capable of *forming* carbon–carbon bonds in a highly stereoselective manner, belong to the class of lyases and are gaining increasing attention in view of their potential in *synthesis*. Since these enzymes are involved in the biosynthesis and biodegradation of sugars, lyase-catalyzed reactions are generally equilibrium-controlled. The following strategies can be applied to drive C–C bond forming reactions towards completion:

- The primary hydroxycarbonyl products often spontaneously cyclize to yield a stable hemiacetal.
- C–C bond forming steps are often embedded in a reaction cascade, which pulls out the formed product from the equilibrium (Sect 3.2, cascade reactions).
- Some C-donor molecules, such as pyruvate, undergo decarboxylation, which provides a strong driving force.

For the sake of clarity, the donor representing the umpolung reagent is drawn with bold C–C bonds (blue) throughout this chapter.

- Aldol reactions catalyzed by aldolases are useful for the elongation of aldehydes by a two- or three-carbon unit yielding β-hydroxy compounds.
- Aldehydes of various size can be coupled in a head-to-head fashion to furnish α-hydroxycarbonyl compounds (acyloins, benzoins).
- A hydroxyacetyl C_2-fragment (equivalent to hydroxyacetaldehyde) is transferred via transketolase reactions.
- For the addition of the C_1-synthon cyanide to aldehydes by hydroxynitrile lyases see Sect. 2.5.3.

2.4.1 Aldol Reactions

Asymmetric C–C bond formation based on catalytic aldol addition reactions remains a challenging subject in synthetic organic chemistry. Although many successful nonbiological strategies have been developed [1496, 1497], most of them are not without drawbacks. They are often stoichiometric in auxiliary reagent and require the use of a chiral metal or organocatalytic enolate complex to achieve stereoselectivity [1498–1501]. Due to the instability of such complexes in aqueous solutions, aldol reactions usually must be carried out in organic solvents at low temperature. Thus, for compounds containing polar functional groups, the employment of conventional aldol reactions requires extensive protection protocols in order to make them lipophilic and to avoid undesired cross-reactions, which limits the application of conventional aldol reactions in aqueous solution. On the other hand, enzymatic aldol reactions catalyzed by aldolases, which are performed in aqueous solution at neutral pH, can be achieved without extensive protection methodology and have therefore attracted increasing interest [1502–1519].

Aldolases were first recognized some 70 years ago. At that time, it was believed that they form an ubiquitous class of enzymes that catalyze a key step in glycolysis by interconversion of hexoses into two three-carbon subunits [1520]. It is now known that aldolases operate on a wide range of substrates including carbohydrates, amino acids and hydroxy acids. A variety of enzymes has been described that add a two- or three-carbon (donor) fragment onto a carbonyl group of an aldehyde or a ketone with high stereospecificity. Since glycolysis and glyconeogenesis are a fundamental pillar of life, almost all organisms possess aldolase enzymes.

Two distinct groups of aldolases, acting via different mechanisms during formation of the (donor) carbanion, have been recognized [1521]. Both of the mechanisms are closely related to conventional aldol reactions, i.e., carbanion formation (umpolung) is achieved via an enolate- or enamine species (Schemes 2.177 and 2.178).

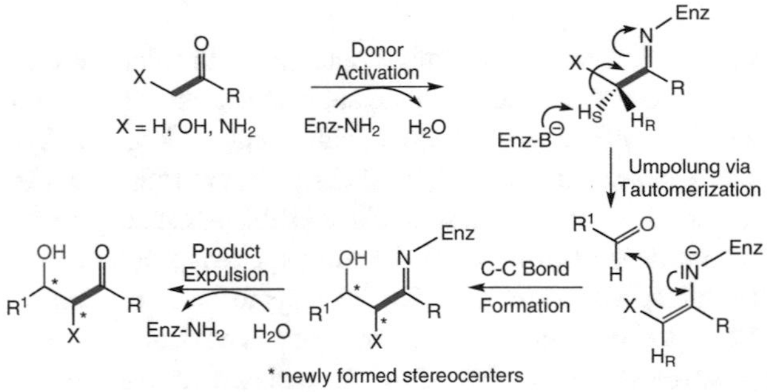

Scheme 2.177   Mechanism of type I aldolases

Type-I aldolases, found predominantly in higher plants and animals, require no metal cofactor. They catalyze the aldol reaction through a Schiff-base intermediate, which tautomerizes to an enamine species (Scheme 2.177) [1522]. First, the donor is covalently linked to the enzyme via the ε-amino group of a conserved lysine residue to form a Schiff base. Next, base-catalyzed abstraction of H_s leads to the formation of an enamine species, which performs a nucleophilic attack on the carbonyl group of the aldehydic acceptor in an asymmetric fashion. Consequently,

the two new chiral centers are formed stereospecifically in a *threo-* or *erythro-*
configuration depending on the enzyme. Finally, hydrolysis of the Schiff base
liberates the aldol product and regenerates the enzyme.

Type II aldolases are found predominantly in bacteria and fungi, and are Zn^{2+}-
dependent enzymes[41] (Scheme 2.178) [1523]. Their mechanism of action proceeds
through a metal-enolate [1524]: an essential Zn^{2+} atom in the active site (coordinated by three nitrogen atoms of histidine residues [1525]) binds the donor via the
hydroxyl and carbonyl groups. This facilitates *pro-(R)*-proton abstraction from the
donor (presumably by a glutamic acid residue acting as base), rendering an enolate,
which launches a nucleophilic attack onto the aldehydic acceptor.

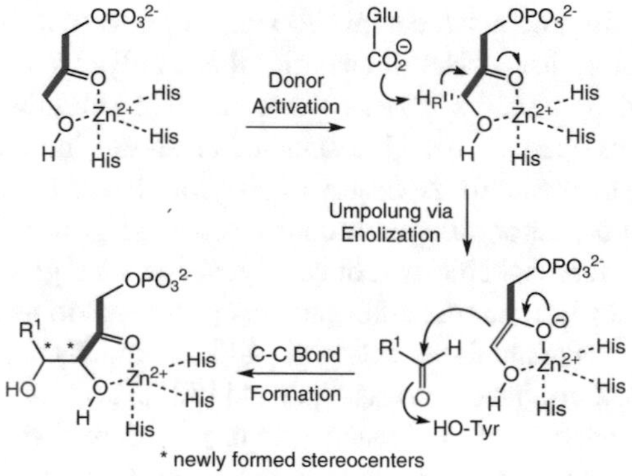

Scheme 2.178 Mechanism of metal-dependent type II aldolases

With few exceptions, the stereochemical outcome of the aldol reaction is controlled
by the enzyme and does not depend on the substrate structure (or on its stereochemistry). Therefore, the configuration of the carbon atoms undergoing C–C bond formation is highly predictable. Furthermore, most aldolases are rather restricted concerning
their donor (the nucleophile), but possess relaxed substrate specificities with respect to
the acceptor (the electrophile), which is the carbonyl group of an aldehyde or ketone.
This is understandable, bearing in mind that the enzyme has to perform an umpolung on
the donor, which is a sophisticated task in an aqueous environment!

To date approximately 50 aldolases have been classified, the most useful and more
readily available enzymes are described in this chapter. Bearing in mind that the
natural substrates of aldolases are carbohydrates, most successful enzyme-catalyzed
aldol reactions have been performed with carbohydrate-like (poly)hydroxy compounds as substrates. Depending on the donor, the carbon-chain elongation generally
involves a two- or three-carbon unit (Scheme 2.179, donors are shown in bold).

---

[41]Some enzymes use Mg^{2+} or Mn^{2+}.

Aldolases are most conveniently classified into five groups according to their donor molecule. The best studied group I uses dihydroxyacetone phosphate (DHAP) as donor, resulting in the formation of a ketose 1-phosphate (Scheme 2.181). Within this group, enzymes capable of forming *all four possible stereoisomers* of the newly generated stereogenic centers in a complementary fashion are available (Scheme 2.180). Group II transfers (non-phosphorylated) short-chain hydroxycarbonyl donors, such as hydroxyacetaldehyde, (di)hydroxacetone (DHA) and 1-hydroxy-2-butanone (Scheme 2.188). Group III uses pyruvate (or phosphoenol pyruvate) as donor to yield 3-deoxy-2-keto acids (Scheme 2.189) [1526]. The fourth group consists of only one enzyme – 2-deoxyribose-5-phosphate aldolase (DERA) – which requires acetaldehyde (or close analogs) as donor to form 2-deoxy aldoses (Scheme 2.190). Finally, group V aldolases couple glycine (as donor) with an acceptor aldehyde to yield α-amino-β-hydroxy acids (Scheme 2.192).

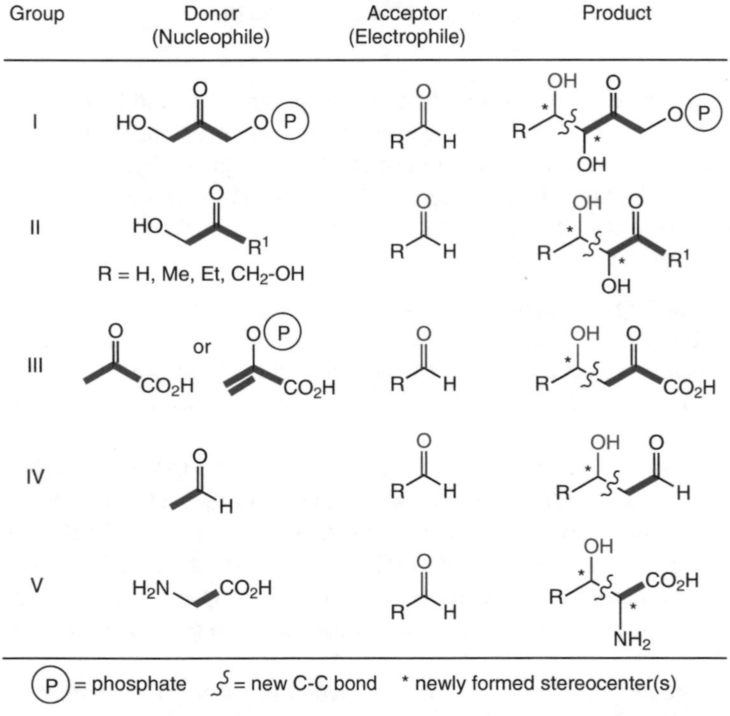

Scheme 2.179 Main groups of aldolases according to donor type

Group I: Dihydroxyacetone Phosphate-Dependent Aldolases

The exploitation of the full synthetic potential of DHAP-dependent aldolases into a general and efficient methodology for asymmetric aldol additions largely depends on the availability of the complete tetrad of enzymes, which allows to create all four possible stereoisomers at will, by simply selecting the correct biocatalyst.

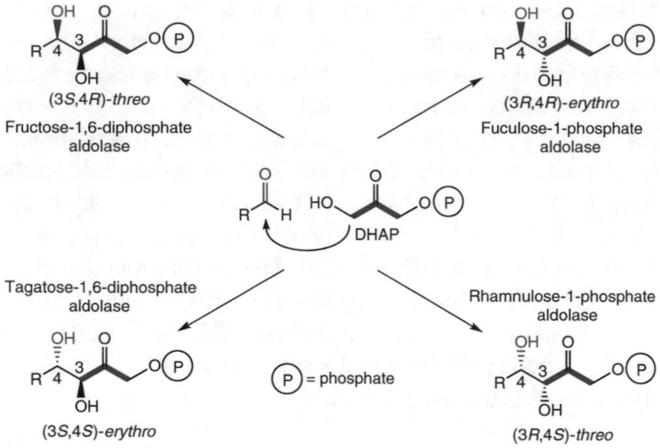

Scheme 2.180 Stereocomplementary DHAP-dependent aldolases

As shown in Scheme 2.180, all four stereocomplementary aldolases occurring in carbohydrate metabolism, which generate the four possible stereoisomeric diol products emerging from the addition of DHAP onto an aldehyde, have been made available by cloning and overexpression. The reaction proceeds with complete stereospecificity with respect to the configuration on carbon 3 and with slightly decreased specificity on carbon 4.

Fructose-1,6-Diphosphate Aldolase Fructose-1,6-diphosphate (FDP) aldolase from rabbit muscle, also commonly known as 'rabbit muscle aldolase' (RAMA), catalyzes the addition of dihydroxyacetone phosphate (DHAP) to D-glyceraldehyde-3-phosphate to form fructose-1,6-diphosphate (Scheme 2.181) [548, 1527].

The equilibrium of the reaction is predominantly on the product side and the specificity of substituent orientation at C-3 and C-4 adjacent to the newly formed vicinal diol bond is always *threo* (Scheme 2.181). However, if the α-carbon atom in the aldehyde component is chiral (C-5 in the product), only low chiral recognition of this remote stereocenter takes place. Consequently, if an α-substituted aldehyde is employed in racemic form, a pair of diastereomeric products will be obtained.

RAMA accepts a wide range of aldehydes in place of its natural substrate (D-glyceraldehyde 3-phosphate), allowing the synthesis of carbohydrates [1528–1531] and analogs such as nitrogen- [547, 1532] and sulfur-containing sugars [1533], deoxysugars [1534], fluoro-sugars, and rare eight- and nine-carbon sugars [1535]. As depicted in Scheme 2.181, numerous aldehydes which are structurally quite unrelated to the natural acceptor are freely accepted [1536–1539].

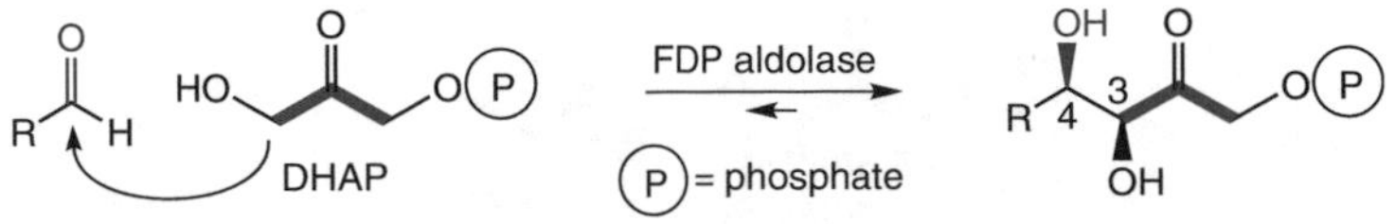

R (natural substrates)	R (non-natural substrates)
D- or L-threose	H, Me, Et, n-Pr, i-Pr, CH=O, CO₂H, CH₂F, CH₂Cl,
D- or L-erythrose	CH₂OCH₂Ph, (CH₂)₂OH, CH(OCH₃)CH₂OH,
L-arabinose, D-ribose,	(CH₂)₂OMe, CHOHMe, CHOHCH₂OMe,
D-lyxose, D-xylose	CH₂Ph, COPh, (CH₂)₃CH=O,
D-glyceraldehyde-3-Ⓟ	CH₂P(O)(OEt)₂, 2- and 3-pyridyl,
	4-cyclohexenyl, CH₂OⓅ, 

Scheme 2.181 Aldol reactions catalyzed by FDP aldolase from rabbit muscle

For RAMA the following rules apply to the aldehyde component:

- In general, unhindered aliphatic, α-heterosubstituted, and protected alkoxy aldehydes are accepted as substrates.
- Sterically hindered aliphatic aldehydes such as pivaldehyde do not react with RAMA, nor do α,β-unsaturated aldehydes or compounds that can readily be eliminated to form α,β-unsaturated aldehydes.
- Aromatic aldehydes are either poor substrates or are unreactive.
- ω-Hydroxy acceptors that are phosphorylated at the terminal hydroxyl group are accepted at enhanced rates relative to the nonphosphorylated species.

In contrast to the relaxed specificity for the acceptor, the requirement for DHAP as the donor is much more stringent. Several isosteric analogs which are more resistant towards spontaneous hydrolysis have been successfully tested as substitutes for DHAP (Scheme 2.182) [1540], however the reaction rates were reduced by about one order of magnitude [1541–1544].

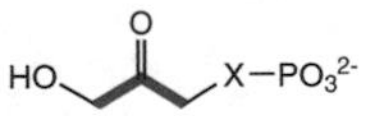

X = O, NH, S, CH₂

Scheme 2.182 Nonnatural DHAP substitutes for fructose-1,6-diphosphate aldolase (RAMA)

Within group-I aldolases, FDP aldolase from rabbit muscle has been extensively used for the synthesis of biologically active sugar analogs on a preparative scale (Scheme 2.183). For example, nojirimycin and derivatives thereof, which have been shown to be potent anti-AIDS agents with no cytotoxicity, have been obtained by a chemoenzymatic approach using RAMA in the key step. As expected, the recognition of the α-hydroxy stereocenter in the acceptor aldehyde was low and gave diastereomers with respect to C₅ [1545, 1546].

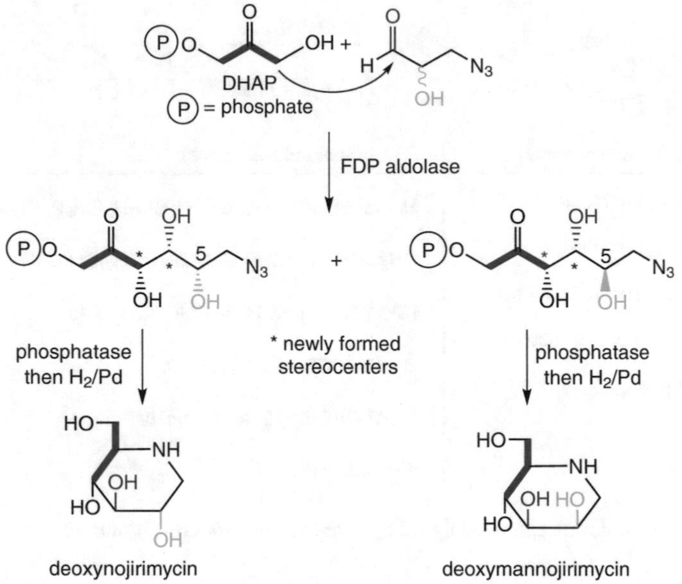

Scheme 2.183 Synthesis of aza-sugar analogs

An elegant synthesis of (+)-*exo*-brevicomin, the sex pheromone of bark beetles made use of FDP-aldolase (Scheme 2.184) [1547]. RAMA-catalyzed addition of DHAP to a δ-keto-aldehyde gave, after enzymatic dephosphorylation, a *threo*-keto-diol, which was cyclized to form a precursor of the pheromone. Finally, the side chain was de-functionalized in four subsequent steps to give (+)-*exo*-brevicomin.

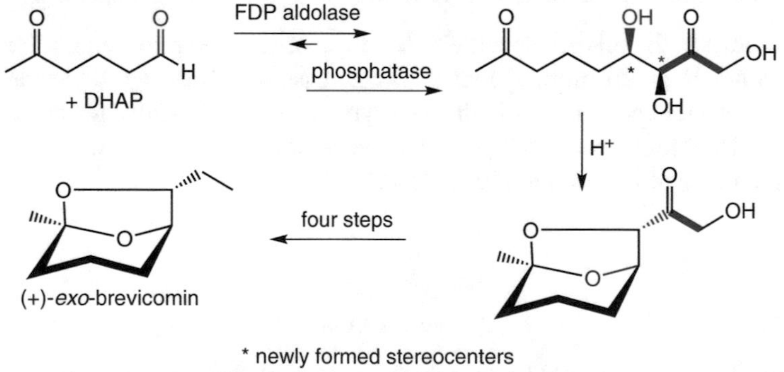

* newly formed stereocenters

Scheme 2.184 Synthesis of (+)-*exo*-brevicomin

Despite the fact that enzymatic aldol reactions are becoming useful in synthetic carbohydrate chemistry, the preparation of aldehyde substrates containing chiral centers remains a problem. Many α-substituted aldehydes racemize in aqueous solution, which would result in the production of a diastereomeric mixture, which is not always readily separable.

The following methods have been used to avoid the (often tedious) separation of diastereomeric products [546].

- Efficient kinetic resolution of α-hydroxyaldehydes can be achieved by inserting a negative charge (such as phosphate or carboxylate) at a distance of four to five atoms from the aldehydic center in order to enhance the binding of the acceptor substrate [1548].
- In some cases, a diastereoselective aldol reaction can be accomplished in a kinetically controlled process via kinetic resolution of the racemic α-substituted aldehyde. Thus, if the reaction is stopped before it reaches equilibrium, a single diastereomer is predominantly formed. However, as mentioned above, the selectivities of aldolases for such kinetic resolutions involving recognition of the (remote) chirality on the α-carbon atom of the aldehyde are usually low.
- In cases wherein one diastereomer of the product is more stable than the other, one can utilize a thermodynamically controlled process (Scheme 2.185). For example, in the aldol reaction of *rac*-2-allyl-3-hydroxypropanal, two diastereomeric products are formed. Due to the hemiacetal ring-formation of the aldol product and because of the reversible nature of the aldol reaction, only the more stable product positioning the 5-allyl substituent in the favorable equatorial position is produced when the reaction reaches equilibrium.
- Another solution to the problem of formation of diastereomeric products is to subject the mixture to the action of glucose isomerase, whereby the D-ketose is converted into the corresponding D-aldose leaving the L-ketose component unchanged [1549].

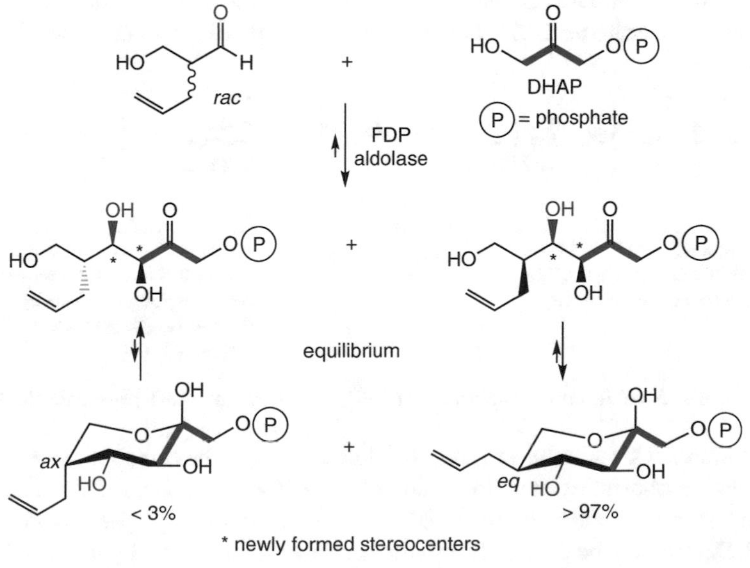

Scheme 2.185 Thermodynamic control in aldolase reactions

A potential limitation on the use of FDP aldolases for the synthesis of monosaccharides is that the products are always *ketoses* with fixed stereochemistry at the newly generated chiral centers on C-3 and C-4. One way to overcome this limitation and to obtain aldoses instead of ketoses makes use of a monoprotected dialdehyde as the acceptor substrate (Scheme 2.186). After the RAMA-catalyzed aldol reaction, the resulting ketone is reduced in a diastereoselective fashion with polyol dehydrogenase. The remaining masked aldehyde is then deprotected to yield a new *aldose*.

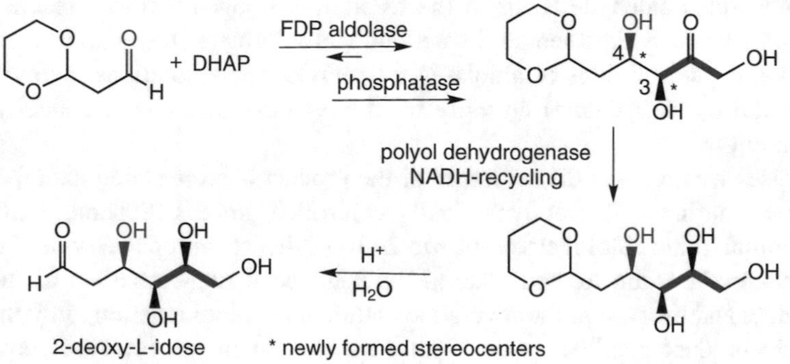
+ DHAP
FDP aldolase
phosphatase
polyol dehydrogenase
NADH-recycling
H+
H2O
2-deoxy-L-idose
* newly formed stereocenters

Scheme 2.186 Synthesis of aldoses using FDP aldolase

Aldol additions catalyzed by DHAP-dependent aldolases which exhibit a complementary stereospecificity to RAMA have been used to a lesser extent (Schemes 2.180 and 2.187) [1550, 1551]. Although the selectivity with respect to the center on carbon 3 is absolute, in some cases the corresponding C-4 diastereomer was formed in minor amounts depending on the R substitutent on the aldehyde. In those cases shown in Scheme 2.187, however, only a single diastereomer was obtained.

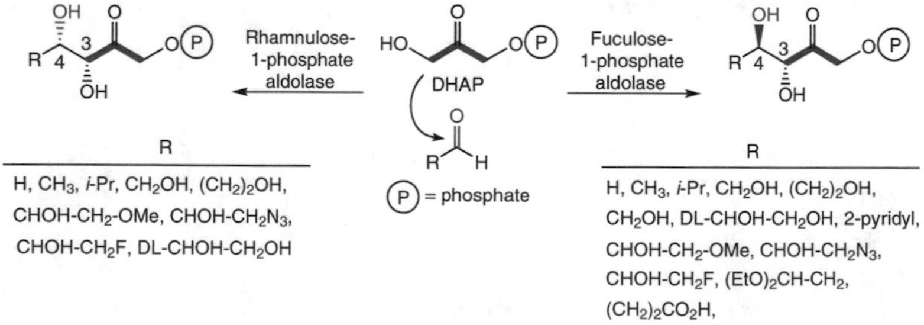
Rhamnulose-1-phosphate aldolase
DHAP
Fuculose-1-phosphate aldolase
P = phosphate
R
H, CH3, i-Pr, CH2OH, (CH2)2OH, CHOH-CH2-OMe, CHOH-CH2N3, CHOH-CH2F, DL-CHOH-CH2OH
R
H, CH3, i-Pr, CH2OH, (CH2)2OH, CH2OH, DL-CHOH-CH2OH, 2-pyridyl, CHOH-CH2-OMe, CHOH-CH2N3, CHOH-CH2F, (EtO)2CH-CH2, (CH2)2CO2H,

Scheme 2.187 Aldol reactions catalyzed by fuculose- and rhamnulose-1-phosphate aldolase

One drawback common to group I aldolases is that they require the expensive and sensitive phosphorylated donor dihydroxyacetone phosphate. This molecule is not very stable in solution ($t_{1/2}$ ~ 20 h at pH 7), and its synthesis is not trivial [1552]. DHAP may be obtained from the hemiacetal dimer of dihydroxyacetone by chemical phosphorylation with $POCl_3$ [1553, 1554], or by enzymatic phosphorylation of dihydroxyacetone at the expense of ATP and glycerol kinase [588] (Sect. 2.1.4). Probably the most elegant and convenient method is the in situ generation of DHAP from fructose-1,6-diphosphate (FDP) using FDP aldolase, forming one molecule of DHAP as well as glyceraldehyde-3-phosphate. The latter can be rearranged by triosephosphate isomerase to give a second DHAP molecule [583]. This two-enzyme protocol has been further extended into a highly integrated 'artificial metabolism' derived from glycolysis to obtain DHAP from inexpensive feedstocks, such as glucose or fructose (yielding two equivalents of DHAP) and sucrose (four equivalents) via an enzymatic cascade consisting of up to seven enzymes [1555]. The use of DHAP can be circumvented by in-situ formation of a DHA borate ester, which is able to mimic DHAP with rhamnulose 1,6-diphosphate aldolase [1556, 1557].

The presence of the phosphate group in the aldol adducts facilitates their purification by precipitation as the corresponding barium salts or via ion-exchange chromatography. Cleavage of phosphate esters is usually accomplished by enzymatic hydrolysis using acid or alkaline phosphatase (Sect. 2.1.4).

Group II: Dihydroxyacetone Dependent Aldolases
The recent discovery of D-fructose-6-phosphate aldolase from *E. coli* and the structurally related transaldolase B variant F178Y has opened an elegant solution to avoid the necessity for phosphorylated DHAP. Since the stereospecificity of these enzymes is identical to that of fructose-1,6-diphosphate aldolase, it represents a useful alternative to RAMA by avoiding the futile phosphorylation of the donor and the dephosphorylation of the aldol product.

Fructose-6-phosphate aldolase and variants thereof accept (non-phosphorylated) dihydroxyacetone and several structural analogs, such as hydroxyacetone, 1-hydroxy-2-butanone and hydroxyacetaldehyde (glycolaldehyde) (Scheme 2.188). In addition to this remarkable donor-flexibility, they also possess a broad tolerance for substituted acceptor aldehydes [1558–1563]. The synthetic applicability of this method was demonstrated by coupling of dihydroxy-acetone with *N*-protected 3-aminopropanal to yield the (3*S*,4*R*)-*threo*-diol, which was reductively deprotected and cyclized to furnish the rare aza-sugar D-fagomine. The latter acts as glycosidase inhibitor and shows antifungal and antibacterial activity [1564].

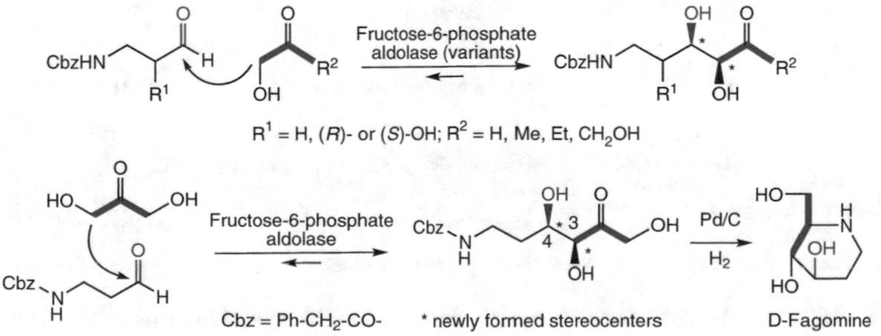

R^1 = H, (*R*)- or (*S*)-OH; R^2 = H, Me, Et, CH$_2$OH

Cbz = Ph-CH$_2$-CO-   * newly formed stereocenters   D-Fagomine

Scheme 2.188 Aldol reaction catalyzed by fructose-6-phosphate aldolase (variants) using non-phosphorylated 1-hydroxy-2-alkanones as donors

Group III: Pyruvate-Dependent Aldolases

For thermodynamic reasons, pyruvate-dependent aldolases have catabolic functions in vivo, whereas their counterparts employing (energy-rich) phosphoenol pyruvate as the donor are involved in the biosynthesis of keto-acids. However, both types of enzymes can be used to synthesize α-keto-γ-hydroxy acids in vitro. In these reactions, the equilibrium is less favorable and usually requires an excess of pyruvate to achieve a reasonable conversion. However, product isolation is facilitated by enzymatic decomposition of excess pyruvate by pyruvate decarboxylase to yield volatile CO_2 and acetaldehyde.

Sialic Acid Aldolase N-Acetylneuraminic acid (NeuAc, also termed sialic acid) aldolase catalyzes the reversible addition of pyruvate onto N-acetylmannosamine to form N-acetylneuraminic acid (Scheme 2.189) [1565, 1566]. Since the equilibrium for this reaction is near unity, an excess of pyruvate must be used in synthetic reactions to drive the reaction towards completion. NeuAc was previously isolated from natural sources such as cow's milk, but increasing demand prompted the development of a two-step synthesis from N-acetylglucosamine using chemical or enzymatic epimerization to N-acetylmannosamine, followed by coupling of pyruvate catalyzed by sialic acid aldolase on a multi-ton scale for the production of a precursor of the anti-viral drug Zanamivir [1567–1570]. Besides NeuAc, the production of structural analogs is of significance since neuraminic acid derivatives play an important role in cell adhesion and biochemical recognition processes [1571]. The cloning of the enzyme has reduced its cost [1572].

In line with the substrate requirements of FDP aldolase, the specificity of sialic acid aldolase appears to be absolute for pyruvate (the donor), but relaxed for the aldehydic acceptor. As may be seen from Scheme 2.189, a range of mannosamine derivatives have been used to synthesize derivatives of NeuAc [1573–1578]. Substitution at C-2 of N-acetylmannosamine is tolerated, and the enzyme exhibits only a slight preference for defined stereochemistry at other centers.

Other group III aldolases of preparative value are 3-deoxy-D-manno-octulosonate (KDO) aldolase [1579, 1580], macrophomate synthase [1581] as well as 2-keto-3-deoxy-6-phosphogluconate and -galactonate aldolases [1526, 1582].

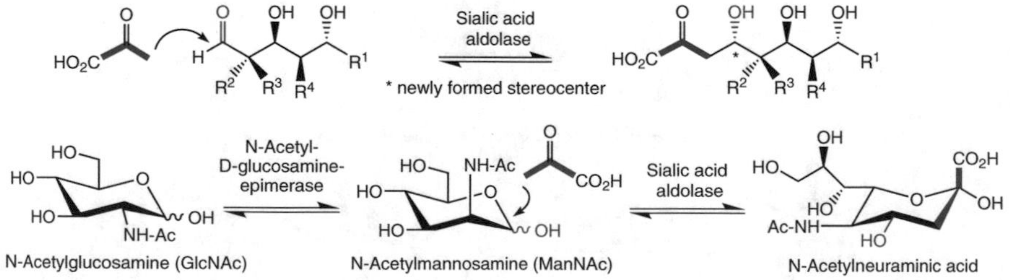

Scheme 2.189 Aldol reactions catalyzed by sialic acid aldolase and industrial-scale synthesis of N-acetylneuraminic acid using a two-enzyme system

Group IV: Acetaldehyde-Dependent Aldolases

2-Deoxyribose-5-Phosphate Aldolase One of the rare aldolases, which accepts acetaldehyde as donor is 2-deoxyribose-5-phosphate (DER) aldolase. In vivo, DER aldolase catalyzes the reversible aldol reaction of acetaldehyde and D-glyceraldehyde-3-phosphate to form 2-deoxyribose-5-phosphate (Scheme 2.190; $R^1 = H$, $R^2 = $ phosphate). This aldolase is unique in that it condenses *two aldehydes* (instead of a ketone and an aldehyde) in a self- or cross-aldol reaction to form aldoses (Schemes 2.179 and 2.190). Like fructose 6-phosphate aldolase, the enzyme (which has been overproduced [1583]) shows a relaxed substrate specificity not only on the acceptor side, but also on the donor side. Thus, besides acetaldehyde it accepts also acetone, fluoroacetone and propionaldehyde as donors, albeit at a much slower rate. Like other aldolases, it transforms a variety of aldehydic acceptors in addition to D-glyceraldehyde-3-phosphate. DERA provides access to β-hydroxy aldehydes and -ketones with generation of a chiral center.

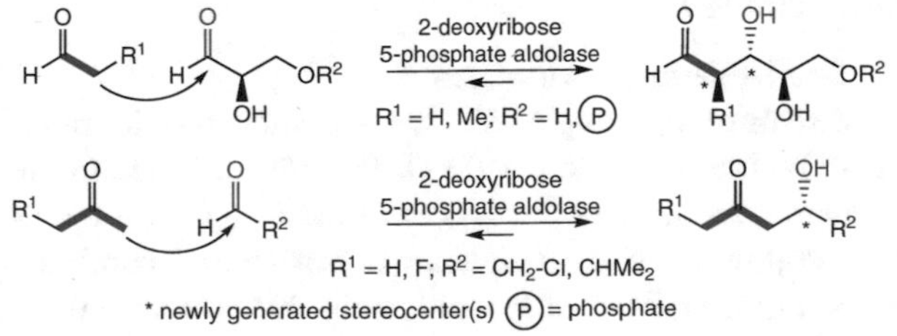

Scheme 2.190 Aldol reactions catalyzed by 2-deoxyribose-5-phosphate aldolase

An elegant method for sequential aldol reactions performed in a one-pot reaction has been discovered for 2-deoxyribose-5-phosphate aldolase (Scheme 2.191) [1584]. When a (substituted) aldehyde was used as acceptor, coupling of acetaldehyde (as donor) led to the corresponding β-hydroxy aldehyde as

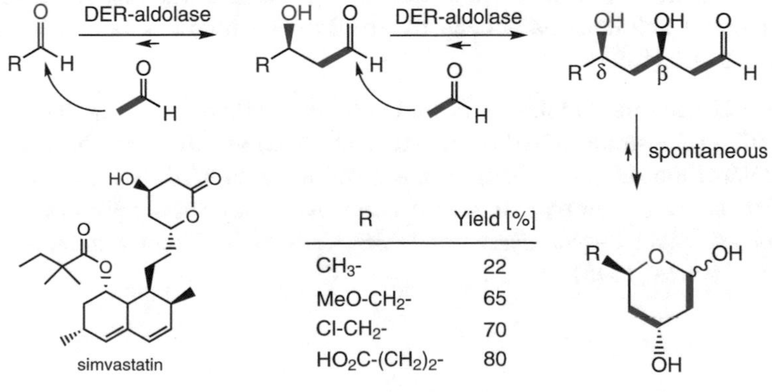

R	Yield [%]
CH$_3$-	22
MeO-CH$_2$-	65
Cl-CH$_2$-	70
HO$_2$C-(CH$_2$)$_2$-	80

Scheme 2.191 Sequential aldol reactions catalyzed by DER aldolase for synthesis of statin side chains

intermediate product. The latter, however, can undergo a second aldol reaction with another acetaldehyde donor, forming a β,δ-dihydroxy aldehyde. At this stage, this aldol cascade (which would lead to the formation of a polymeric product if uninterrupted) is terminated by the (spontaneous) formation of a stable hemiacetal (lactol). The latter does not possess a free aldehydic group and therefore cannot serve as acceptor any more.

The dihydroxylactols thus obtained can be oxidized by NaOCl to the corresponding lactones, which represent the chiral side chains of several cholesterol-lowering 3-hydroxy-3-methylglutaryl-(HMG)-CoA reductase inhibitors, collectively denoted as 'statins' [1585].[42] Several derivatives thereof are produced on industrial scale using DER-aldolase mutants at product concentrations exceeding 100 g/L [1586, 1587].

This concept provides rapid access to polyfunctional complex products from cheap starting materials in a one-pot reaction. It has recently been extended by combining various types of aldolases together to perform three- and four-substrate cascade reactions [1588, 1589].

Group V: Glycine-Dependent Aldolases
One remarkable feature of group V aldolases is their requirement for an amino acid as donor – glycine (Scheme 2.192) [1590, 1591]. Since the donor bears an amino group, their mechanism of action is related to Type I aldolases, with the difference that umpolung of the (glycine) donor to an enamine species is not effected by an ε-amino-moiety of lysine within the protein, but via Schiff-base formation with the aldehyde group of a pyridoxal-5′-phosphate cofactor (PLP, Sect. 2.6.2, Scheme 2.221). However, the nucleophilic attack of Cα onto an aldehyde acceptor forming an α-amino-β-hydroxy acid is essentially the same. Since two new stereocenters are formed, four possible stereoisomers can by formally obtained. However, in contrast to DHAP-dependent aldolases (Scheme 2.180), the complete set of stereo-complementary threonine aldolases has not yet been found [1592, 1593]. β-Hydroxyamino acids are multifunctional compounds with numerous applications in the synthesis of complex bioactive structures, such as peptide mimetics, protease inhibitors and antibiotics. It is thus not surprising, that threonine aldolases have been frequently used for their synthesis, also on industrial scale [1594].

D- and L-Threonine Aldolases These enzymes are involved in the biosynthesis/degradation of α-amino-β-hydroxyamino acids, such as threonine and they exquisitely control the stereochemistry of the α-amino configuration, which is either D or L, depending on the type of enzyme. However, they show only low-moderate specificities for the β-hydroxy-center, which leads to diastereomeric *threo*/*erythro* product mixtures [1595].

---

[42]For instance, atorvastatin (Lipitor™), rosuvastatin (Crestor™) or simvastatin.

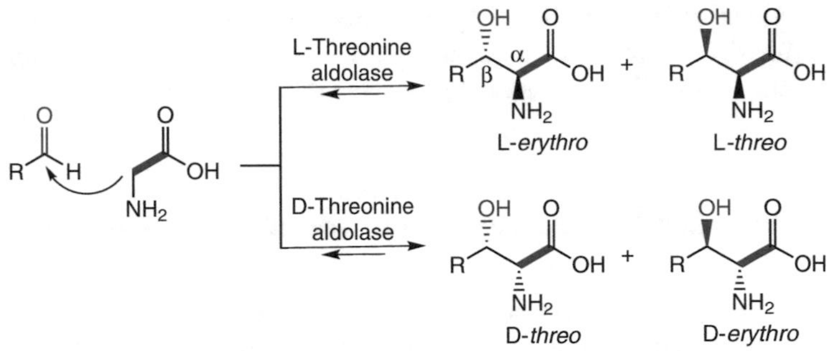

R	Enzyme	D : L	threo : erythro
Me	L-ThrA	<1 : 99	1 : 99
Me	D-ThrA	>99 : 1	~50 : 50
i-Pr	L-ThrA	<1 : 99	6 : 94
i-Pr	D-ThrA	>99 : 1	93 : 7

Scheme 2.192 Aldol reactions catalyzed by L-threonine aldolase

For example, recombinant D- and L-threonine aldolases from *E. coli* and *Xanthomonas oryzae*, respectively, were very faithful with respect to the formation of D- or L-configurated centers at Cα, but their diastereoselectivity for the β-hydroxy group was less pronounced. In particular, the D-enzyme gave a 1:1 *threo-erythro* mixture of the natural amino acid threonine (R = Me), which was improved in case of the sterically more demanding non-canonical analog (R = *i*−Pr).

For biocatalytic applications, several threonine aldolases show broad substrate tolerance for various acceptors, including aromatic aldehydes [1596, 1597]; however, conjugated enals were not accepted. The L-enzyme from *Candida humicola* was used in the synthesis of multifunctional α-amino-β-hydroxy acids, which possess interesting biological properties (Scheme 2.192) [1597]. A number of benzyloxy- and alkyloxy aldehydes were found to be good acceptors. Although the stereoselectivity of the newly generated α-center was absolute (providing only L-amino acids), the selectivity for the β-position bearing the hydroxyl group was less pronounced, leading to *threo-* and *erythro*-configurated products.

As with other aldolases, efforts were undertaken to overcome the limitation of threonine aldolases for their donor glycine. Screening uncovered novel aldolases, which were able to accept D- or L-alanine as donor, which opens the possibility to synthesize α-amino-α,α-dialkyl-β-hydroxy acids containing a quaternary center, which is difficult to obtain by conventional methods [1598, 1599]. Compounds of this type are important conformational modifiers of physiologically active peptides and building blocks for protease inhibitors.

Unfortunately, the position of the equilibrium does not favor synthesis, which requires to push the reaction by employing either an excess of the donor glycine (which is difficult to separate from the product) or the acceptor aldehyde (which at high concentrations may deactive the enzyme). A recently developed protocol relies on pulling of the equilibrium by (irreversible) decarboxylation of the formed α-amino-β-hydroxycarboxylic acid catalyzed by a decarboxylase to yield the corresponding aminoalcohols as final products [1600].

2.4.2   Thiamine-Dependent Acyloin and Benzoin Reactions

In the aldol reaction, C–C coupling always takes place in a head-to-tail fashion between the umpoled Cα atom of an enolate- or enamine-species (acting as donor) and the carbonyl C of an acceptor forming a β-hydroxycarbonyl product. In contrast, head-to-head coupling of two aldehydic species involving both carbonyl C atoms would lead to α-hydroxycarbonyl compounds, such as acyloins or benzoins. For this reaction, one aldehyde has to undergo umpolung at the carbonyl C, which is accomplished with the aid of an intriguing cofactor: thiamine diphosphate (ThDP, Scheme 2.193) [1601–1604]. This cofactor is an essential element for the formation/cleavage of C–C, C–N, C–O, C–P, and C–S bonds and plays a vital role as vitamin B_1 [1605–1607]. A schematic representation of the mechanism of enzymatic carboligation by ThDP-dependent enzymes is depicted in Scheme 2.193 [1608]. In a first step, ThDP is deprotonated at the iminium carbon, leading to a resonance-stabilized carbanion. The latter performs a nucleophilic attack on an aldehyde (R^1–CH=O), which is converted into the donor by forming a covalently bound carbinol species bearing a negative charge, equivalent to an enamine. This umpoled species attacks an acceptor aldehyde (R^2–CH=O) going in hand with C–C bond formation. Tautomerization of the diolate intermediate releases the α-ketol (acyloin/benzoin) product and regenerates the cofactor. Overall, the net reaction constitutes the transfer or an acyl anion equivalent to an aldehyde.

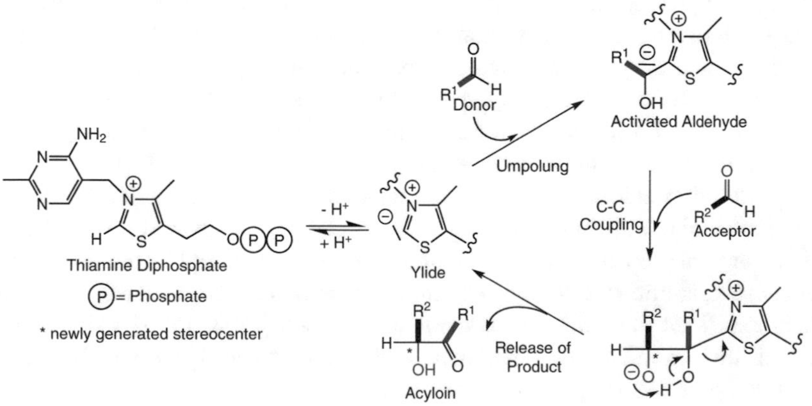

Scheme 2.193 Thiamine diphosphate-dependent carboligation of aldehydes (acyloin reaction)

In the enzymatic aldol reaction, the role of the donor and acceptor is strictly determined by the high specificity of the enzyme for the donor, hence only a single coupling product can be obtained. In contrast, the possible product range is more complex in acyloin and benzoin reactions: If only a single aldehyde species is used as substrate, only one product is obtained via homocoupling; however, a pair of regioisomeric α-hydroxyketones can be obtained via cross-coupling, when two different aldehydes are used, the ratio of which is determined by the preference of the enzyme for the donor versus acceptor, e.g., acetaldehyde versus benzaldehyde or vice versa (Schemes 2.194 and 2.200).

Stereocontrol in mixed acyloin and benzoin reactions is high only if the carboligation encompasses at least one (large) aromatic aldehyde, whereas with two (small) aliphatic aldehydes only moderate e.e.s are generally obtained.

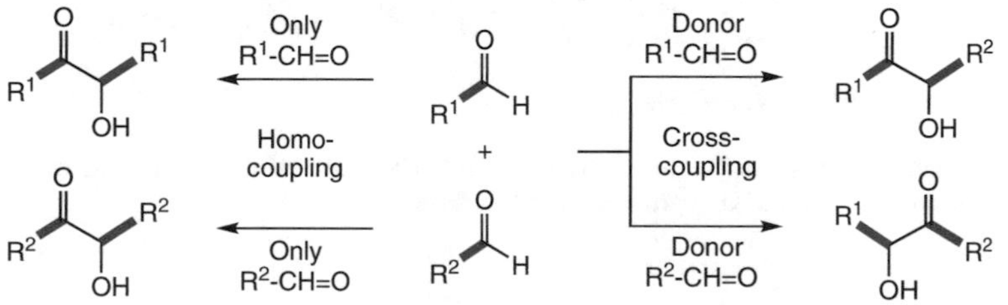

Scheme 2.194 Regioisomeric α-hydroxyketones obtained from homo- and cross-coupling of aldehydes

Acyloin and Benzoin Reactions[43]

Historically, the biocatalytic acyloin reaction was first observed by Liebig in 1913 during studies on baker's yeast [1609]. A few years later, Neuberg and Hirsch reported the formation of 3-hydroxy-3-phenylpropan-2-one (phenyl acetyl carbinol, PAC) from benzaldehyde by fermenting baker's yeast [1610]. Without knowledge on the actual enzyme(s) involved, this biotransformation assumed early industrial importance when it was shown that the acyloin thus obtained could be converted into (−)-ephedrine by diastereoselective reductive amination, a process which is operated on a scale of ~500 t/a [1611, 1612] (Scheme 2.195). Subsequent studies revealed that this yeast-based protocol can be extended to a broad range of aldehydes [1613, 1614].

---

[43]For the sake of clarity, the redox-neutral acyloin formation from two aldehydes is referred to as 'acyloin reaction', as opposed to the 'acyloin condensation', which constitutes the reductive condensation of two esters.

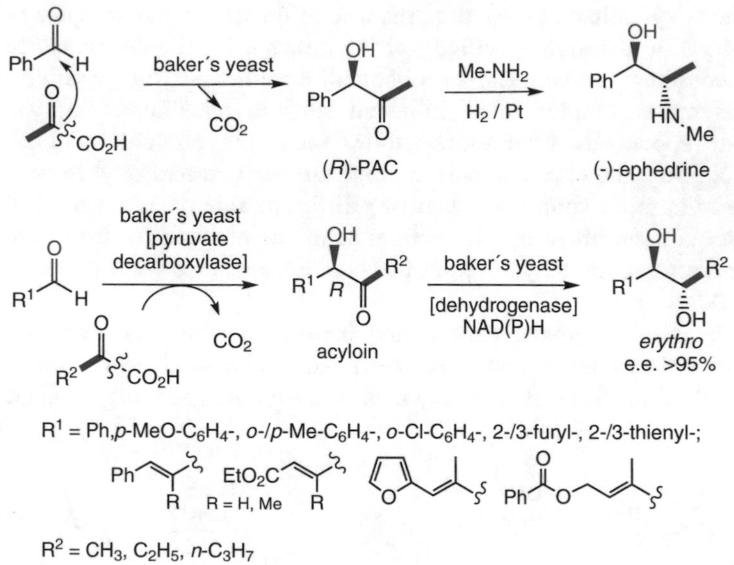

Scheme 2.195 Synthesis of (−)-ephedrine via baker's yeast catalyzed acyloin reaction and acyloin formation catalyzed by pyruvate decarboxylase

Despite its important history, it was during the early 1990s, that the reaction pathway was elucidated in detail [1615] and it turned out that the enzyme responsible for this reaction is pyruvate decarboxylase (PDC) [1616]. The C_2-unit (equivalent to acetaldehyde) originates from the decarboxylation of pyruvate and is transferred to the si-face of the aldehydic substrate to form an (R)-α-hydroxyketone (acyloin) with the aid of the cofactor TDP [1617]. Since pyruvate decarboxylase accepts α-ketoacids other than pyruvate, C_2-through C_4-equivalents can be transferred onto a large variety of aldehydes [1618–1620]. In whole-cell (yeast) transformations, the resulting acyloin is often reduced in a subsequent step by yeast alcohol dehydrogenase to give the erythro-diol. The latter reaction is a common feature of baker's yeast whose stereochemistry is guided by Prelog's rule (see Sect. 2.2.2 and 2.2.3, Scheme 2.112). The optical purity of the diols is usually better than 90% [1621–1624].

It must be mentioned, however, that for baker's yeast-catalyzed acyloin reactions the yields of chiral diols are usually in the range of 10–35%, but this is offset by the ease of the reaction and the low price of the reagents used. Depending on the substrate structure, the reduction of the acceptor aldehyde to give the corresponding primary alcohol (catalyzed by yeast alcohol dehydrogenases, see Sect. 2.2.2) and saturation of the α,β-double bond (catalyzed by ene-reductases, see Sect. 2.2.5) are the major competing reactions. To avoid low yields associated with yeast-catalyzed acyloin- [1625–1627] and benzoin-reactions [1625], isolated enzymes are nowadays used [1629–1631].

Pyruvate Decarboxylase
In vivo, pyruvate decarboxylase [EC 4.1.1.1] catalyzes the nonoxidative decarboxylation of pyruvate to acetaldehyde and is thus a key enzyme in the fermentative production of ethanol. The most well-studied PDCs are obtained from baker's yeast [1625, 1632, 1633] and from *Zymomonas mobilis* [1634].

From a synthetic viewpoint, however, its carboligation activity is more important [1635–1637]: All PDCs investigated so far prefer small aliphatic aldehydes as donors, used either directly or applied in the form of the respective α-ketocarboxylic acids [1638]. The latter are decarboxylated during the course of the reaction, which drives the equilibrium towards carboligation. Straight-chain α-ketoacids up to C-6 are good donors, whereas branched and aryl-aliphatic analogs are less suitable. On the acceptor side, aromatic aldehydes are preferred, leading to PAC-type acyloins (Schemes 2.195 and 2.196). Self-coupling of small aldehydes yielding acetoin-type products may occur.

Benzoylformate Decarboxylase
BFD [EC 4.1.1.7] is derived from mandelate catabolism, where it catalyzes the nonoxidative decarboxylation of benzoyl formate to yield benzaldehyde. Again, the reverse carboligation reaction is more important [1639–1641]. As may be deduced from its natural substrate, is exhibits a strong preference for large aldehydes as donor substrates encompassing a broad range of aromatic, heteroaromatic, cyclic aliphatic and olefinic aldehydes [1628]. With acetaldehyde as acceptor, it yields the complementary regio-isomeric product to PDC (Scheme 2.196).

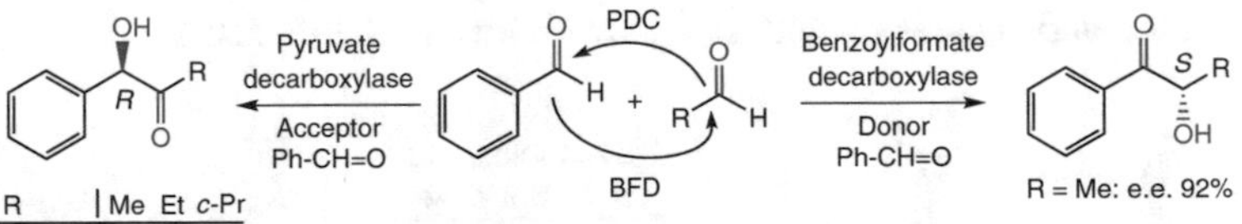

Scheme 2.196 Regiocomplementary carboligation of aldehydes catalyzed by pyruvate and benzoylformate decarboxylase

Benzaldehyde Lyase
Benzaldehyde lyase (BAL) [EC 4.1.2.38] from *Pseudomonas fluorescens*, which was able to grow on lignin-degradation products, such as benzoin, is a powerful biocatalyst for the homo- and cross-carboligation of various aromatic and aliphatic aldehydes. In contrast to PDC and BFD, BAL shows only negligible decarboxylation activity, while C–C lyase- and carboligation are dominant [1642–1644]. Especially the self-ligation of benzaldehyde yields benzoin with high activity and stereoselectivity (e.e. >99%), making this enzyme very interesting for industrial processes [1645]. For benzoin formation, *o*-, *m*-, and *p*-substituted aromatic aldehydes are widely accepted as donors [1646]. Cross-coupling of aromatic and aliphatic aldehydes (acting as acceptor) result in the formation of (*R*)-2-hydroxypropiophenone derivatives in analogy to BFD. On the acceptor side, formaldehyde, acetaldehyde and close derivatives, such as phenyl-, mono-, or dimethoxyacetaldehyde are tolerated.

The remarkable synthetic potential of BAL is demonstrated by the regiocomplementary benzoin reaction of α,β-unsaturated aldehydes acting as donor or acceptor, respectively. While large aldehydes acted as donors (product type A), small counterparts served as acceptors leading to isomeric olefinic acyloins B in high e.e.s [1647] (Scheme 2.197).

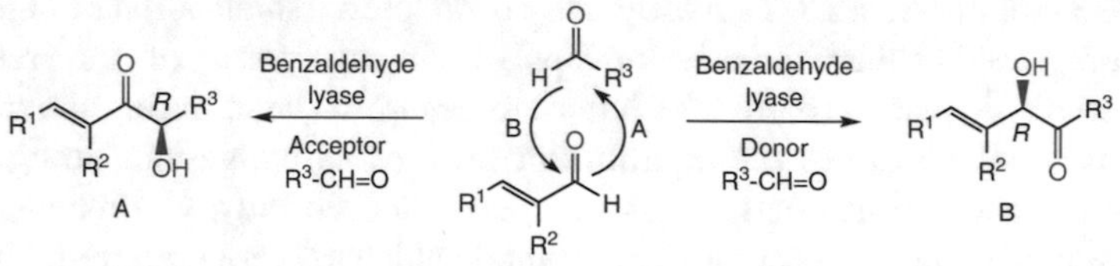

Scheme 2.197 Regiocomplementary carboligation of aldehydes catalyzed by benzaldehyde lyase

α-Ketoglutarate decarboxylases

A useful extention to the set of acyloin-forming enzymes is the use of α-ketoglutarate decarboxylases [1648]. Like pyruvate decarboxylase, they catalyze the decarboxylative carboligation between an α-keto acid and aldehyde, but they use α-ketoglutarate as donor (Scheme 2.198). As a key molecule in the Krebs-cycle, the latter is abundantly available from glutamate. Among the enzymes tested, SucA from *E. coli* showed excellent stereoselectivities for aliphatic acceptors, whereas MenD (from *Mycobacterium tuberculosis*) was best for (substituted) benzaldehydes [1649]. This strategy allows acyloin formation from aldehydes going in hand with extension by a (succinoyl) C_4-unit. Concomitant decarboxylation provides a strong driving force and ensures quantitative conversions (Scheme 2.198).

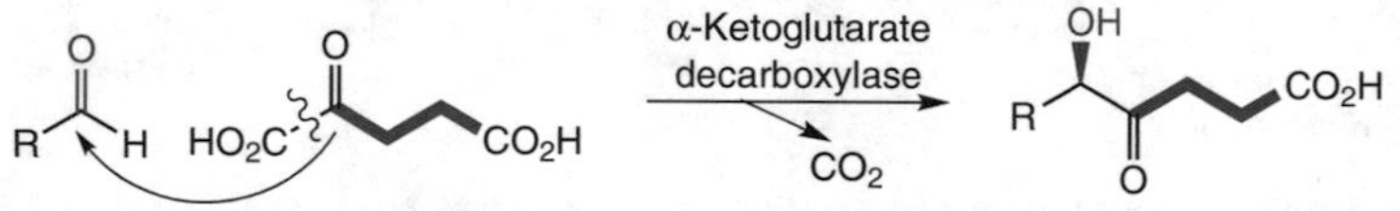

R	Enzyme	E.e. [%]
Me, Et	SucA	94
n-Bu	SucA	90
n-Pent	SucA	82
o-F-, *m*-I-C$_6$H$_4$-, Ph	MenD	94–96

Scheme 2.198 Acyloin formation with C_4-extention using α-ketoglutarate decarboxylase SucA

Transketolase

In the oxidative pentose phosphate pathway, ThDP-dependent transketolase[44] catalyzes the reversible interconversion of phosphorylated aldoses and ketoses via transfer of a terminal 2-carbon hydroxyacetyl-unit (Scheme 2.199) [1650]. Its

---

[44]Correctly, this enzyme has the charming name 'D-seduheptulose-7-phosphate: D-glyceraldehyde-3-phosphate glycoaldehyde transferase'.

mechanism resembles a classical acyloin reaction mediated by ThDP [1651, 1652]. Fortunately, the natural phosphorylated substrate(s) can be replaced by the donor hydroxypyruvate [1653], which is decarboxyled to furnish a hydroxyacetaldehyde unit thereby driving the reaction towards completion. The C-2 fragment is transferred onto an aldehyde acceptor yielding an acyloin possessing a *threo*-diol configuration. This method has allowed the synthesis of a number of monosaccharide-like acyloins on a preparative scale [1654–1657]. Transketolases were initially be obtained from yeast [1658] and spinach [1659], their overexpression has opened the way for large-scale production [1660, 1661].

Transketolases from various sources have been shown to possess a broad acceptor spectrum yielding products with complete (*S*)-stereospecificity for the newly formed C-3 stereocenter [1662]. Generic aldehydes are usually converted with full stereocontrol and even α,β-unsaturated aldehydes are accepted to some degree.

α-Hydroxy aldehydes show enhanced rates by mimicking the natural substrate [1651]. Interestingly, transketolase recognizes chirality in the aldehydic acceptor moiety to a greater extent than aldolases. Thus, when (stereochemically stable) racemic α-hydroxyaldehydes are employed as acceptors, an efficient kinetic resolution of the α-center is achieved (Scheme 2.199). Only the (α*R*)-enantiomer is transformed into the corresponding keto-triol leaving the (α*S*)-counterpart behind [1663]. In a related manner, when (±)-3-azido-2-hydroxypropionaldehyde was chosen as acceptor, only the D-(*R*)-isomer reacted and the L-(*S*)-enantiomer remained unchanged [1546].

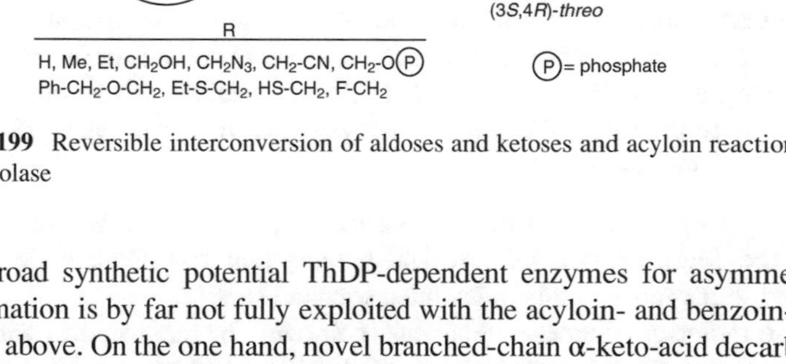
Transketolase
Aldose
Ketose
rac
Transketolase
CO_2
(3S,4R)-threo
R
H, Me, Et, CH_2OH, CH_2N_3, CH_2-CN, CH_2-O(P)
Ph-CH_2-O-CH_2, Et-S-CH_2, HS-CH_2, F-CH_2
(P)= phosphate

Scheme 2.199 Reversible interconversion of aldoses and ketoses and acyloin reaction catalyzed by transketolase

The broad synthetic potential ThDP-dependent enzymes for asymmetric C–C bond formation is by far not fully exploited with the acyloin- and benzoin-reactions discussed above. On the one hand, novel branched-chain α-keto-acid decarboxylases favorably extend the limited substrate tolerance of traditional enzymes, such as PDC, by accepting sterically hindered α-ketoacids as donors [1664]. On the other hand, the acceptor range may be significantly widened to encompass ketones, α-ketoacids and even CO_2, which leads to novel types of products (Scheme 2.200).

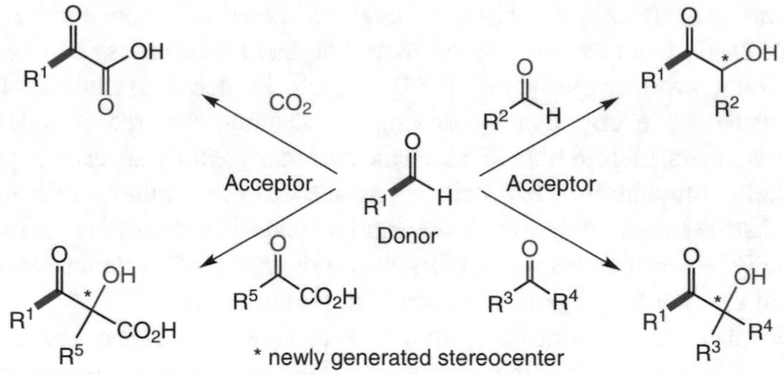

Scheme 2.200 Future potential of thiamine-dependent C–C bond formation

2.5 Addition and Elimination Reactions

Among the various types of transformations used in organic synthesis, addition reactions are the 'cleanest' since two components are combined into a single product with 100% atom efficiency [1665, 1666].

The asymmetric addition of (small) molecules, such as water, ammonia and C-H acidic carbon nucleophiles (such as hydrogen cyanide, nitroalkanes, β-dicarbonyl compounds) onto C=C or C=O bonds is typically catalyzed by lyases. Depending on the substitution pattern of the substrate, up to two chiral centers are created from a prochiral substrate via desymmetrization.

2.5.1 Addition of Water

The asymmetric addition of water onto olefins is one of the 'dream-reactions' in organic synthesis and represents one of the (largely unsolved) problems of catalysis. Enzymes called hydratases [EC 4.2.1.X][45] can catalyze this reaction [1667].

In analogy to the rules of chemical catalysis, two different types of enzymatic hydration mechanisms exist:

– Hydration of electron-rich (isolated) alkenes proceeds via acid-catalysis and obeys the Markovnikov rule, which dictates that the nucleophile [OH⁻] is attached to the more highly substituted carbon [1668].
– Electron-deficient alkenes, which are polarized by an electron-withdrawing (carbonyl) group, are hydrated via Michael-type addition with nucleophilic attack at Cβ. If the corresponding proton to be added is in the α-position to a carboxyl group, hydration occurs in an *anti*-fashion (Scheme 2.202), in case of a (coenzyme A) thioester, the *syn*-product is preferred [1669, 1670].

---

[45]Occasionally also called 'hydro-lyases'.

For biotransformations, hydratases acting on isolated olefins are developed to a lesser extent, although a few have gained industrial importance.

Oleate hydratases from different sources are employed for the production of (*R*)-10-hydroxystearic acid starting from oleate [(9*Z*)-octadecenoic acid] with high volumetric productivities (12 g L^{-1} h^{-1}) (Scheme 2.201). Although regio- and stereoselectivities are very high, the substrate tolerance is somewhat limited to various long-chain unsaturated fatty acids possessing a (9*Z*)-olefinic bond, such as palmitoleic [(9*Z*)-hexadecenoic], γ-linoleic [(all-*Z*)-6,9,12-octadecatrienoic], linoleic [(all-*Z*)-9,12-octadecadienoic), myristoleic [(9*Z*)-tetradecenoic] and α-linoleic acid [(all-*Z*)-9,12,15-octadecatrienoic acid) in decreasing order of reactivity relative to oleic acid [1671].

In nature, limonene hydratase is involved in the biodegradation of the monoterpene limonene, which is available in large amounts as waste-product from citrus fruit processing. Regio- and enantioselective hydration of (*R*)-limonene yields (*R*)-α-terpineol, which is a popular olfactory component exerting a strong lilac-like smell [1672]. In contrast, the (*S*)-enantiomer displays a conifer-like odor.

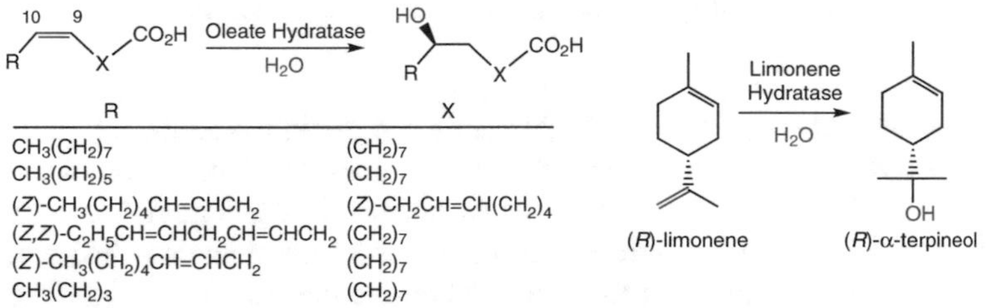

Scheme 2.201 Regio- and stereoselective hydration of non-activated isolated olefins

Fumarase and malease catalyze the stereospecific addition of water onto C=C bonds conjugated to a carboxylic acid [1673]. Both reactions are mechanistically related and proceed in a *anti*-fashion [1674, 1675], with protonation occurring from the *re*-side (Scheme 2.202). Both enzymes are complementary with respect to the (*E*)- or (*Z*)-configuration of their substrates and show exceptional stereoselectivities for the nucleophilic attack, but their substrate tolerance is rather narrow.

The addition of water onto fumaric acid by fumarase leads to (*S*)-malic acid (Scheme 2.202). The latter is used as an acidulant in fruit juices, carbonated soft drinks, and candies, and is performed at a capacity of ~2000 t/year [1676, 1677]. While fumarate and chlorofumaric acid are well accepted, the corresponding (sterically more encumbered) bromo-, iodo-, and methyl derivatives are transformed at exceedingly low rates, albeit with excellent stereoselectivities. Replacement of one of the carboxylic groups or changing the stereochemistry of the double bond from (*E*) to (*Z*) is not tolerated by fumarase [1678].

The analogous hydration of the stereoisomeric (Z)-isomer (maleic acid) is catalysed by malease (maleate hydratase) and produces the mirror-image (R)-malate [1679, 1680]. The latter enzyme also accepts 2-methylmaleate (citraconate) to form (R)-2-hydroxy-2-methylsuccinate (2-methyl-maleate) [1681].

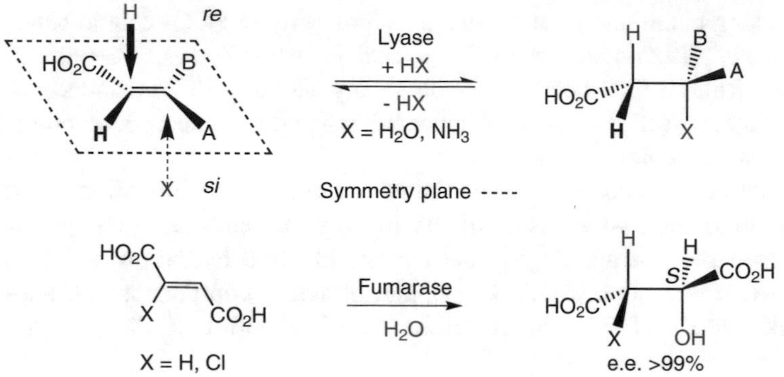

Scheme 2.202 *anti*-Selective asymmetric hydration and hydroamination of activated conjugated C=C bonds and fumarase-catalyzed formation of (S)-malate from fumaric acid

In contrast to the hydration of highly activated olefinic *di*acids, α,β-unsaturated *mono*carboxylic acids have to be activated via a thioester linkage onto the cofactor Coenzyme A (Scheme 2.203). The latter is catalyzed by an enoyl-CoA synthetase and requires ATP as energy source. The enoyl-CoA intermediate is hydrated by an enoyl-CoA hydratase in a *syn*-fashion yielding the corresponding β-hydroxyacyl-CoA as product, which is finally hydrolyzed by a thioesterase to liberate the β-hydroxycarboxylic acid and CoA, which re-enters the catalytic cycle. Due to the complexity of this multienzyme-system requiring ATP and CoA, hydration of acrylic acid derivatives is always performed using whole cells [1423, 1682, 1683].

An elegant example for this biotransformation is the asymmetric hydration of crotonobetaine yielding the 'nutraceutical' (R)-carnitine (Scheme 2.203), which is used as an additive in baby food, geriatric nutrition and health sport. In order to avoid the undesired degradation of the product by the whole-cell biocatalyst, mutant strains lacking carnitine dehydrogenase have been developed to produce (R)-carnitine at a capacity of >100 t/year [1684–1686].

The capacity of microbial cells of different origin to perform an asymmetric hydration of C=C bonds has only been poorly investigated but they show a promising synthetic potential. For instance, *Fusarium solani* cells are capable of hydrating the non-activated 'inner' (E)-double bond of terpene alcohols (e.g., nerolidol) or – ketones (e.g., geranyl acetone) in a highly selective manner [1687]. However, side reactions such as hydroxylation, ketone-reduction, or degradation of the carbon skeleton represent a major drawback. On the other hand, resting cells of *Rhodococcus rhodochrous* catalyzed the asymmetric addition of water onto the C=C bond

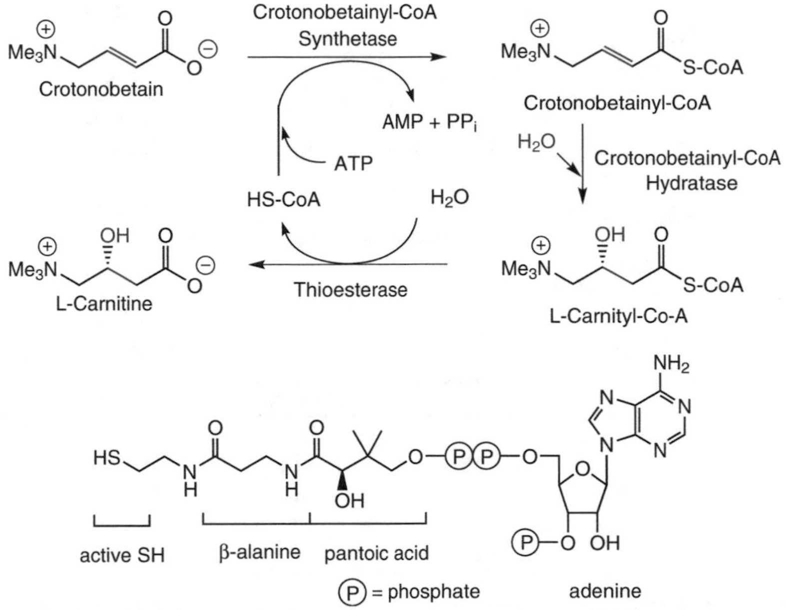

Scheme 2.203   Asymmetric hydration of crotonobetaine to carnitine via a multienzyme-system

of α,β-unsaturated butyrolactones with high enantioselectivity, furnishing β-hydroxylactones in moderate yields [1688]. Furthermore, C=C hydration reactions are often not far from equilibrium, which requires sophisticated process engineering, such as selective crystallization or the use of membrane technology.

2.5.2   Addition of Ammonia

The addition of ammonia across C=C bonds is equivalent to a 'hydroamination' and is catalyzed by ammonia lyases [1689], such as aspartase [EC 4.3.1.1], 3-methylaspartase [1690], and phenylalanine ammonia lyase [EC 4.3.1.5]. The equilibrium can be shifted by high concentrations of ammonia (~4–6 M) [1691].

Enzymatic amination of fumaric acid using aspartase is used for the production of L-aspartic acid at a capacity of ~10,000 t/year (Scheme 2.204) [1692–1695]. Although aspartase is one of the most specific enzymes known by accepting only its natural substrate [1696–1698], some variations concerning the N-nucleophile are tolerated: Hydroxylamine, hydrazine, methoxyamine, and methylamine are accepted and furnish the corresponding N-substituted L-aspartate derivatives [1699–1701].

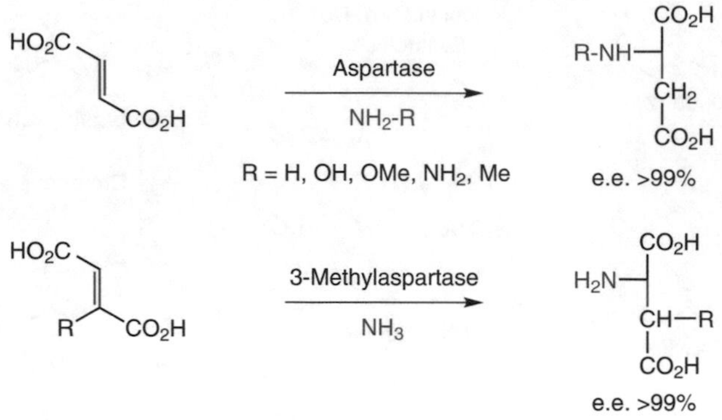

HO₂C
‖
CO₂H

Aspartase
—————→
NH₂-R

R = H, OH, OMe, NH₂, Me

$$R-NH-\overset{\displaystyle CO_2H}{\underset{\displaystyle CO_2H}{\underset{\displaystyle |}{\overset{|}{C}H_2}}}$$

e.e. >99%

HO₂C
‖
R   CO₂H

3-Methylaspartase
—————→
NH₃

$$H_2N-\overset{\displaystyle CO_2H}{\underset{\displaystyle CO_2H}{\underset{\displaystyle |}{\overset{|}{C}H-R}}}$$

e.e. >99%

R	Yield [%]
H	90
Me, Et, Cl	60-61
i-Pr	54
n-Pr	49
n-Bu	0

Scheme 2.204 Amination of fumarate derivatives by aspartase and 3-methylaspartase

In contrast to aspartase, some structural variations are tolerated by the related 3-methylaspartase (Scheme 2.204). For instance, the methyl group in the natural substrate may be replaced by a chlorine atom or by small alkyl moieties [1702], but the fluoro- and the iodo-analog are not good substrates. Although the bromo-derivative is accepted, it irreversibly inhibits the enzyme [1703].

In a related fashion, asymmetric amination of (*E*)-cinnamic acid yields L-phenyl-alanine using L-phenylalanine ammonia lyase [EC 4.3.1.5] at a capacity of 10,000 t/ year [1423, 1704]. The enzyme tolerates a wide variety of halogen substituents on the phenyl ring and also accepts heterocyclic analogs, such as pyridyl and thienyl derivatives [1705]. A fascinating variant of this biotransformation consists in the use of phenylalanine aminomutase from *Taxus chinensis* (yew tree), which interconverts α- to β-phenylalanine in the biochemical route leading to the side chain of taxol [1706]. In contrast to the majority of the cofactor-independent C–O and C–N lyases discussed above, its activity depends on the protein-derived internal cofactor 5-methylene-3,5-dihydroimidazol-4-one (MIO) [1707]. Since the reversible α,β-isomerization proceeds via (*E*)-cinnamic acid as achiral intermediate, the latter can be used as substrate for the amination reaction. Most remarkably, the ratio of α- versus β-amino acid produced (which is 1:1 for the natural substrate, R = H) strongly depends on the type and the position of substituents on the aryl moiety: While *o*-substituents favor the formation of α-phenylalanine derivatives, *p*-substituted substrates predominantly lead to β-amino analogs. A gradual switch between both pathways occurred with *m*-substituted compounds. With few exceptions, the stereoselectivity remained excellent (Scheme 2.205) [1708, 1709].

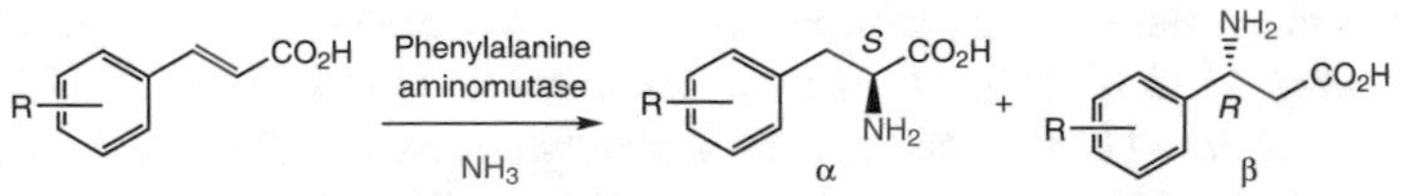

R	Ratio α/β	α E.e. [%]	β E.e. [%]
H	1 : 1	>99	>99
o-F, o-Cl, o-Br, o-Me	>98 : 2	>99	n.d.
m-F	86 : 14	92	n.d.
m-Me	20 : 80	>99	>99
p-n-Pr	12 : 88	n.d.	>99
p-Et	9 : 91	n.d.	>99
p-Me	4 : 96	>99	>99

Scheme 2.205 Formation of α- and β-phenylalanine derivatives using phenylalanine ammonia mutase

2.5.3 Cyanohydrin Formation and Henry-Reaction

Hydroxynitrile lyase enzymes (HNLs) catalyze the asymmetric addition of hydrogen cyanide onto a carbonyl group of an aldehyde or a ketone thus forming a chiral cyanohydrin [1710–1715], [46] a reaction which was used for the first time as long ago as 1908 [1716]. In nature, HNLs catalyze the cleavage of cyanohydrins derived from cyanoglucosides and cyanolipids (Scheme 2.95) to liberate HCN as a defence mechanism against herbivores and microbial attack. This activity is not only widespread in plants, but is also found in bacteria, fungi, lichens and insects. Although they catalyze the same reaction, HNLs belong to (at least) four different structural folds, i.e. α/β-hydrolase proteins, oxidoreductases, cupins[47] and alcohol dehydrognases, which indicates a perfect example of convergent evolution [1717].

   Cyanohydrins are rarely used as products per se, but they represent versatile starting materials for the synthesis of various types of compounds [1718]. Most prominent, chiral cyanohydrins constitute the alcohol moieties of several commercial pyrethroid insecticides (see Scheme 2.208) (see below) [1719].

   Since only a single enantiomer is produced during the reaction – through desymmetrization of a prochiral substrate – the availability of enzymes possessing opposite stereochemical preference is of importance to gain access to both (R)- and (S)-cyanohydrins (Scheme 2.207). Fortunately, an impressive number of hydroxy-nitrile lyases can be isolated from cyanogenic plants to meet this requirement [1720–1724].

   (R)-Specific enzymes are obtained predominantly from the Rosacea family (almond, plum, cherry, apricot) and they have been thoroughly investigated [1725–1728]. They contain FAD in its oxidized form as a prosthetic group located near (but not in) the active site, but this moiety does not participate in catalysis and seems to be an evolutionary relict.

---

[46]Outdated terms for hydroxynitrile lyases are 'oxynitrilases' or 'hydroxynitrilases'.

[47]A family of small barrel-shaped proteins, named after 'cupa' (Latin) = small barrel.

On the other hand, (S)-hydroxynitrile lyases [1729–1732] were found in *Sorghum bicolor* [1733] (millet), *Hevea brasiliensis* [1734, 1735] (rubber tree), *Ximenia americana* [1736] (sandalwood), *Sambucus niger* [1737] (elder), *Manihot esculenta* [1731, 1738] (cassava), flax, and clover. They do not contain FAD and they exhibit a more narrow substrate tolerance, as aliphatic aldehydes are not always accepted. Furthermore, the reaction rates and optical purities are sometimes lower than those which are obtained when the (R)-enzyme is used. Based on X-ray structures [1739, 1740], the mechanism of enzymatic cyanohydrin formation has been elucidated as follows (Scheme 2.206) [1741]: The substrate is positioned in the active site with its carbonyl group bound through a network of hydrogen bonds involving His/Cys/Tyr or Ser/Thr-moieties, while the lipophilic residue is accommodated in a hydrophobic pocket. Nucleophilic addition of cyanide anion occurs from opposite sides from cyanide-binding pockets, which are made of positively charged Arg/Lys- or His/Lys-residues [394].

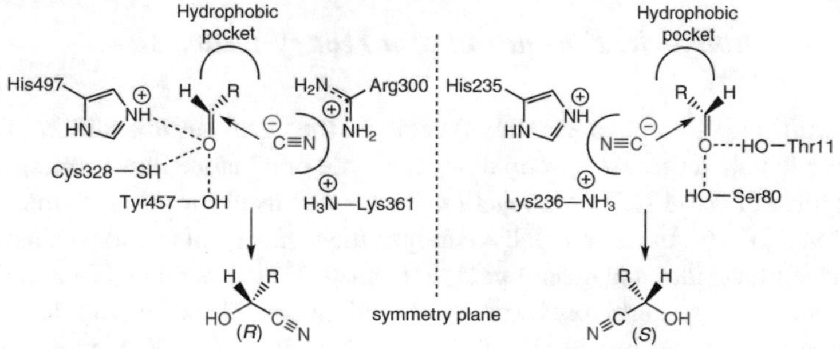

Scheme 2.206 Mechanism of (R)- and (S)-hydroxynitrile formation by HNLs from almond and *Hevea brasiliensis*, respectively

The following set of rules for the substrate-acceptance of (R)-hydroxynitrile lyase was delineated [1742].

• The best substrates are aromatic aldehydes, which may be substituted in the *meta*- or *para*-position; also heteroaromatics such as furan and thiophene derivatives are well accepted [1743–1746].
• Straight-chain aliphatic aldehydes are transformed as long as they are not longer than six carbon atoms; the α-position may be substituted with a methyl group. It is noteworthy, that also α,β-unsaturated aliphatic aldehydes were transformed into the corresponding cyanohydrins in a clean reaction. No formation of saturated β-cyano aldehydes through Michael-type addition of hydrogen cyanide across the C=C double bond occurred. The latter is a common side reaction using traditional methodology.
• Methyl ketones are transformed into cyanohydrins [1747], while ethyl ketones are impeded by low yields [1748].
• For large or sterically demanding aldehydes, such as *o*-chlorobenzaldehyde, (R)-HNL mutants possessing a more spacious active site were constructed [1749, 1750]. The (R)-*o*-chloromandelonitrile thus obtained represents the chiral core of the blockbuster clopidogrel (Plavix) to prevent heart attack or stroke (Scheme 2.207, Table 2.5).

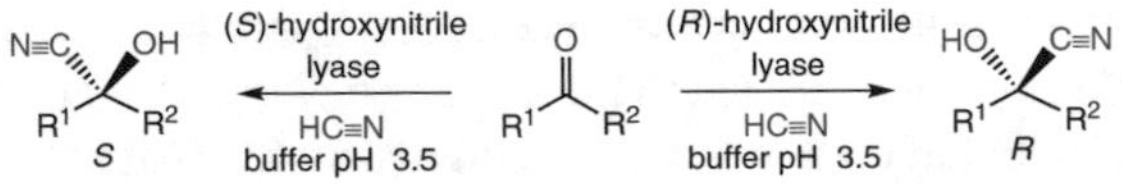

Scheme 2.207 Stereocomplementary asymmetric cyanohydrin formation

Table 2.5   Synthesis of (R)-cyanohydrins from aldehydes and ketones

R^1	R^2	e.e. (%)
Ph–	H	94
p-MeO–C$_6$H$_4$–	H	93
2-furyl–	H	98
n-C$_3$H$_7$–	H	92–96
t-Bu–	H	73
(E)-CH$_3$–CH=CH–	H	69
n-C$_3$H$_7$–	Me	95
n-C$_4$H$_9$–	Me	98
(CH$_3$)$_2$CH–(CH$_2$)$_2$–	Me	98
CH$_2$=CH(CH$_3$)–	Me	94
Cl–(CH$_2$)$_3$–	Me	84

The (S)-hydroxynitrile lyase from *Hevea brasiliensis* has been made available in sufficient quantities by cloning and overexpression to allow industrial-scale applications [1751]. Of particular interest is the synthesis of the (S)-cyanohydrin from *m*-phenoxybenzaldehyde (Table 2.6), which is an important intermediate for synthetic pyrethroids.

Table 2.6   Synthesis of (S)-cyanohydrins from aldehydes

R^1	R^2	e.e. (%)
Ph–	H	96–98
p-HO–C$_6$H$_4$–	H	94–99
m-PhO–C$_6$H$_4$–	H	96
3-thienyl–	H	98
n-C$_5$H$_{11}$–	H	84
n-C$_8$H$_{17}$–	H	85
CH$_2$=CH–	H	84
(E)-CH$_3$–CH=CH–	H	92
(E)-n-C$_3$H$_7$–CH=CH–	H	97
(Z)-n-C$_3$H$_7$–CH=CH–	H	92
n-C$_5$H$_{11}$–	CH$_3$	92
(CH$_3$)$_2$CH–CH$_2$–	CH$_3$	91

Synthetic pyrethroids comprise a class of potent insecticides with structural similarities to a number of naturally occurring chrysanthemic acid esters found in the extract of pyrethrum flowers (*Chrysanthemum cinerariaefolium*) (Scheme 2.208). These natural products constitute highly potent insecticides, but their instability (inherent to the cyclopentenone moiety) precludes their broad application in agriculture. This fact has led to the development of a range of closely related analogs, which retain the high insecticidal activity of their natural ancestors but are more stable. All of these synthetic pyrethroids contain asymmetric carbon atoms and their insecticidal activity resides predominantly in one particular isomer. In order to reduce the environmental burden during pest control, single isomers are marketed [1752].

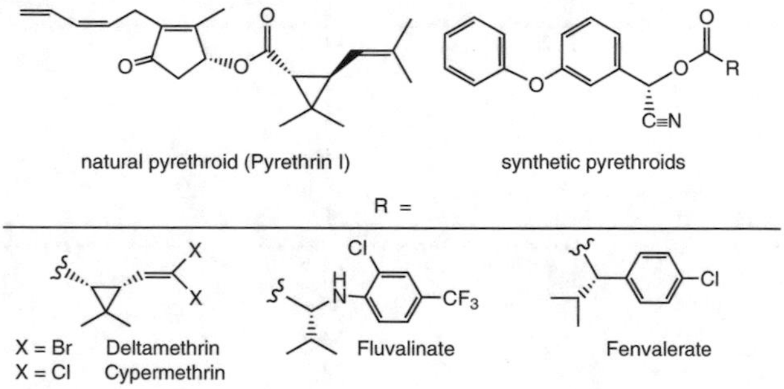

Scheme 2.208 Natural and synthetic pyrethroids

Two particular problems which are often encountered in hydroxynitrile lyase-catalyzed reactions are the spontaneous nonenzymatic formation of racemic cyanohydrin and racemization of the product due to equilibration of the reaction. As a result, the optical purity of the product is decreased. Bearing in mind that both the chemical formation and the racemization of cyanohydrins are pH-dependent and require water, three different techniques have been developed in order to suppress the depletion of the optical purity of the product.

- Adjusting the pH of the medium to a value below 3.5, which is the lower operational pH-limit for most hydroxynitrile lyases.
- Lowering the water-activity of the medium [1753] by using water-miscible organic cosolvents such as ethanol or methanol. Alternatively, the reaction can be carried out in a biphasic aqueous-organic system or in a monophasic organic solvent (e.g., ethyl acetate, di-*i*-propyl, or methyl *t*-butyl ether) which contains only traces of water to preserve the enzyme's activity.
- In order to avoid the use of hazardous hydrogen cyanide, *trans*-cyanation reactions were developed using acetone cyanohydrin [1754, 1755] as donor for hydrogen cyanide. The latter is considerably more easy to handle due to its higher boiling point (82 °C) compared to HCN (26 °C). Using this technique, the competing chemical cyanohydrin formation is negligible due to the low concentration of free hydrogen cyanide.

A fascinating variant of the enzymatic cyanohydrin formation consists in the use of nitroalkanes (as nonnatural nucleophiles) instead of cyanide (Scheme 2.209), which constitutes a biocatalytic equivalent to the Henry-reaction, which is not known in nature thus far. It produces vicinal nitro-alcohols, which are valuable precursors for amino alcohols. Using (*S*)-HNL from *Hevea brasiliensis* or *Arabidopsis thaliana*

[1756], the asymmetric addition of nitromethane to *p*-methyl- and halogen-substituted benzaldehydes gave the nitroalcohols in high e.e.s, while for *p*-nitro- and *m*-hydroxybenzaldehyde the stereoselectivity dropped sharply [1757, 1758]. With nitroethane, two stereocenters are created: Whereas the stereoselectivity for the alcoholic center at C_1 was high (e.e. 95%), the recognition for the adjacent center bearing the nitro moiety was modest and diastereomers were formed.

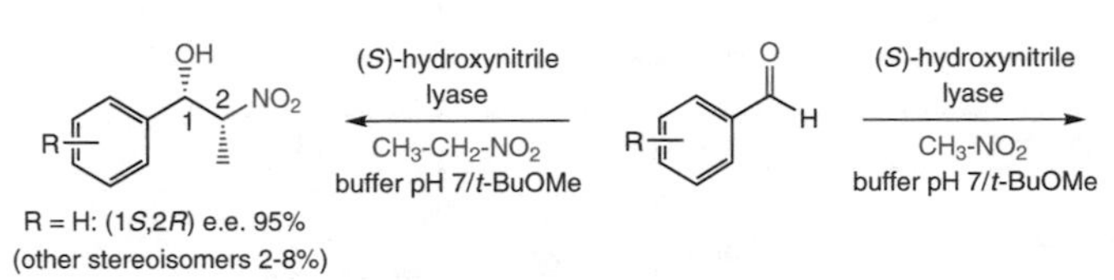

R = H: (1*S*,2*R*) e.e. 95%
(other stereoisomers 2-8%)

R	e.e. [%]
m-OH	18
p-NO$_2$	28
H	97
o-, *m*-, or *p*-Cl	95-98

Scheme 2.209 Asymmetric Henry-reaction catalyzed by (*S*)-hydroxynitrile lyases

2.5.4   Michael-Type Additions

Since its serendipitous discovery in 1887 [1759], the nucleophilic addition of 1,3-dicarbonyl donors onto α,β-unsaturated carbonyl acceptors – known as the Michael-addition – became a central tool for the (stereoselective) C–C bond formation [1760, 1761]. Despite its simplicity based on base-catalysis, it is a puzzling fact that nature did not evolve an analogous enzyme and a 'Michael-lyase' as such does not exist.

Enzyme-catalyzed C–C bond forming Michael-type additions of umpoled car-bonyl species onto α,β-unsaturated carbonyl acceptors are extremely rare and are found in polyketide pathways, e.g. in the biosynthesis of rhizoxin [1762]. An analogous reaction where a ThDP-bound carbanion performs a 1,4-addition onto an enal (instead of the usual 1,2-addition onto a carbonyl group to form an acyloin, Schemes 2.193 and 2.197) are catalyzed by PigD and MenD and are equivalent to the Stetter reaction [1763, 1764].

Early studies in search for enzyme-catalyzed Michael-type additions focused on the exploitation of the catalytic promiscuity of well-characterized hydrolytic enzymes, such as lipases, acylases and proteases using 1,3-dicarbonyl donors and enals or nitroalkenes as acceptors. Although encouraging catalytic activities translating into yields of up to 90% were found, stereoselectivities remained disappointingly low and e.e.s typically ranged within ~20–40% [1765–1767]. In the majority of cases, the product was (near) racemic [1768] or were not reported at all. Attempts to rationally design a 'Michael-lyase' from a lipase-scaffold gave disappointingly low stereoselectivities [1769–1772]. Consequently, these seminal studies are interesting from an evolutionary and mechanistic standpoint, but they are synthetically irrelevant.

As a typical example for a stereoselective Michael-type addition catalyzed by hydrolytic enzymes is depicted in Scheme 2.210. When α-trifluoromethyl propenoic acid was subjected to the action of various proteases, lipases and esterases in the presence of a nucleophile (NuH), such as water, amines, and thiols, chiral propanoic acids were obtained in moderate optical purity [1773]. The reaction mechanism probably involves the formation of an acyl enzyme interme-diate (Sect. 1.1, Scheme 2.1). Being an activated derivative, the latter is more electrophilic than the 'free' carboxylate and undergoes an asymmetric Michael addition by the nucleophile, directed by the chiral environment of the enzyme.

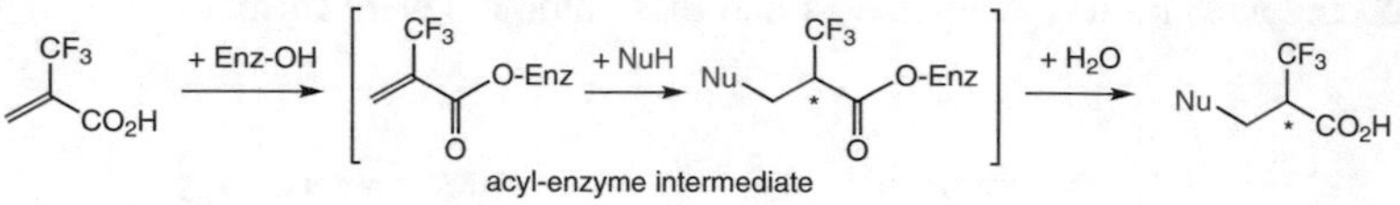

Nucleophile	Enzyme	e.e. [%]
H_2O	*Candida rugosa* lipase	70
Et_2NH	*Candida rugosa* lipase	71
H_2O	pig liver esterase	60
Et_2NH	pig liver esterase	69
PhSH	pig liver esterase	50

Scheme 2.210 Asymmetric Michael addition catalyzed by hydrolytic enzymes

The first successful attempt to unearth a highly stereoselective enzyme-catalyzed Michael-type addition employed 4-oxalocrotonate tautomerase (4-OT),[48] which cata-lyzes the tautomerization of 2-hydroxymuconate to 2-oxo-3-hexenedioate (4-oxalocrotonate), a step in the oxidative biodegradation of alkylbenzenes, such as toluene and xylene (Scheme 2.211). 4-OT contains a rare N-terminal proline residue (Pro-1) in its active site, which acts as acid/base in the 'natural' tautomerization. In contrast, the mechanism of the Michael-addition presumably involves a nucleophilic

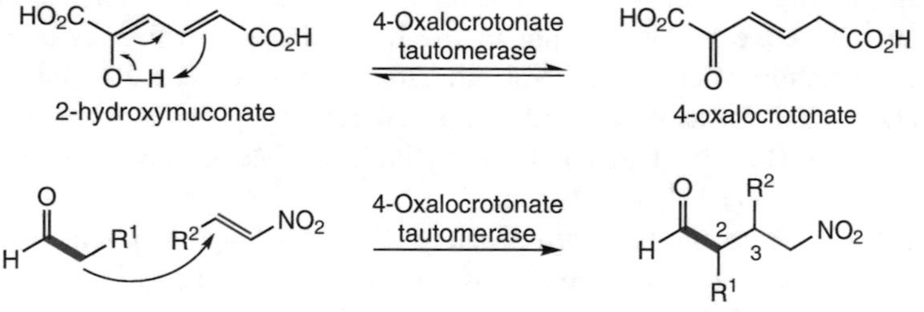

R^1	R^2	e.e. [%]	d.e. [%]	yield [%]
H	Ph	89 (S)	—	46
H	p-Cl-C_6H_4	84 (S)	—	64
H	(E)-CH=CH-Ph	95 (S)	—	38
H	i-Bu	98 (R)	—	74
Me	Ph	50 (2R,3S)	86	64

Scheme 2.211 Tautomerization and Michael-type addition catalyzed by 4-oxalocrotonate tautomerase

---

[48][EC 5.3.2.6] also called 2-hydroxymuconate tautomerase or 4-oxalocrotonate isomerase.

enamine intermediate (derived from a Schiff-base between Pro-1 and the aldehyde donor), a reminiscent to the organocatalytic analog. With acetaldehyde as donor, e.e.s of up to 98% were achieved, the stereoselectivities were lower with propanal [1774–1776].

2.6   Transfer Reactions

2.6.1   Glycosyl Transfer Reactions

Glycosidic molecules in the form of oligo- or polysaccharides represent about two-thirds of the carbon found in the biosphere, largely in the form of (hemi) cellulose and chitin.[49] While β-linked polysaccharides provide structural support, α-linked glycans, such as starch and glycogen, are more easily cleaved and serve as energy storage [1777]. As a consequence, glycosyl transfer is certainly one of the most important biochemical reactions [1778].

While polysaccharides are the basis for material sciences [1779], oligo-sugars play a vital role in intracellular migration and secretion of glycoproteins, cell–cell interactions, oncogenesis, and interaction of cell surfaces with pathogens [1780–1782]. The building blocks are monosaccharides which (theoretically) occur in an enormous number of stereoisomers, which results in a structural diversity far greater than that possible with peptides of comparable size [1783].[50] Fortunately, Nature is using almost exclusively pentoses and hexoses for in vivo synthesis.

The ready availability of such oligosaccharides of well-defined structure is critical for the synthesis of drug candidates. Isolation of these materials from natural sources is a complex task and is not economical on a large scale due to their low concentration in carbohydrate mixtures obtained from natural sources. Chemical synthesis of complex oligosaccharides is one of the greatest challenges facing synthetic organic chemistry since it requires many protection and deprotection steps which result in low overall yields [1784]. In this context, biocatalysts are attractive as they allow the regio- and stereospecific synthesis of oligosaccharides with a minimum of protection and deprotection steps [1504–1506, 1785–1793]. There are four groups of enzymes which can be used for the synthesis of oligosaccharides[51] (Scheme 2.212) [1794]. However, differences are not always clear-cut and mixed activities are sometimes observed.

Glycosyl transferases are responsible for the *biosynthesis* of oligosaccharides in vivo. They require that the sugar donor is activated on the anomeric center by

---

[49]Cellulose is the most abundant organic carbon in the ecosphere and its global standing crop has been estimated as 9.2×10^{11} tons, with an annual production of 0.85×10^{11} tons; the annual production of marine chitin was estimated as 2.3×10^9 tons.

[50]The possible number of linear and branched oligosaccharide isomers for a reducing hexasaccharide was calculated to encompass 1.05×10^{12} structures, see [1782].

[51]http://www.cazy.org.

phosphorylation prior to the condensation step. The activating group acts as a leaving group (LG) and is either a nucleoside diphosphate (in the Leloir pathway [1795]) or a simple phosphate (in non-Leloir enzymes [1796]).

Trans-glycosidases interconvert carbohydrate chains by transferring one sugar unit from a (di)saccharide (acting as donor) onto an acceptor and they represent an alternative pathway for gluconeogenesis, which is independent on phosphorylated donors. Due to their role in biosynthesis, glycosyl transferases and trans-glycosidases are generally rather specific with respect to their substrate(s) and the nature of the glycosidic bond to be formed [1797].

Glycosidases belong to the class of hydrolytic enzymes and have a *catabolic* function in vivo as they hydrolyze glycosidic linkages to form mono- or oligosaccharides from polysugars. Consequently, they are generally less specific when compared to glycosyl transferases. Glycosidases can be used for glycoside synthesis in the reverse (condensation) direction.

Glycosyl phosphorylases use phosphate (instead of water) for the breakdown of a glycosidic bond, which yields a glycosyl phosphate rather than a non-activated monosugar. The phosphorolytic degradation of oligosaccharides is a more energy-efficient alternative to hydrolysis by glycosidases.

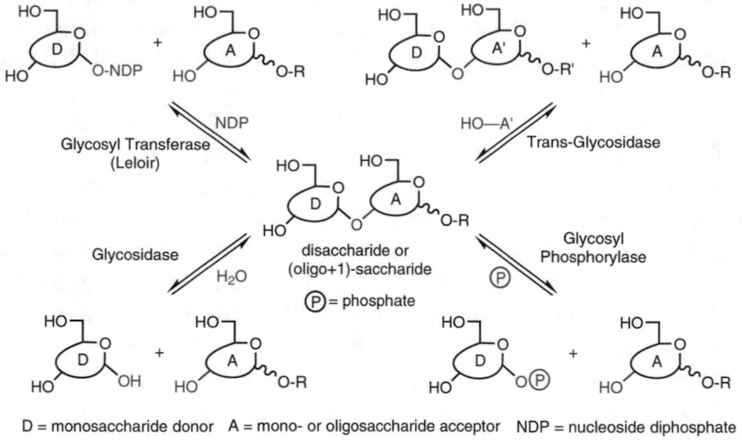

Scheme 2.212 Four different types of enzymatic glycosylation reactions

2.6.1.1 Glycosyl Transferases

In the Leloir-pathway, a sugar is phosphorylated in a first step by a kinase to give a sugar-1-phosphate. This activated sugar subsequently reacts with a nucleoside triphosphate under catalysis of a nucleoside transferase and forms a chemically activated nucleoside diphosphate sugar (NDP, Scheme 2.212) [1798]. These key nucleoside diphosphate sugars constitute the activated 'donors' in the subsequent condensation with the hydroxyl group 'acceptors' a mono- or oligosaccharide, a protein or a lipid, which is catalyzed by a glycosyl transferase. The key building blocks serving as donors are UDP-Glc, UDP-Gal, UDP-GlcNAc, UDP-GalNAc, UDP-Xyl, GDP-Man, GDP-Fuc and CMP-sialic acid [1799]. Since each NDP-sugar requires a distinct group of glycosyl transferases, a large number of highly specific glycosyl transferases are neccessary. More than one hundred glycosyl transferases have been identified to date and each one appears to specifically catalyze the formation of a unique glycosidic linkage [1800].

Chemists employ Leloir-enzymes to the synthesis of oligosaccharides and the majority of synthetic reactions are performed by UDP-glycosyl transferases [1801–1806]. Two requirements are currently limiting large-scale applications, namely the availability of the sugar nucleoside phosphates at reasonable costs and the matching glycosyl transferases. Only a few of these enzymes are commercially available, because isolation of these membrane-bound (unstable) proteins is difficult, since they are present only in low concentrations [1807, 1808]. The availability of NDP-sugars is ensured by in-situ regeneration of the sugar nucleotide from the released nucleoside phosphate via utilizing phosphorylating enzymes, which avoids co-product inhibition caused by the released nucleoside diphosphate [1809] (Scheme 2.213).

The point of interest to synthetic chemists is the range of acceptors and donors that can be used in glycosyl transferase-catalyzed reactions. Fortunately, the specificity of glycosyl transferases is high but not absolute.

UDP-galactosyl (UDP-Gal) transferase is the best-studied transferase in terms of specificity for the acceptor sugar. It has been demonstrated that this enzyme catalyzes the transfer of UDP-Gal to a remarkable range of acceptor substrates of the carbohydrate-type [1800, 1810–1813] (Table 2.7). Other glycosyl transferases, although less well-studied than UDP-Gal transferase, also appear to tolerate various acceptors as substrates [1814–1817].

Table 2.7 Glycoside synthesis using β-galactosyl transferase from the Leloir pathway (donor = UDP-Gal, Scheme 2.212)

Acceptor	Product
Glc–OH	β-Gal–(1→4)-Glc–OH
GlcNAc–OH	β-Gal–(1→4)-GlcNAc–OH
β-GlcNAc–(1→4)-Gal–OH	β-Gal–(1→4)-β-GlcNAc–(1→4)-Gal–OH
β-GlcNAc–(1→6)-Gal–OH	β-Gal–(1→4)-β-GlcNAc–(1→6)-Gal–OH
β-GlcNAc–(1→3)-Gal–OH	β-Gal–(1→4)-β-GlcNAc–(1→3)-Gal–OH

The use of the multienzyme systems, which arise due to the need to prepare the activated UDP-donor sugar in situ, is exemplified with the synthesis of *N*-acetyllactosamine [1818] (Scheme 2.213). Glucose-6-phosphate is isomerized to its 1-phosphate by phosphoglucomutase. Transfer of the activating group (UDP) from UTP is catalyzed by UDP-glucose pyrophosphorylase liberating pyrophosphate, which is destroyed by inorganic pyrophosphatase. Then, the center at carbon 4 is epimerized by UDP-galactose epimerase in order to drive the process out of the equilibrium. Finally, using galactosyl transferase, UDP-galactose is linked to *N*-acetylglucosamine to yield *N*-acetyllactosamine. The liberated UDP is recycled back to the respective triphosphate by pyruvate kinase at the expense of phosphoenol pyruvate. The overall yield of this sequence was in the range of 70% when performed on a scale greater than 10 g.

An elegant method for the recycling of UDP-glucose from UDP employs sucrose synthase for the transfer of a glucose unit from (cheap) sucrose onto UDP. Process optimization allowed UDP-Glc synthesis at ~100 g scale [1819].

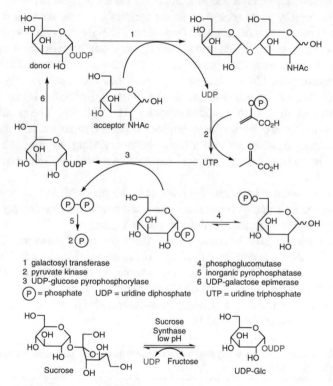

1 galactosyl transferase
2 pyruvate kinase
3 UDP-glucose pyrophosphorylase
(P) = phosphate   UDP = uridine diphosphate

4 phosphoglucomutase
5 inorganic pyrophosphatase
6 UDP-galactose epimerase
UTP = uridine triphosphate

Scheme 2.213 Synthesis of *N*-acetyllactosamine using a six-enzyme system (Leloir pathway) and recycling of UDP-glucose at the expense of sucrose

2.6.1.2 Glycoside Phosphorylases and Trans-glycosidases

In order to avoid the need for activated nucleoside diphosphate sugar donors, attention has been drawn to oligosaccharide synthesis via non-Leloir transferases (Schemes 2.212 and 2.214). In this case, the activated donor is a more simple sugar-1-phosphate, which can be transferred by a single glycoside phosphorylase. The latter catalyzes the reversible cleavage/formation of a glycosidic bond using phosphate as nucleophile/leaving group, respectively [1820, 1821]. Retaining and inverting disaccharide phosphorylases for the conversion of sucrose, maltose, cellobiose and α,α-trehalose are known, which opens the way to use these naturally abundant disaccharides as cheap glycosyl donors [1822].

For example, trehalose, one of the major storage carbohydrates in plants, fungi, and insects, was synthesized from glucose and its 1-phosphate using trehalose phosphorylase as the catalyst in the reverse (condensation) reaction [1823].

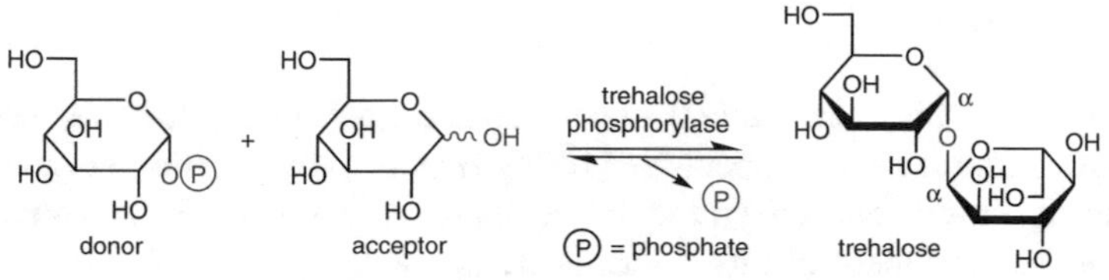

Scheme 2.214 Synthesis of trehalose via the non-Leloir pathway

Glycoside synthesis is even more simple using trans-glycosidases, which use cheap non-phosphorylated (di)saccharides as donor substrates. Interestingly, the energy of the glycosidic bond in sucrose is of the same level as that of nucleotide-activated sugars, such as UDP-Glu. The major disadvantage is the small number of available specificities, which is mainly limited to the transfer of α-glycosyl- and β-fructosyl-groups [1824].

Sucrose phosphorylase is also capable of catalyzing glycosyl-transfer reactions. This bacterial transglucosidase catalyzes the cleavage of the disaccharide sucrose using phosphate as nucleophile to yield α-D-glucose-1-phosphate and D-fructose (Scheme 2.215). In the absence of phosphate, the enzyme-glucosyl intermediate can be intercepted by various nucleophiles bearing an alcoholic group to yield the corresponding α-D-glucosides in high yields [1825, 1826]. Aryl alcohols and polyhydroxylated compounds, such as sugars and sugar alcohols are often glycosylated in a highly selective fashion. The major advantage of this system is the weak hydrolase activity of sucrose phosphorylase and the high-energy content of the cheap glucosyl donor sucrose. Several of these products constitute biocompatible solutes, which regulate the water-balance of the cell, prevent protein denaturation and stabilize membranes and are thus used as natural osmolytes and moisturising agents for cosmetics [1827].

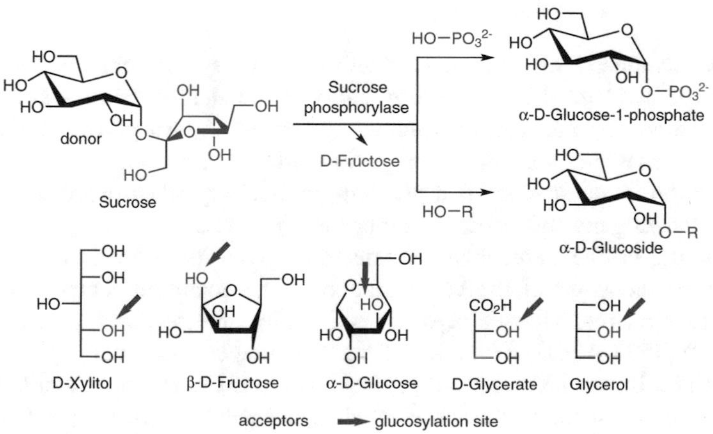

Scheme 2.215 Synthesis of α-D-glycosides using sucrose phosphorylase via trans-glycosylation

2.6.1.3 Glycosidases

A glycosidic bond is more stable than a phosphate ester or peptide bond, with a half-life of $\sim 10^7$ years at room temperature [1828]. Hence, its hydrolytic cleavage requires exceptionally proficient enzymes: Glycosidases (also termed 'glycohydrolases' [1829]) are independent of any cofactor and show k_{cat} values of $\sim 10^2$ s^{-1}, which translates into a rate acceleration of $\sim 10^{17}$. In general, glycosidases show high (but not absolute) specificity for both the glycosyl moiety and the nature of the glycosidic linkage, but little if any specificity for the aglycone component which acts as a leaving group ([LG-H], Scheme 2.216) [1830]. It has long been recognized that the nucleophile (NuH, which is water in the 'normal' hydrolytic pathway) can be replaced by other nucleophiles, such as another sugar or a primary or secondary (nonnatural) acceptor alcohol. This allows to turn the degradative nature of glycosyl hydrolysis towards the more useful *synthetic* direction [1785, 1831–1835]. Interestingly, this potential was already recognized as early as 1913! [1836].

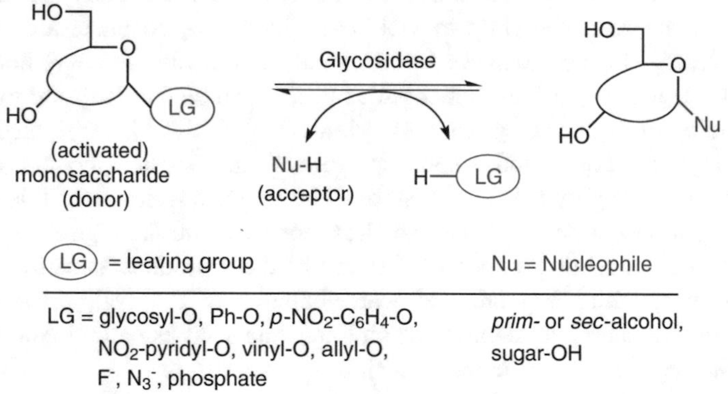

Glycosidase

(activated)
monosaccharide
(donor)

Nu-H
(acceptor)

H—LG

LG = leaving group

Nu = Nucleophile

LG = glycosyl-O, Ph-O, *p*-NO$_2$-C$_6$H$_4$-O,
NO$_2$-pyridyl-O, vinyl-O, allyl-O,
F$^-$, N$_3^-$, phosphate

prim- or *sec-*alcohol,
sugar-OH

Scheme 2.216 Glycoside synthesis using glycosidases

Major advantages of glycosidase-catalyzed glycosyl transfer are that there is minimal (or zero) need for protection and that the stereochemistry at the newly formed anomeric center can be completely controlled through the choice of the appropriate enzyme, i.e., an α- or β-glucosidase. However, regiocontrol with respect to the acceptor remains a problem, particularly when mono- or oligosaccharides carrying multiple hydroxy groups are involved.

Depending on the stereochemical course of glycoside formation, i.e., whether *retention* or *inversion* of the configuration at the anomeric center is observed, glycosidases operate via two separate and distinct mechanisms (Schemes 2.217 and 2.218) [1837–1841]. Examples of the retaining enzymes are β-galactosidase, invertase and lysozyme. Inverting glycosidases, such as trehalase and β-amylase, have been used for the synthesis of alkyl glycosides to a lesser extent. In recent years, a number of thermostable glycosidases have been identified and characterized. The most remarkable among them are the β-glucosidase [1842] and the β-galactosidase from the hyperthermophilic archean *Pyrococcus furiosus*.

Although the first proposal for the mechanism of retaining glycosidases in 1953 has undergone some refinements, it is still valid in its sense (Scheme 2.217) [633, 1843, 1844]: The active site contains two glutamic acid residues (Glu^1 and Glu^2), which can act as an acid or a base, respectively. In the first step, Glu^1 acts as an acid by protonation of the anomeric oxygen, making the (oligo)saccharide moiety [RO] a good leaving group, while the glycosyl residue is bound to the enzyme via Glu^2 as oxonium ion [1845, 1846]. Then, the leaving group ROH is displaced by the incoming nucleophile NuH (usually water) via diffusion. In the second step, the nucleophile is deprotonated by Glu^1 and attacks the glycosyl-enzyme intermediate from the same face from which the leaving group R-OH was expelled. Since both steps constitute an S_N2-reaction, *double inversion* results in *net retention* of configuration.

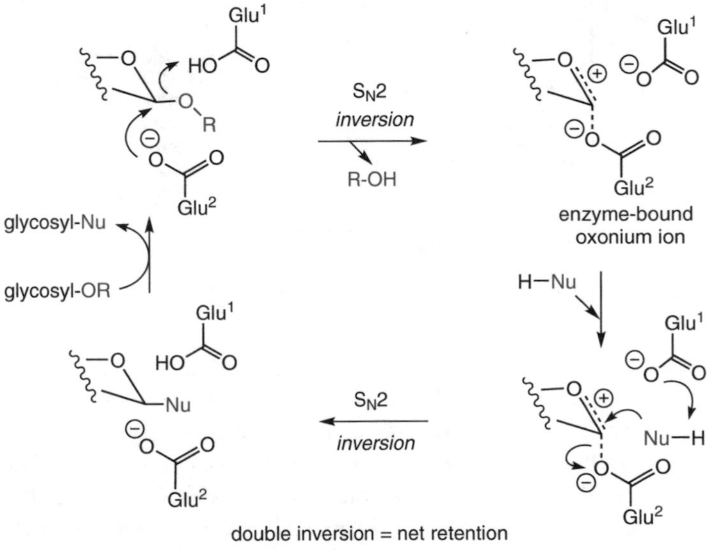

Scheme 2.217 Mechanism of retaining glycosidases

In contrast, *inverting* glycosidases act via a single step: Direct nucleophilic displacement of the aglycone moiety (ROH) by a nucleophile (NuH) via S_N2 leads to *inversion* of anomeric configuration (Scheme 2.218).

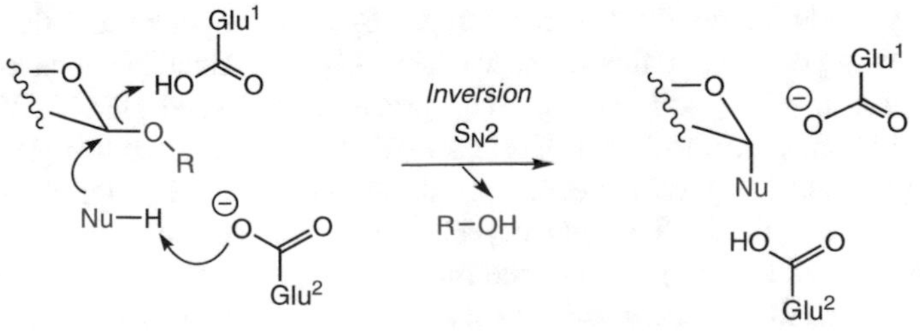

Scheme 2.218 Mechanism of inverting glycosidases

Reverse Hydrolysis

Glycosidases can be used for the synthesis of glycosides in two modes. The *thermodynamic* approach is the reversal of glycoside hydrolysis by shifting the equilibrium of the reaction from hydrolysis to synthesis. This procedure uses a free (nonactivated) monosaccharide as substrate and it has been referred to as 'direct glycosylation' or 'reverse hydrolysis' (Fig. 2.19, pathway A) [1847–1850]. Since in an aqueous environment the equilibrium constant for this reaction lies strongly in favor of hydrolysis, high concentrations of both the monosaccharide and the nucleophilic component (carbohydrate or alcohol) must be used. As a consequence, yields in these reactions are generally low and reaction mixtures comprised of thick syrups up to 75% by weight are not amenable to scale-up.

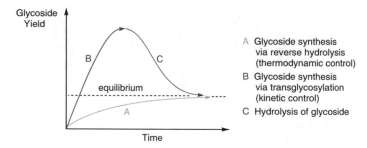

Fig. 2.19 Glycosylation via thermodynamic and kinetic control

Other methods to improve such procedures make use of aqueous-organic two-phase systems [1851, 1852] and polyethylene-glycol-modified glycosidases [1853]. However, the direct enzymatic synthesis of alkyl glycosides is generally hampered by the low solubility of carbohydrates in organic media. More polar solvents, such as DMF, DMSO or pyridine, are inapplicable because the products are often intended for use in food and personal care products. Alternatively, the reaction can be performed at temperatures below 0 °C or the glycoside formed can be removed from the reaction medium by selective adsorption [1854]. In summary, glycoside synthesis via the reverse hydrolysis approach is less than ideal.

Transglycosylation

The second strategy – the *kinetic* approach – utilizes a preformed activated glycoside as donor, which is coupled onto the nucleophile acceptor by an appropriate glycosidase and is referred to as 'transglycosylation' (Fig. 2.19, pathway B) [1855, 1856]. The enzyme-glycoside intermediate is then trapped by a nucleophile other than water to yield a new glycoside. In this case, activated glycosyl donors which possess an aglycone moiety with good leaving group properties are used [1857, 1858]. Good donors are, for instance, glycosyl fluorides [578, 1859, 1860], -azides [1861, 1862], (hetero)aryl- (usually *p*-nitrophenyl- or nitropyridyl- [1863]), vinyl- and allyl-glycosides [1864, 1865]. Transglycosylation gives higher yields as compared to reverse hydrolysis and is generally the method of choice [1866, 1867]. Since the glycoside formed during the reaction is also a substrate for the enzyme in hydrolysis causing its degradation (Fig. 2.19, pathway C), the success of this procedure as a preparative method depends on the following crucial parameters:

- Transglycosylation must be faster than glycoside hydrolysis
- The rate of hydrolysis of the product being slower than that of the glycosyl donor

In practice these conditions can be attained readily. It should be emphasized that an analogous situation can be found in enzymatic peptide synthesis using proteases (Sect. 3.1.4). The primary advantages of using glycosidases in comparison to glycosyl transferases is that expensive activated sugar nucleosides are not required and glycosidases generally are more readily available than glycosyl transferases. Furthermore, there is total control over the α/β-configuration at the newly generated anomeric center.

The major drawbacks, however, are incomplete yields and the frequent formation of product mixtures due to the limited selectivity of glycosidases with respect to the glycosidic acceptor, in particular due to the formation of undesired 1,6-linkages. The regio- and stereoselectivity of transglycosylation reactions is influenced by a number of parameters such as reaction temperature [1868], concentration of organic cosolvent, the reactivity of the activated donor [1869], the nature of the aglycone [1870, 1871], and the anomeric configuration of the acceptor glycoside [1872] (Table 2.8).

Table 2.8   Transglycosylation catalyzed by glycosidases (Scheme 2.216)

Enzyme	Donor/glycoside	Acceptor/nucleophile	Product(s)
α-Galactosidase	α-Gal-O-p-C$_6$H$_4$-NO$_2$	α-Gal-O-allyl	α-Gal-$(1{\to}3)$-α-Gal-O-allyl
α-Galactosidase	α-Gal-O-p-C$_6$H$_4$-NO$_2$	α-Gal-O-Me	α-Gal-$(1{\to}3)$-α-Gal-O-Me[a]
α-Galactosidase	α-Gal-O-p-C$_6$H$_4$-NO$_2$	β-Gal-O-Me	α-Gal-$(1{\to}6)$-β-Gal-O-Me[b]
β-Galactosidase	β-Gal-O-o-C$_6$H$_4$-NO$_2$	α-Gal-O-Me	β-Gal-$(1{\to}6)$-α-Gal-O-Me
β-Galactosidase	β-Gal-O-o-C$_6$H$_4$-NO$_2$	β-Gal-O-Me	β-Gal-$(1{\to}3)$-β-Gal-O-Me[c]

[a]α-Gal-$(1{\to}6)$-α-Gal-O-Me
[b]α-Gal-$(1{\to}3)$-β-Gal-O-Me
[c]β-Gal-$(1{\to}6)$-β-Gal-O-Me are formed as side products

This latter fact has been used as a convenient tool to modulate the regioselectivity of glycosylation by switching the configuration at the anomeric center of the glycosidic acceptor. This technique has been denoted as 'anomeric control' (Scheme 2.219).

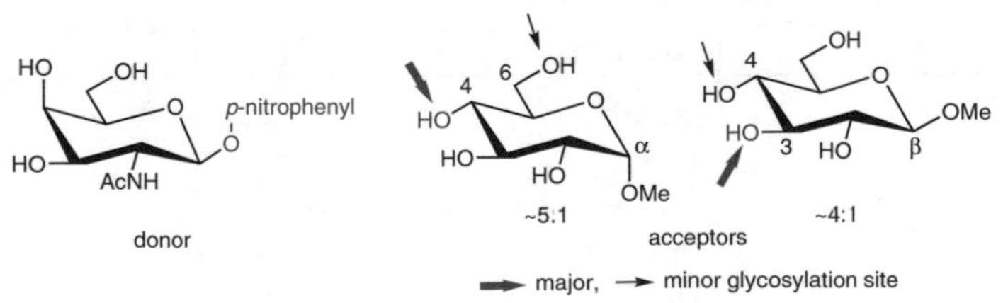

donor   acceptors

major, → minor glycosylation site

Scheme 2.219 Anomeric control in N-acetylglucosaminyl transfer onto α- and β-D-methylglucosides by β-galactosidase

For instance, when the α-anomer of methyl-D-glucoside was used as acceptor and p-nitrophenyl-β-N-acetyl-D-galactosaminide as donor in a *trans*-glycosylation reaction catalyzed by β-galactosidase from *Aspergillus oryzae*, two transfer products possessing a 1,4- and 1,6-linkage were formed in a ratio of ~5:1, respectively. On the other hand, when using the β-anomer of the acceptor, the corresponding 1,3- and 1,4-glucosides were formed instead (ratio ~4:1) [1873].

Besides the synthesis of natural glycosides, a considerable number of nonnatural alcohols have been employed as nucleophiles in transglycosylation reactions (Table 2.9) [1874, 1875]. The types of transformation include the desymmetrization of *meso*-diols and the kinetic resolution of racemic primary and secondary alcohols. In discussing enantioselection towards a (chiral) nonnatural acceptor, it should be kept in mind that the donor carbohydrate moiety is chiral and, as a consequence, the glycosylation products are *diastereomers* rather than enantiomers. In general, the stereocontrol during desymmetrization of prochiral or the kinetic resolution of racemic alcohols by glycosidases performs much worse than, e.g., lipases and alcohol dehydrogenases.

Cyclic *meso*-1,2-diols have been transformed into the corresponding monoglycosides in good diastereoselectivity using β-galactosidase from *Escherichia coli*, which is readily available from the dairy industry. As may be seen from Table 2.9, the selectivity strongly depends on the structure of the aglycone component [1876].

In some cases, the kinetic resolution of racemic primary and secondary alcohols was feasible. On the one hand, the enantioselectivity of glycosidases involving the glycosylation of primary alcohol moieties in 1,2-propanediol, glycerol or glycidol was negligible [1877, 1878], however, better results were obtained for *sec*-alcohols (Table 2.9) [1879–1881]. This fact is understandable if one considers the rules for chiral recognition for carboxyl ester hydrolases (see Schemes 2.20 and 2.45), where the distance of the center of chirality to the point of reaction should be a minimum. It is apparent that stereoselective glycosylation of alcohols is inferior compared to ester hydrolysis / esterification using standard hydrolases.

In contrast, regio-selective glycosylation of (poly)hydroxy compounds, such as steroids (digoxin, digitoxin), terpenoids (geraniol), hydroquinones (arbutin), vitamins (α-tocopherol), flavonoids (quercetin) and (poly)phenols (resveratrol) is a valuable technique to enhance the stability and water-solubility of these compounds or to modulate their (bio)activity.

Table 2.9 Desymmetrization of *meso*-diols and kinetic resolution of alcohols by glycosylation using β-galactosidase from *Escherichia coli*

Donor/glycoside	Acceptor/nucleophile	Product	d.e. [%]
β-Gal-*O*-Ph	OH, 1,2 OH	O-β-Gal, 1,2 OH	90–96
β-Gal-OPh	OH, OH *rac*	OH, O-β-Gal	95[a]
β-Gal-OC₆H₄-*o*-NO₂	Ph, *rac*, OH	Ph, *rac*, OH	98

[a]The β-galactosidase from *Sulfolobus solfataricus* was used

Glyco-Synthases

A major improvement in the use of glycosidases for glycoside synthesis was the rational re-design of the catalytic site to disable the undesired hydrolysis of the glycoside product, while maintaining glycoside synthesis activity (Scheme 2.220).

Replacement of the Glu^2-residue acting as base in the native enzyme by a Ser residue allowed to bind an activated glycosyl fluoride as donor. The latter is attacked by the acceptor nucleophile, which is deprotonated by Glu^1, forming the glycoside product. In the native enzyme, the latter would undergo subsequent hydrolysis by a water molecule activated by Glu^2 but this is impossible in the Ser-mutant. Such active-site mutants of glycosidases (aptly denoted as 'glycosynthases' [1882–1887]) show greatly enhanced yields of glycosides due to the elimination of their undesired hydrolysis.

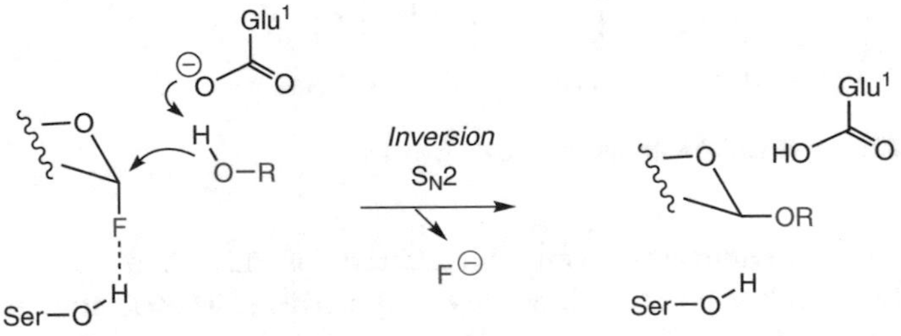

Scheme 2.220 Mechanistic principle of an inverting glycos*ynthase*

2.6.2 Amino Transfer Reactions

Transaminases [EC 2.6.1.X][52] catalyze the redox-neutral transfer of ammonia between an amine donor and a carbonyl acceptor group (Scheme 2.221) [97, 1888–1893]. Since free ammonia is highly toxic to living cells, this reaction is mediated via an 'activated benzaldehyde' (pyridoxal-5'-phosphate, PLP, vitamin B_6) as cofactor, which functions as a molecular shuttle for the $[NH_3]$-moiety. In a first step, PLP forms an aldimine Schiff base with the amine-donor. Tautomerization of the C=N bond catalyzed by a conserved Lys residue yields a ketimine, which is hydrolyzed to yield the cofactor in its aminated form (pyridoxamine, PMP). In the second step, the latter reacts (in reversed order) through the same events with the carbonyl group of the substrate to yield the amine product with regeneration of PLP [1894, 1895].

---

[52]Also denoted as amino transferases.

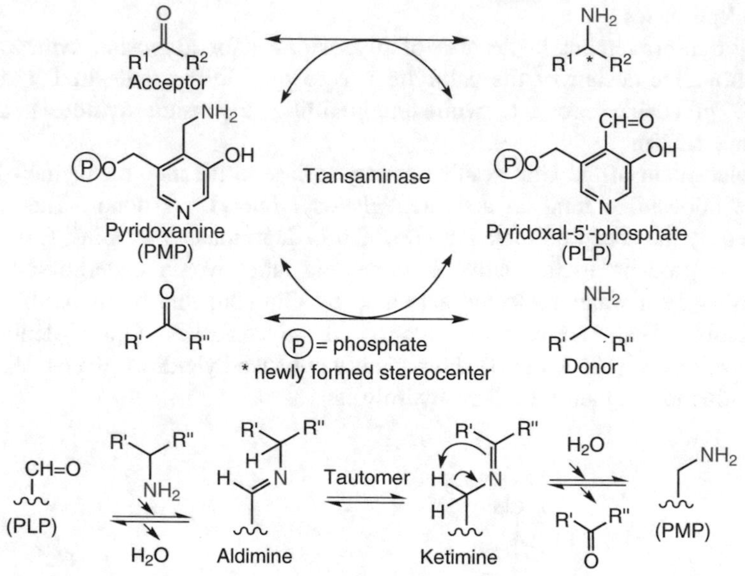

Scheme 2.221 Transaminase-catalyzed amino-transfer

Although transaminases were discovered already half a century ago [1896, 1897], their use for the biocatalytic synthesis of (nonnatural) amines was initially impeded by two obstacles: (i) The majority of transaminases available were only active on α-amino/α-ketoacids as substrates and (ii) techniques to shift the equilibrium towards amine formation had to be developed. The first significant advances in transamination for organic synthesis were achieved by Celgene Co., who employed transaminases for the synthesis of nonracemic amines, preferentially via (less efficient) kinetic resolution of *rac*-amines by enantioselective de-amination [1898, 1899]. Within the last decade, several breakthroughs with respect to the (commercial) availability of stereo-complementary transaminases possessing a broad substrate spectrum and a set of techniques to shift the equilibrium of transamination in favor of amine synthesis were accomplished, which make enzymatic transamination nowadays a reliable technique for the industrial-scale asymmetric synthesis of amines [1900, 1901].

On a genomic level, transaminases are classified into subgroups [1902–1904]. α-Transaminases are very specific, as they only act on α-amino acids yielding the corresponding α-keto acids. In contrast, ω-transaminases (ω-TA) are more flexible and accept substrates bearing a distant carboxylate moiety (such as lysine, ornithine, β-alanine, and ω-aminobutyrate). Since they also accept *prim*-amines lacking a carboxylate group (e.g. i-Pr-NH$_2$, 2-Bu-NH$_2$), they are also referred to as 'amine transaminases [1905].

In view to access both stereoisomers of a chiral amine via transamination by choice of an appropriate (*R*)- or (*S*)-selective ω-TA, screening studies were undertaken which revealed an impressive number of stereo-complementary ω-TAs [1906–1909]. The most widely used enzymes are obtained from *Vibrio fluvialis* [1910], *Chromobacterium violaceum* [1911, 1912], *Pseudomonas aeruginosa* [1913], *Bacillus megaterium* [1914], and *Alcaligenes denitrificans* [1915]. Thermostable mutants were derived from an ω-TA from *Arthrobacter citreus* [1916].

Because the transamination is reversible, the synthesis of nonracemic amines using ω-transaminases can be operated in two modes (Scheme 2.222):

- Enantioselective deamination starts from a racemic amine via kinetic resolution, where one enantiomer is converted into the corresponding ketone, leaving the desired amine enantiomer untouched, which can be recovered in 50% theoretical yield. For thermodynamic reasons, pyruvate was employed as preferred amine acceptor yielding D- or L-alanine as by-product, depending on the stereopreference of the enzyme.
- Desymmetrization of prochiral ketones via asymmetric amination is preferred for its superior efficiency. Depending on the substrate preference of the employed transaminase, sacrificial amine donors derived from the α-aminoacid pool (e.g. Ala, Phe, Glu, Asp) or simple amines (i-Pr-NH$_2$, 2-Bu-NH$_2$) are commonly employed. It should be kept in mind that the absolute configuration of a chiral amine-donor has to match the stereospecificity of the ω-TA in order to be accepted.

In transamination, equilibrium constants are close to unity at best and the amino transfer from an α-amino acid to a ketone is strongly disfavored.[53] To even worsen the situation, ω-TAs often show cosubstrate and/or coproduct inhibition at elevated concentrations, which prevents to *push* amine formation by employing an excess of amine donor [1917]. In contrast, *pulling* the equilibrium by co-product removal is much more effective. The following strategies have been developed (Scheme 2.222) [1918]:

- The most simple approach is to use i-Pr-NH$_2$ as amine donor and to remove the coproduct acetone at elevated temperature by evaporation [1919, 1920].
- Non-volatile coproducts are usually removed by an additional enzymatic step: For instance, decarboxylation of an α-ketoacid (e.g., pyruvate or phenylpyruvate, formed from alanine or phenylalanine, respectively) using pyruvate or phenylpyruvate decarboxylase, yields an aldehyde and CO$_2$ [1921, 1922]. Although this provides a strong driving force, the aldehyde thus formed is usually a good substrate and gets aminated.
- Carbonyl-reduction of the keto-coproduct by a suitable dehydrogenase in presence of NAD(P)H-recycling yields the corresponding alcohol. For instance, pyruvate can be conveniently reduced to lactate by lactate dehydrogenase [1923].

---

[53]The equilibrium constant between acetophenone and alanine was reported to be 8.8×10^{-4}, see [1916].

- The most efficient approach is probably the direct recycling of alanine from pyruvate via NADH-dependent reduction in presence of ammonia catalyzed by alanine dehydrogenase. Overall, this sequence is equivalent to a metal-free reductive amination [1924]. Using ammonium formate and formate dehydrogenase for NAD(P)H-recycling, this resembles a biocatalytic equivalent of the Leuckart-Wallach reaction.
- The use of amine donors, which form an unstable keto co-product [1925]. For instance, α,ω-diamino acids, such as ornithine (n = 2) or lysine (n = 3) yield α-amino-ω-ketoacids, which (nonenzymatically) cyclize to the corresponding Δ^2-pyrroline-5-carboxylate and Δ^1-piperidine-2-carboxylate, respectively, as dead-end products [1926, 1927]. In a related approach, o-xylylene diamine gives an amino aldehyde, which spontaneously undergoes 5-exo-trig cyclization, followed by tautomerization to yield iso-indol. The latter forms coloured polymers, which may serve as indicator for positive hits in mutant libraries, but complicates downstream-processing in prep-scale reactions [1928].

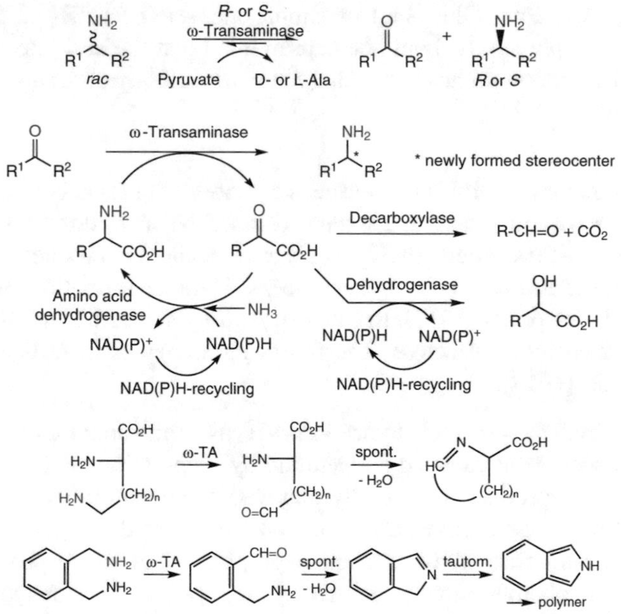

Scheme 2.222 Enantioselective de-amination of rac-amines (kinetic resolution) and asymmetric transamination of ketones (desymmetrization) with amine donors for equilibrium shift

To date, a broad range of wild-type (R)- and (S)-ω-transaminases are available, which accept ketones bearing a large and small group, while mutants accepting sterically demanding substrates bearing two bulky groups were obtained by directed evolution [1929]. Together with efficient techniques to shift the equilibrium, the stage was set for the large-scale synthesis of nonracemic amines from the corresponding ketones. Considering the importance of amino groups in active pharmaceutical ingredients (APIs)[54] it is not surprising that the asymmetric enzymatic transamination on industrial scale was predominantly developed for pharma-applications, which is illustrated by the following examples (Scheme 2.223).

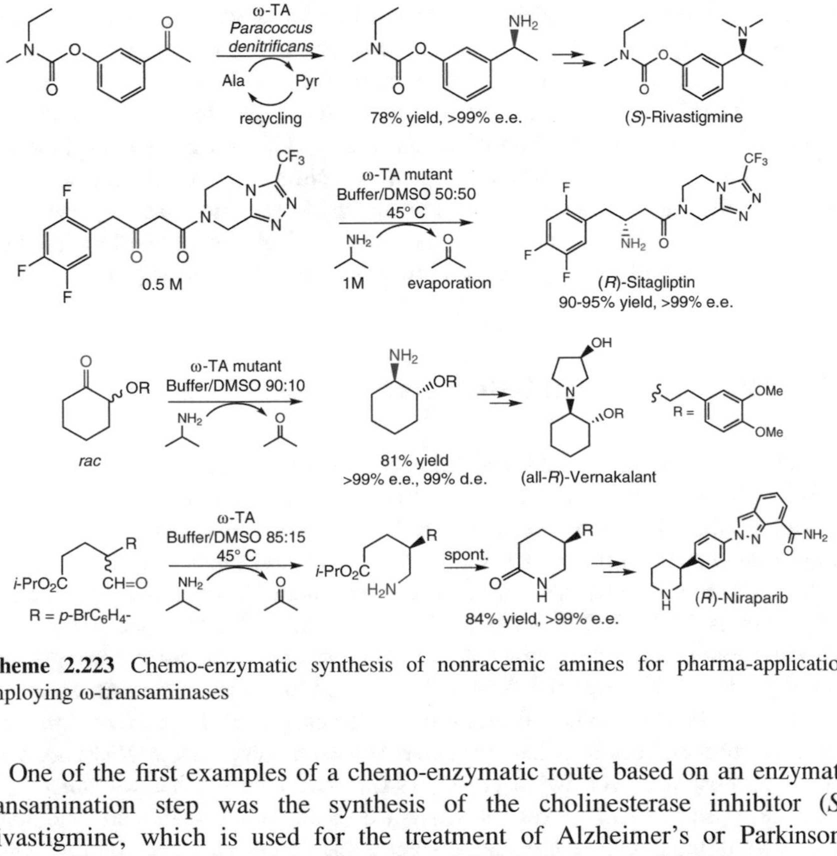

Scheme 2.223 Chemo-enzymatic synthesis of nonracemic amines for pharma-applications employing ω-transaminases

One of the first examples of a chemo-enzymatic route based on an enzymatic transamination step was the synthesis of the cholinesterase inhibitor (S)-Rivastigmine, which is used for the treatment of Alzheimer's or Parkinson's disease. The introduction of the (S)-amino moiety was accomplished by asymmetric amination of the aryl-methyl ketone precursor in 78% isolated yield and >99% e.e. using an ω-TA from *Paracoccus denitrificans* [1930]. The most prominent transamination-based process implemented on industrial scale – a benchmark in biocatalytic synthesis – was the production of the anti-diabetic drug (R)-Sitagliptin. Since wild-type ω-TAs were barely able to convert the sterically demanding ketone precursor (~4% conversion at best), elaborated directed evolution over 11 rounds was neccessary to provide a 27-mutant enzyme with improved catalytic performance and enhanced tolerance towards process conditions (~50% DMSO, 45 °C).

---

[54]The relative abundance of functional groups in APIs (in decreasing order) is hydroxy (~40%), carboxy (~22%), amino (~16%), sulfoxide (~3%), others (19%).

Eventually, (R)-sitagliptin was obtained in 90–95% yield and >99% e.e. at 200 g/L substrate concentration using i-Pr-NH$_2$ (1M) as amine donor [1931]. Due to its superior efficiency, the biocatalytic route replaced the Rh[Josiphos]-catalysed asymmetric enamine hydrogenation process.

An ω-TA was also engineered to adapt it to a process for the synthesis of the antiarrhythmic agent (all-R)-vernakalant. In this case, enzyme evolution was directed to provide a transaminase variant with inverted diastereoselectivity with respect to the chiral center adjacent to the carbonyl moiety. Careful choice of the reaction conditions allowed the in-situ racemization of the starting ketone, providing a single *trans*-diastereomer in 81% yield, 99% d.e. and >99% e.e. via dynamic resolution [1932]. In analogy, dynamic resolution is also feasible for aldehydes bearing a configurationally unstable center at Cα. This strategy was exploited for the preparation of the anti-cancer agent (R)-Niraparib, which relies on the enantioselective amination of a racemic aldehyde precursor as key step. The δ-amino-ester thus formed undergoes spontaneous ring-closure yielding a lactam, which efficiently pulls the equilibrium towards product formation [1933].

2.7 Halogenation and Dehalogenation Reactions

Halogen-containing compounds are not only produced by man, but also by Nature [1934–1936]. A brominated indole derivative – Tyrian purple dye[55] – was isolated from the mollusc *Murex brandaris* by the Phoenicians. Since that time, more than 5000 halogenated natural products of various structural types have been isolated from sources such as bacteria, fungi, algae, higher plants, marine molluscs, insects, and mammals [1937, 1938]. Whereas fluorinated and iodinated species are rather rare, chloro and bromo derivatives are found more often. The former are predominantly produced by terrestrial species [1939] and the latter in marine organisms [1940]. For instance, about 10^7 tons of bromoalkanes such as bromoform and methylene bromide are released from coastal brown algae *Ascophyllum nodosum* into the atmosphere worldwide [1941, 1942]. Although the natural function of halogenating enzymes is not yet known, they do seem to be involved in the defence mechanism of their hosts. For instance, some algae produce halometabolites, which makes them inedible to animals [1943]. In contrast to hydrolytic or redox enzymes, which have been investigated since about a century, halogen-converting enzymes are a more recent subject of research after the first halogenase was reported in 1966 [1944–1950].

---

[55]6,6'-Dibromoindigo.

2.7.1 Halogenation

Despite the impressive number of halometabolites identified so far, only a few types of halogenating enzymes have been characterized to date [1951–1954]: Whereas flavin- [1955] and α-ketoglutarate-dependent (nonheme) iron-halogenases [1956] are rather substrate specific, halo*peroxidases* show a broad substrate scope and thus had a dominant impact in biotransformations [1451, 1471, 1957–1960]. These enzymes are widely distributed in nature and enable a multitude of electrophilic halogenation reactions following the general equation shown in Scheme 2.224, where X stands for halide (Cl^-, Br^- and I^-, but not F^-,[56] [1961]). The individual enzymes are called chloro-, bromo-, and iodoperoxidase. The name reflects the smallest halide ion that they can oxidize, in correlation to the corresponding redox potential. For redox reactions catalyzed by haloperoxidases which do not involve a halide (such as hydroxylation, epoxidation, or sulfoxidation) see Sect. 2.3.4.

Despite their mechanistic differences, the overall net reaction of haloperoxidases consists of a two-electron oxidation of halide anion at the expense of H_2O_2 yielding $[X^+]$ (Scheme 2.224).[57] Depending on the type of halide and the reaction conditions, such as pH, the electrophilic species may be halonium (X^+, X_3^+), halogen (X_2) or hypohalous acid/hypohalite (HOX/OX^-). Two major types of haloperoxidases depend on a catalytic metal:

- The mechanism of heme-iron-dependent enzymes is closely related to that of peroxygenases, i.e. H_2O_2-dependent oxidation of Fe^{3+} in the enzyme's resting state yields the Fe^{4+}-species Compound I (Scheme 2.171). The latter oxidizes halide in a two-electron transfer step (Scheme 2.224) [1962, 1963].
- In contrast, vanadium-depending haloperoxidases do not change the oxidation state of V^{5+} during the catalytic cycle, but switch between vanadate and peroxo-vanadate [1964, 1965].

For both enzymes, the fate of X^+ generated and the existence of a metal-bound hypohalite adduct is under debate, for heme-dependent haloperoxidases this elusive species is ironically denoted as 'Compound X' [1966]. The actual halogenation reaction is believed to take place outside of the active site and consequently, any asymmetric induction observed in haloperoxidase-catalyzed reactions is usually low.

---

[56]The high electronegativity of fluorine renders the formation of F^+ energetically prohibitive. Enzymatic (nucleophilic) fluorination is extremely rare and requires S-adenosylmethionine (SAM) as cofactor.

[57]In a side reaction, X^+ may react with H_2O_2 to form singlet oxygen ($X^+ + H_2O_2 \rightarrow {}^1O_2 + X^- + H^+$).

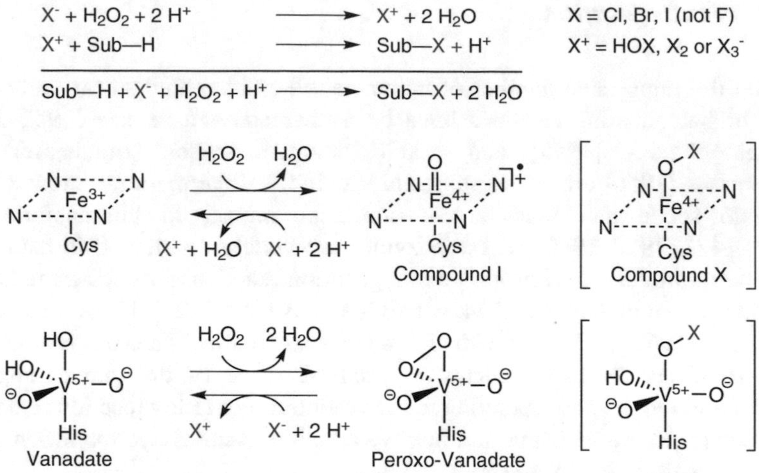

$$X^- + H_2O_2 + 2\ H^+ \longrightarrow X^+ + 2\ H_2O \qquad X = Cl,\ Br,\ I\ (not\ F)$$
$$X^+ + Sub\text{—}H \longrightarrow Sub\text{—}X + H^+ \qquad X^+ = HOX,\ X_2\ or\ X_3^-$$

$$Sub\text{—}H + X^- + H_2O_2 + H^+ \longrightarrow Sub\text{—}X + 2\ H_2O$$

Scheme 2.224 Enzymatic halogenation catalyzed by haloperoxidases

Bearing in mind their unique position as halogenating enzymes and the large variety of structurally different halometabolites produced by them, it is not surprising that the majority of haloperoxidases are characterized by a low product selectivity and wide substrate tolerance.

The most intensively studied haloperoxidases are the chloroperoxidase from the mold *Caldariomyces fumago* [1473] and bromoperoxidases from algae [1967] and bacteria such as *Pseudomonas aureofaciens* [1968], *Ps. pyrrocinia* [1969], and *Streptomyces* sp. [1970]. The only iodoperoxidase of preparative use is isolated from horseradish root [1971].

Halogenation of Alkenes
Haloperoxidases transform alkenes by a formal addition of hypohalous acid to produce halohydrins. The mechanism is identical to that of chemical halohydrin formation and proceeds via a halonium intermediate [1972–1974], (Scheme 2.225).

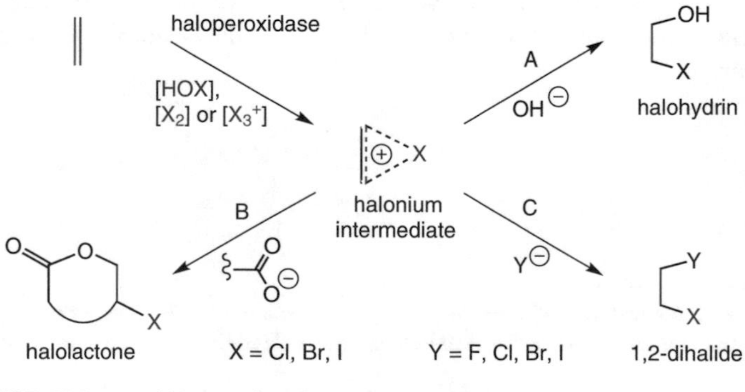

Scheme 2.225 Haloperoxidase-catalyzed transformation of alkenes

Functional groups present in the alkene can lead to products other than the expected halohydrin (pathway A) by competing with hydroxyl anion for the halonium intermediate. Unsaturated carboxylic acids, for instance, are transformed into the corresponding halolactones due to the nucleophilicity of the carboxylate group (pathway B) affording a halolactonization [1975, 1976]. Similarly, in presence of elevated concentrations of halide, 1,2-dihalides are formed (pathway C) [1977]. This latter transformation offers the unique possibility of introducing fluorine, which is not oxidized by haloperoxidases, into the substrate. Furthermore, migration of functional groups such as halogen [1978] and loss of carbon-containing units such as acetate and formaldehyde may occur, particularly when an oxygen substituent is attached to the C=C bond [1979, 1980].

All types of carbon–carbon double bonds – isolated (e.g., propene), conjugated (e.g., butadiene) and cumulative (e.g., allene) – are reactive (Scheme 2.226) [1981]. The size of the substrate seems to be of little importance since steroids [1982] and sterically demanding bicyclic alkenes [1983] are accepted equally well. Any regioselectivity observed reflects the (predominant) chemical and nonenzymatic nature of halohydrin formation. The same holds for diastereoselectivity on (bi)cyclic structures, where attack of the halonium species preferably occurs from the less hindered *exo*-side, followed by nucleophilic ring opening in a *trans*-fashion. Geraniol was halogenated on the (electronically favored) terminal C=C bond, the corresponding bromonium ion underwent intramolecular 6-*exo-tet* cyclization yielding a cyclohexane carbenium ion, which upon deprotonation gave a mixture of regio-isomeric alkenes in racemic form bearing the Br and CH$_2$-OH substituents in the stereochemically preferred diaxial position (plus additional side products) [1984].

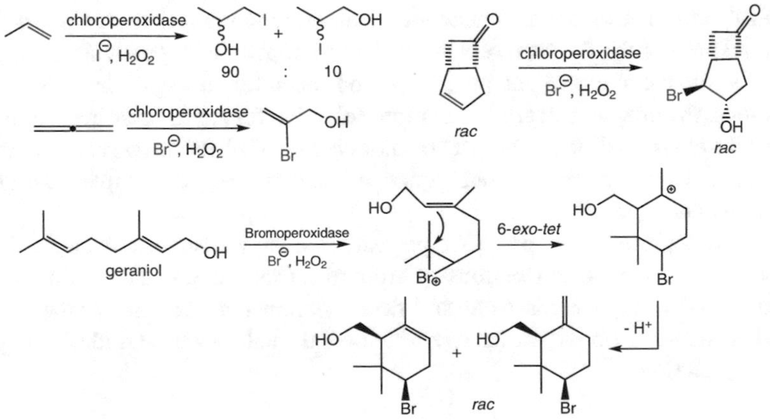

Scheme 2.226 Regio- and diastereoselective formation of halohydrins from alkenes

Halogenation of Alkynes

With alkyne substrates, haloperoxidase-catalyzed reactions yield α-haloketones (Scheme 2.227) [1985]. As with alkenes, the product distribution depends on the halide ion concentration. Both homogeneous and mixed dihalides can be formed, dependent upon whether a single halide species or a mixture of halide ions are present.

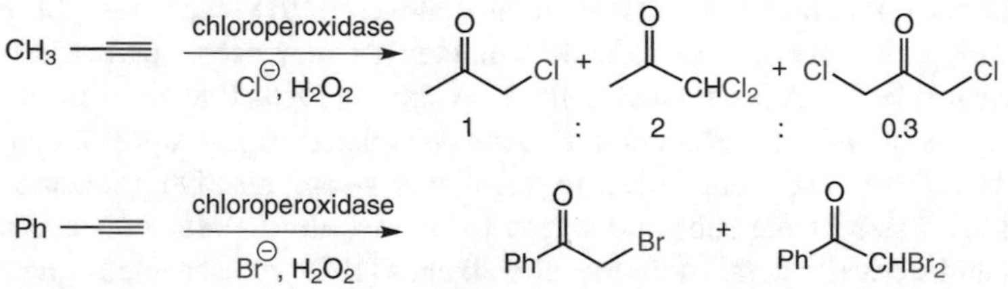

Scheme 2.227 Haloperoxidase-catalyzed reactions of alkynes

Halogenation of Aromatic Compounds

A wide range of electron-rich aromatic and heteroaromatic compounds are readily halogenated by haloperoxidases [1986–1988]. Bearing in mind the electrophilic character of the halogenating species, electron-rich phenols [1989, 1990] and anilines [1991] as well as their respective *O*- and *N*-alkyl derivatives are particularly well accepted. As in chemical electrophilic halogenation, the regioselectivity is dominated by the *ortho*- and *para*-directing effect of the substituent (Scheme 2.228) [1992]. In a comparative study, the bromination of phenol was performed with the V-depending bromoperoxidase from *Ascophyllum nodosum* and with different chemical brominating agents under identical reaction conditions. Two key points can be taken: Firstly, the ratio of *ortho*/*para* bromophenol is somewhat comparable in the chemical and enzymatic processes, and secondly, owing to the mild reaction conditions, the enzymatic reaction is more selective for *mono*-bromination than the chemical transformations [1993]. The color change of phenolic dyes such as phenol red or fluorescein upon halogenation serves as a simple assay for haloperoxidases [1994].

Since haloperoxidases are also peroxidases, they also can catalyze halide-independent peroxidation reactions of aromatics (Sect. 2.3.4). Thus, dimerization, polymerization, oxygen insertion and de-alkylation reactions are encountered as undesired side-reactions, particularly whenever the halide ion is omitted or depleted from the reaction mixture.

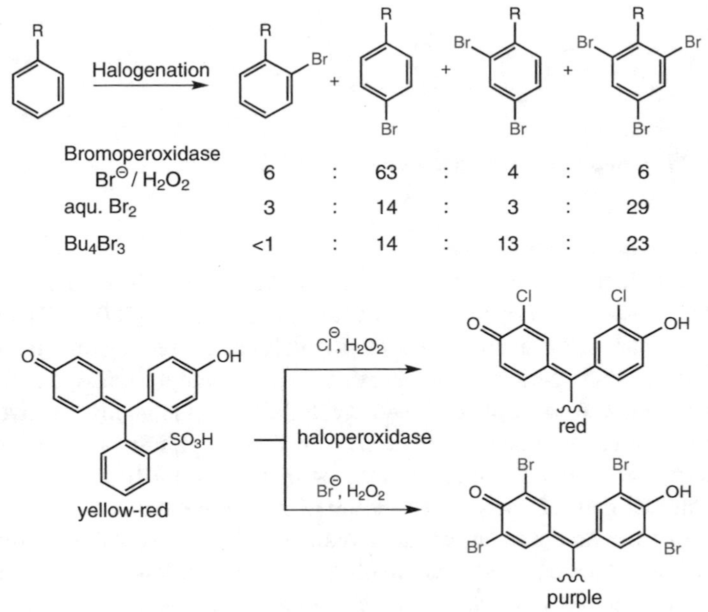

Scheme 2.228   Halogenation of aromatic compounds

Halogenation of C–H Groups

Similar to the chemical process, enzymatic halogenation of C–H groups is only possible if they are activated by adjacent electron-withdrawing substituents, for example carbonyl groups, which facilitate enolization. Since the reactivity depends on the enol content of the substrate, simple ketones like 2-heptanone are unreactive [1995], but highly enolized 1,3-diketones are readily halogenated to give the corresponding 2-*mono*- or 2,2-dihalo derivatives (Scheme 2.229) [1996]. For instance, monochloro dimedone has been used extensively to detect chlorinating and brominating haloperoxidases due to a hypsochromic shift of its absorbance maximum from 290 nm (owing to its enol content) to shorter wavelengths. In addition, it served as mimic in the elucidation of the biosynthesis of the intriguing highly chlorinated metabolite caldariomycin, which is formed by the (haloperoxidase-producing) fungus *Caldariomyces fumago*.

As with the formation of halohydrins from alkenes, stereoselectivities are low and the reactivity of the substrate is independent of its size. For example, monocyclic compounds such as barbituric acid derivatives [1997] and sterically demanding polycyclic steroids are equally well accepted [1998]. β-Ketoacids are also halogenated, but the spontaneous decarboxylation of the intermediate α-halo-β-ketoacid affords the corresponding α-haloketones [1999]. The chloroperoxidase-catalyzed halogenation of oximes was shown to proceed via a two-step sequence through a halonitroso intermediate which is further oxidized to furnish an α-halonitro product [2000].

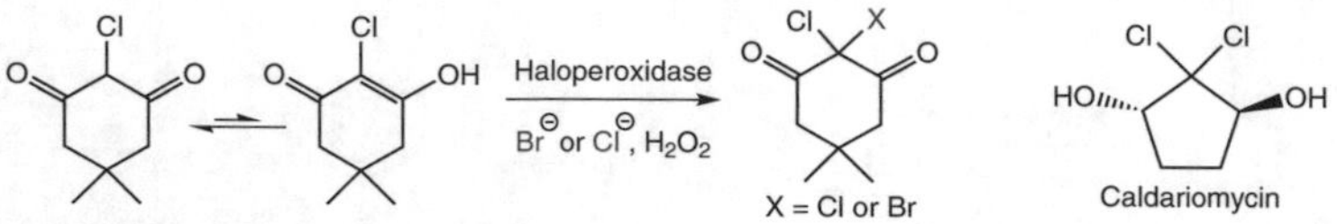

Scheme 2.229 Halogenation of electronically activated C–H groups

Halogenation of N- and S-Atoms

Amines are halogenated by haloperoxidases to form unstable haloamines, which readily deaminate or decarboxylate, liberating the halogen [2001]. This pathway constitutes a part of the natural mammalian defence system against microorganisms, parasites and, perhaps, tumor cells. In an analogous fashion, thiols are oxidized to yield the corresponding sulfenyl halides. These highly reactive species are prone to undergo nucleophilic attack by hydroxyl ion or by excess thiol [2002, 2003]. As a result, sulfenic acids or disulfides are formed, respectively. Due to a lack of control, these reactions are of no synthetic use.

In view of the predominant chemical nature of biohalogenation, it seems that enzymatic halogenation reactions involving haloperoxidases do not show any significant advantage over the usual chemical reactions due to their lack of stereoselectivity. A benefit, however, lies in the mild reaction conditions employed.

2.7.2 Dehalogenation

The concentrations of haloorganic compounds in the ecosphere has remained reasonably constant due to the establishment of an equilibrium between biosynthesis and biodegradation. Due to man's recent activities, a large number of halogen-containing compounds – most of which are recalcitrant – are liberated either by intent (e.g., insecticides), or because of poor practice (lead scavengers in gasoline) or through abuse (dumping of waste) into the ecosystem. These halogenated compounds would rapidly pollute the earth if there were no microbial dehalogenation pathways [2004, 2005]. Five major pathways for enzymatic degradation of halogenated compounds have been discovered (Table 2.10) [2006–2009].

Table 2.10   Major biodegradation pathways of halogenated compounds

Reaction type	Starting material		Products
Reductive dehalogenation	C–X	→	C–H + X⁻
Oxidative degradation	H–C–X	→	C=O + HX
Dehydrohalogenation	H–C–C–X	→	C=C + HX
Hydrolysis	C–X + H_2O	→	C–OH + HX
Epoxide formation	HO–C–C–X	→	epoxide + HX

X = Cl, Br, I

Redox enzymes are responsible for the replacement of the halogen by a hydrogen atom via reductive dehalogenation [2010, 2011] and oxidative dehalogenation yields a carbonyl group [2012]. Elimination of hydrogen halide leads to the formation of an alkene [2013], which is further degraded by oxidation. Since all of these pathways proceed either with a loss of a functional group or through removal of a chirality center, they are of little use for the biocatalytic synthesis of organic compounds. On the other hand, the enzyme-catalyzed hydrolytic replacement of a halide by a hydroxy group and the formation of an epoxide from a halohydrin take place in a stereocontrolled fashion and are therefore of synthetic interest.

Dehalogenases
Hydrolytic dehalogenation catalyzed by dehalogenases [EC 3.8.1.X, formally hydrolases] proceeds by formal nucleophilic substitution of the halogen atom with a hydroxyl ion going in hand with *inversion* of configuration [2014]. The mechanism has close similarities to that of epoxide hydrolases (Sect. 2.1.5), i.e. the carboxyl moiety of an aspartate residue attacks the halide by forming an 'alkyl-enzyme intermediate' (Scheme 2.230). Being a carboxyl ester, the latter is hydrolyzed by a hydroxyl ion which is provided from water by the aid of a histidine [631, 2015].

To date, two types of dehalogenases have gained importance for preparative biotransformations due to their stereospecificities on haloalkanes and α-haloacids.

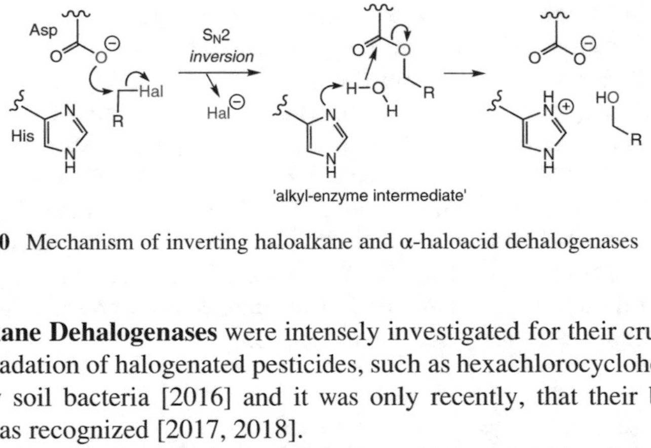

'alkyl-enzyme intermediate'

Scheme 2.230 Mechanism of inverting haloalkane and α-haloacid dehalogenases

Haloalkane Dehalogenases were intensely investigated for their crucial role in the biodegradation of halogenated pesticides, such as hexachlorocyclohexane (Lindane),[58] by soil bacteria [2016] and it was only recently, that their biocatalytic potential was recognized [2017, 2018].

Typical substrates for haloalkane dehalogenases DhaA, LinB and DbjA are *prim* or *sec* chloro-, bromo- and iodoalkanes. Halogens attached to olefinic or aromatic carbons are unreactive, as well as CX_2, CX_3 or C-F moieties. Enantioselectivities on 2-bromoalkanes with chain lengths of C_4 - C_7 ranged from poor to good [2019]. However, more polar substrates, such as α-bromo esters and -amides were resolved with excellent E-values with a strong preference for the (*R*)-enantiomer [2020, 2021]. Due to inversion of configuration during hydrolysis, the hydroxy product and remaining non-converted substrate are both (*S*)-configured (homochiral) (Scheme 2.231).

---

[58]The use of Lindane was banned in 2009.

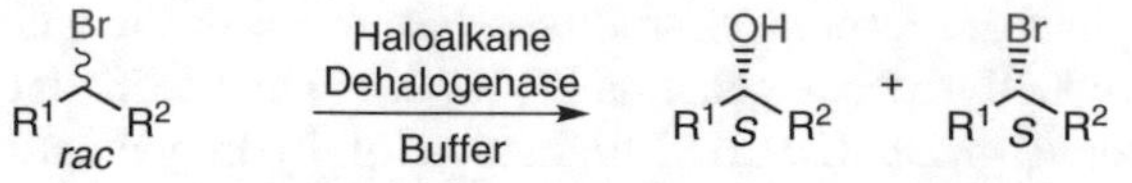

R^1	R^2	Enzyme	*E*	R^1	R^2	Enzyme	*E*
Me	Et	DhaA	2	Me	CO_2Me	DbjA	>200
Me	*n*-Pr	DbjA	145	Et	CO_2Me	DbjA	>200
Me	*n*-Bu	DbjA	68	Me	CO-NHCH$_2$Ph	DbjA	>200
Me	*n*-Pent	DbjA	28	Et	CO-NHCH$_2$Ph	LinB	>200

Scheme 2.231 Kinetic resolution of 2-bromoalkanes, α-bromoesters and amides using haloalkane dehalogenases

α-Haloacid Dehalogenases are specific for short-medium chain 2-halocarboxylic acids [2022, 2023]. In contratst to haloalkane dehalogenases, they can also convert α-fluoroalkanoates,[59] but they are inactive on (nonactivated) haloalkanes. Interestingly, the reactivity of halides with α-haloacid dehalogenases depends on the source of enzyme. In some cases it increases from iodine to fluorine derivatives, which is in sharp contrast to the corresponding chemical reactivity with nucleophiles such as hydroxyl ion [2024]. The most intriguing aspect of α-haloacid dehalogenases is their enantiospecificity [2025]. Depending on the growth conditions, the microbial production of stereo-complementary (*R*)- or (*S*)-specific enzymes may be induced [2026–2028].

(*S*)-2-Chloropropionic acid is a key chiral synthon required for the synthesis of a range of important α-aryl- and α-aryloxypropionic acids used as anti-inflammatory agents and herbicides, respectively (Scheme 2.35). Several strategies to resolve racemic 2-chloropropionic acid via enzymatic ester hydrolysis using 'classic' hydrolases proceed with varying degrees of selectivity [2029]. An elegant approach makes use of an (*R*)-specific α-haloacid dehalogenase from *Pseudomonas putida* NCIMB 12018 (Scheme 2.232) [2030–2032]. Thus, from a racemic mixture of α-haloacid, the (*R*)-enantiomer is converted into the (*S*)-hydroxyacid product via *inversion* of configuration leaving the (*S*)-α-haloacid behind. Some minor structural variations of the substrate are tolerated. This process has been scaled-up to industrial production at a capacity of 2000 t/year [2033].

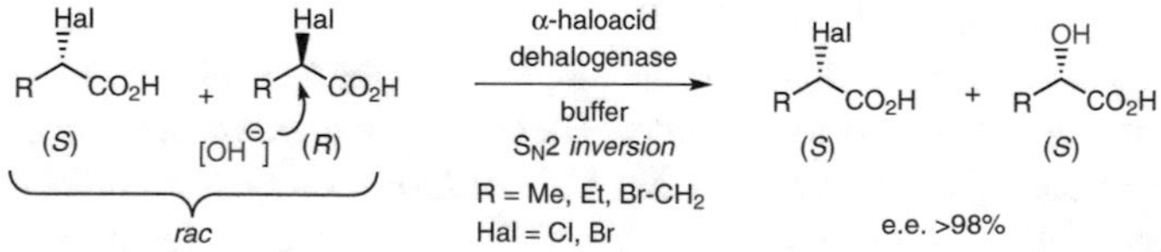

Scheme 2.232 Resolution of 2-chloropropanoic acid derivatives by α-haloacid dehalogenase

[59] Fluoroacetate is extremely toxic.

The hydrolytic instability of α-bromoacids in aqueous solvent systems and the limited solubility of long-chain analogs can be overcome by using organic solvents [2034]. Thus, long-chain α-haloacids (which were not accepted as substrates in water) were successfully transformed with good specificity in presence of toluene, acetone or DMSO.

Halohydrin Dehalogenases

The biodegradation of halohydrins proceeds through a two-step mechanism involving epoxide-formation catalysed by halohydrin dehalogenases [EC 4.5.1.X, formally lyases],[60] followed by epoxide hydrolase-mediated formation of vic-diols (Sect. 2.1.5), which are oxidatively degraded. A number of organisms possessing halohydrin dehalogenase and epoxide hydrolase activity, respectively, were found among bacteria (*Flavo-* [2035, 2036], *Corynebacteria* [2037], *Arthrobacter erithrii* [2038, 2039], *Pseudomonas* sp. [2040]), fungi (*Caldariomyces fumago*), and algae (*Laurencia pacifica*).

First hints on the stereoselectivity of halohydrin dehalogenases were obtained from studies on the desymmetrization of prochiral 1,3-dichloropropan-2-ol yielding epichlorohydrin using resting cells of *Corynebacterium* sp. (Scheme 2.233) [2041]. In a two-step sequence, (*R*)-3-chloropropane-1,2-diol was formed in 74% e.e. via epichlorohydrin through the sequential action of an (unspecified) halohydrin dehalogenase and an epoxide hydrolase [2042]. Further studies revealed that these activities are widespread among bacteria [2043–2047].

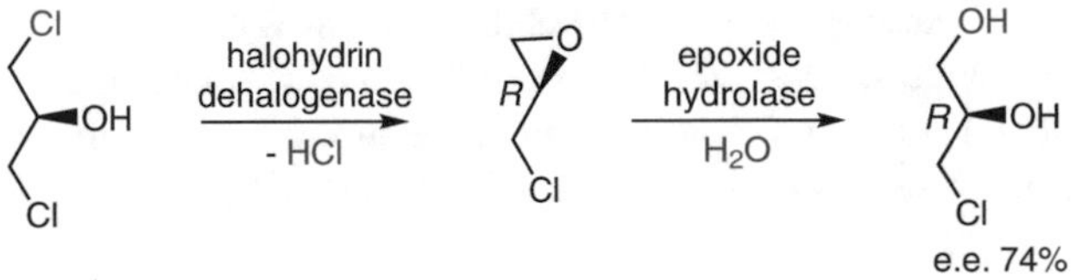

Scheme 2.233 Asymmetric microbial degradation of prochiral halohydrin by a *Corynebacterium* sp

A breakthrough was achieved by cloning and overexpression of halohydrin dehalogenases from *Agrobacterium radiobacter*, which allowed the preparative-scale application of these enzymes under well-defined conditions [2048].

The mechanism of halohydrin dehalogenase was shown to proceed in a reversible fashion via nucleophilic attack of halide (provided by a lipophilic halide binding site) with simultaneous activation of the epoxide through protonation by a Tyr residue within a conserved catalytic triad of Ser-Tyr-Arg (Scheme 2.234) [2049, 2050].

---

[60]Halohydrin dehalogenases were also (ambiguously) termed 'haloalcohol dehalogenases' or 'halohydrin epoxidases'.

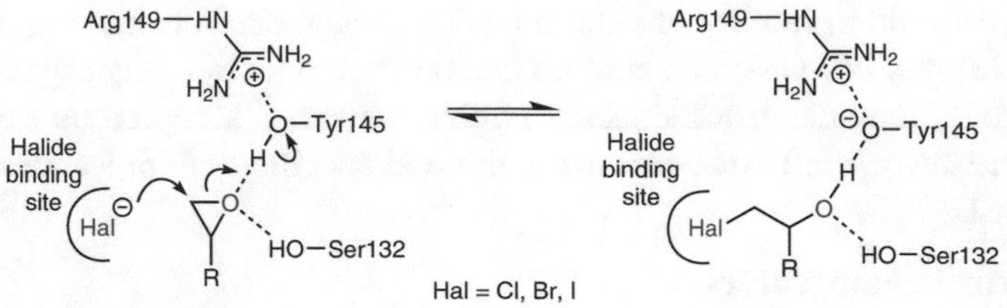

Scheme 2.234 Catalytic mechanism of halohydrin dehalogenase from *Agrobacterium radiobacter*

Using pure halohydrin dehalogenase (HheC), competing activities observed in whole-cell preparations were eliminated and halohydrins could be resolved via enantioselective ring-closure with excellent selectivities yielding (*R*)-epoxides and nonreacted (*S*)-halohydrins (Scheme 2.235) [2051, 2052]. A stereo-complementary enzyme (HheA) showing opposite stereoselectivity could be identified [2053].

Subsequent studies revealed that the natural nucleophile halide (Cl, Br, I) could be replaced by nonnatural analogs, such as azide [2054], nitrite [2055], cyanide [2056], (thio)cyanate and formate by maintining the exquisite regioselectivity of nucleophilic attack at the less hindered oxirane carbon atom. Whereas the reaction rates observed with cyanide, (thio)cyanate, and formate were comparable to those using halide, azide and nitrite proved to be much better nucleophiles [2057]. Nonlinear and nonanionic nucleophiles, such as H_2S, acetate, PO_4^{3-}, CO_3^{2-}, BO_3^{3-}, and F^- were unreactive. The use of *N*-nucleophiles opened the way to prepare 1,2- and 1,3-aminoalcohols using azide or cyanide via the corresponding 1-azido-2-ols and 1-cyano-2-ols, respectively (Scheme 2.236).

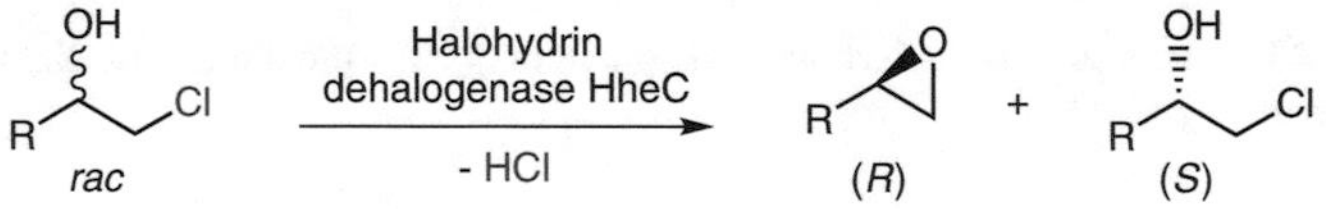

R	Enantioselectivity (*E*)
(*E*)-Me-CH=CH-	>200
(*E*)-Et-CH=CH-	177
(*E*)-*n*-C$_4$H$_9$-CH=CH-	>200
(*E*)-Ph-CH=CH-	>200
2-Furyl	>200
2-Thiophenyl	65

Scheme 2.235 Kinetic resolution of halohydrins using halohydrin dehalogenase

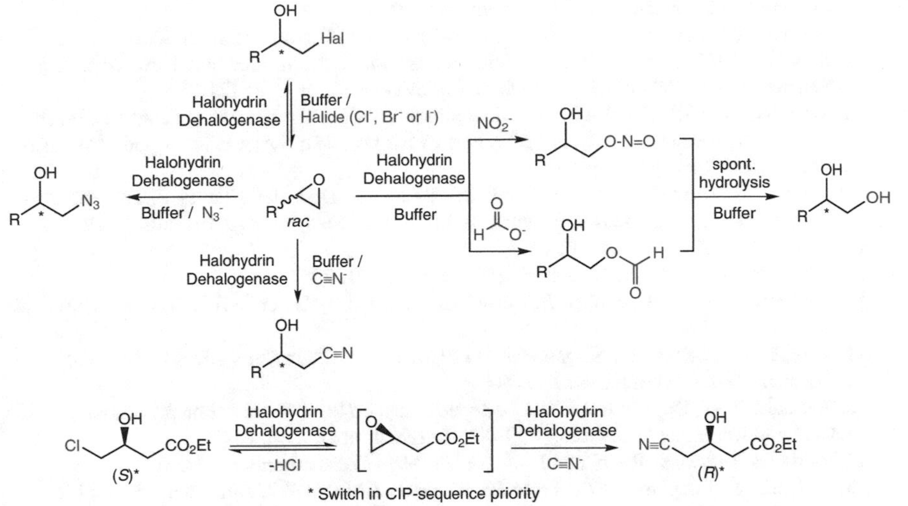

Scheme 2.236 Regio- and enantioselective ring-opening of epoxides using nonnatural nucleophiles catalysed by halohydrin dehalogenase

The mono-nitrite (or formate) esters of *vic*-diols obtained via enzymatic ring-opening of epoxides in presence of nitrite (or formate) are unstable and undergo spontaneous (nonenzymatic) hydrolysis to furnish the corresponding diols. This protocol offers a useful complement to the asymmetric hydrolysis of epoxides. Depending on the type of substrate and the enzymes used, enantio-complementary epoxide hydrolysis can be achieved [2058].

A one-pot two-step transformation of ethyl (*S*)-4-chloro-3-hydroxybutanoate (obtained via asymmetric bioreduction of the corresponding β-ketoester) via (reversible) epoxide-formation followed by ring-opening with cyanide was accomplished on a kg-scale using a halohydrin dehalogenase mutant. Ethyl (*R*)-4-cyano-4-butanoate was thus obtained in a highly chemoselective fashion without formation of byproducts, which plagued the chemical process. The latter product is a key intermediate for the synthesis of antihypocholesterolemic 'statin' agents [2059] (Scheme 2.236).

References

References to Sect. 2.1

1. Bornscheuer UT, Kazlauskas RJ (2006) Hydrolases in Organic Synthesis, 2nd ed. Wiley-VCH, Weinheim
2. Boland W, FRößL C, Lorenz M (1991) Synthesis 1049
3. Lee HC, Ko YH, Baek SB, Kim DH (1998) Bioorg. Med. Chem. Lett. 8: 3379

4. Otto H-H, Schirmeister T (1997) Chem. Rev. 97: 133
5. Harrison MJ, Burton NA, Hillier IH, Gould IR (1996) Chem. Commun. 2769
6. Fersht A (1985) Enzyme Structure and Mechanism, 2nd edn. Freeman, New York, p 405
7. Kazlauskas RJ, Weber HK (1998) Curr. Opinion Chem. Biol. 2: 121
8. Jones JB, Beck JF (1976) Asymmetric Syntheses and Resolutions using Enzymes. In: Jones JB, Sih CJ, Perlman D (eds) Applications of Biochemical Systems in Organic Chemistry. Wiley, New York, part I, p 107
9. Brady L, Brzozowski AM, Derewenda ZS, Dodson E, Dodson G, Tolley S, Turkenburg JP, Christiansen L, Huge-Jensen B, Norskov L, Thim L, Menge U (1990) Nature 343: 767
10. Blow D (1990) Nature 343: 694
11. Schrag JD, Li Y, Wu S, Cygler M (1991) Nature 351: 761
12. Sussman JL, Harel M, Frolow F, Oefner C, Goldman A, Toker L, Silman I (1991) Science 253: 872
13. Kirchner G, Scollar MP, Klibanov AM (1986) J. Am. Chem. Soc. 107: 7072
14. Faber K, Riva S (1992) Synthesis 895
15. Starmans WAJ, Doppen RG, Thijs L, Zwanenburg B (1998) Tetrahedron Asymmetry 9: 429
16. Garcia MJ, Rebolledo F, Gotor V (1993) Tetrahedron Lett. 34: 6141
17. Puertas S, Brieva R, Rebolledo F, Gotor V (1993) Tetrahedron 49: 6973
18. Kitaguchi H, Fitzpatrick PA, Huber JE, Klibanov AM (1989) J. Am. Chem. Soc. 111: 3094
19. Gotor V (1992) Enzymatic aminolysis, hydrazinolysis and oximolysis reactions. In: Servi S (ed) Microbial Reagents in Organic Synthesis, Nato ASI Series C, vol 381. Kluwer, Dordrecht, p 199
20. Abernethy JL, Albano E, Comyns J (1971) J. Org. Chem. 36: 1580
21. Silver MS (1966) J. Am. Chem. Soc. 88: 4247
22. Björkling F, Frykman H, Godtfredsen SE, Kirk O (1992) Tetrahedron 48: 4587
23. Öhrner N, Orrenius C, Mattson A, Norin T, Hult K (1996) Enzyme Microb. Technol. 19: 328
24. Chen CS, Sih CJ (1989) Angew. Chem. Int. Ed. 28: 695
25. Schoffers E, Golebiowski A, Johnson CR (1996) Tetrahedron 52: 3769
26. Matsumoto K, Tsutsumi S, Ihori T, Ohta H (1990) J. Am. Chem. Soc. 112: 9614
27. Björkling F, Boutelje J, Gatenbeck S, Hult K, Norin T, Szmulik P (1985) Tetrahedron 41: 1347
28. Krisch K (1971) Carboxyl ester hydrolases. In: Boyer PD (ed) The Enzymes, 3rd edn, vol 5. Academic Press, New York, p 43
29. Ramos-Tombo GM, Schär H-P, Fernandez i Busquets X, Ghisalba O (1986) Tetrahedron Lett. 27: 5707
30. Gais HJ, Lukas KL (1984) Angew. Chem. Int. Ed. 23: 142
31. Kasel W, Hultin PG, Jones JB (1985) J. Chem. Soc., Chem. Commun. 1563
32. Wang YF, Chen CS, Girdaukas G, Sih CJ (1984) J. Am. Chem. Soc. 106: 3695
33. Kroutil W, Kleewein A, Faber K (1997) Tetrahedron Asymmetry 8: 3251
34. Wang YF, Chen CS, Girdaukas G, Sih CJ (1985) Extending the applicability of esterases of low enantioselectivity in asymmetric synthesis. In: Porter R, Clark S (eds) Enzymes in Organic Synthesis, Ciba Foundation Symposium 111. Pitman, London, p 128
35. Sih CJ, Wu SH (1989) Topics Stereochem. 19: 63
36. Dakin HD (1903) J. Physiol. 30: 253
37. Gruber CC, Lavandera I, Faber K, Kroutil W (2006) Adv. Synth. Catal. 348: 1789
38. Klempier N, Faber K, Griengl H (1989) Synthesis 933
39. Chen CS, Fujimoto Y, Girdaukas G, Sih CJ (1982) J. Am. Chem. Soc. 104: 7294
40. Martin VS, Woodard SS, Katsuki T, Yamada Y, Ikeda M, Sharpless KB (1981) J. Am. Chem. Soc. 103: 6237
41. Bredig G, Fajans K (1908) Ber. dtsch. chem. Ges. 41: 752
42. Janes LE, Kazlauskas RJ (1997) J. Org. Chem. 62: 4560
43. Zmijewski MJ, Sullivan G, Persichetti R, Lalonde J (1997) Tetrahedron Asymmetry 8: 1153

44. Straathof AJJ, Jongejan JA (1997) Enzyme Microb. Technol. 21: 559
45. Gawley RE (2006) J. Org. Chem. 71: 2411
46. Faber K (1997) Enantiomer 2: 411
47. Straathof AJJ, Rakels JLL, Heijnen JJ (1992) Biocatalysis 7: 13
48. van Tol JBA, Jongejan JA, Geerlof A, Duine JA (1991) Recl. Trav. Chim. Pays-Bas 110: 255
49. Lu Y, Zhao X, Chen ZN (1995) Tetrahedron Asymmetry 6: 1093
50. Rakels JLL, Straathof AJJ, Heijnen JJ (1993) Enzyme Microb. Technol. 15: 1051
51. Faber K, Hönig H, Kleewein A (1995) Recent developments: determination of the selectivity of biocatalytic kinetic resolution of enantiomers – the 'enantiomeric ratio'. In: Roberts SM (ed) Preparative Biotransformations. Wiley, New York
52. Oberhauser T, Bodenteich M, Faber K, Penn G, Griengl H (1987) Tetrahedron 43: 3931
53. Chen CS, Wu, SH, Girdaukas G, Sih CJ (1987) J. Am. Chem. Soc. 109: 2812
54. Langrand G, Baratti J, Buono G, Triantaphylides C (1988) Biocatalysis 1: 231
55. Kroutil W, Kleewein A, Faber K (1997) Tetrahedron Asymmetry 8: 3263
56. Guo ZW, Wu SH, Chen CS, Girdaukas G, Sih CJ (1990) J. Am. Chem. Soc. 112: 4942
57. Caron G, Kazlauskas R (1993) Tetrahedron Asymmetry 4: 1995
58. Kazlauskas RJ (1989) J. Am. Chem. Soc. 111: 4953
59. Wu SH, Zhang LQ, Chen CS, Girdaukas G, Sih CJ (1985) Tetrahedron Lett. 26: 4323
60. Alfonso I, Astorga C, Rebolledo F, Gotor V (1996) Chem. Commun. 2471
61. Macfarlane ELA, Roberts SM, Turner NJ (1990) J. Chem. Soc., Chem. Commun. 569
62. Chen CS, Liu YC (1991) J. Org. Chem. 56: 1966
63. Faber K (1992) Indian J. Chem. 31B: 921
64. Stecher H, Faber K (1997) Synthesis 1
65. Strauss UT, Felfer U, Faber K (1999) Tetrahedron Asymmetry 10: 107
66. Faber K (2001) Chem. Eur. J. 7: 5004
67. Diaz-Rodriguez A, Lavandera I, Gotor V (2015) Curr. Green Chem. 2: 192.
68. Rachwalski M, Vermue N, Rutjes FPJT (2013) Chem. Soc. Rev. 42: 9268.
69. Guo ZW (1993) J. Org. Chem. 58: 5748
70. Ebbers EJ, Ariaans GJA, Houbiers JPM, Bruggink A, Zwanenburg B (1997) Tetrahedron 53: 9417
71. Adams E (1976) Adv. Enzymol. Relat. Areas Mol. Biol. 44: 69
72. Yagasaki M, Ozaki A (1997) J. Mol. Catal. B 4: 1
73. Danda H, Nagatomi T, Maehara A, Umemura T (1991) Tetrahedron 47: 8701
74. Takano S, Suzuki M, Ogasawara K (1993) Tetrahedron Asymmetry 4: 1043
75. Kanerva LT (1996) Acta Chem. Scand. 50: 234
76. Pedragosa-Moreau S, Morisseau C, Baratti J, Zylber J, Archelas A, Furstoss R (1997) Tetrahedron 53: 9707
77. El Gihani MT, Williams JMJ (1999) Curr. Opinion Chem. Biol. 3: 11
78. Kitamura M, Tokunaga M, Noyori R (1993) Tetrahedron 49: 1853
79. Ward RS (1995) Tetrahedron Asymmetry 6: 1475
80. Verho O, Baeckvall J-E (2015) J. Am. Chem. Soc. 137: 3996.
81. Takizawa S, Groeger H, Sasai H (2015) Chem. Eur. J. Org. 21: 8992.
82. Turner NJ (2010) Curr. Opin. Chem. Biol. 14: 115.
83. Eliel EL, Wilen SH, Mander LN (eds) (1994) Stereochemistry of Organic Compounds. Wiley, New York, p 315
84. Morrisson JD (ed) (1983) Asymmetric Synthesis. Academic Press, New York, vol 1, p 1
85. Fülling G, Sih CJ (1987) J. Am. Chem. Soc. 109: 2845
86. Brinksma J, van der Deen H, van Oeveren A, Feringa BL (1998) J. Chem. Soc., Perkin Trans. 1, 4159
87. Um P-J, Drueckhammer DG (1998) J. Am. Chem. Soc. 120: 5605
88. Dinh PM, Williams JMJ, Harris W (1999) Tetrahedron Lett. 40: 749

89. Taniguchi T, Ogasawara K (1997) Chem. Commun. 1399
90. Wegman MA, Hacking MAPJ, Rops J, Pereira P, van Rantwijk F, Sheldon RA (1999) Tetrahedron Asymmetry 10: 1739
91. Inagaki M, Hiratake J, Nishioka T, Oda J (1992) J. Org. Chem. 57: 5643
92. Kitamura M, Tokunaga M, Noyori R (1993) J. Am. Chem. Soc. 115: 144
93. Tan DS, Günter MM, Drueckhammer DG (1995) J. Am. Chem. Soc. 117: 9093
94. Williams RM (1989) Synthesis of Optically Active α-Amino Acids. Pergamon Press, Oxford
95. http://www.prweb.com/pdfdownload/8151116.pdf
96. Leuchtenberger W, Huthmacher K, Drauz K (2005) Appl. Microbiol. Biotechnol. 69: 1.
97. Taylor PP, Pantaleone DP, Senkpeil RF, Fotheringham IG (1998) Trends Biotechnol. 16: 412
98. Yoshimura T, Jhee KH, Soda K (1996) Biosci. Biotechnol. Biochem. 60: 181
99. Ohshima T, Soda K (1989) Trends Biotechnol. 7: 210
100. Schmidt-Kastner G, Egerer P (1984) Amino acids and peptides. In: Rehm HJ, Reed G (eds) Biotechnology. Verlag Chemie, Weinheim, vol 6a, p 387
101. Kamphuis J, Boesten WHJ, Kaptein B, Hermes HFM, Sonke T, Broxterman QB, van den Tweel WJJ, Schoemaker HE (1992) The production and uses of optically pure natural and unnatural amino acids. In: Collins AN, Sheldrake GN, Crosby J (eds) Chirality in Industry. Wiley, Chichester, p 187
102. Enei H, Shibai H, Hirose Y (1982) Amino acids and related compounds. In: Tsao GT (ed) Annual Reports on Fermentation Processes. Academic Press, New York, vol 5, p 79
103. Abbott BJ (1976) Adv. Appl. Microbiol. 20: 203
104. Rozzell D, Wagner F (eds) (1992) Biocatalytic Production of Amino Acids and Derivatives. Hanser Publ., Munich
105. Yonaha K, Soda K (1986) Adv. Biochem. Eng. Biotechnol. 33: 95
106. Soda K, Tanaka H, Esaki N (1983) Amino acids. In: Rehm HJ, Reed G (eds) Biotechnology, vol 3. Verlag Chemie, Weinheim, p 479
107. Wagner I, Musso H (1983) Angew. Chem. Int. Ed. 22: 816
108. Kaptein B, Boesten WHJ, Broxterman QB, Peters PJH, Schoemaker HE, Kamphuis J (1993) Tetrahedron Asymmetry 4: 1113
109. Lu W, Rey P, Benezra A (1995) J. Chem. Soc., Perkin Trans. 1, 553
110. Yamada S, Hongo C, Yoshioka R, Chibata I (1983) J. Org. Chem. 48: 843
111. Kamphuis J, Boesten WHJ, Broxterman QB, Hermes HFM, van Balken JAM, Meijer EM, Schoemaker HE (1990) Adv. Biochem. Eng. Biotechnol. 42: 134
112. Warburg O (1905) Ber. dtsch. chem. Ges. 38: 187
113. Jones M, Page MI (1991) J. Chem. Soc. 316
114. Miyazawa T, Takitani T, Ueji S, Yamada T, Kuwata S (1988) J. Chem. Soc., Chem. Commun. 1214
115. Miyazawa T, Iwanaga H, Ueji S, Yamada T, Kuwata S (1989) Chem. Lett. 2219
116. Chenevert R, Bel Rhlid R, Letourneau M, Gagnon R, D'Astous L (1993) Tetrahedron Asymmetry 4: 1137
117. Jones JB, Kunitake T, Niemann C, Hein GE (1965) J. Am. Chem. Soc. 87: 1777
118. Jones JB, Beck JF (1976) Applications of chymotrypsin in resolutions and asymmetric synthesis. In: Jones JB, Sih CJ, Perlman D (eds) Applications of Biochemical Systems in Organic Chemistry. Wiley, New York, part I, p 137
119. Dirlam NC, Moore BS, Urban FJ (1987) J. Org. Chem. 52: 3287
120. Berger A, Smolarsky M, Kurn N, Bosshard HR (1973) J. Org. Chem. 38: 457
121. Cohen SG (1969) Trans. NY Acad. Sci. 31: 705
122. Hess PD (1971) Chymotrypsin – chemical properties and catalysis. In: Boyer GP (ed) The Enzymes, vol 3. Academic Press, New York, p 213
123. Chenevert R, Letourneau M, Thiboutot S (1990) Can. J. Chem. 68: 960
124. Roper JM, Bauer DP (1983) Synthesis 1041

125. Izquierdo MC, Stein RL (1990) J. Am. Chem. Soc., 112: 6054
126. Chen ST, Wang KT, Wong CH (1986) J. Chem. Soc., Chem. Commun. 1514
127. Glänzer BI, Faber K, Griengl H (1987) Tetrahedron 43: 771
128. Chen ST, Huang WH, Wang KT (1994) J. Org. Chem. 59: 7580
129. Greenstein JP, Winitz M (1961) Chemistry of the Amino Acids. Wiley, New York, p 715
130. Boesten WHJ, Dassen BHN, Kerkhoffs PL, Roberts MJA, Cals MJH, Peters PJH, van Balken JAM, Meijer EM, Schoemaker HE (1986) Efficient enzymic production of enantiomerically pure amino acids. In: Schneider MP (ed) Enzymes as Catalysts in Organic Synthesis. Reidel, Dordrecht, p 355
131. Sonke T, Kaptein B, Boesten WHJ, Broxterman QB, Schoemaker HE, Kamphuis J, Formaggio F, Toniolo T, Rutjes FPJT (2000) In: Patel RN (ed) Stereoselective Biocatalysis. Marcel Dekker, New York, p 23
132. Eichhorn E, Roduit J-P, Shaw N, Heinzmann K, Kiener A (1997) Tetrahedron Asymmetry 8: 2533
133. Kamphuis J, Boesten WHJ, Broxterman QB, van Balken JAM, Meijer EM, Schoemaker HE (1990) Adv. Biochem. Eng. Biotechnol. 42: 133
134. Yasukawa K, Asano Y (2012) Adv. Synth. Catal. 354: 3332.
135. Sambale C, Kula MR (1987) Biotechnol. Appl. Biochem. 9: 251
136. Chibata I, Tosa T, Sato T, Mori T (1976) Methods Enzymol. 44: 746
137. Greenstein JP (1957) Methods Enzymol. 3: 554
138. Chibata I, Ishikawa T, Tosa T (1970) Methods Enzymol. 19: 756
139. Bommarius AS, Drauz K, Klenk H, Wandrey C (1992) Ann. New York Acad. Sci. 672: 126
140. Tosa T, Mori T, Fuse N, Chibata I (1967) Biotechnol. Bioeng. 9: 603
141. Chenault HK, Dahmer J, Whitesides GM (1989) J. Am. Chem. Soc. 111: 6354
142. Mori K, Otsuka T (1985) Tetrahedron 41: 547
143. Baldwin JE, Christie MA, Haber SB, Kruse LI (1976) J. Am. Chem. Soc. 98: 3045
144. Cobley CJ, Hanson CH, Lloyd MC, Simmonds S, Peng WJ (2011) Org. Proc. Res. Dev. 15: 284.
145. Verseck S, Bommarius A S, Kula MR (2001) Appl. Microbiol. Biotechnol. 55: 345
146. May O, Verseck S, Bommarius A S, Drauz K (2002) Org. Proc. Res. Dev. 6: 452
147. Baxter S, Royer S, Grogan G, Brown F, Holt-Tiffin KE, Taylor IN, Fotheringham IG, Campopiano DJ (2012) J. Am. Chem. Soc. 134: 19310
148. Tokuyama S, Hatano K (1996) Appl. Microbiol. Biotechnol. 44: 774
149. Liu J, Asano Y, Ikoma K, Yamashita S, Hirose Y, Shimoyama T, Takahashi S, Nakayama T, Nishino T (2012) J. Biosci. Bioeng. 114: 391.
150. Wang W, Xi H, Bi Q, Hu Y, Zhang Y, Ni M (2013) Microbiol. Res. 168: 360.
151. Arima J, Isoda Y, Hatanaka T, Mori N (2013) World J. Microbiol. Biotechnol. 29: 899.
152. Hambardzumyan AA, Mkhitaryan AV, Paloyan AM, Dadayan SA, Saghyan AS (2016) Appl. Biochem. Microbiol. 52: 250.
153. Shaw NM, Robins K, Kiener A (2003) Adv. Synth. Catal. 345: 425
154. Petersen M, Sauter M (1999) Chimia 53: 608
155. Solodenko VA, Kasheva TN, Kukhar VP, Kozlova EV, Mironenko DA, Svedas VK (1991) Tetrahedron 47: 3989
156. Bücherer HT, Steiner W (1934) J. Prakt. Chem. 140: 291
157. Ogawa J, Shimizu S (1997) J. Mol. Catal. B 2: 163
158. Yamada H, Takahashi S, Kii Y, Kumagai H (1978) J. Ferment. Technol. 56: 484
159. Ogawa J, Shimizu S (2000) In: Patel RN (ed) Stereoselective Bioocatalysis. Marcel Dekker, New York, p 1
160. Runser S, Chinski N, Ohleyer E (1990) Appl. Microbiol. BIotechnol. 33: 382
161. Möller, A, Syldatk C, Schulze M, Wagner F (1988) Enzyme Microb. Technol. 10: 618
162. Nam S-H, Park H-S, Kim H-S (2005) Chem. Rec. 5: 298.
163. Wallach DP, Grisolia S (1957) J. Biol. Chem. 226: 277

164. Olivieri R, Fascetti E, Angelini L (1981) Biotechnol. Bioeng. 23: 2173
165. Yamada H, Shimizu S, Shimada H, Tani Y, Takahashi S, Ohashi T (1980) Biochimie 62: 395
166. Guivarch M, Gillonnier C, Brunie JC (1980) Bull. Soc. Chim. Fr. 91
167. Syldatk C, Müller R, Pietzsch M, Wagner F (1992) Biocatalytic Production of Amino Acids and Derivatives. Hanser Publ., Munich, p 129
168. Yamada H, Takahashi S, Yoshiaki K, Kumagai H (1978) J. Ferment. Technol. 56: 484
169. Martinez-Rodriguez S, Martinez-Gomez AI, Rodriguez-Vico F, Clemente-Jimenez JM, Las Heras-Vazquez FJ (2010) Appl. Microbiol. Biotechnol. 85: 441.
170. Watabe K, Ishikawa T, Mukohara Y, Nakamura H (1992) J. Bacteriol. 174: 7989
171. Pietzsch M, Syldatk C, Wagner F (1992) Ann. New York Acad. Sci. 672: 478
172. Assaf Z, Eger E, Vitnik Z, Fabian WMF, Ribitsch D, Guebitz GM, Faber K, Hall M (2014) ChemCatChem 6: 2517.
173. Evans C, McCague R, Roberts SM, Sutherland AG (1991) J. Chem. Soc., Perkin Trans. 1, 656
174. Fukumura T, Talbot G, Misono H, Teramura Y, Kato K, Soda K (1978) FEBS Lett. 89: 298.
175. Okazaki S, Suzuki A, Mizushima T, Kawano T, Komeda H, Asano Y, Yamane T (2009) Biochemistry 48: 941.
176. Asano Y, Yamaguchi S (2005) J. Mol. Catal. B: Enzym. 36: 22.
177. Payoungkiattikun W, Okazaki S, Nakano S, Ina A, H-Kittikun A, Asano Y (2015) Appl. Biochem. Biotechnol. 176: 1303.
178. Jones M, Page MI (1991) J. Chem. Soc., Chem. Commun. 316
179. Brieva R, Crich JZ, Sih CJ (1993) J. Org. Chem. 58: 1068
180. Forro E, Fülöp F (2006) Chem. Eur. J. 12: 2587
181. Evans C, McCague R, Roberts SM, Sutherland AG, Wisdom R (1991) J. Chem. Soc., Perkin Trans. 1, 2276
182. Ohno M, Otsuka M (1989) Org. React. 37: 1
183. Pearson AJ, Bansal HS, Lai YS (1987) J. Chem. Soc., Chem. Commun. 519
184. Johnson CR, Penning TD (1986) J. Am. Chem. Soc. 108: 5655
185. Suemune H, Harabe T, Xie ZF, Sakai K (1988) Chem. Pharm. Bull. 36: 4337
186. Dropsy EP, Klibanov AM (1984) Biotechnol. Bioeng. 26: 911
187. Chenevert R, Martin R (1992) Tetrahedron Asymmetry 3: 199
188. Kotani H, Kuze Y, Uchida S, Miyabe T, Limori T, Okano K, Kobayashi S, Ohno M (1983) Agric. Biol. Chem. 47: 1363
189. Lambrechts C, Galzy P (1995) Biosci. Biotech. Biochem. 59: 1464
190. Jackson MA, Labeda DP, Becker LA (1995) Enzyme Microb. Technol. 17: 175
191. Quax WJ, Broekhuizen CP (1994) Appl. Microbiol. Biotechnol. 41: 425
192. Bornscheuer UT, Kazlauskas R J (2006) Hydrolases in Organic Synthesis, 2nd ed. Wiley-VCH, Weinheim, p 179
193. Zocher F, Krebsfänger N, Yoo OJ, Bornscheuer UT (1998) J. Mol. Catal. B 5: 199
194. Schlacher A, Stanzer T, Soelkner B, Klingsbichel E, Petersen E, Schmidt M, Klempier N, Schwab H (1997) J. Mol. Catal. B 3: 25
195. Jones JB (1980) In: Dunnill P, Wiseman A, Blakeborough N (eds) Enzymic and Non-enzymic Catalysis. Horwood/Wiley, New York, p 54
196. Schubert Wright C (1972) J. Mol. Biol. 67: 151
197. Philipp M, Bender ML (1983) Mol. Cell. Biochem. 51: 5
198. Fruton JS (1971) Pepsin. In: Boyer PD (ed) The Enzymes, vol 3. Academic Press, London, p 119
199. Bianchi D, Cabri W, Cesti P, Francalanci F, Ricci M (1988) J. Org. Chem. 53: 104
200. Fancetic O, Deretic V, Marjanovic N, Glisin V (1988) Biotechnol. Forum 5: 90
201. Baldaro E, Fuganti C, Servi S, Tagliani A, Terreni M (1992) The use of immobilized penicillin acylase in organic synthesis. In: Servi S (ed) Microbial Reagents in Organic Synthesis. NATO ASI Ser. C, vol 381. Kluwer, Dordrecht, p 175

202. Bender ML, Killheffer JV (1973) Crit. Rev. Biochem. 1: 149
203. Barnier JP, Blanco L, Guibe-Jampel E, Rousseau G (1989) Tetrahedron 45: 5051
204. Schultz M, Hermann P, Kunz H (1992) Synlett. 37
205. Moorlag H, Kellogg RM, Kloosterman M, Kaptein B, Kamphuis J, Schoemaker HE (1990) J. Org. Chem. 55: 5878
206. Kallwass HKW, Yee C, Blythe TA, McNabb TJ, Rogers EE, Shames SL (1994) Bioorg. Med. Chem. 2: 557
207. Chen ST, Fang JM (1997) J. Org. Chem. 62: 4349
208. Henke E, Bornscheuer UT, Schmid RD, Pleiss J (2003) ChemBioChem 4: 485
209. Henke E, Pleiss J, Bornscheuer UT (2002) Angew. Chem. Int. Ed. 41: 3211
210. Krishna S H, Persson M, Bornscheuer UT (2002) Tetrahedron Asymmetry 13: 2693
211. Heymann E, Junge W (1979) Eur. J. Biochem. 95: 509
212. Öhrner N, Mattson A, Norin T, Hult K (1990) Biocatalysis 4: 81
213. Lam LKP, Brown CM, De Jeso B, Lym L, Toone EJ, Jones JB (1988) J. Am. Chem. Soc. 110: 4409
214. Polla A, Frejd T (1991) Tetrahedron 47: 5883
215. Seebach D, Eberle M (1986) Chimia 40: 315
216. Reeve DC, Crout DHG, Cooper K, Fray MJ (1992) Tetrahedron Asymmetry 3: 785
217. Senayake CH, Bill TJ, Larsen RD, Leazer J, Reiter PJ (1992) Tetrahedron Lett. 33: 5901
218. De Jeso B, Belair N, Deleuze H, Rascle MC, Maillard B (1990) Tetrahedron Lett. 31: 653
219. Tanyeli C, Sezen B, Demir AS, Alves RB, Arseniyadis S (1999) Tetrahedron Asymmetry 10: 1129
220. Jongejan JA, Duine JA (1987) Tetrahedron Lett. 28: 2767
221. Burger U, Erne-Zellweger D, Mayerl CM (1987) Helv. Chem. Acta 70: 587
222. Hazato A, Tanaka T, Toru T, Okamura N, Bannai K, Sugiura S, Manabe K, Kurozumi S (1983) Nippon Kagaku Kaishi 9: 1390
223. Hazato A, Tanaka T, Toru T, Okamura N, Bannai K, Sugiura S, Manabe K, Kurozumi S (1984) Chem. Abstr. 100: 120720q
224. Papageorgiou C, Benezra C (1985) J. Org. Chem. 50: 1145
225. Sicsic S, Leroy J, Wakselman C (1987) Synthesis 155
226. Schirmeister T, Otto HH (1993) J. Org. Chem. 58: 4819
227. Schneider M, Engel N, Boensmann H (1984) Angew. Chem. Int. Ed. 23: 66
228. Luyten M, Müller S, Herzog B, Keese R (1987) Helv. Chim. Acta 70: 1250
229. Huang FC, Lee LFH, Mittal RSD, Ravikumar PR, Chan JA, Sih CJ, Capsi E, Eck CR (1975) J. Am. Chem. Soc. 97: 4144
230. Herold P, Mohr P, Tamm C (1983) Helv. Chim. Acta 76: 744
231. Adachi K, Kobayashi S, Ohno M (1986) Chimia 40: 311
232. Ohno M, Kobayashi S, Iimori T, Wang Y-F, Izawa T (1981) J. Am. Chem. Soc. 103: 2405
233. Mohr P, Waespe-Sarcevic, Tamm C, Gawronska K, Gawronski JK (1983) Helv. Chim. Acta 66: 2501
234. Cohen SG, Khedouri E (1961) J. Am. Chem. Soc. 83: 1093
235. Cohen SG, Khedouri E (1961) J. Am. Cherm. Soc. 83: 4228
236. Roy R, Rey AW (1987) Tetrahedron Lett. 28: 4935
237. Santaniello E, Chiari M, Ferraboschi P, Trave S (1988) J. Org. Chem. 53: 1567
238. Gopalan AS, Sih CJ (1984) Tetrahedron Lett. 25: 5235
239. Mohr P, Waespe-Sarcevic N, Tamm C, Gawronska K, Gawronski JK (1983) Helv. Chim. Acta 66: 2501
240. Schregenberger C, Seebach D (1986) Liebigs Ann. Chem. 2081
241. Sabbioni G, Jones JB (1987) J. Org. Chem. 52: 4565
242. Björkling F, Boutelje J, Gatenbeck S, Hult K, Norin T (1985) Appl. Microbiol. Biotechnol. 21: 16
243. Bloch R, Guibe-Jampel E, Girard G (1985) Tetrahedron Lett. 26: 4087

244. Laumen K, Schneider M (1984) Tetrahedron Lett. 25: 5875
245. Harre M, Raddatz P, Walenta R, Winterfeldt E (1982) Angew. Chem. Int. Ed. 21: 480
246. Hummel A, Brüsehaber E, Böttcher D, Trauthwein H, Doderer K, Bornscheuer UT (2007)
     Angew. Chem. Int. Ed. 46: 8492
247. Deardorff DR, Mathews AJ, McMeekin DS, Craney CL (1986) Tetrahedron Lett. 27: 1255
248. Laumen K, Schneider MP (1986) J. Chem. Soc., Chem. Commun. 1298
249. Johnson CR, Bis SJ (1992) Tetrahedron Lett. 33: 7287
250. Sih CJ, Gu QM, Holdgrün X, Harris K (1992) Chirality 4: 91
251. Wang YF, Sih CJ (1984) Tetrahedron Lett. 25: 4999
252. Iriuchijima S, Hasegawa K, Tsuchihashi G (1982) Agric. Biol. Chem. 46: 1907
253. Mohr P, Rösslein L, Tamm C (1987) Helv. Chim. Acta 70: 142
254. Ramaswamy S, Hui RAHF, Jones JB (1986) J. Chem. Soc., Chem. Commun. 1545
255. Alcock NW, Crout DHG, Henderson CM, Thomas SE (1988) J. Chem. Soc., Chem.
     Commun. 746
256. Sicsic S, Ikbal M, Le Goffic F (1987) Tetrahedron Lett. 28: 1887
257. Klunder AJH, Huizinga WB, Hulshof AJM, Zwanenburg B (1986) Tetrahedron Lett. 27:
     2543
258. Crout DHG, Gaudet VSB, Laumen K, Schneider MP (1986) J. Chem. Soc., Chem.
     Commun. 808
259. Lange S, Musidlowska A, Schmidt-Dannert C, Schmitt J, Bornscheuer UT (2001) Chem-
     BioChem 2: 576
260. Musidlowska A, Lange S, Bornscheuer UT (2001) Angew. Chem. Int. Ed. 40: 2851
261. Dominguez de Maria P, Garcia-Burgos CA, Bargeman G, van Gemert RW (2007) Synthesis
     1439
262. Hermann M, Kietzmann MU, Ivancic M, Zenzmaier C, Luiten RGM, Skranc W,
     Wubbolts M, Winkler M, Birner-Gruenberger R, Pichler H, Schwab H (2008)
     J. Biotechnol. 133: 301
263. May O (2009) Green chemistry with biocatalysis for production of pharmaceuticals. In:
     Tao J, Lin GQ, Liese A (eds) Biocatalysis for the Pharmaceutical Industry. Wiley Asia,
     Singapore, p 310
264. Sugai T, Kuwahara S, Hishino C, Matsuo N, Mori K (1982) Agric. Biol. Chem. 46: 2579
265. Oritani T, Yamashita K (1980) Agric. Biol. Chem. 44: 2407
266. Ohta H, Miyamae Y, Kimura Y (1989) Chem. Lett. 379
267. Ziffer H, Kawai K, Kasai K, Imuta M, Froussios C (1983) J. Org. Chem. 48: 3017
268. Takaishi Y, Yang YL, DiTullio D, Sih CJ (1982) Tetrahedron Lett. 23: 5489
269. Glänzer BI, Faber K, Griengl H (1987) Tetrahedron 43: 5791
270. Mutsaers JHGM, Kooreman HJ (1991) Recl. Trav. Chim. Pays-Bas 110: 185
271. Smeets JWH, Kieboom APG (1992) Recl. Trav. Chim. Pays-Bas 111: 490
272. Stahly GP, Starrett RM (1997) Production methods for chiral non-steroidal anti-inflamma-
     tory profen drugs. In: Collins AN, Sheldrake GN, Crosby J (eds) Chirality in Industry
     II. Wiley, Chichester, pp 19–40
273. Jones JB, Beck JF (1976) Applications of chymotrypsin in resolution and asymmetric
     synthesis. In: Jones JB, Sih CJ, Perlman D (eds) Applications of Biochemical Systems in
     Organic Synthesis. Wiley, New York, p 137
274. Uemura A, Nozaki K, Yamashita J, Yasumoto M (1989) Tetrahedron Lett. 30: 3819
275. Kvittingen L, Partali V, Braenden JU, Anthonsen T (1991) Biotechnol. Lett. 13: 13
276. Pugniere M, San Juan C, Previero A (1990) Tetrahedron Lett. 31: 4883–4886
277. Shin C, Seki M, Takahashi N (1990) Chem. Lett. 2089
278. Miyazawa T, Iwanaga H, Yamada T, Kuwata S (1994) Biotechnol. Lett. 16: 373
279. Harrison FG, Gibson ED (1984) Proc. Biochem.19: 33.
280. Pathak T, Waldmann H (1998) Curr. Opinion Chem. Biol. 2: 112
281. Waldmann H, Reidel A (1997) Angew. Chem. Int. Ed. 36: 647
282. Francetic O, Deretic V, Marjanovic N, Glisin V (1988) Biotech-Forum 5: 90

283. Waldmann H (1988) Liebigs Ann. Chem. 1175
284. Fuganti C, Grasselli P, Servi S, Lazzarini A, Casati P (1988) Tetrahedron 44: 2575
285. Waldmann H (1989) Tetrahedron Lett. 30: 3057
286. Fernandez-Lafuente R, Guisan JM, Pregnolato M, Terreni M (1997) Tetrahedron Lett. 38: 4693
287. Pohl T, Waldmann H (1995) Tetrahedron Lett. 36: 2963
288. Baldaro E, Faiardi D, Fuganti C, Grasselli P, Lazzarini A (1988) Tetrahedron Lett. 29: 4623
289. Berger B, de Raadt A, Griengl H, Hayden W, Hechtberger P, Klempier N, Faber K (1992) Pure Appl. Chem. 64: 1085
290. Dale JA, Dull DL, Mosher HS (1969) J. Org. Chem. 34: 2543
291. Feichter C, Faber K, Griengl H (1991) J. Chem. Soc., Perkin Trans. 1, 653
292. Faber K, Ottolina G, Riva S (1993) Biocatalysis 8: 91
293. Kamezawa M, Raku T, Tachibana H, Ohtani T, Naoshima Y (1995) Biosci. Biotechnol. Biochem. 59: 549
294. Naoshima Y, Kamezawa M, Tachibana H, Munakata Y, Fujita T, Kihara K, Raku T (1993) J. Chem. Soc., Perkin Trans. 1, 557
295. Guanti G, Banfi L, Narisano E (1992) Asymmetrized tris(hydroxymethyl)methane and related synthons as new highly versatile chiral building blocks. In: Servi S (ed) Microbial Reagents in Organic Synthesis. Nato ASI Series C, vol 381. Kluwer, Dordrecht, pp 299–310
296. Morgan B, Oehlschlager AC, Stokes TM (1992) J. Org. Chem. 57: 3231
297. Nguyen BV, Nordin O, Vörde C, Hedenström E, Högberg HE (1997) Tetrahedron Asymmetry 8: 983
298. Nordin O, Hedenström E, Högberg H-E (1994) Tetrahedron Asymmetry 5: 785
299. Björkling F, Boutelje J, Gatenbeck S, Hult K, Norin T (1986) Bioorg. Chem. 14: 176
300. Guanti G, Banfi L, Narisano E, Riva R, Thea S (1986) Tetrahedron Lett. 27: 4639
301. Santaniello E, Ferraboschi P, Grisenti P, Aragozzini F, Maconi E (1991) J. Chem. Soc., Perkin Trans. 1, 601
302. Björkling F, Boutelje J, Gatenbeck, S, Hult K, Norin T (1985) Tetrahedron Lett. 26: 4957
303. Guo Z-W, Sih CJ (1989) J. Am. Chem. Soc. 111: 6836
304. Barton MJ, Hamman JP, Fichter KC, Calton GJ (1990) Enzyme Microb. Technol. 12: 577
305. Grandjean D, Pale P, Chuche J (1991) Tetrahedron Lett. 32: 3043
306. Sugai T, Hamada K, Akeboshi T, Ikeda H, Ohta H (1997) Synlett. 983
307. Liu YY, Xu JH, Xu QG, Hu Y (1999) Biotechnol. Lett. 21: 143
308. Lam LKP, Hui RAHF, Jones JB (1986) J. Org. Chem. 51: 2047
309. Phillips RS (1992) Enzyme Microb. Technol. 14: 417
310. Phillips RS (1996) Trends Biotechnol. 14: 13
311. Keinan E, Hafeli EK, Seth KK, Lamed R (1986) J. Am. Chem. Soc. 108: 162
312. Holmberg E, Hult K (1991) Biotechnol. Lett. 13: 323
313. Boutelje J, Hjalmarsson M, Hult K, Lindbäck M, Norin T (1988) Bioorg. Chem. 16: 364
314. Sakai T, Kawabata I, Kishimoto T, Ema T, Utaka M (1997) J. Org. Chem. 62: 4906
315. Sakai T, Kishimoto T, Tanaka Y, Ema T, Utaka M (1998) Tetrahedron Lett. 39: 7881
316. Sakai T, Mitsutomi H, Korenaga T, Ema T (2005) Tetrahedron Asymmetry 16: 1535
317. Varma RS (1999) Green Chem. 43
318. Obermayer D, Gutmann B, Kappe CO (2009) Angew. Chem. Int. Ed. 48: 8321
319. de la Hoz A, Diaz-Ortiz A, Moreno A (2005) Chem. Soc. Rev. 34: 164
320. Kappe CO (2013) Angew. Chem. Int. Ed. 52: 7924.
321. Rosana MR, Tao Y, Stiegman AE, Dudley BD (2012) Chem. Sci. 3: 1240.
322. Loupy A, Petit A, Hamelin J, Texier-Boullet F, Jacquault P, Mathe D (1998) Synthesis 1213
323. Parker MC, Besson T, Lamare S, Legoy MD (1996) Tetrahedron Lett. 37: 8383
324. Carrillo-Munoz JR, Bouvet D, Guibe-Jampel E, Loupy A, Petit A (1996) J. Org. Chem. 61: 7746
325. Bornscheuer UT, Kazlauskas RJ (2006) Hydrolases in Organic Synthesis, 2nd ed. Wiley-VCH Weinheim, p 43

326. Grunwald P (2009) Biocatalysis, Imperial College Press, p 777
327. Bommarius AS, Riebel BR (2004) Biocatalysis. Wiley-VCH, Weinheim, p 61
328. Penning TM, Jez JM (2001) Chem. Rev. 101: 3027
329. Reetz MT, Zonta A, Schimossek K, Liebeton K, Jaeger KE (1997) Angew. Chem. Int. Ed. 36: 2830
330. Renata H, Wang ZJ, Arnold FH (2015) Angew. Chem. Int. Ed. 54: 3351.
331. Cheng F, Zhu L, Schwaneberg U (2015) Chem. Commun. 51: 9760.
332. Reetz MT (2011) Angew. Chem. Int. Ed. 50: 138.
333. Toscano MD, Woycechowsky KJ, Hilvert D (2007) Angew. Chem. Int. Ed. 46: 3212
334. Gerlt JA, Babbitt PC (2009) Curr. Opinion Chem. Biol. 13: 10
335. Morley KL, Kazlauskas RJ (2005) Trends Biotechnol. 23: 231
336. Eigen M (1971) Naturwissenschaften 58: 465.
337. Shivange AV, Marienhagen J, Mundhada H, Schenk A, Schwaneberg U (2009) Curr. Opinion Chem. Biol. 13: 19
338. Baumann M, Stürmer R, Bornscheuer UT (2001) Angew. Chem. Int. Ed. 40: 4201
339. Reetz MT, Becker MH, Kühling KM, Holzwarth A (1998) Angew. Chem. Int. Ed. 37: 2647
340. Turner NJ (2009) Nature Chem. Biol. 5: 567
341. Reetz MT (2006) Adv. Catal. 49: 1
342. Demirjian DC, Shah PC, Moris-Varas F (1999) Top. Curr. Chem. 200: 1
343. Wahler D, Reymond JL (2001) Curr. Opinion Chem. Biol. 5: 152
344. Guo HH, Choe J, Loeb LA (2004) Proc. Natl. Acad. Sci. USA 101: 9205.
345. Chen KQ, Arnold FH (1993) Proc. Natl. Acad. Sci. USA 90: 5618.
346. Reetz MT (2002) Tetrahedron 58: 6595
347. Kourist R, Bartsch S, Bornscheuer UT (2007) Adv. Synth. Catal. 349: 1391
348. Henke E, Pleiss J, Bornscheuer UT (2003) ChemBioChem 4: 485
349. Bartsch S, Kourist R, Bornscheuer UT (2008) Angew. Chem. Int. Ed. 47: 1508
350. Böttcher D, Bornscheuer UT (2006) Nature Protocols 1: 2340
351. Jansonius JN (1987) Enzyme mechanism: what X-ray crystallography can(not) tell us. In: Moras D, Drenth J, Strandlberg B, Suck D, Wilson K (eds) Crystallography in Molecular Biology. Plenum Press, New York, p 229
352. Rubin B (1994) Struct. Biol. 1: 568
353. Heliwell JR, Helliwell M (1996) Chem. Commun. 1595
354. Alternatively, protein structures can be determined in solution by NMR spectroscopy, provided that the enzyme is not too large ($\leq$40 kDa).
355. Fersht A (1977) Enzyme Structure and Mechanism. Freeman, San Francisco, p 15
356. Schrag JD, Cygler M (1993) J. Mol. Biol. 230: 575
357. Grochulski P, Li Y, Schrag JD, Bouthillier F, Smith P, Harrison D, Rubin B, Cygler M (1993) J. Biol. Chem. 268: 12843
358. Uppenberg J, Hansen MT, Patkar S, Jones TA (1994) Structure 2: 293
359. Noble MEM, Cleasby A, Johnson LN, Egmond MR, Frenken LGJ (1993) FEBS Lett. 331: 123
360. Burkert U, Allinger NL (1982) Molecular mechanics; In: Caserio MC (ed) ACS Monograph, vol 177. ACS, Washington
361. Zhang Y (2009) Curr. Opin. Struct. Biol. 19: 145.
362. Wijma HJ, Janssen DB (2013) FEBS J. 280: 2948.
363. Gerlt JA, Allen KN, Almo SC, Armstrong RN, Babbitt PC, Cronan JE, Dunaway-Mariano D, Imker HJ, Jacobson MP, Minor W, Poulter CD, Raushel FM, Sali A, Shoichet BK, Sweedler JV (2011) Biochemistry 50: 9950.
364. Pleiss J (2011) Curr. Opin. Biotechnol. 22: 611.
365. Norin M, Hult K, Mattson A, Norin T (1993) Biocatalysis 7: 131
366. Fitzpatrick PA, Ringe D, Klibanov AM (1992) Biotechnol. Bioeng. 40: 735
367. Ortiz de Montellano PR, Fruetel JA, Collins JR, Camper DL, Loew GH (1991) J. Am. Chem. Soc. 113: 3195

368. Trott O, Ohlson AJ (2010) J. Comput. Chem. 31: 455.; http://vina.scripps.edu.
369. Oberhauser T, Faber K, Griengl H (1989) Tetrahedron 45: 1679
370. Kazlauskas RJ, Weissfloch ANE, Rappaport AT, Cuccia LA (1991) J. Org. Chem. 56: 2656
371. Karasaki Y, Ohno M (1978) J. Biochem. (Tokyo) 84: 531
372. Ahmed SN, Kazlauskas RJ, Morinville AH, Grochulski P, Schrag JD, Cygler M (1994) Biocatalysis 9: 209
373. Provencher L, Wynn H, Jones JB, Krawczyk AR (1993) Tetrahedron Asymmetry 4: 2025
374. Desnuelle P (1972) The lipases. In: Boyer PO (ed) The Enzymes, vol 7. Academic Press, New York, p 575
375. Wooley P, Petersen SB (eds) (1994) Lipases, their Structure, Biochemistry and Applications. Cambridge UP, Cambridge
376. Macrae AR (1983) J. Am. Oil Chem. Soc. 60: 291
377. Hanson M (1990) Oils Fats Int. 5: 29
378. Nielsen T (1985) Fette Seifen Anstrichmittel 87: 15
379. Jaeger KE, Reetz MT (1998) Trends Biotechnol. 16: 396
380. Schmid RD, Verger R (1998) Angew. Chem. Int. Ed. 37: 1608
381. Alberghina L, Schmidt RD, Verger R (eds) (1991) Lipases: Structure, Mechanism and Genetic Engineering. GBF Monographs, vol 16. Verlag Chemie, Weinheim
382. Ransac S, Carriere F, Rogalska E, Verger R, Marguet F, Buono G, Pinho Melo E, Cabral JMS, Egloff MPE, van Tilbeurgh H, Cambillau C (1996) The Kinetics, Specificites and Structural Features of Lipases. In: Op den Kamp AF (ed) Molecular Dynamics of Biomembranes. NATO ASI Ser, vol H 96. Springer, Heidelberg, pp 265–304
383. Sarda L, Desnuelle P (1958) Biochim. Biophys. Acta 30: 513
384. Schonheyder F, Volqvartz K (1945) Acta Physiol. Scand. 9: 57
385. Verger R (1997) Trends Biotechnol. 15: 32
386. Theil F (1995) Chem. Rev. 95: 2203
387. Miranda AS, Miranda LSM, de Souza ROMA (2015) Biotechnol. Adv. 33: 372.
388. Bianchi D, Cesti P (1990) J. Org. Chem. 55: 5657
389. Iriuchijima S, Kojima N (1981) J. Chem. Soc., Chem. Commun. 185
390. Derewenda ZS, Wei Y (1995) J. Am. Chem. Soc. 117: 2104
391. Nagao Y, Kume M, Wakabayashi RC, Nakamura T, Ochiai M (1989) Chem. Lett. 239
392. Kazlauskas RJ, Weissfloch ANE (1997) J. Mol. Catal. B 3: 65
393. Kato K, Gong Y, Tanaka S, Katayama M, Kimoto H (1999) Biotechnol. Lett. 21: 457
394. Mugford P, Wagner U, Jiang Y, Faber K, Kazlauskas RJ (2008) Angew. Chem. Int. Ed. 47: 8782
395. Effenberger F, Gutterer B, Ziegler T, Eckhart E, Aichholz R (1991) Liebigs Ann. Chem. 47
396. Brockerhoff H (1968) Biochim. Biophys. Acta 159: 296
397. Brockmann HL (1981) Methods Enzymol. 71: 619
398. Desnuelle P (1961) Adv. Enzymol. 23: 129
399. Servi S (1999) Top. Curr. Chem. 200: 127
400. D'Arrigo P, Servi S (1997) Trends Biotechnol. 15: 90
401. Hultin PG, Jones JB (1992) Tetrahedron Lett. 33: 1399
402. Wimmer Z (1992) Tetrahedron 48: 8431
403. Claßen A, Wershofen S, Yusufoglu A, Scharf HD (1987) Liebigs Ann. Chem. 629
404. Jones JB, Hinks RS (1987) Can. J. Chem. 65: 704
405. Kloosterman H, Mosmuller EWJ, Schoemaker HE, Meijer EM (1987) Tetrahedron Lett. 28: 2989
406. Shaw JF, Klibanov AM (1987) Biotechnol. Bioeng. 29: 648
407. Sweers HM, Wong CH (1986) J. Am. Chem. Soc. 108: 6421
408. Hennen WJ, Sweers HM, Wang YF, Wong CH (1988) J. Org. Chem. 53: 4939
409. Waldmann H, Sebastian D (1994) Chem. Rev. 94: 911
410. Ballesteros A, Bernabé M, Cruzado C, Martin-Lomas M, Otero C (1989) Tetrahedron 45: 7077

411. Guibé-Jampel E, Rousseau G, Salaun J (1987) J. Chem. Soc., Chem. Commun. 1080
412. Cohen SG, Milovanovic A (1968) J. Am. Chem. Soc. 90: 3495
413. Laumen K, Schneider M (1985) Tetrahedron Lett. 26: 2073
414. Hemmerle H, Gais HJ (1987) Tetrahedron Lett. 28: 3471
415. Banfi L, Guanti G (1993) Synthesis 1029
416. Guanti G, Banfi L, Narisano E (1990) Tetrahedron Asymmetry 1: 721
417. Guanti G, Narisano E, Podgorski T, Thea S, Williams A (1990) Tetrahedron 46: 7081
418. Patel RN, Robison RS, Szarka LJ (1990) Appl. Microbiol. Biotechnol. 34: 10
419. Gao Y, Hanson RM, Klunder JM, Ko SY, Masamune H, Sharpless KB (1987) J. Am. Chem. Soc. 109: 5765
420. Marples BA, Roger-Evans M (1989) Tetrahedron Lett. 30: 261
421. Ladner WE, Whitesides GM (1984) J. Am. Chem. Soc. 106: 7250
422. Ramos-Tombo GM, Schär HP, Fernandez i Busquets X, Ghisalba O (1986) Tetrahedron Lett. 27: 5707
423. Bornemann S, Crout DHG, Dalton H, Hutchinson DW (1992) Biocatalysis 5: 297
424. Palomo JM, Segura RL, Mateo C, Terreni M, Guisan JM, Fernandez-Lafuente R (2005) Tetrahedron Asymmetry 16: 869
425. Cotterill IC, Sutherland AG, Roberts SM, Grobbauer R, Spreitz J, Faber K (1991) J. Chem. Soc., Perkin Trans. 1, 1365
426. Cotterill IC, Dorman G, Faber K, Jaouhari R, Roberts SM, Scheinmann F, Spreitz J, Sutherland AG, Winders JA, Wakefield BJ (1990) J. Chem. Soc., Chem. Commun. 1661
427. De Jersey J, Zerner B (1969) Biochemistry 8: 1967
428. Crich JZ, Brieva R, Marquart P, Gu RL, Flemming S, Sih CJ (1993) J. Org. Chem. 58: 3252
429. Cotterill IC, Finch H, Reynolds DP, Roberts SM, Rzepa HS, Short KM, Slawin AMZ, Wallis CJ, Williams DJ (1988) J. Chem. Soc., Chem. Commun. 470
430. Naemura K, Matsumura T, Komatsu M, Hirose Y, Chikamatsu H (1988) J. Chem. Soc., Chem. Commun. 239
431. Pearson AJ, Lai YS (1988) J. Chem. Soc., Chem. Commun. 442
432. Pearson AJ, Lai YS, Lu W, Pinkerton AA (1989) J. Org. Chem. 54: 3882
433. Gautier A, Vial C, Morel C, Lander M, Näf F (1987) Helv. Chim. Acta 70: 2039
434. Pawlak JL, Berchtold GA (1987) J. Org. Chem. 52: 1765
435. Sugai T, Kakeya H, Ohta H (1990) J. Org. Chem. 55: 4643
436. Kitazume T, Sato T, Kobayashi T, Lin JT (1986) J. Org. Chem. 51: 1003
437. Abramowicz DA, Keese CR (1989) Biotechnol. Bioeng. 33: 149
438. Hoshino O, Itoh K, Umezawa B, Akita H, Oishi T (1988) Tetrahedron Lett. 29: 567
439. Sugai T, Kakeya H, Ohta H, Morooka M, Ohba S (1989) Tetrahedron 45: 6135
440. Pottie M, Van der Eycken J, Vandevalle M, Dewanckele JM, Röper H (1989) Tetrahedron Lett. 30: 5319
441. Dumortier L, Van der Eycken J, Vandewalle M (1989) Tetrahedron Lett. 30: 3201
442. Klempier N, Geymeyer P, Stadler P, Faber K, Griengl H (1990) Tetrahedron Asymmetry 1: 111
443. Yamaguchi Y, Komatsu O, Moroe T (1976) J. Agric. Chem. Soc. Jpn. 50: 619
444. Cygler M, Grochulski P, Kazlauskas RJ, Schrag JD, Bouthillier F, Rubin B, Serreqi AN, Gupta AK (1994) J. Am. Chem. Soc. 116: 3180
445. Hönig H, Seufer-Wasserthal P (1990) Synthesis 1137
446. Pai YC, Fang JM, Wu SH (1991) J. Org. Chem. 59: 6018
447. Saf R, Faber K, Penn G, Griengl H (1988) Tetrahedron 44: 389
448. Königsberger K, Faber K, Marschner C, Penn G, Baumgartner P, Griengl H (1989) Tetrahedron 45: 673
449. Hult K, Norin T (1993) Indian J. Chem. 32B: 123
450. Wu SH, Guo ZW, Sih CJ (1990) J. Am. Chem. Soc. 112: 1990
451. Allenmark S, Ohlsson A (1992) Biocatalysis 6: 211

452. Uppenberg J, Patkar S, Bergfors T, Jones TA (1994) J. Mol. Biol. 235: 790
453. Patkar SA, Björkling F, Zyndel M, Schulein M, Svendsen A, Heldt-Hansen HP, Gormsen E (1993) Indian J. Chem. 32B: 76
454. Rogalska E, Cudrey C, Ferrato F, Verger R (1993) Chirality 5: 24
455. Hoegh I, Patkar S, Halkier T, Hansen MT (1995) Can. J. Bot. 73 (Suppl 1): S869
456. Anderson EM, Larsson KM, Kirk O (1998) Biocatalysis Biotrans. 16: 181
457. Martinelle M, Hult K (1995) Biochim. Biophys. Acta 1251: 191
458. Uppenberg J, Öhrner N, Norin M, Hult K, Kleywegt GJ, Patkar S, Waagen V, Anthonsen T, Jones A (1995) Biochemistry 34: 16838
459. Holla EW, Rebenstock HP, Napierski B, Beck G (1996) Synthesis 823
460. Ohtani T, Nakatsukasa H, Kamezawa M, Tachibana H, Naoshima Y (1997) J. Mol. Catal. B 4: 53
461. Sanchez VM, Rebolledo F, Gotor V (1997) Tetrahedron Asymmetry 8: 37
462. Konegawa T, Ohtsuka Y, Ikeda H, Sugai T, Ohta H (1997) Synlett. 1297
463. Adam W, Diaz MT, Saha-Möller CR (1998) Tetrahedron Asymmetry 9: 791
464. Mulvihill MJ, Gage JL, Miller MJ (1998) J. Org. Chem. 63: 3357
465. Kingery-Wood J, Johnson JS (1996) Tetrahedron Lett. 37: 3975
466. Hansen TV, Waagen V, Partali V, Anthonsen HW, Anthonsen T (1995) Tetrahedron Asymmetry 6: 499
467. Waagen V, Hollingsaeter I, Partali V, Thorstad O, Anthonsen T (1993) Tetrahedron Asymmetry 4: 2265
468. Kato K, Katayama M, Fujii S, Kimoto H (1996) J. Ferment. Bioeng. 82: 355
469. Xie ZF (1991) Tetrahedron Asymmetry 2: 733
470. Xie ZF, Nakamura I, Suemune H, Sakai K (1988) J. Chem. Soc., Chem. Commun. 966
471. Xie ZF, Sakai K (1989) Chem. Pharm. Bull. 37: 1650
472. Seemayer R, Schneider MP (1990) J. Chem. Soc., Perkin Trans. 1, 2359
473. Laumen K, Schneider MP (1988) J. Chem. Soc., Chem. Commun. 598
474. Kalaritis P, Regenye RW, Partridge JJ, Coffen DL (1990) J. Org. Chem. 55: 812
475. Klempier N, Geymayer P, Stadler P, Faber K, Griengl H (1990) Tetrahedron Asymmetry 1: 111
476. Kloosterman M, Kierkels JGT, Guit RPM, Vleugels LFW, Gelade ETF, van den Tweel WJJ, Elferink VHM, Hulshof LA, Kamphuis J (1991) Lipases: biotransformations, active site models and kinetics. In: Alberghina L, Schmidt RD, Verger R (eds) Lipases: Structure, Mechanism and Genetic Engineering GBF Monographs, vol 16. Verlag Chemie, Weinheim, p 187
477. Lemke K, Lemke M, Theil F (1997) J. Org. Chem. 62: 6268
478. Johnson CR, Adams JP, Bis SJ, De Jong RL, Golebiowski A, Medich JR, Penning TD, Senanayake CH, Steensma DH, Van Zandt MC (1993) Indian J. Chem. 32B: 140
479. Schrag JD, Li Y, Cygler M, Lang D, Burgdorf T, Hecht HJ, Schmid R, Schomburg D, Rydel TJ, Oliver JD, Strickland LC, Dunaway CM, Larson SB, Day J, McPherson A (1997) Structure 5: 187
480. Kim KK, Song HK, Shin DH, Hwang KY, Suh SW (1996) Structure 5: 173
481. Hughes DL, Bergan JJ, Amato JS, Bhupathy M, Leazer JL, McNamara JM, Sidler DR, Reider PJ, Grabowski EJJ (1990) J. Org. Chem. 55: 6252
482. Mizuguchi E, Takemoto M, Achiwa K (1993) Tetrahedron Asymmetry 4: 1961
483. Ors M, Morcuende A, Jimenez-Vacas MI, Valverde S, Herradon B (1996) Synlett. 449
484. Soriente A, Laudisio G, Giorgano M, Sodano G (1995) Tetrahedron Asymmetry 6: 859
485. Burgess K, Henderson I (1989) Tetrahedron Lett. 30: 3633
486. Itoh T, Takagi Y, Nishiyama S (1991) J. Org. Chem. 56: 1521
487. Allen JV, Williams JMJ (1996) Tetrahedron Lett. 37: 1859
488. Kim M-J, Ahn Y, Park J (2005) Bull. Korean Chem. Soc. 26: 515
489. Hoyos P, Pace V, Alcantara A R (2012) Adv. Synth. Catal. 354: 2585
490. Sugiyama K, Oki Y, Kawanishi S, Kato K, Ikawa T, Egi M, Akai S (2016) Catal. Sci. Technol. 6: 5023
491. Verho O, Baeckvall J-E (2015) J. Am. Chem. Soc. 137: 3996

492. Scilimati A, Ngooi T K, Sih CJ (1998) Tetrahedron Lett. 29: 4927
493. Bänzinger M, Griffiths GJ, McGarrity JF (1993) Tetrahedron Asymmetry 4: 723
494. Liang S, Paquette L A (1990) Tetrahedron Asymmetry 1: 445
495. Waldinger C, Schneider M, Botta M, Corelli F, Summa V (1996) Tetrahedron Asymmetry 7: 1485
496. Patel RN, Banerjee A, Ko RY, Howell JM, Li WS, Comezoglu FT, Partyka RA, Szarka L (1994) Biotechnol. Appl. Biochem. 20: 23
497. Takano S, Yamane T, Takahashi M, Ogasawara K (1992) Tetrahedron Asymmetry 3: 837
498. Takano S, Yamane T, Takahashi M, Ogasawara K (1992) Synlett 410
499. Gaucher A, Ollivier J, Marguerite J, Paugam R, Salaun J (1994) Can. J. Chem. 72: 1312
500. Xie ZF, Suemune H, Sakai K (1993) Tetrahedron Asymmetry 4: 973
501. Huge Jensen B, Galluzzo DR, Jensen RG (1987) Lipids 22: 559
502. Chan C, Cox PB, Roberts SM (1988) J. Chem. Soc., Chem. Commun. 971
503. Estermann H, Prasad K, Shapiro MJ (1990) Tetrahedron Lett. 31: 445
504. Hilgenfeld R, Saenger W (1982) Top. Curr. Chem. 101: 1
505. Naemura K, Takahashi N, Chikamatsu H (1988) Chem. Lett. 1717
506. Tinapp P (1971) Chem. Ber. 104: 2266
507. Satoh T, Suzuki S, Suzuki Y, Miyaji Y, Imai Z (1969) Tetrahedron Lett. 10: 4555
508. Effenberger F (1994) Angew. Chem. Int. Ed. 33: 1555
509. Mitsuda S, Yamamoto H, Umemura T, Hirohara H, Nabeshima S (1990) Agric. Biol. Chem. 54: 2907
510. Effenberger F, Stelzer U (1993) Chem. Ber. 126: 779
511. Mitsuda S, Nabeshima S, Hirohara H (1989) Appl. Microbiol. Biotechnol. 31: 334
512. Matsumae H, Shibatani T (1994) J. Ferment. Bioeng. 77: 152.
513. Shibatani T, Omori K, Akatsuka H, Kawai E, Matsumae H (2000) J. Mol. Catal. B: Enzym. 10: 141
514. Martinez C A, Hu S, Dumond Y, Tao J, Kelleher P, Tully L (2008) Org. Proc. Res. Dev. 12: 392.
515. Watts S (1989) New Scientist July 1, p. 45.; https://www.novonordisk.com/content/dam/ Denmark/HQ/aboutus/documents/HistoryBook_UK.pdf
516. Sheldon RA (2007) Green Chem. 9: 1273
517. Okumura S, Iwai M, Tsujisaka Y (1979) Biochim. Biophys. Acta 575: 156
518. Jensen RG (1974) Lipids 9: 149
519. Chen S, Su L, Chen J, Wu J (2013) Biotechnol. Adv. 31: 1754
520. Martinez C, Abergel C, Cambillau C, de Geus P, Lauwereys M (1991) Crystallographic study of a recombinant cutinase from *Fusarium solani pisi*. In: Alberghina L, Schmidt RD, Verger R (eds) Lipases: Structure, Mechanism and Genetic Engineering. GBF Monographs, vol 16. Verlag Chemie, Weinheim, p 67
521. Martinez C, Nicholas A, van Tilbeurgh H, Egloff MP, Cudrey C, Verger R, Cambillau C (1994) Biochemistry 33: 83
522. Martinez C, De Geus P, Lauwereys M, Matthysens G, Cambillau C (1992) Nature 356: 615
523. Dumortier L, Liu P, Dobbelaere S, Van der Eycken J, Vandewalle M (1992) Synlett. 243
524. Dimarogona M, Nikolaivits E, Kanelli M, Christakopoulos P, Sandgren M, Topakas E (2015) Biochim. Biophys. Acta 1850: 2308
525. Perz V, Zumstein M T, Sander M, Zitzenbacher S, Ribitsch D, Guebitz G M (2015) Biomacromolecules 16: 3889
526. Holmberg E, Hult K (1991) Biotechnol. Lett. 323
527. Holmberg E, Szmulik P, Norin T, Hult K (1989) Biocatalysis 2: 217
528. Itoh T, Ohira E, Takagi Y, Nishiyama S, Nakamura K (1991) Bull. Chem. Soc. Jpn. 64: 624
529. Bamann E, Laeverenz P (1930) Ber. dtsch. chem. Ges. 63: 394
530. Shimizu S, Kataoka M (1996) Chimia 50: 409
531. Novick NJ, Tyler ME (1982) J. Bacteriol. 149: 364

532. Dong YH, Wang LH, Xu JL, Zhang HB, Zhang XF, Zhang LH (2001) Nature 411: 813
533. Gutman AL, Zuobi K, Guibe-Jampel E (1990) Tetrahedron Lett. 31: 2037
534. Kataoka M, Honda K, Shimizu S (2000) Eur. J. Biochem. 267: 3
535. Honda K, Kataoka K, Shimizu S (2002) Appl. Microbiol. Biotechnol. 60: 288
536. Onakunle OA, Knowles CJ, Bunch AW (1997) Enzyme Microb. Technol. 21: 245
537. Honda K, Tsubioi H, Minetoki T, Nose H, Sakamoto K, Kataoka M, Shimizu S (2005) Appl. Microbiol. Biotechnol. 66: 520
538. Kesseler M, Friedrich T, Höffken H W, Hasuer B (2002) Adv. Synth. Catal. 344: 1103
539. Vekiru E, Fruhauf S, Hametner C, Schatzmayr G, Krska R, Moll WD, Schuhmacher R (2016) World Mycotox. J. 9: 353.
540. Zhang Y, An J, Yang G-Y, Bai A, Zheng B, Lou Z, Wu G, Ye W, Chen H-F, Feng Y, Manco G (2015) PLoS One 10: e0115130/1-e0115130/16.
541. Remy B, Plener L, Poirier L, Elias M, Daude D, Chabriere E (2016) Sci. Rep. 6: 37780.
542. Zawilska JB, Wojcieszak J, Olejniczak AB (2013) Pharmacol. Rep. 65: 1.
543. Hecker SJ, Erion MD (2008) J. Med. Chem. 51: 2328.
544. Gauss D, Schönenberger B, Molla GS, Kinfu BM, Chow J, Liese A, Streit WR, Wohlgemuth R (2016) Biocatalytic phosphorylation of metabolites, in: Applied Biocatalysis: From Fundamental Science to Industrial Applications, Liese A, Hilterhaus L, Kettling U, Antranikian G (eds.) Wiley-VCH, pp. 147.
545. Fujii H, Koyama T, Ogura K (1982) Biochim. Biophys. Acta 712: 716
546. Durrwachter JR, Wong CH (1988) J. Org. Chem. 53: 4175
547. Straub A, Effenberger F, Fischer P (1990) J. Org. Chem. 55: 3926
548. Bednarski MD, Simon ES, Bischofberger N, Fessner WD, Kim MJ, Lees W, Saito T, Waldmann H, Whitesides GM (1989) J. Am. Chem. Soc. 111: 627
549. Schultz M, Waldmann H, Kunz H, Vogt W (1990) Liebigs Ann. Chem. 1019
550. Scollar MP, Sigal G, Klibanov AM (1985) Biotechnol. Bioeng. 27: 247
551. van Henk T, Hartog AF, Ruijssenaars HJ, Kerkman R, Schoemaker HE, Wever R (2007) Adv. Synth. Catal. 349: 1349
552. Herdewijn P, Balzarini J, De Clerq E, Vanderhaege H (1985) J. Med. Chem. 28: 1385
553. Borthwick AD, Butt S, Biggadike K, Exall AM, Roberts SM, Youds PM, Kirk BE, Booth BR, Cameron JM, Cox SW, Marr CLP, Shill MD (1988) J. Chem. Soc., Chem. Commun. 656
554. Langer RS, Hamilton BC, Gardner CR, Archer MD, Colton CC (1976) AIChE J. 22: 1079
555. Chenault HK, Simon ES, Whitesides GM (1988) Biotechnol. Gen. Eng. Rev. 6: 221
556. Andexer JN, Richter M (2015) ChemBioChem 16: 380.
557. Alissandratos A, Caron K, Loan TD, Hennessy JE, Easton CJ (2016) ACS Chem. Biol. 11: 3289.
558. Ayuso-Fernandez I, Galmes MA, Bastida A, Garcia-Junceda E (2014) ChemCatChem 6: 1059.
559. Shih Y-S, Whitesides GM (1977) J. Org. Chem. 42: 4165.
560. Berke W, Schüz H-J, Wandrey C, Morr M, Denda G, Kula M-R (1988) Biotechnol. Bioeng. 32: 130.
561. Wong CH, Haynie SL, Whitesides GM (1983) J. Am. Chem. Soc. 105: 115
562. Hirschbein BL, Mazenod FP, Whitesides GM (1982) J. Org. Chem. 47: 3765
563. Simon ES, Grabowski S, Whitesides GM (1989) J. Am. Chem. Soc. 111: 8920
564. Crans DC, Whitesides GM (1983) J. Org. Chem. 48: 3130
565. Bolte J, Whitesides GM (1984) Bioorg. Chem. 12: 170
566. Augé C, Mathieu C, Mérienne C (1986) Carbohydr. Res. 151: 147
567. Wong CH, Haynie SL, Whitesides GM (1982) J. Org. Chem. 5416
568. Zhang J, Wu B, Zhang Y, Kowal P, Wang PG (2003) Org. Lett. 5: 2583.
569. Shiba T, Tsutsumi K, Ishige K, Noguchi T (2000) Biochemistry (Moscow) 65: 315

570. Brown MRW, Kornberg A (2008) Trends Biochem. Sci. 33: 284.
571. Rao NN, Gomez-Garcia MR, Kornberg A (2009) Ann. Rev. Biochem. 78: 605.
572. Kameda A, Shiba T, Kawazoe Y, Satoh Y, Ihara Y, Munekata M, Ishige K, Noguchi T (2001) J. Biosci. Bioeng. 91: 557
573. Haeusler PA, Dieter L, Rittle KJ, Shepler LS, Paszkowski AL, Moe OA (1992) Biotechnol. Appl. Biochem. 15: 125.
574. Kazlauskas RJ, Whitesides GM (1985) J. Org. Chem. 50: 1069
575. Marshall DL (1973) Biotechnol. Bioeng. 15: 447
576. Cantoni GL (1975) Ann. Rev. Biochem. 44: 435
577. Baughn RL, Adalsteinsson O, Whitesides GM (1978) J. Am. Chem. Soc. 100: 304
578. Drueckhammer DG, Wong CH (1985) J. Org. Chem. 50: 5912
579. Chenault HK, Mandes RF, Hornberger KR (1997) J. Org. Chem. 62: 331
580. Chenault HK, Mandes RF, Hornberger KR (1997) J. Org. Chem. 62: 331.
581. Pollak A, Baughn RL, Whitesides GM (1977) J. Am. Chem. Soc. 77: 2366
582. Wong CH, Whitesides GM (1981) J. Am. Chem. Soc. 103: 4890
583. Wong CH, Whitesides GM (1983) J. Org. Chem. 48: 3199
584. van Herk T, Hartog AF, Schoemaker H E, Wever R (2006) J. Org. Chem. 71: 6244
585. Gross A, Abril O, Lewis JM, Geresh S, Whitesides GM (1983) J. Am. Chem. Soc. 105: 7428
586. Thorner JW, Paulus H (1973) In: The Enzymes; Boyer PD (ed) Academic Press, New York, vol 8, p 487
587. Rios-Mercadillo VM, Whitesides GM (1979) J. Am. Chem. Soc. 101: 5829
588. Crans DC, Whitesides GM (1985) J. Am. Chem. Soc. 107: 7019
589. Crans DC, Whitesides GM (1985) J. Am. Chem. Soc. 107: 7008
590. Chenault HK, Chafin LF, Liehr S (1998) J. Org. Chem. 63: 4039
591. Eibl H (1980) Chem. Phys. Lipids 26: 405
592. Vasilenko I, Dekruijff B, Verkleij A (1982) Biochim. Biophys. Acta 685: 144
593. Axelrod B (1948) J. Biol. Chem. 172: 1.
594. Appleyard J (1948) Biochem. J. 42: 596.
595. Wever R, Babich L, Hartog AF (2015) Transphosphorylation, in: Science of Synthesis, Biocatalysis in Organic Synthesis, Faber K, Fessner W-D, Turner NJ (eds.), vol. 1, pp. 223.
596. Suzuki E, Ishikawa K, Mihara Y, Shimba N, Asano Y (2007) Bull. Chem. Soc. Jpn. 80: 276.
597. Hemrika W, Renirie R, Dekker HL, Barnett P, Wever R (1997) Proc. Natl. Acad. Sci. USA 94: 2145.
598. van Herk T, Hartog AF, van der Burg AM, Wever R (2005) Adv. Synth. Catal. 347: 1155
599. Tanaka N, Hasan Z, Hartog AF, van Herk T, Wever R (2003) Org. Biomol. Chem. 1: 2833.
600. Mihara Y, Ishikawa K, Suzuki E, Asano Y (2004) Biosci. Biotechnol. Biochem. 68: 1046.
601. Babich L, Hartog AF, van der Horst MA, Wever R (2012) Chem. Eur. J. 18: 6604.
602. Ishikawa K, Mihara Y, Shimba N, Ohtsu N, Kawasaki H, Suzuki E, Asano Y (2002) Prot. Eng. 15: 539.
603. Pradines A, Klaébé A, Périé J, Paul F, Monsan P (1991) Enzyme Microb. Technol. 13: 19
604. Schoevaart R, van Rantwijk F, Sheldon RA (2000) J. Org. Chem. 65: 6940.
605. van Herk T, Hartog AF, van der Burg AM, Wever R (2005) Adv. Synth. Catal. 347: 1155.
606. Dissing K, Uerkvitz W (2006) Enzyme Microb. Technol. 38: 683.
607. Hartog AF, van Herk T, Wever R (2011) Adv. Synth. Catal. 353: 2339.
608. Tasnadi G, Lukesch M, Zechner M, Jud W, Hall M, Ditrich K, Baldenius K, Hartog AF, Wever R, Faber K (2016) Eur. J. Org. Chem. 45.
609. Asano Y, Mihara Y, Yamada H (1999) J. Mol. Catal. B: Enzym. 6: 271.
610. Mihara Y, Ishikawa K, Suzuki E, Asano Y (2004) Biosci. Biotechnol. Biochem. 68: 1046
611. Schurig V, Betschinger F (1992) Chem. Rev. 92: 873
612. Kolb HC, Van Nieuenhze MS, Sharpless KB (1994) Chem. Rev. 94: 2483
613. Scott JW (1984) Chiral carbon fragments and their use in synthesis. In: Scott JW, Morrison JD (eds) Asymmetric Synthesis, vol 4. Academic Press, Orlando, p 5

614. Finn MG, Sharpless KB (1985) On the mechanism of asymmetric epoxidation with titanium-tartrate catalysts. In: Scott JW, Morrison JD (eds) Asymmetric Synthesis, vol 5. Academic Press, Orlando, p 247
615. Jacobsen EN, Zhang W, Muci AR, Ecker JR, Deng L (1991) J. Am. Chem. Soc. 113: 7063
616. Pedragosa-Moreau S, Archelas A, Furstoss R (1995) Bull. Chim. Soc. Fr. 132: 769
617. de Bont JAM (1993) Tetrahedron Asymmetry 4: 1331
618. Onumonu AN, Colocoussi N, Matthews C, Woodland MP, Leak DJ (1994) Biocatalysis 10: 211
619. Besse P, Veschambre H (1994) Tetrahedron 50: 8885
620. Hartmans S (1989) FEMS Microbiol. Rev. 63: 235
621. Tokunaga M, Larrow JF, Kakiuchi F, Jacobsen EN (1997) Science 277: 936
622. Schaus SE, Brandes BD, Larrow JF, Tokunaga M, Hansen KB, Gould AE, Furrow ME, Jacobsen EN (2002) J. Am. Chem. Soc. 124: 1307.
623. Lu AYH, Miwa GT (1980) Ann. Rev. Pharmacol. Toxicol. 20: 513
624. Seidegard J, De Pierre JW (1983) Biochim. Biophys. Acta 695: 251
625. Armstrong RN (1987) Crit. Rev. Biochem. 22: 39
626. Hanzlik RP, Heidemann S, Smith D (1978) Biochem. Biophys. Res. Commun. 82: 310
627. Arand M, Wagner H, Oesch F (1996) J. Biol. Chem. 271: 4223
628. Nardini M, Ridder IS, Rozeboom HJ, Kalk KH, Rink R, Janssen DB, Dijkstra BW (1999) J. Biol. Chem. 274: 14579
629. Tzeng HF, Laughlin LT, Lin S, Armstrong RN (1996) J. Am. Chem. Soc. 118: 9436
630. DuBois GC, Appella E, Levin W, Lu AYH, Jerina DM (1978) J. Biol. Chem. 253: 2932
631. Janssen DB, Pries F, van der Ploeg JR (1994) Annu. Rev. Microbiol. 48: 163
632. Verschueren KHG, Seljee F, Rozeboom HJ, Kalk KH, Dijkstra BW (1993) Nature 363: 693
633. Withers SG, Warren RAJ, Street IP, Rupitz K, Kempton JB, Aebersold R (1990) J. Am. Chem. Soc. 112: 5887
634. Arand M, Hallberg BM, Zou J, Bergfors T, Oesch F, van der Werf M, de Bont JAM, Jones TA, Mowbray SL (2003) EMBO J. 22: 2583
635. Bellucci G, Chiappe C, Cordoni A, Marioni F (1994) Tetrahedron Lett. 35: 4219
636. Escoffier B, Prome JC (1989) Bioorg. Chem. 17: 53
637. Mischitz M, Mirtl C, Saf R, Faber K (1996) Tetrahedron Asymmetry 7: 2041
638. Moussou P, Archelas A, Baratti J, Furstoss R (1998) Tetrahedron Asymmetry 9: 1539
639. Daiboun T, Elalaoui MA, Thaler-Dao H, Chavis C, Maury G (1993) Biocatalysis 7: 227
640. Lu AYH, Levin W (1978) Methods Enzymol. 52: 193
641. Berti G (1986) Enantio- and diastereoselectivity of microsomal epoxide hydrolase: potential applications to the preparation of non-racemic epoxides and diols. In: Schneider MP (ed) Enzymes as Catalysts in Organic Synthesis, NATO ASI Series, vol 178. Reidel, Dordrecht, p 349
642. Bellucci G, Chiappe C, Marioni F (1989) J. Chem. Soc., Perkin Trans. 1, 2369
643. Bellucci G, Capitani I, Chiappe C, Marioni F (1989) J. Chem. Soc., Chem. Commun. 1170
644. Wu S, Li A, Chin YS, Li Z (2013) ACS Catal. 3: 752.
645. Weijers CAGM (1997) Tetrahedron: Asymm. 8: 639.
646. Zheng H, Reetz MT (2010) J. Am. Chem. Soc. 132: 15744.
647. Watabe T, Suzuki S (1972) Biochem. Biophys. Res. Commun. 46: 1120
648. El-Sherbeni AA, El-Kadi AOS (2014) Arch. Toxicol. 88: 2013.
649. Weijers CAGM, De Haan A, De Bont JAM (1988) Appl. Microbiol. Biotechnol. 27: 337
650. Yamada Y, Kikuzaki H, Nakatani N (1992) Biosci. Biotechnol. Biochem. 56: 153
651. Kolattokudy PE, Brown L (1975) Arch. Biochem. Biophys. 166: 599
652. Pedragosa-Moreau S, Archelas A, Furstoss R (1996) Tetrahedron Lett. 37: 3319
653. Misawa E, Chan Kwo Chion CKC, Archer IV, Woodland MC, Zhou NY, Carter SF, Widdowson DA, Leak DJ (1998) Eur. J. Biochem. 253: 173
654. Morisseau C, Archelas A, Guitton C, Faucher D, Furstoss R (1999) Eur. J. Biochem. 263: 386
655. van der Werf MJ, Overkamp KM, de Bont JAM (1998) J. Bacteriol. 180: 5052
656. Mischitz M, Faber K, Willetts A (1995) Biotechnol. Lett. 17: 893

657. Kroutil W, Genzel Y, Pietzsch M, Syldatk C, Faber K (1998) J. Biotechnol. 61: 143
658. Svaving J, de Bont JAM (1998) Enzyme Microb. Technol. 22: 19
659. Archer IVJ (1997) Tetrahedron 53: 15617
660. Weijers CAGM, de Bont JAM (1999) J. Mol. Catal. B 6: 199
661. Orru RVA, Archelas A, Furstoss R, Faber K (1999) Adv. Biochem. Eng. Biotechnol. 63: 145
662. Faber K, Mischitz M, Kroutil W (1996) Acta Chem. Scand. 50: 249
663. de Vries EJ, Janssen DB (2003) Curr. Opin. Chem. Biol. 14: 414.
664. Bala N, Chimni SS (2010) Tetrahedron: Asymm. 21: 2879.
665. Reetz MT, Bocola M, Wang L-W, Sanchis J, Cronin A, Arand M, Zhou J, Archelas A, Bottalla A-L, Naworyta A, Mowbray SL (2009) J. Am. Chem. Soc. 131: 7334.
666. Kotik M, Archelas A, Wohlgemuth R (2012) Curr. Org. Chem. 16: 451.
667. Orru RVA, Faber K (1999) Curr. Opinion Chem. Biol. 3: 16
668. Botes AL, Steenkamp JA, Letloenyane MZ, van Dyk MS (1998) Biotechnol. Lett. 20: 427; the microorganism used in this study (*Chryseomonas luteola*) causes inner ear infections in infants and belongs to safety Class II.
669. Weijers CAGM (1997) Tetrahedron Asymmetry 8: 639
670. Botes AL, Weijers CAGM, van Dyk MS (1998) Biotechnol. Lett. 20: 421
671. Weijers CAGM, Botes AL, van Dyk MS, de Bont JAM (1998) Tetrahedron Asymmetry 9: 467
672. Morisseau C, Nellaiah H, Archelas A, Furstoss R, Baratti JC (1997) Enzyme Microb. Technol. 20: 446
673. Spelberg JHL, Rink R, Kellogg RM, Janssen DB (1998) Tetrahedron Asymmetry 9: 459
674. Grogan G, Rippe C, Willetts A (1997) J. Mol. Catal. B 3: 253
675. Pedragosa-Moreau S, Archelas A, Furstoss R (1996) Tetrahedron 52: 4593
676. Mischitz M, Kroutil W, Wandel U, Faber K (1995) Tetrahedron Asymmetry 6: 1261
677. Moussou P, Archelas A, Furstoss R (1998) Tetrahedron 54: 1563
678. Kroutil W, Mischitz M, Plachota P, Faber K (1996) Tetrahedron Lett. 37: 8379
679. Mischitz M, Faber K (1996) Synlett. 978
680. Archer IVJ, Leak DJ, Widdowson DA (1996) Tetrahedron Lett. 37: 8819
681. Steinreiber A, Mayer SF, Saf R, Faber K (2001) Tetrahedron Asymmetry 12: 1519
682. Steinreiber A, Mayer SF, Faber K (2001) Synthesis 2035.
683. Edegger K, Mayer SF, Steinreiber A, Faber K (2004)Tetrahedron 60: 583.
684. Archer IVJ, Leak DJ, Widdowson DA (1996) Tetrahedron Lett. 37: 8819.
685. Imai K, Marumo S, Mori K (1974) J. Am. Chem. Soc. 96: 5925
686. Pedragosa-Moreau S, Archelas A, Furstoss R (1993) J. Org. Chem. 58: 5533
687. Orru RVA, Mayer SF, Kroutil W, Faber K (1998) Tetrahedron 54: 859
688. Kroutil W, Mischitz M, Faber K (1997) J. Chem. Soc., Perkin Trans. 1, 3629
689. Goswami A, Totleben MJ, Singh AK, Patel RN (1999) Tetrahedron: Asymm. 10: 3167.
690. Kong X-D, Ma Q, Zhou J, Zeng B-B, Xu J-H (2014) Angew. Chem. Int. Ed. 53: 6641.
691. Legras JL, Chuzel G, Arnaud A, Galzy P (1990) World Microbiol. Biotechnol. 6: 83
692. Solomonson LP (1981) Cyanide as a metabolic inhibitor. In: Vennesland B et al. (eds) Cyanide in Biology. Academic Press, London, p 11
693. Legras JL, Jory M, Arnaud A, Galzy P (1990) Appl. Microbiol. Biotechnol. 33: 529
694. Jallageas JC (1980) Adv. Biochem. Eng. 14: 1
695. Meth-Cohn O, Wang MX (1995) Tetrahedron Lett. 36: 9561
696. Hjort CM, Godtfredsen SE, Emborg C (1990) J. Chem. Technol. Biotechnol. 48: 217
697. Thompson LA, Knowles CJ, Linton EA, Wyatt JM (1988) Chem. Brit. 900
698. Nagasawa T, Yamada H (1989) Trends Biotechnol. 7: 153
699. Asano Y (2002) J. Biotechnol. 94: 65.
700. Wang M-X (2011) Topics Organomet. Chem. 36: 105.
701. Prasad S, Bhalla TC (2010) Biotechnol. Adv. 28: 725.
702. Arnaud A, Galzy P, Jallageas JC (1976) Folia Microbiol. 21: 178
703. Nagasawa T, Takeuchi K, Yamada H (1988) Biochem. Biophys. Res. Commun. 155: 1008

704. Nagasawa T, Nanba H, Ryuno K, Takeuchi K, Yamada H (1987) Eur. J. Biochem. 162: 691
705. Brennan BA, Alms G, Nelson MJ, Durney LT, Scarrow RC (1996) J. Am. Chem. Soc. 118: 9194
706. Sugiura Y, Kuwahara J, Nagasawa T, Yamada H (1987) J. Am. Chem. Soc. 109: 5848
707. Nagume T (1991) J. Mol. Biol. 220: 221
708. Kuhn ML, Martinez S, Gumataotao N, Bornscheuer U, Liu D, Holz RC (2012) Biochem. Biophys. Res. Commun. 424: 365.
709. Yamanaka Y, Hashimoto K, Ohtaki A, Noguchi K, Yohda M, Odaka M (2010) J. Biol. Inorg. Chem. 15: 655.
710. Song L, Wang M, Shi J, Xue Z, Wang MX, Qian S (2007) Biochem. Biophys. Res. Commun. 362: 319
711. Nagashima S, Nakasako M, Dohmae N, Tsujimura M, Takio K, Odaka M, Yohda M, Kamiya N, Endo I (1998) Nature Struct. Biol. 5: 347
712. Huang W, Jia J, Cummings J, Nelson M, Schneider G, Lindqvist Y (1997) Structure 5: 691
713. Miyanaga A, Fushinobu S, Ito K, Wakagi T (2001) Biochem. Biophys. Res. Commun. 288: 1169
714. Endo I, Odaka M, Yohda M (1999) Trends Biotechnol. 17: 244
715. Odaka M, Fujii K, Hoshino M, Noguchi T, Tsujimura M, Nagashima S, Yohda M, Nagamune T, Inoue Y, Endo I (1997) J. Am. Chem. Soc. 119: 3785
716. Desai LV, Zimmer M (2004) J. Chem. Soc. Dalton Trans. 872
717. Kobayashi M, Shimizu S (1998) Nature Biotechnol. 16: 733
718. Kobayashi M, Shimizu S (2000) Curr. Opinion Chem. Biol. 4: 95
719. Martinkova L, Uhnakova B, Patek M, Nesvera J, Kren V (2009) Environ. Int. 35: 162.
720. Asano Y, Fujishiro K, Tani Y, Yamada H (1982) Agric. Biol. Chem. 46: 1165
721. Layh N, Parratt J, Willetts A (1998) J. Mol. Catal. B 5: 476
722. Brenner C (2002) Curr. Opinion Struct. Biol. 12: 775
723. Kobayashi M, Goda M, Shimizu S (1998) Biochem. Biophys. Res. Commun. 253: 662
724. Kobayashi M, Shimizu S (1994) FEMS Microbiol. Lett. 120: 217.
725. Ingvorsen K, Yde B, Godtfredsen SE, Tsuchiya RT (1988) Microbial hydrolysis of organic nitriles and amides. In: Cyanide Compounds in Biology. Ciba Foundation Symp. 140. Wiley, Chichester, p 16
726. Ohta H (1996) Chimia 50: 434
727. Wyatt JM, Linton EA (1988) The industrial potential of microbial nitrile biochemistry. In: Cyanide Compounds in Biology. Ciba Foundation Symp. 140, Wiley, Chichester, p 32
728. Nagasawa T, Yamada H (1990) Pure Appl. Chem. 62: 1441
729. De Raadt A, Klempier N, Faber K, Griengl H (1992) Microbial and enzymatic transformation of nitriles. In: Servi S (ed) Microbial Reagents in Organic Synthesis., NATO ASI Series C, vol 381. Kluwer, Dordrecht, p 209
730. Nagasawa T, Yamada H (1990) Large-scale bioconversion of nitriles into useful amides and acids. In: Abramowicz DA (ed) Biocatalysis. Van Nostrand Reinhold, New York, p 277
731. Sugai T, Yamazaki T, Yokohama M, Ohta H (1997) Biosci. Biotechnol. Biochem. 61: 1419
732. Crosby J, Moilliet J, Parratt JS, Turner NJ (1994) J. Chem. Soc., Perkin Trans. 1, 1679
733. Nazly N, Knowles CJ, Beardsmore AJ, Naylor WT, Corcoran EG (1983) J. Chem. Technol. Biotechnol. 33: 119
734. Knowles CJ, Wyatt JM (1988) World Biotechnol. Rep. 1: 60
735. Wyatt JM (1988) Microbiol. Sci. 5: 186
736. Ingvorsen K, Hojer-Pedersen B, Godtfredsen SE (1991) Appl. Environ. Microbiol. 57: 1783
737. Maestracci M, Thiéry A, Arnaud A, Galzy P (1988) Indian J. Microbiol. 28: 34
738. Bui K, Arnaud A, Galzy P (1982) Enzyme Microb. Technol. 4: 195
739. Lee CY, Chang HN (1990) Biotechnol. Lett. 12: 23
740. Asano Y, Yasuda T, Tani Y, Yamada H (1982) Agric. Biol. Chem. 46: 1183
741. Ryuno K, Nagasawa T, Yamada H (1988) Agric. Biol. Chem. 52: 1813
742. Nagasawa T, Yamada H (1988) Pure Appl. Chem. 62: 1441
743. Asano Y (2015) Hydrolysis of Nitriles to Amides, in: Science of Synthesis, Biocatalysis in Organic Synthesis, 1: 255.

744. Mauger J, Nagasawa T, Yamada H (1989) Tetrahedron 45: 1347
745. Mauger J, Nagasawa T, Yamada H (1988) J. Biotechnol. 8: 87
746. Nagasawa T, Mathew CD, Mauger J, Yamada H (1988) Appl. Environ. Microbiol. 54: 1766
747. Petersen M, Kiener A (1999) Green Chem. 4: 99
748. Kobayashi M, Nagasawa T, Yanaka N, Yamada H (1989) Biotechnol. Lett. 11: 27
749. Kobayashi M, Yanaka N, Nagasawa T, Yamada H (1990) J. Antibiot. 43: 1316
750. Mathew CD, Nagasawa T, Kobayashi M, Yamada H (1988) Appl. Environ. Microbiol. 54: 1030
751. Vaughan PA, Cheetham PSJ, Knowles CJ (1988) J. Gen. Microbiol. 134: 1099
752. Nagasawa T, Yamada H, Kobayashi M (1988) Appl. Microbiol. Biotechnol. 29: 231
753. Bengis-Garber C, Gutman AL (1988) Tetrahedron Lett. 29: 2589
754. Kobayashi M, Nagasawa T, Yamada H (1988) Appl. Microbiol. Biotechnol. 29: 231
755. Nishise H, Kurihara M, Tani Y (1987) Agric. Biol. Chem. 51: 2613
756. Meth-Cohn O, Wang MX (1997) J. Chem. Soc., Perkin Trans. 1, 3197
757. Taylor SK, Chmiel NH, Simons LJ, Vyvyan JR (1996) J. Org. Chem. 61: 9084
758. Shen Y, Wang M, Li X, Zhang J, Sun H, Luo J (2012) J. Chem. Technol. Biotechnol. 87: 1396.
759. Hann EC, Eisenberg A, Fager SK, Perkins NE, Gallagher FG, Cooper S, Gavagan JE, Stieglitz B, Hennessy SM, DiCosimo R (1999) Bioorg. Med. Chem. 7: 2239.
760. Kieny-L'Homme MP, Arnaud A, Galzy P (1981) J. Gen. Appl. Microbiol. 27: 307
761. Yokoyama M, Sugai T, Ohta H (1993) Tetrahedron Asymmetry 6: 1081
762. Stolz A, Trott S, Binder M, Bauer R, Hirrlinger B, Layh N, Knackmuss HJ (1998) J. Mol. Catal. B 5: 137
763. Maddrell SJ, Turner NJ, Kerridge A, Willetts AJ, Crosby J (1996) Tetrahedron Lett. 37: 6001
764. Robertson DE, Chaplin JA, DeSantis G, Podar M, Madden M, Chi E, Richardson T, Milan A, Miller M, Weiner DP, Wong K, McQuaid J, Farwell B, Preston L A, Tan X, Snead MA, Keller M, Mathur E, Kretz PL, Burk MJ, Short JM (2004) Appl. Environ. Microbiol. 70: 2429
765. Robertson DE, Steer BA (2004) Curr. Opinion Chem. Biol. 8: 141
766. DeSantis G, Zhu Z, Greenberg A, Wong K, Chaplin J, Hanson SR, Farwell B, Nicholson LW, Rand CL, Weiner DP, Robertson DE, Burk MJ (2002) J. Am. Chem. Soc. 124: 9024
767. Furuhashi K, et al. (1992) Appl. Microbiol. Biotechnol. 37: 184
768. Fukuda Y, Harada T, Izumi Y (1973) J. Ferment. Technol. 51: 393
769. Yamamoto K, Oishi K, Fujimatsu I, Komatsu I (1991) Appl. Environ. Microbiol. 57: 3028
770. Yamamoto K, Ueno Y, Otsubo K, Kawakami K, Komatsu K (1990) Appl. Environ. Microbiol. 56: 3125
771. Gröger H (2001) Adv. Synth. Catal. 343: 547
772. Macadam AM, Knowles CJ (1985) Biotechnol. Lett. 7: 865
773. Arnaud A, Galzy P, Jallageas JC (1980) Bull. Soc. Chim. Fr. II: 87
774. Bhalla TC, Miura A, Wakamoto A, Ohba Y, Furuhashi K (1992) Appl. Microbiol. Biotechnol. 37: 184
775. Choi SY, Goo YM (1986) Arch. Pharm. Res. 9: 45
776. Effenberger F, Böhme J (1994) Bioorg. Med. Chem. 2: 715
777. Martinkova L, Stolz A, Knackmuss HJ (1996) Biotechnol. Lett. 18: 1073
778. Yamamoto K, Komatsu KI (1991) Agric. Biol. Chem. 55: 1459
779. Fallon R D, Stieglitz B, Turner I (1997) Appl. Microbiol. Biotechnol. 47: 156
780. Kakeya H, Sakai N, Sugai T, Ohta H (1991) Tetrahedron Lett. 32: 1343
781. Bianchi D, Bosetti A, Cesti P, Franzosi G, Spezia S (1991) Biotechnol. Lett. 13: 241
782. Osprian I, Jarret C, Strauss U, Kroutil W, Orru RVA, Felfer U, Willetts AJ, Faber K (1999) J. Mol. Catal. B 6: 555
783. Layh N, Willetts A (1998) Biotechnol. Lett. 20: 329
784. May SW, Padgette SR (1983) Biotechnology 677

References to Sect. 2.2

785. Jones JB, Beck JF (1976) Asymmetric syntheses and resolutions using enzymes. In: Jones JB, Sih CJ, Perlman D (eds) Applications of Biochemical Systems in Organic Chemistry. Wiley, New York, p 236
786. Hummel W, Kula MR (1989) Eur. J. Biochem. 184: 1
787. Willner I, Mandler D (1989) Enzyme Microb. Technol. 11: 467
788. Chenault HK, Whitesides GM (1987) Appl. Biochem. Biotechnol. 14: 147
789. Wichmann R, Vasic-Racki D (2005) Adv. Biochem. Eng. Biotechnol. 92: 225
790. van der Donk WA, Zhao HM (2003) Curr. Opinion Biotechnol. 14: 421
791. Jones JB, Sneddon DW, Higgins W, Lewis AJ (1972) J. Chem. Soc., Chem. Commun. 856
792. Jensen MA, Elving PJ (1984) Biochim. Biophys. Acta 764: 310
793. Wienkamp R, Steckhan E (1982) Angew. Chem. Int. Ed. 21: 782
794. Simon H, Bader J, Günther H, Neumann S, Thanos J (1985) Angew. Chem. Int. Ed. 24: 539
795. Mandler D, Willner I (1986) J. Chem. Soc., Perkin Trans. 2, 805
796. Jones JB, Taylor KE (1976) Can. J. Chem. 54: 2069
797. Legoy MD, Laretta-Garde V, LeMoullec JM, Ergan F, Thomas D (1980) Biochimie 62: 341
798. Julliard M, Le Petit J, Ritz P (1986) Biotechnol. Bioeng. 28: 1774
799. van Eys J (1961) J. Biol. Chem. 236: 1531
800. Gupta NK, Robinson WG (1966) Biochim. Biophys. Acta 118: 431
801. Wang SS, King CK (1979) Adv. Biochem. Eng. Biotechnol. 12: 119
802. Karabatsos GL, Fleming JS, Hsi N, Abeles RH (1966) J. Am. Chem. Soc. 88: 849
803. Stampfer W, Kosjek B, Moitzi C, Kroutil W, Faber K (2002) Angew. Chem. Int. Ed. 41: 1014
804. Kara S, Spickermann D, Schrittwieser JH, Leggewie C, van Berkel WJH, Arends IWCE, Hollmann F (2013) Green Chem. 15: 330.
805. Zuhse R, Leggewie C, Hollmann F, Kara S (2015) Org. Proc. Res. Dev. 19: 369.
806. Levy HR, Loewus FA, Vennesland B (1957) J. Am. Chem. Soc. 79: 2949
807. Tischer W, Tiemeyer W, Simon H (1980) Biochimie 62: 331
808. Wichmann R, Wandrey C, Bückmann AF, Kula MR (1981) Biotechnol. Bioeng. 23: 2789
809. Slusarczyk H, Pohl M, Kula MR (1998) In: Ballesteros A, Plou FJ, Iborra JL, Halling P (eds) Stability and Stabilisation of Biocatalysts. Elsevier: Amsterdam, p 331
810. Shaked Z, Whitesides GM (1980) J. Am. Chem. Soc. 102: 7104
811. Hummel W, Schütte H, Schmidt E, Wandrey C, Kula MR (1987) Appl. Microbiol. Biotechnol. 26: 409
812. Weuster-Botz D, Paschold H, Striegel B, Gieren H, Kula M-R, Wandrey C (1994) Chem. Ing. Tech. 17: 131
813. Seelbach K, Riebel B, Hummel W, Kula MR, Tishkov VI, Egorov AM, Wandrey C, Kragl U (1996) Tetrahedron Lett. 37: 1377
814. Tishkov VI, Galkin AG, Fedorchuk VV, Savitsky PA, Rojkova AM, Gieren H, Kula MR (1999) Biotechnol. Bioeng. 64: 187
815. Tishkov VI, Galkin AG, Marchenko GN, Tsygankov YD, Egorov AM (1993) Biotechnol. Appl. Biochem. 18: 201
816. Vandecasteele J-P, Lemal J (1980) Bull. Soc. Chim. Fr. 101
817. Wong C-H, Drueckhammer DG, Sweers HM (1985) J. Am. Chem. Soc. 107: 4028
818. Pollak A, Blumenfeld H, Wax M, Baughn RL, Whitesides GM (1980) J. Am. Chem. Soc. 102: 6324
819. Hirschbein BL, Whitesides GM (1982) J. Am. Chem. Soc. 104: 4458
820. Wong CH, Gordon J, Cooney CL, Whitesides GM (1981) J. Org. Chem. 46: 4676
821. Guiseley KB, Ruoff PM (1961) J. Org. Chem. 26: 1248
822. Johannes TW, Woodyer RD, Zhao H (2007) Biotechnol. Bioeng. 96: 18
823. Vrtis JM, White AK, Metcalf WW, van der Donk WA (2002) Angew. Chem. Int. Ed. 41: 3257

824. Costas AM, White A K, Metcalf WW (2001) J. Biol. Chem. 276: 17429
825. Woodyer RD, van der Donk W A, Zhao HM (2003) Biochemistry 42: 11604
826. Woodyer R, D van der Donk W A, Zhao HM (2006) Comb. Chem. High Throughput Screen. 9: 237
827. Johannes TW, Woodyer RD, Zhao HM (2005) Appl. Environ. Microbiol. 71: 5728
828. Dodds DR, Jones JB (1982) J. Chem. Soc., Chem. Commun. 1080
829. Wang SS, King C-K (1979) Adv. Biochem. Eng. 12: 119
830. Mansson MO, Larsson PO, Mosbach K (1982) Methods Enzymol. 89: 457
831. Wong CH, Whitesides GM (1983) J. Am. Chem. Soc. 105: 5012
832. Utaka M, Yano T, Ema T, Sakai T (1996) Chem. Lett. 1079
833. Danielsson B, Winquist F, Malpote JY, Mosbach K (1982) Biotechnol. Lett. 4: 673
834. Payen B, Segui M, Monsan P, Schneider K, Friedrich CG, Schlegel HG (1983) Biotechnol. Lett. 5: 463
835. Holzer AK, Hiebler K, Mutti FG, Simon RC, Lauterbach L, Lenz O, Kroutil W (2015) Org. Lett. 17: 2431.
836. Schneider K, Schlegel HG (1976) Biochim. Biophys. Acta 452: 66
837. Lee LG, Whitesides GM (1986) J. Org. Chem. 51: 25
838. Carrea G, Bovara R, Longhi R, Riva S (1985) Enzyme Microb. Technol. 7: 597
839. Matos JR, Wong CH (1986) J. Org. Chem. 51: 2388
840. Bednarski MD, Chenault HK, Simon ES, Whitesides GM (1987) J. Am. Chem. Soc. 109: 1283
841. Morokutti A, Lyskowski A, Sollner S, Pointner E, Fitzpatrick TB, Kratky C, Gruber K, Macheroux P (2005) Biochemistry 44: 13724
842. Riebel BR, Gibbs PR, Wellborn WB, Bommarius AS (2003) Adv. Synth. Catal. 345: 707
843. Ross R P, Claiborne A (1992) J. Mol. Biol. 227: 658
844. Ward DE, Donelly CJ, Mullendore ME, van der Oost J, de Vos WM, Crane EJ (2001) Eur. J. Biochem. 268: 5816
845. Riebel BR, Gibbs PR, Wellborn WB, Bommarius AS (2002) Adv. Synth. Catal. 3454: 1156
846. Lemière GL, Lepoivre JA, Alderweireldt FC (1985) Tetrahedron Lett. 26: 4527
847. Lemière GL (1986) Alcohol dehydrogenase catalysed oxidoreduction reactions in organic chemistry. In: Schneider MP (ed) Enzymes as Catalysts in Organic Synthesis, NATO ASI Series C, vol 178. Reidel, Dordrecht, p 19
848. Devaux-Basseguy R, Bergel A, Comtat M (1997) Enzyme Microb. Technol. 20: 248
849. Hummel W (1997) Adv. Biochem. Eng. Biotechnol. 58: 145
850. Hummel W (1999) New alcohol dehydrogenases for the synthesis of chiral compounds. In: Scheper T (ed) New Enzymes for Organic Synthesis. Springer, Heidelberg, pp 145–184
851. Prelog V (1964) Pure Appl. Chem. 9: 119
852. Peters J, Minuth T, Kula MR (1993) Biocatalysis 8: 31
853. Hou CT, Patel R, Barnabe N, Marczak I (1981) Eur. J. Biochem. 119: 359
854. Bradshaw CW, Hummel W, Wong CH (1992) J. Org. Chem. 57: 1532
855. Hummel W (1990) Appl. Microbiol. Biotechnol. 34: 15
856. Prelog V (1964) Colloqu. Ges. Physiol. Chem. 14: 288
857. Bradshaw CW, Fu H, Shen GJ, Wong CH (1992) J. Org. Chem. 57: 1526
858. Kula MR, Kragl U (2000) Dehydrogenases in the synthesis of chiral compounds. In: Patel RN (ed) Stereoselective Biocatalysis. Marcel Dekker, New York, p 839
859. Roberts SM (1988) Enzymes as catalysts in organic synthesis; In: Cooper A, Houben JL, Chien LC (eds) The Enzyme Catalysis Process. NATO ASI Series A, vol 178. Reidel, Dordrecht, p 443
860. Nakamura K, Miyai T, Kawai J, Nakajima N, Ohno A (1990) Tetrahedron Lett. 31: 1159
861. MacLeod R, Prosser H, Fikentscher L, Lanyi J, Mosher HS (1964) Biochemistry 3: 838
862. Lepoivre JA (1984) Janssen Chim. Acta 2: 20
863. Plant A (1991) Pharm. Manufact. Rev., March: 5
864. Cedergen-Zeppezauer ES, Andersson I, Ottonello S (1985) Biochemistry 24: 4000

865. Ganzhorn AJ, Green DW, Hershey AD, Gould RM, Plapp BV (1987) J. Biol. Chem. 262: 3754
866. Jones JB, Schwartz HM (1981) Can. J. Chem. 59: 1574
867. Van Osselaer TA, Lemière GL, Merckx EM, Lepoivre JA, Alderweireldt FC (1978) Bull. Soc. Chim. Belg. 87: 799
868. Jones JB, Takemura T (1984) Can. J. Chem. 62: 77
869. Davies J, Jones JB (1979) J. Am. Chem. Soc. 101: 5405
870. Lam LKP, Gair IA, Jones JB (1988) J. Org. Chem. 53: 1611
871. Park DH, Plapp BV (1991) J. Biol. Chem. 266: 13296.
872. Krawczyk AR, Jones JB (1989) J. Org. Chem. 54: 1795
873. Irwin AJ, Jones JB (1976) J. Am. Chem. Soc. 98: 8476
874. Sadozai SK, Merckx EM, Van De Val AJ, Lemière GL, Esmans EL, Lepoivre JA, Alderweireldt FC (1982) Bull. Soc. Chim. Belg. 91: 163
875. Nakazaki M, Chikamatsu H, Fujii T, Sasaki Y, Ao S (1983) J. Org. Chem. 48: 4337
876. Nakazaki M, Chikamatsu H, Naemura K, Suzuki T, Iwasaki M, Sasaki Y, Fujii T (1981) J. Org. Chem. 46: 2726
877. Nakazaki M, Chikamatsu H, Sasaki Y (1983) J. Org. Chem. 48: 2506
878. Matos JR, Smith MB, Wong CH (1985) Bioorg. Chem. 13: 121
879. Takemura T, Jones JB (1983) J. Org. Chem. 48: 791
880. Haslegrave JA, Jones JB (1982) J. Am. Chem. Soc. 104: 4666
881. Fries RW, Bohlken DP, Plapp BV (1979) J. Med. Chem. 22: 356
882. Dodds DR, Jones JB (1988) J. Am. Chem. Soc. 110: 577
883. Yamazaki Y, Hosono K (1988) Tetrahedron Lett. 29: 5769
884. Yamazaki Y, Hosono K (1989) Tetrahedron Lett. 30: 5313
885. Lemière GL, Van Osselaer TA, Lepoivre JA, Alderweireldt FC (1982) J. Chem. Soc., Perkin Trans. 2, 1123
886. Secundo F, Phillips RS (1996) Enzyme Microb. Technol. 19: 487
887. Keinan E, Sinha SC, Sinha-Bagchi A (1991) J. Chem. Soc., Perkin Trans. 1, 3333
888. Keinan E, Seth KK, Lamed R, Ghirlando R, Singh SP (1990) Biocatalysis 3: 57
889. Rothig TR, Kulbe KD, Buckmann F, Carrea G (1990) Biotechnol. Lett. 12: 353
890. Keinan E, Sinha SC, Singh SP (1991) Tetrahedron 47: 4631
891. Edegger K, Stampfer W, Seisser B, Faber K, Mayer SF, Oehrlein R, Hafner A, Kroutil W (2006) Eur. J. Org. Chem. 1904
892. Kosjek B, Stampfer W, Pogorevc M, Goessler W, Faber K, Kroutil W (2004) Biotechnol. Bioeng. 86: 55
893. Stampfer W, Kosjek B, Faber K, Kroutil W (2003) J. Org. Chem. 68: 402
894. Drueckhammer DG, Sadozai SK, Wong CH, Roberts SM (1987) Enzyme Microb. Technol. 9: 564
895. Keinan E, Seth KK, Lamed R (1986) J. Am. Chem. Soc. 108: 3474
896. De Amici M, De Micheli C, Carrea G, Spezia S (1989) J. Org. Chem. 54: 2646
897. Drueckhammer DG, Barbas III CF, Nozaki K, Wong CH (1988) J. Org. Chem. 53: 1607
898. Kelly DR, Lewis JD (1991) J. Chem. Soc., Chem. Commun. 1330
899. Schubert T, Hummerl W, Kula MR, Müller M (2001) Eur. J. Org. Chem. 4181
900. Kim MJ, Whitesides GM (1988) J. Am. Chem. Soc. 110: 2959
901. Kim M-J, Kim JY (1991) J. Chem. Soc., Chem. Commun. 326
902. Luyten MA, Bur D, Wynn H, Parris W, Gold M, Frieson JD, Jones JB (1989) J. Am. Chem. Soc. 111: 6800
903. Casy G, Lee TV, Lovell H (1992) Tetrahedron Lett. 33: 817
904. Schütte H, Hummel H, Kula MR (1984) Appl. Microbiol. Biotechnol. 19: 167
905. Hummel W, Schütte H, Kula MR (1985) Appl. Microbiol. Biotechnol. 21: 7
906. De Amici M, De Micheli C, Molteni G, Pitrè D, Carrea G, Riva S, Spezia S, Zetta L (1991) J. Org. Chem. 56: 67

907. Butt S, Davies HG, Dawson MJ, Lawrence GC, Leaver J, Roberts SM, Turner MK, Wakefield BJ, Wall WF, Winders JA (1987) J. Chem. Soc., Perkin Trans. 1, 903
908. Riva S, Ottolina G, Carrea G, Danieli B (1989) J. Chem. Soc., Perkin Trans. 1, 2073
909. Carrea G, Colombi F, Mazzola G, Cremonesi P, Antonini E (1979) Biotechnol. Bioeng. 21: 39
910. Butt S, Davies HG, Dawson MJ, Lawrence GC, Leaver J, Roberts SM, Turner MK, Wakefield BJ, Wall WF, Winders JA (1985) Tetrahedron Lett. 26: 5077
911. Leaver J, Gartenmann TCC, Roberts SM, Turner MK (1987) In: Laane C, Tramper J, Lilly MD (eds) Biocatalysis in Organic Media. p 411, Elsevier, Amsterdam
912. Man H, Kedziora K, Kulig J, Frank A, Lavandera I, Gotor-Fernandez V, Rother D, Hart S, Turkenburg JP, Grogan G (2014) Topics Catal. 57: 356
913. Lavandera I, Kern A, Ferreira-Silva B, Glieder A, de Wildeman S, Kroutil W (2008) J. Org. Chem. 73: 6003
914. Nakamura K, Yoneda T, Miyai T, Ushio K, Oka S, Ohno A (1988) Tetrahedron Lett. 29: 2453
915. Fontana A (1984) Thermophilic Enzymes and Their Potential Use in Biotechnology. Dechema, Weinheim, p 221
916. Bryant FO, Wiegel J, Ljungdahl LG (1988) Appl. Environ. Microbiol. 54: 460
917. Pham VT, Phillips RS, Ljungdahl LG (1989) J. Am. Chem. Soc. 111: 1935
918. Daniel RM, Bragger J, Morgan HW (1990) Enzymes from extreme environments. In: Abramowicz DA (ed) Biocatalysis. Van Nostrand Reinhold, New York, p 243
919. Willaert JJ, Lemière GL, Joris LA, Lepoivre JA, Alderweideldt FC (1988) Bioorg. Chem. 16: 223
920. Pham VT, Phillips RS (1990) J. Am. Chem. Soc. 112: 3629
921. Chen CS, Zhou BN, Girdaukas G, Shieh WR, VanMiddlesworth F, Gopalan AS, Sih CJ (1984) Bioorg. Chem. 12: 98
922. Shieh WR, Gopalan AS, Sih CJ (1985) J. Am. Chem. Soc. 107: 2993
923. Hoffmann RW, Ladner W, Helbig W (1984) Liebigs Ann. Chem. 1170
924. Nakamura K, Ushio K, Oka S, Ohno A (1984) Tetrahedron Lett. 25: 3979
925. Fuganti C, Grasselli P, Casati P, Carmeno M (1985) Tetrahedron Lett. 26: 101
926. Nakamura K, Kawai Y, Oka S, Ohno A (1989) Tetrahedron Lett. 30: 2245
927. Christen M, Crout DHG (1987) Bioreactors and Biotransformations, In: Moody GW, Baker PB (eds) Elsevier, London, p 213
928. Sakai T, Nakamura T, Fukuda K, Amano E, Utaka M, Takeda A (1986) Bull. Chem. Soc. Jpn. 59: 3185
929. Buisson D, Azerad R, Sanner C, Larcheveque M (1991) Tetrahedron Asymmetry 2: 987
930. Ushio K, Inoue K, Nakamura K, Oka S, Ohno A (1986) Tetrahedron Lett. 27: 2657
931. Kometani T, Kitatsuji E, Matsuno R (1991) Agric. Biol. Chem. 55: 867
932. Dahl AC, Madsen JO (1998) Tetrahedron Asymmetry 9: 4395
933. Buisson D, Azerad R (1986) Tetrahedron Lett. 27: 2631
934. Seebach D, Züger MF, Giovannini F, Sonnleitner B, Fiechter A (1984) Angew. Chem. Int. Ed. 23: 151
935. Servi S (1990) Synthesis 1
936. Kometani T, Yoshii H, Matsuno R (1996) J. Mol. Catal. B 1: 45
937. Sih CJ, Chen CS (1984) Angew. Chem. Int. Ed. 23: 570
938. Ward OP, Young CS (1990) Enzyme Microb. Technol. 12: 482
939. Csuk R, Glänzer B (1991) Chem. Rev. 91: 49
940. Neuberg C, Lewite A (1918) Biochem. Z. 91: 257
941. Neuberg C (1949) Adv. Carbohydr. Chem. 4: 75
942. Ticozzi C, Zanarotti A (1988) Tetrahedron Lett. 29: 6167
943. Dondoni A, Fantin G, Fogagnolo M, Mastellari A, Medici A, Nefrini E, Pedrini P (1988) Gazz. Chim. Ital. 118: 211

944. Bucchiarelli M, Forni A, Moretti I, Prati F, Torre G, Resnati G, Bravo P (1989) Tetrahedron 45: 7505
945. Bernardi R, Bravo P, Cardillo R, Ghiringhelli D, Resnati G (1988) J. Chem. Soc., Perkin Trans. 1: 2831
946. Kitazume T, Kobayashi T (1987) Synthesis 87
947. Kitazume T, Nakayama Y (1986) J. Org. Chem. 51: 2795
948. Takano S, Yanase M, Sekiguchi Y, Ogasawara K (1987) Tetrahedron Lett. 28: 1783
949. Bucchiarelli M, Forni A, Moretti I, Torre G (1983) Synthesis 897
950. Kitazume T, Lin JT, (1987) J. Fluorine Chem. 34: 461
951. Fujisawa T, Hayashi H, Kishioka Y (1987) Chem. Lett. 129
952. Seebach D, Roggo S, Maetzke T, Braunschweiger H, Cerkus J, Krieger M (1987) Helv. Chim. Acta 70: 1605
953. Nakamura K, Inoue Y, Shibahara J, Oka S, Ohno A (1988) Tetrahedron Lett. 29: 4769
954. Guette JP, Spassky N (1972) Bull. Soc. Chim. Fr. 4217
955. Levene PA, Walti A (1943) Org. Synth., Coll. Vol. II: 545
956. Takano S (1987) Pure Appl. Chem. 59: 353
957. Itoh T, Yoshinaka A, Sato T, Fujisawa T (1985) Chem. Lett. 1679
958. Kozikowski AP, Mugrage BB, Li CS, Felder L (1986) Tetrahedron Lett. 27: 4817
959. Bernardi R, Ghiringhelli D (1987) J. Org. Chem. 52: 5021
960. Top S, Jaouen G, Gillois J, Baldoli C, Maiorana S (1988) J. Chem. Soc., Chem. Commun. 1284
961. Yamazaki Y, Hosono K (1988) Agric. Biol. Chem. 52: 3239
962. Syldatk C, Andree H, Stoffregen A, Wagner F, Stumpf B, Ernst L, Zilch H, Tacke R (1987) Appl. Microbiol. Biotechnol. 27: 152
963. Yamazaki Y, Kobayashi H (1993) Chem. Express 8: 97
964. Le Drian C, Greene AE (1982) J. Am. Chem. Soc. 104: 5473
965. Belan A, Bolte J, Fauve A, Gourcy JG, Veschambre H (1987) J. Org. Chem. 52: 256
966. Hirama M, Nakamine T, Ito S (1984) Chem. Lett. 1381
967. Deshong P, Lin M-T, Perez JJ (1986) Tetrahedron Lett. 27: 2091
968. Tschaen DM, Fuentes LM, Lynch JE, Laswell WL, Volante RP, Shinkai I (1988) Tetrahedron Lett. 29: 2779
969. Mori K (1989) Tetrahedron 45: 3233
970. Kramer A, Pfader H (1982) Helv. Chim. Acta 65: 293
971. Sih CJ, Zhou B, Gopalan AS, Shieh WR, VanMiddlesworth F (1983) Strategies for Controlling the Stereochemical Course of Yeast Reductions. In: Bartmann W, Trost BM (eds) Selectivity—a Goal for Synthetic Efficiency. Proc. 14th Workshop Conference Hoechst. Verlag Chemie, Weinheim, p 250
972. Heidlas J, Engel KH, Tressl R (1991) Enzyme Microb. Technol. 13: 817
973. Nakamura K, Inoue K, Ushio K, Oka S, Ohno A (1987) Chem. Lett. 679
974. Nakamura K, Kawai Y, Oka S, Ohno A (1989) Bull. Chem. Soc. Jpn. 62: 875
975. Nakamura K, Higaki M, Ushio K, Oka S, Ohno A (1985) Tetrahedron Lett. 26: 4213
976. Chibata I, Tosa T, Sato T (1974) Appl. Microbiol. 27: 878
977. Nakamura K, Kawai Y, Ohno A (1990) Tetrahedron Lett. 31: 267
978. Hayakawa R, Nozawa K, Kimura K, Shimizu M (1999) Tetrahedron 55: 7519
979. Ushio K, Hada J, Tanaka Y, Ebara K (1993) Enzyme Microb. Technol. 15: 222
980. Miya H, Kawada M, Sugiyama Y (1996) Biosci. Biotechnol. Biochem. 60: 95
981. Arnone A, Biagnini G, Cardillo R, Resnati G, Begue JP, Bonnet-Delpon D, Kornilov A (1996) Tetrahedron Lett. 37: 3903
982. Deol B, Ridley D, Simpson G (1976) Aust. J. Chem. 29: 2459
983. Nakamura K, Miyai T, Nozaki K, Ushio K, Ohno A (1986) Tetrahedron Lett. 27: 3155
984. Fujisawa T, Itoh T, Sato T (1984) Tetrahedron Lett. 25: 5083
985. Buisson D, Henrot S, Larcheveque M, Azerad R (1987) Tetrahedron Lett. 28: 5033
986. Akita H, Furuichi A, Koshoji H, Horikoshi K, Oishi T (1983) Chem. Pharm. Bull. 31: 4376

987. Frater G, Müller U, Günther W (1984) Tetrahedron 40: 1269
988. VanMiddlesworth F, Sih CJ (1987) Biocatalysis 1: 117
989. Hoffmann RW, Helbig W, Ladner W (1982) Tetrahedron Lett. 23: 3479
990. Chenevert R, Thiboutot S (1986) Can. J. Chem. 64: 1599
991. Ohta H, Ozaki K, Tsuchihashi G (1986) Agric. Biol. Chem. 50: 2499
992. Nakamura K, Kawai Y, Ohno A (1991) Tetrahedron Lett. 32: 2927
993. Nakamura K, Miyai T, Nozaki K, Ushio K, Oka S, Ohno A (1986) Tetrahedron Lett. 27: 3155
994. Buisson D, Azerad R, Sanner C, Larcheveque M (1990) Biocatalysis 3: 85
995. Buisson D, Sanner C, Larcheveque M, Azerad R (1987) Tetrahedron Lett. 28: 3939
996. Cabon O, Buisson D, Lacheveque M, Azerad R (1995) Tetrahedron Asymmetry 6: 2199
997. Nishida T, Matsumae H, Machida I, Shibatani T (1995) Biocatalysis Biotrans. 12: 205
998. Sato T, Tsurumaki M, Fujisawa T (1986) Chem. Lett. 1367
999. Besse P, Veschambre H (1993) Tetrahedron Asymmetry 4: 1271
1000. Soukup M, Wipf B, Hochuli E, Leuenberger HGW (1987) Helv. Chim. Acta 70: 232
1001. Nakamura K, Inoue K, Ushio K, Oka S, Ohno A (1988) J. Org. Chem. 53: 2598
1002. Brooks DW, Mazdiyasni H, Chakrabarti S (1984) Tetrahedron Lett. 25: 1241
1003. Brooks DW, Woods KW (1987) J. Org. Chem. 52: 2036
1004. Brooks DW, Mazdiyasni H, Grothaus PG (1987) J. Org. Chem. 52: 3223
1005. Brooks DW, Mazdiyasni H, Sallay P (1985) J. Org. Chem. 50: 3411
1006. Fujisawa T, Kojima E, Sato T (1987) Chem. Lett. 2227
1007. Takeshita M, Sato T (1989) Chem. Pharm. Bull. 37: 1085
1008. Kieslich K (1976) Microbial Transformations of Non-Steroid Cyclic Compounds. Thieme, Stuttgart
1009. Besse P, Sokoltchik T, Veschambre H (1998) Tetrahedron Asymmetry 9: 4441
1010. Tidswell EC, Salter GJ, Kell DB, Morris JG (1997) Enzyme Microb. Technol. 21: 143
1011. Wipf B, Kupfer E, Bertazzi R, Leuenberger HGW (1983) Helv. Chim. Acta 66: 485
1012. Bernardi R, Cardillo R, Ghiringhelli D, de Pavo V (1987) J. Chem. Soc., Perkin Trans. 1, 1607
1013. Ikeda H, Sato E, Sugai T, Ohta H (1996) Tetrahedron 52: 8113
1014. Fujisawa T, Onogawa Y, Sato A, Mitsuya T, Shimizu M (1998) Tetrahedron 54: 4267
1015. Wei ZL, Li ZY, Lin GQ (1998) Tetrahedron 54: 13059
1016. Fantin G, Fogagnolo M, Giovannini PP, Medici A, Pedrini P, Gardini F, Lanciotti R (1996) Tetrahedron 52: 3547
1017. Akakabe Y, Naoshima Y (1993) Phytochemistry 32: 1189
1018. Bruni R, Fantin G, Maietti S, Medici A, Pedrini P, Sacchetti G (2006) Tetrahedron Asymmetry 17: 2287
1019. Vicenzi JT, Zmijewski MJ, Reinhard MR, Landen BE, Muth WL, Marler PG (1997) Enzyme Microb. Technol. 20: 494
1020. Hummel W (1997) Biochem. Eng. 58: 145
1021. Patel RN, Banerjee A, McNamee CG, Brzozowski D, Hanson RL, Szarka LJ (1993) Enzyme Microb. Technol. 15: 1014
1022. Haberland J, Hummel W, Daußmann T, Liese A (2002) Org. Proc. Res. Dev. 6: 458
1023. Schmidt E, Ghisalba O, Gygax D, Sedelmeier G (1992) J. Biotechnol. 24: 315
1024. Kataoka M, Kita K, Wada M, Yasohara Y, Hasegawa J, Shimizu S (2003) Appl. Microbiol. Biotechnol. 62: 437
1025. Liang J, Lalonde J, Borup B, Mitchell V, Mundorff E, Trinh N, Kochrekar DA, Ramachandran NC, Pai GG (2010) Org. Proc. Res. Dev. 14: 193.
1026. Azerad R, Buisson D (1992) In: Servi S (ed) Microbial Reagents in Organic Synthesis. NATO ASI Series C, vol 381, p. 421, Kluwer, Dordrecht
1027. Buisson D, Azerad R, Sanner C, Lacheveque M (1992) Biocatalysis 5: 249
1028. Gadler P, Glueck S M, Kroutil W, Nestl BM, Larissegger-Schnell B, Ueberbacher BT, Wallner SR, Faber K (2006) Biochem. Soc. Trans 34: 296

1029. Nakamura K, Inoue Y, Matsuda T, Ohno A (1995) Tetrahedron Lett. 36: 6263
1030. Fantin G, Fogagnolo M, Giovannini PP, Medici A, Pedrini, P (1995) Tetrahedron Asymmetry 6: 3047
1031. Carnell AJ (1999) Adv. Biochem. Eng. Biotechnol. 63: 57
1032. Voss CV, Gruber CC, Kroutil W (2010) Synlett 991
1033. Hasegawa J, Ogura M, Tsuda S, Maemoto S, Kutsuki H, Ohashi T (1990) Agric. Biol. Chem. 54: 1819
1034. Setyahadi S, Harada E, Mori N, Kitamoto Y (1998) J. Mol. Catal. B 4: 205
1035. Azerad R, Buisson D (1992) Stereocontrolled reduction of β-ketoesters with *Geotrichum candidum*. In: Servi S (ed) Microbial Reagents in Organic Synthesis. NATO ASI Series C, vol 381. Kluwer, Dordrecht, pp 421–440
1036. Voss CV, Gruber CC, Faber K, Knaus T, Macheroux P, Kroutil W (2008) J. Am. Chem. Soc. 130: 13969
1037. Voss CV, Gruber CC, Kroutil W (2008) Angew. Chem. Int. Ed. 47: 741
1038. Ogawa J, Xie SX, Shimizu S (1999) Biotechnol. Lett. 21: 331
1039. Brown GM (1971) In: Florkin M, Stotz EH (eds) Comprehensive Biochemistry, vol. 21. Elsevier, New York, p 73.
1040. Shimizu S, Hattori S, Hata H, Yamada Y (1987) Appl. Environ. Microbiol. 53: 519
1041. Shimizu S, Hattori S, Hata H, Yamada H (1987) Enzyme Microb. Technol. 9: 411
1042. Page PCB, Carnell AJ, McKenzie MJ (1998) Synlett. 774
1043. Matsumura S, Kawai Y, Takahashi Y, Toshima K (1994) Biotechnol. Lett. 16: 485
1044. Carnell AJ, Iacazio G, Roberts SM, Willetts AJ (1994) Tetrahedron Lett. 35: 331
1045. Moon Kim B, Guare JP, Hanifin CM, Arford-Bickerstaff DJ, Vacca JP, Ball RG (1994) Tetrahedron Lett. 35: 5153
1046. Ghislieri D, Turner N (2014) Top. Catal. 57: 284.
1047. Nugent TC, El-Shazly M (2010) Adv. Synth. Catal. 352: 753.
1048. Breuer M, Ditrich K, Habicher T, Hauer B, Keßeler M, Stürmer R, Zelinski T (2004) Angew. Chem. Int. Ed. 43: 788.
1049. Schrittwieser JH, Velikogne S, Kroutil W (2015) Adv. Synth. Catal. 357: 1655.
1050. Mangas-Sanchez J, France SP, Montgomery SL, Aleku GA, Man H, Sharma M, Ramsden JI, Grogan G, Turner NJ (2017) Curr. Opin. Chem. Biol. 37: 19.
1051. Grogan G, Turner NJ (2016) Chem. Eur. J. 22: 1900
1052. Chimni SS, Singh RJ (1997) World J. Microbiol. Biotechnol. 14: 247.
1053. Li H, Williams P, Micklefield J, Gardiner JM, Stephens G (2004) Tetrahedron 60: 753.
1054. Vaijayanthi T, Chadha A (2008) Tetrahedron: Asymm. 19: 93.
1055. Mitsukura K, Suzuki M, Tada K, Yoshida T, Nagasawa T (2010) Org. Biomol. Chem. 8: 4533.
1056. Mitsukura K, Kuramoto T, Yoshida T, Kimoto N, Yamamoto H, Nagasawa T (2013) Appl. Microbiol. Biotechnol. 97: 8079.
1057. Leipold F, Hussain S, Ghislieri D, Turner NJ (2013) ChemCatChem 5: 3505.
1058. Hussain S, Leipold F, Man H, Wells E, France SP, Mulholland KR, Grogan G, Turner NJ (2015) ChemCatChem 7: 579.
1059. Imine reductase Engineering Database, https://ired.biocatnet.de/
1060. Scheller PN, Fademrecht S, Hofzelter S, Pleiss J, Leipold F, Turner NJ, Nestl BM, Hauer B (2014) ChemBioChem 15: 2201.
1061. Scheller PN, Lenz M, Hammer SC, Hauer B, Nestl B (2016) ChemCatChem 7: 3239.
1062. Huber T, Schneider L, Präg A, Gerhardt S, Einsle O, Müller M (2014) ChemCatChem 6: 2248.
1063. Wetzl D, Gand M, Ross A, Müller H, Matzel P, Hanlon SP, Müller M, Wirz B, Höhne M, Iding H (2016) ChemCatChem 8: 2023.
1064. Ohshima T, Soda K (2000) Amino acid dehydrogenases and their applications. In: Patel RN (ed) Stereoselective Biocatalysis. Marcel Dekker, New York, p. 877

1065. Bommarius AS, Au SK (2015) Amino acid and amine dehydrogenases, in: Science of Synthesis, Biocatalysis in Organic Synthesis (Faber K, Fessner W-D, Turner NJ, eds.), 2: 335.
1066. Moore JC, Savile CK, Pannuri S, Kosjek B, Janey JM (2012) Comprehensive Chirality (Carreira EM, Yamamoto H, eds.) 9: 318.
1067. Engel PC (2011) Biochem. Soc. Trans. 39: 425.
1068. Hanson RL (2009) ACS Symposium Series 1009: 306.
1069. Vedha-Peters K, Gunawardana M, Rozzell JD, Novick SJ (2006) J. Am. Chem. Soc. 128: 10923.
1070. Akita H, Suzuki H, Doi K, Ohshima T (2014) Appl. Microbiol. Biotechnol. 98: 1135.
1071. Hanson RL, Johnston RM, Goldberg SL, Parker WL, Goswami A (2013) Org. Proc. Res. Dev. 17: 693.
1072. Sekimoto T, Matsuyama T, Fukui T, Tanizawa K (1993) J. Biol. Chem. 268: 27039
1073. Bommarius AS, Schwarm M, Drauz K (1998) J. Mol. Catal. B: Enzym. 5: 1
1074. Krix G, Bommarius AS, Drauz K, Kottenhahn M, Schwarm M, Kula MR (1997) J. Biotechnol. 53: 29
1075. Kragl U, Vasic-Racki D, Wandrey C (1996) Bioproc. Eng. 14: 291
1076. Itoh N, Yachi C, Kudome T (2000) J. Mol. Catal. B: Enzym. 10: 281.
1077. Abrahamson MJ, Vazquez-Figueroa E, Woodall NB, Moore JC, Bommarius AS (2012) Angew. Chem. Int. Ed. 51: 3969.
1078. Abrahamson MJ, Wong JW, Bommarius AS (2013) Adv. Synth. Catal. 355: 1780.
1079. http://nobelprize.org/nobel_prizes/chemistry/laureates/2001/knowles-lecture.html   and http://nobelprize.org/nobel_prizes/chemistry/laureates/2001/noyori-lecture.html.
1080. Yang JW, Hechavarria Fonseca MT, Vignola N, List B (2005) Angew. Chem. Int. Ed. 44:108
1081. Williams RE, Bruce NC (2002) Microbiology 148: 1607
1082. Steinbacher S, Stumpf M, Weinkauf S, Rohdich F, Bacher A, Simon H (2002) Enoate reductase family. In: Chapman SK, Perham RN, Scrutton NS (eds) Flavins and Flavoproteins. Weber, p 941
1083. Warburg O, Christian W (1933) Biochem. Z. 266: 377
1084. Barna T, Messiha HL, Petosa C, Bruce NC, Scrutton NS, Moody PCE (2002) J. Biol. Chem. 277: 30976
1085. Schaller F, Biesgen C, Müssig C, Altmann T, Weiler EW (2000) Planta 210: 979
1086. Vaz ADN, Chakraborty S, Massey V (1995) Biochemistry 34: 4246
1087. Snape Jr, N. Walkley A, Morby AP, Nicklin S, White GF (1997) J. Bacteriol. 179: 7796
1088. Nishino SF, Spain JC (1993) Appl. Environ. Microbiol. 59: 2520
1089. Barna TM, Khan H, Bruce NC, Barsukov I, Scrutton NS, Moody PCE (2001) J. Mol. Biol. 310: 433
1090. Steinbacher S, Stumpf M, Weinkauf S, Rohdich F, Bacher A, Simon H (2002) Enoate Reductase Family. In: Flavins and Flavoproteins; Stephen K, Perham RN, Scrutton NS (eds.) Chapman, Cambridge, UK, pp. 941-949.
1091. Simon H (1991) Chem. Biochem. Flavoenzymes 2: 317.
1092. Kohli RM, Massey V (1998) J. Biol. Chem. 273: 32763
1093. Shimoda K, Ito DI, Izumi S, Hirata T (1996) J. Chem. Soc. Perkin Trans. 1, 4: 355
1094. Schlieben NH, Niefind K, Müller J, Riebel B, Hummel W, Schomburg D (2005) J. Mol. Biol. 349: 801
1095. Kurata A, Kurihara T, Kamachi H, Esaki N (2004) Tetrahedron Asymmetry 15: 2837
1096. Kataoka M, Kotaka A, Hasegawa A, Wada M, Yoshizumi A, Nakamori S, Shimizu S (2002) Biosci. Biotechnol. Biochem. 66: 2651
1097. Williams RE, Rathbone DA, Scrutton NS, Bruce NC (2004) Appl. Environ. Microbiol. 70: 3566
1098. Fuganti C, Grasselli P (1979) J. Chem. Soc. Chem. Commun. 995
1099. Fischer FG, Wiedemann O (1934) Liebigs Ann. Chem. 513: 260

1100. Desrut M, Kergomard A, Renard MF, Veschambre H (1981) Tetrahedron 37: 3825
1101. Simon H, White H, Lebertz H, Thanos I (1987) Angew. Chem. Int. Ed. 26: 785
1102. Simon H (1993) Indian J. Chem. 32B: 170
1103. Hauer B, Stuermer R, Hall M, Faber K (2007) Curr. Opinion Chem. Biol. 11: 203
1104. Hall M, Stueckler C, Kroutil W, Macheroux P, Faber K (2007) Angew. Chem. Int. Ed. 46: 3934
1105. Toogood HS, Gardiner JM, Scrutton NS (2010) ChemCatChem 2: 892
1106. Hall M, Stueckler C, Ehammer H, Pointner E, Oberdorfer G, Gruber K, Hauer B, Stuermer R, Kroutil W, Macheroux P, Faber K (2008) Adv. Synth. Catal. 350: 411
1107. Hall M, Stueckler C, Hauer B, Stuermer R, Friedrich T, Breuer M, Kroutil W, Faber K (2008) Eur. J. Org. Chem. 1511
1108. Mueller NJ, Stueckler C, Hauer B, Baudendistel N, Housden H, Bruce NC, Faber K (2010) Adv. Synth. Catal. 352: 387
1109. Swiderska MA, Stewart JD (2006) J. Mol. Catal. B: Enzym. 42: 52
1110. Swiderska MA, Stewart JD (2006) Org. Lett. 8: 6131
1111. Toogood HS, Scrutton NS (2014) Curr. Opin. Chem. Biol. 19: 107.
1112. Fuganti C, Grasselli P (1989) Baker's yeast mediated synthesis of natural products. In: Whitaker JR, Sonnet PE (eds) Biocatalysis in Agricultural Biotechnology, ACS Symp. Ser. 389. ACS, Washington, p 359
1113. Suemune H, Hayashi N, Funakoshi K, Akita H, Oishi T, Sakai K (1985) Chem. Pharm. Bull. 33: 2168
1114. Gil G, Ferre E, Barre M, Le Petit J (1988) Tetrahedron Lett. 29: 3797
1115. Mueller A, Stuermer R, Hauer B, Rosche B (2007) Angew. Chem. Int. Ed. 46: 3316
1116. Fuganti C, Grasselli P (1982) J. Chem. Soc., Chem. Commun. 205
1117. Fuganti C, Grasselli P, Servi S (1983) J. Chem. Soc., Perkin Trans. 1, 241
1118. Sato T, Hanayama K, Fujisawa T (1988) Tetrahedron Lett. 29: 2197
1119. Kergomard A, Renard MF, Veschambre H (1982) J. Org. Chem. 47: 792
1120. Sih CJ, Heather JB, Sood R, Price P, Peruzzotti G, Lee HFH, Lee SS (1975) J. Am. Chem. Soc. 97: 865
1121. Durchschein K, Ferreira-da Silva B, Wallner S, Macheroux P, Kroutil W, Glueck S M, Faber K (2010) Green Chem. 12: 616
1122. Ohta H, Ozaki K, Tsuchihashi G (1987) Chem. Lett. 191
1123. Kitazume T, Ishikawa N (1984) Chem. Lett. 587
1124. Kosjek B, Fleitz FJ, Dormer PG, Kuethe JT, Devine PN (2008) Tetrahedron: Asymmetry 19: 1403
1125. Oberdorfer G, Gruber K, Faber K, Hall M (2012) Synlett 1857.
1126. Gramatica P, Manitto P, Monti D, Speranza G (1987) Tetrahedron 43: 4481
1127. Gramatica P, Manitto P, Ranzi BM, Delbianco A, Francavilla M (1983) Experientia 38: 775
1128. Gramatica P, Manitto P, Monti D, Speranza G (1988) Tetrahedron 44: 1299
1129. Leuenberger HG, Boguth W, Widmer E, Zell R (1976) Helv. Chim. Acta 59: 1832
1130. Cho E, Hankinson SE, Rosner B, Willet WC, Colditz GA (2008) Am. J. Clin. Nutr. 87: 1837
1131. Ohta H, Kobayashi N, Ozaki K (1989) J. Org. Chem. 54: 1802
1132. Hall M, Stueckler C, Ehammer H, Pointner E, Kroutil W, Macheroux P, Faber K (2007) Org. Lett. 9: 5409
1133. Mangan D, Miskelly I, Moody TS (2012) Adv. Synth. Catal. 354: 2185.
1134. Utaka M, Konishi S, Mizuoka A, Ohkubo T, Sakai T, Tsuboi S, Takeda A (1989) J. Org. Chem. 54: 4989
1135. Utaka M, Konishi S, Okubo T, Tsuboi S, Takeda A (1987) Tetrahedron Lett. 28: 1447
1136. Tasnadi G, Winkler CK, Clay D, Sultana N, Fabian WMF, Hall M, Ditrich K, Faber K (2012) Chem. Eur. J.: 18: 10362.
1137. Tasnadi G, Winkler CK, Clay D, Hall M, Faber K (2012) Catal. Sci. Technol. 2: 1509.
1138. Takabe K, Hiyoshi H, Sawada H, Tanaka M, Miyazaki A, Yamada T, Kitagiri T, Yoda H (1992) Tetrahedron Asymmetry 3: 1399

1139. Turrini NG, Hall M, Faber K (2015) Adv. Synth. Catal. 357: 1861.
1140. Vaz ADN, Chakraborty S, Massey V (1995) Biochemistry 34: 4246.
1141. Winkler CK, Clay D, Entner M, Plank M, Faber K (2014) Chem. Eur. J.: 20: 1403.

References to Sect. 2.3

1142. Glueck SM, Gümüs S, Fabian WMF, Faber K (2010) Chem. Soc. Rev. 39: 313
1143. Fang JM, Lin CH, Bradshaw CW, Wong CH (1995) J. Chem. Soc., Perkin Trans. 1, 967
1144. Schmid RD, Urlacher VB (eds) (2007) Modern Biooxidation Methods. Wiley-VCH, Weinheim
1145. Hollmann F, Arends IWCE, Buehler K, Schallmey A, Bühler B (2011) Green Chem. 13: 226.
1146. Fonken GS, Johnson RA (1972) Chemical oxidations with microorganisms. In: Belew JS (ed) Oxidation in Organic Chemistry, vol 2. Marcel Dekker, New York, p 185
1147. Reichstein T (1934) Helv. Chim. Acta 17: 996
1148. Sato K, Yamada Y, Aida K, Uemura T (1967) Agric. Biol. Chem. 31: 877
1149. Touster O, Shaw DRD (1962) Physiol. Rev. 42: 181
1150. Bernhauer K, Knobloch H (1940) Biochem. Z. 303: 308
1151. Kaufmann H, Reichstein T (1967) Helv. Chim. Acta 50: 2280
1152. Jones JB, Taylor KE (1976) Can. J. Chem. 54: 2969 and 2974
1153. Drueckhammer DG, Riddle VW, Wong CH (1985) J. Org. Chem. 50: 5387
1154. Irwin AJ, Jones JB (1977) J. Am. Chem. Soc. 99: 1625
1155. Jones JB, Jacovac IJ (1990) Org. Synth., coll. vol. 7: 406
1156. Bisogno FR, Garcia-Urdiales E, Valdes H, Lavandera I, Kroutil W, Suarez D, Gotor V (2010) Chem. Eur. J. 16: 11012.
1157. Wong C-H, Matos JR (1985) J. Org. Chem. 50: 1992
1158. Könst P, Merkens H, Kara S, Kochius S, Vogel A, Zuhse R, Holtmann D, Arends IWCE, Hollmann F (2012) Angew. Chem. Int. Ed. 51: 9914.
1159. Irwin AJ, Jones JB (1977) J. Am. Chem. Soc. 99: 556
1160. Jones JB, Lok KP (1979) Can. J. Chem. 57: 1025
1161. Ng GSY, Yuan LC, Jakovac IJ, Jones JB (1984) Tetrahedron 40: 1235
1162. Lok KP, Jakovac IJ, Jones JB (1985) J. Am. Chem. Soc. 107: 2521
1163. Pickl M, Fuchs M, Glueck SM, Faber K (2015) Appl. Microbiol. Biotechnol. 99: 6617.
1164. van Hellemond EW, Leferink NGH, Heuts PHM, Fraaije MW, van Berkel WJH (2000) Adv. Appl. Microbiol. 60: 17.
1165. Dijkman WP, Binda C, Fraaije MW, Mattevi A (2015) ACS Catal. 5: 1833.
1166. Ghisleri D, Turner NJ (2014) Top. Catal. 57: 284.
1167. Heath RS, Pontini M, Bechi B, Turner NJ (2014) ChemCatChem 6: 996.
1168. Ghislieri D, Green AP, Pontini M, Willies SC, Rowles I, Frank A, Grogan G, Turner NJ (2013) J. Am. Chem. Soc. 135: 10863.
1169. Rowles I, Malone KJ, Etchells LL, Willies SC, Turner NJ (2012) ChemCatChem 4: 1259.
1170. Schrittwieser JH, Groenendaal B, Resch V, Ghislieri D, Wallner S, Fischereder E-M, Fuchs E, Grischek B, Sattler JH, Macheroux P, Turner NJ, Kroutil W (2014) Angew. Chem. Int. Ed. 53: 3731.
1171. Koehler V, Bailey KR, Znabet A, Raftery J, Helliwell M, Turner NJ (2010) Angew. Chem. Int. Ed. 49: 2182.
1172. Li T, Liang J, Ambrogelly A, Brennan B, Gloor G, Huisman G, Lalonde J, Lekhal A, Mijts B, Muley S, Newman L, Tobin M, Wong G, Zaks A, Zhang X (2012) J. Am. Chem. Soc. 134: 6467.
1173. Holland HL (1992) Organic Synthesis with Oxidative Enzymes. Verlag Chemie, Weinheim

1174. Walsh C (1979) Enzymatic Reaction Mechanisms. Freeman, San Francisco, p 501
1175. Dalton H (1980) Adv. Appl. Microbiol. 26: 71
1176. Hayashi O (ed) (1974) The Molecular Mechanism of Oxygen Activation. Academic Press, New York
1177. Gunsalus IC, Pederson TC, Sligar SG (1975) Ann. Rev. Microbiol. 377
1178. Dawson JH (1988) Science 240: 433
1179. Hou CT (1986) Biotechnol. Gen. Eng. Rev. 4: 145
1180. Dunford HB (1982) Adv. Inorg. Biochem. 4: 41
1181. Dix TA, Benkovic SS (1988) Acc. Chem. Res. 21: 101
1182. Massey V (2000) Biochem. Soc. Trans. 28: 283
1183. Ziegler DM (1990) Trends Pharmacol. Sci. 11: 321
1184. Massey V (1994) J. Biol. Chem. 269: 22459
1185. Sato R, Omura T (eds) (1978) Cytochrome P-450. Academic Press, New York
1186. Ortiz de Montellano PR (ed) (1986) Cytochrome P-450. Plenum Press, New York
1187. Takemori S (1987) Trends Biochem. Sci. 12: 118
1188. Dawson JH, Sono M (1987) Chem. Rev. 87: 1255
1189. Müller HG (1990) Biocatalysis 4: 11
1190. Alexander LS, Goff HM (1982) J. Chem. Educ. 59: 179
1191. Poulos TL, Finzel BC, Howard AJ (1986) Biochemistry 25: 5314
1192. Poulos TL, Finzel BC, Gunsalus IC, Wagner GC, Kraut J (1985) J. Biol. Chem. 260: 16122
1193. Green MT, Dawson JH, Gray HB (2004) Science 304: 1653.
1194. McQuarters AB, Wolf MW, Hunt AP, Lehnert N (2014) Angew. Chem. Int. Ed. 53: 4750.
1195. Rittle J, Green MT (2010) Science 330: 933.
1196. Cirino PC, Arnold FH (2003) Angew. Chem. Int. Ed. 42: 3299.
1197. Nelson DR (2004) Hum. Genomics 4: 59.
1198. Grinberg AV, Hannemann F, Schiffler B, Muller J, Heinemann U, Bernhardt R (2000) Proteins 40: 590
1199. Bernhardt R (2006) J. Biotechnol. 124: 128
1200. Ortiz de Montanello PR (ed) (2005) Cytochrome P450: Structure, Mechanism and Biochemistry, 3rd ed. Kluwer/Plenum Press, New York
1201. Li Q-S, Schwaneberg U, Fischer P, Schmid RD (2000) Chem. Eur. J. 6: 1531.
1202. Holtmann D, Hollmann F (2016) ChemBioChem 17: 1391.
1203. Bucko M, Gemeiner P, Schenkmayerova A, Krajcovic T, Rudroff F, Mihovilovic MD (2016) Appl. Microbiol. Biotechnol. 100: 6585.
1204. Roiban G-D, Reetz MT (2015) Chem. Commun. 51: 2208.
1205. Caswell JM, O'Neill M, Taylor SJC, Moody TS (2013) Curr. Opin. Chem. Biol. 17: 271.
1206. Ryerson CC, Ballou DP, Walsh C (1982) Biochemistry 21: 2644
1207. Visser CM (1983) Eur. J. Biochem. 135: 543
1208. Ghisla S, Massey V (1989) Eur. J. Biochem. 181: 1
1209. Huijbers MME, Montersino S, Westphal AH, Tischler D, van Berkel WJH (2014) Arch. Biochem. Biophys. 544: 2.
1210. Entsch B, van Berkel WJ (1995) FASEB J 9: 476
1211. Jadan AP, Moonen MJH, Boeren SA, Golovleva LA, Rietjens IMCM, van Berkel WJH (2004) Adv. Synth. Catal. 346: 376
1212. Schmid A, Vereyken I, Held M, Witholt B (2001) J. Mol. Catal. B: Enzym. 11: 455
1213. Meyer A, Schmid A, Held M, Westphal A H, Rothlisberger M, Kohler HPE, van Berkel WJH, Witholt B (2002) J. Biol. Chem. 277: 5575
1214. Meyer A, Held M, Schmid A, Kohler HPE, Witholt B (2003) Biotechnol. Bioeng. 81: 518
1215. Otto K, Hofstetter K, Rothlisberger M, Witholt B, Schmid A (2004) J. Bacteriol. 186: 5292
1216. Hollmann F, Hofstetter K, Habicher T, Hauer B, Schmid A (2005) J. Am. Chem. Soc. 127: 6540
1217. Dijkman WP, de Gonzalo G, Mattevi A, Fraaije MW (2013) Appl. Microbiol. Biotechnol. 97: 5177.

1218. Fraaije MW, Mattevi A (2008) Nature Chem. Biol. 4: 719.
1219. Gandomkar S, Fischereder E-M, Schrittwieser JH, Wallner S, Habibi Z, Macheroux P, Kroutil W (2015) Ang. Chem. Int. Ed. 54: 15051.
1220. Mansuy D, Battioni P (1989) In: Hill CL (ed) Activation and Functionalisation of Alkanes. Wiley, New York, p 195
1221. Johnson RA (1978) Oxygenations with micro-organisms. In: Trahanovsky WS (ed) Oxidation in Organic Synthesis, part C. Academic Press, New York, p 131
1222. Fonken G, Johnson RA (1972) Chemical Oxidations with Microorganisms. Marcel Dekker, New York
1223. Kieslich K (1984) In: Rehm HJ, Reed G (eds) Biotechnology, vol 6a. Verlag Chemie, Weinheim, p 1
1224. Kieslich K (1980) Bull. Soc. Chim. Fr. 11: 9
1225. Sariaslani FS (1989) Crit. Rev. Biotechnol. 9: 171
1226. Bruce NC, French CE, Hailes AM, Long MT, Rathbone DA (1995) Trends Biotechnol. 13: 200
1227. Breslow R (1980) Acc. Chem. Res. 13: 170
1228. Fossey J, Lefort D, Massoudi M, Nedelec JY, Sorba J (1985) Can. J. Chem. 63: 678
1229. Barton DHR, Kalley F, Ozbalik N, Young E, Balavoine G (1989) J. Am. Chem. Soc. 111: 7144
1230. Holland HL (1999) Curr. Opinion Chem. Biol. 3: 22
1231. Mansuy D (1990) Pure Appl. Chem. 62: 741
1232. Kieslich K (1969) Synthesis 120
1233. Kolot FB (1983) Process Biochem. 19
1234. Marsheck WJ (1971) Progr. Ind. Microbiol. 10: 49
1235. Holland HL (1984) Acc. Chem. Res. 17: 389
1236. Holland HL (1982) Chem. Soc. Rev. 11: 371
1237. Sedlaczek L (1988) Crit. Rev. Biotechnol. 7: 186
1238. Weiler EW, Droste M, Eberle J, Halfmann HJ, Weber A (1987) Appl. Microbiol. Biotechnol. 27: 252
1239. Perlman D, Titius E, Fried J (1952) J. Am. Chem. Soc. 74: 2126
1240. Peterson DH, Murray HC, Eppstein SH, Reineke LM, Weintraub A, Meister PD, Leigh HM (1952) J. Am. Chem. Soc. 74: 5933
1241. Fried J, Thoma RW, Gerke JR, Herz JE, Donin MN, Perlman D (1952) J. Am. Chem. Soc. 74: 3692
1242. Sawada S, Kulprecha S, Nilubol N, Yoshida T, Kinoshita S, Taguchi H (1982) Appl. Environ. Microbiol. 44: 1249
1243. Gbewonyo K, Buckland BC, Lilly MD (1991) Biotechnol. Bioeng. 37: 1101
1244. Cohen N, Eichel WF, Lopersti RJ, Neukom C, Saucy G (1976) J. Org. Chem. 41: 3505
1245. Branca Q, Fischli A (1977) Helv. Chim. Acta 60: 925
1246. Evans DA, Sacks CE, Kleschick WA, Taber TR (1979) J. Am. Chem. Soc. 101: 6789
1247. Ohashi T, Hasegawa J (1992) New preparative methods for optically active β-hydroxycarboxylic acids. In: Collins, AN, Sheldrake GN, Crosby J (eds) Chirality in Industry. Wiley, New York, p 249
1248. Goodhue CT, Schaeffer JR (1971) Biotechnol. Bioeng. 13: 203
1249. Aberhart DJ (1977) Bioorg. Chem. 6: 191
1250. Jung H, Jung K, Kleber HP (1993) Adv. Biochem. Eng. Biotechnol. 50: 21
1251. Ciegler A (1974) Microbial transformations of terpenes. In: CRC Handbook of Microbiology, vol 4. CRC Press, Boca Raton, pp 449–458
1252. Krasnobajew V (1984) Terpenoids. In: Rehm HJ, Reed G (eds) Biotechnology, vol 6a. Verlag Chemie: Weinheim, p 97
1253. Lamare V, Furstoss R (1990) Tetrahedron 46: 4109
1254. Rosazza JPN, Steffens JJ, Sariaslani S, Goswami A, Beale JM, Reeg S, Chapman R (1987) Appl. Environ. Microbiol. 53: 2482

1255. Liu WG, Goswami A, Steffek RP, Chapman RL, Sariaslani FS, Steffens JJ, Rosazza JPN (1988) J. Org. Chem. 53: 5700
1256. Archelas A, Furstoss R, Waegell B, le Petit J, Deveze L (1984) Tetrahedron 40: 355
1257. Johnson RA, Herr ME, Murray HC, Reineke LM (1971) J. Am. Chem. Soc. 93: 4880
1258. Furstoss R, Archelas A, Waegell B, le Petit J, Deveze L (1981) Tetrahedron Lett. 22: 445
1259. Johnson RA, Herr ME, Murray HC, Fonken GS (1970) J. Org. Chem. 35: 622
1260. Archelas A, Fourneron JD, Furstoss R (1988) Tetrahedron Lett. 29: 6611
1261. Fonken GS, Herr ME, Murray HC, Reineke LM (1968) J. Org. Chem. 33: 3182
1262. Archelas A, Fourneron JD, Furstoss R (1988) J. Org. Chem. 53: 1797
1263. Johnson RA, Herr ME, Murray HC, Fonken GS (1968) J. Org. Chem. 33: 3217
1264. Furstoss R, Archelas A, Fourneron JD, Vigne B (1986) A model for the hydroxylation site of the fungus *Beauveria sulfurescens*. In: Schneider MP (ed) Enzymes as Catalysts in Organic Synthesis, NATO ASI Series C, vol 178. Reidel, Dordrecht, p 361
1265. Holland HL, Morris TA, Nava PJ, Zabic M (1999) Tetrahedron 55: 7441
1266. de Raadt A, Griengl H, Petsch M, Plachota P, Schoo N, Weber H, Braunegg G, Kopper I, Kreiner M, Zeiser A (1996) Tetrahedron Asymmetry 7: 491
1267. de Raadt A, Griengl H, Petsch M, Plachota P, Schoo N, Weber H, Braunegg G, Kopper I, Kreiner M, Zeiser A, Kieslich K (1996) Tetrahedron Asymmetry 7: 467
1268. de Raadt A, Griengl H, Petsch M, Plachota P, Schoo N, Weber H, Braunegg G, Kopper I, Kreiner M, Zeiser A (1996) Tetrahedron Asymmetry 7: 473
1269. Braunegg G, de Raadt A, Feichtenhofer S, Griengl H, Kopper I, Lehmann A, Weber H (1999) Angew. Chem. Int. Ed. 38: 2763
1270. Olah GA, Ernst TD (1989) J. Org. Chem. 54: 1204
1271. Komiyam AM (1989) J. Chem. Soc., Perkin Trans. 1, 2031
1272. Zimmer H, Lankin DC, Horgan SW (1971) Chem. Rev. 71: 229
1273. Powlowski JB, Dagley S, Massey V, Ballou DP (1987) J. Biol. Chem. 262: 69
1274. Wiseman A, King DJ (1982) Topics Enzymol. Ferment. Biotechnol. 6: 151
1275. Boyd DR, Campbell RM, Craig HC, Watson CG, Daly JW, Jerina DM (1976) J. Chem. Soc., Perkin Trans. 1, 2438
1276. Vigne B, Archelas A, Furstoss R (1991) Tetrahedron 47: 1447
1277. Yoshioka H, Nagasawa T, Hasegawa R, Yamada H (1990) Biotechnol. Lett. 679
1278. Theriault RJ, Longfield TH (1973) Appl. Microbiol. 25: 606
1279. Glöckler R, Roduit JP (1996) Chimia 50: 413
1280. Watson GK, Houghton C, Cain RB (1974) Biochem. J. 140: 265
1281. Hoeks FWJMM, Meyer HP, Quarroz D, Helwig M, Lehky P (1990) Scale-up of the process for the biotransformation of nicotinic acid into 6-hydroxynicotinic acid. In: Copping LG, Martin RE, Pickett JA, Bucke C, Bunch AW (eds) Opportunities in Biotransformations. Elsevier, London, p 67
1282. Pasutto FM, Singh NN, Jamali F, Coutts RT, Abuzar S (1987) J. Pharm. Sci. 76: 177
1283. Klibanov AM, Berman Z, Alberti BN (1981) J. Am. Chem. Soc. 103: 6263
1284. Doddema HJ (1988) Biotechnol. Bioeng. 32: 716
1285. Kazandjian RZ, Klibanov AM (1985) J. Am. Chem. Soc. 107: 5448
1286. Kolb HC, VanNieuwenhze MS, Sharpless KB (1994) Chem. Rev. 94: 2483
1287. Johnson RA, Sharpless KB (1993) In: Ojima I (ed) Catalytic Asymmetric Synthesis. Verlag Chemie, New York, pp 103–158
1288. Pfenninger A (1986) Synthesis 89
1289. Konishi K, Oda K, Nishida K, Aida T, Inoue S (1992) J. Am. Chem. Soc. 114: 1313
1290. Leak DJ, Aikens PJ, Seyed-Mahmoudian M (1992) Trends Biotechnol. 10: 256
1291. Weijers CAGM, de Haan A, de Bont JAM (1988) Microbiol. Sci. 5: 156
1292. May SW (1979) Enzyme Microb. Technol. 1: 15
1293. Furuhashi K (1986) Econ. Eng. Rev. 18 (7/8): 21
1294. Abraham WR, Stumpf B, Arfmann HA (1990) J. Essent. Oil Res. 2: 251

1295. Habets-Ctützen AQH, Carlier SJN, de Bont JAM, Wistuba D, Schurig V, Hartmans S, Tramper J (1985) Enzyme Microb. Technol. 7: 17
1296. de Smet MJ, Witholt B, Wynberg H (1981) J. Org. Chem. 46: 3128
1297. Furuhashi K (1986) Chem. Econ. Eng. Rev. 18 (7–8): 21
1298. Takahashi O, Umezawa J, Furuhashi K, Takagi M (1989) Tetrahedron Lett. 30: 1583
1299. Furuhashi K (1992) Biological methods to optically active epoxides. In: Collins AN, Sheldrake GN, Crosby J (eds) Chirality in Industry. Wiley, New York, p 167
1300. White RF, Birnbaum J, Meyer RT, ten Broeke J, Chemerda JM, Demain AL (1971) Appl. Microbiol. 22: 55
1301. Peterson JA, Basu D, Coon MJ (1966) J. Biol. Chem. 241: 5162
1302. de Smet MJ, Witholt B, Wynberg H (1983) Enzyme Microb. Technol. 5: 352
1303. Jurtshuk P, Cardini GE (1972) Crit. Rev. Microbiol. 1: 254
1304. Fu H, Newcomb M, Wong CH (1991) J. Am. Chem. Soc. 113: 5878
1305. Katopodis AG, Wimalasena K, Lee J, May SW (1984) J. Am. Chem. Soc. 106: 7928
1306. May SW, Abbott BJ (1973) J. Biol. Chem. 248: 1725
1307. May SW (1976) Catal. Org. Synth. 4: 101
1308. May SW, Schwartz RD, Abbott BJ, Zaborsky OR (1975) Biochim. Biophys. Acta 403: 245
1309. Habets-Crützen AQH, de Bont JAM (1985) Appl. Microbiol. Biotechnol. 22: 428
1310. Brink LES, Tramper J (1987) Enzyme Microb. Technol. 9: 612
1311. Brink LES, Tramper J (1985) Biotechnol. Bioeng. 27: 1258
1312. Hou CT, Patel R, Laskin AI, Barnabe N, Barist I (1983) Appl. Environ. Microbiol. 46: 171
1313. Furuhashi K (1981) Ferment. Ind. 39: 1029
1314. Ohta H, Tetsukawa H (1979) Agric. Biol. Chem. 43: 2099
1315. Johnstone SL, Phillips GT, Robertson BW, Watts PD, Bertola MA, Koger HS, Marx AF (1987) Stereoselective synthesis of (S)-β-blockers via microbially produced epoxide intermediates. In: Laane C, Tramper J, Lilly MD (eds) Biocatalysis in Organic Media. Elsevier, Amsterdam, p 387
1316. Fu H, Shen GJ, Wong CH (1991) Recl. Trav. Chim. Pays-Bas 110: 167
1317. Howe R, Shanks RG (1966) Nature 210: 1336
1318. Weijers CAGM, van Ginkel CG, de Bont JAM (1988) Enzyme Microb. Technol. 10: 214
1319. Schmid A, Hofstetter K, Freiten H-J, Hollmann F, Witholt B (2001) Adv. Synth. Catal. 343: 752
1320. Panke S, Held W, Wubbolts MG, Witholt B, Schmid A (2002) Biotechnol. Bioeng. 80: 33
1321. Bravo P, Resnati G, Viani F (1985) Tetrahedron Lett. 26: 2913
1322. Solladie G (1981) Synthesis 185
1323. Goldberg SI, Sahli MS (1967) J. Org. Chem. 32: 2059
1324. Solladie G, Demailly G, Greck C (1985) J. Org. Chem. 50: 1552
1325. Kagan HB, Dunach E, Nemeck C, Pitchen P, Samuel O, Zhao S (1985) Pure Appl. Chem. 57: 1922
1326. Holland HL (1988) Chem. Rev. 88: 473
1327. Phillips RS, May SW (1981) Enzyme Microb. Technol. 3: 9
1328. Dodson RM, Newman N, Tsuchiya HM (1962) J. Org. Chem. 27: 2707
1329. Auret BJ, Boyd DR, Henbest HB, Watson CG, Balenovic K, Polak V, Johanides V, Divjak S (1974) Phytochemistry 13: 65
1330. Ohta H, Okamoto Y, Tsuchihashi G (1985) Agric. Biol. Chem. 49: 2229
1331. Ohta H, Matsumoto S, Okamoto Y, Sugai T (1989) Chem. Lett., 625
1332. Holland HL, Brown FM, Lakshmaiah G, Larsen BG, Patel M (1997) Tetrahedron Asymmetry 8: 683
1333. Holland HL, Carter IM (1983) Bioorg. Chem. 12: 1
1334. Buist PH, Marecak DM, Partington ET, Skala P (1990) J. Org. Chem. 55: 5667
1335. Beecher J, Brackenridge I, Roberts SM, Tang J, Willetts AJ (1995) J. Chem. Soc., Perkin Trans. 1, 1641
1336. Ohta H, Okamoto Y, Tsuchihashi G (1985) Agric. Biol. Chem. 49: 671

1337. Abushanab E, Reed D, Suzuki F, Sih CJ (1978) Tetrahedron Lett. 19: 3415
1338. Holland HL, Pöpperl H, Ninniss RW, Chenchaiah PC (1985) Can. J. Chem. 63: 1118
1339. Poje M, Nota O, Balenovic K (1980) Tetrahedron 36: 1895
1340. Auret BJ, Boyd DR, Breen F, Greene RME, Robinson PM (1981) J. Chem. Soc., Perkin Trans. 1, 930
1341. Auret BJ, Boyd DR, Cassidy ES, Turley F, Drake AF, Mason SF (1983) J. Chem. Soc., Chem. Commun. 282
1342. Okamoto Y, Ohta H, Tsuchihashi G (1986) Chem. Lett. 2049
1343. Carrea G, Redigolo B, Riva S, Colonna S, Gaggero N, Battistel E, Bianchi D (1992) Tetrahedron: Asymmetry 3: 1063.
1344. Secundo F, Carrea G, Dallavalle S, Franzosi G (1993) Tetrahedron: Asymmetry 4: 1981.
1345. Colonna S, Gaggero N, Carrea G, Pasta P (1997) Chem. Commun. 439.
1346. de Gonzalo G, Pazmino DET, Ottolina G, Fraaije MW, Carrea G (2005) Tetrahedron: Asymmetry 16: 3077.
1347. Kamerbeek NM, Olsthoorn AJJ, Fraaije MW, Janssen DB (2003) Appl. Environ. Microbiol. 69: 419.
1348. Rios-Martinez A, de Gonzalo G, Torres Pazmino DE, Fraaije MW, Gotor V (2010) Eur. J. Org. Chem. 2409.
1349. Krow GR (1981) Tetrahedron 37: 2697
1350. Mimoun H (1982) Angew. Chem. Int. Ed. 21: 734
1351. Criegee R (1948) Liebigs Ann. Chem. 560: 127
1352. Lee JB, Uff BC (1967) Quart. Rev. 21: 429
1353. Schwab JM, Li WB, Thomas LP (1983) J. Am. Chem. Soc. 105: 4800
1354. Gunsalus IC, Peterson TC, Sligar SG (1975) Ann. Rev. Biochem. 44: 377
1355. Walsh CT, Chen YCJ (1988) Angew. Chem. Int. Ed. 27: 333
1356. Roberts SM, Wan PWH (1998) J. Mol. Catal. B 4: 111
1357. Bolm C, Schlingloff G, Weickhardt K (1994) Angew. Chem. Int. Ed. 33: 1848
1358. Abril O, Ryerson CC, Walsh C, Whitesides GM (1989) Bioorg. Chem. 17: 41
1359. Alphand V, Archelas A, Furstoss R (1990) J. Org. Chem. 55: 347
1360. Taschner MJ, Black DJ (1988) J. Am. Chem. Soc. 110: 6892
1361. Taschner MJ, Black DJ, Chen QZ (1993) Tetrahedron Asymmetry 4: 1387
1362. Kelly DR, Knowles CJ, Mahdi JG, Taylor IN, Wright MA (1995) J. Chem. Soc., Chem. Commun. 729
1363. Ottolina G, Carrea G, Colonna S, Rückemann A (1996) Tetrahedron Asymmetry 7: 1123
1364. Ouazzani-Chahdi J, Buisson D, Azerad R (1987) Tetrahedron Lett. 28: 1109
1365. Levitt MS, Newton RF, Roberts SM, Willetts AJ (1990) J. Chem. Soc., Chem. Commun. 619
1366. Alphand V, Archelas A, Furstoss R (1989) Tetrahedron Lett. 30: 3663
1367. Carnell AJ, Roberts SM, Sik V, Willetts AJ (1990) J. Chem. Soc., Chem. Commun. 1438
1368. Doig SD, Simpson H, Alphand V, Furstoss R, Woodley JM (2003) Enzyme Microb. Technol. 32: 347
1369. Alphand V, Carrea G, Wohlgemuth R, Furstoss R, Woodley JM (2003) Trends Biotechnol. 21: 318
1370. Stark D, von Stockar U (2003) Adv. Biochem. Eng. Biotechnol. 80: 149
1371. Vicenci JT, Zmijewski MJ, Reinhard MR, Landen BE, Muth WL, Marler PG (1997) Enzyme Microb. Technol. 29: 494
1372. Alphand V, Furstoss R (1992) J. Org. Chem. 57: 1306
1373. Petit F, Furstoss R (1993) Tetrahedron Asymmetry 4: 1341
1374. Grogan G, Roberts SM, Wan P, Willetts AJ (1993) Biotechnol. Lett. 15: 913
1375. Willetts A (1997) Trends Biotechnol. 15: 55
1376. Kelly DR, Knowles CJ, Mahdi JG, Wright MA, Taylor IN, Roberts SM, Wan PWH, Grogan G, Pedragosa-Moreau S, Willetts AJ (1996) Chem. Commun. 2333
1377. Nealson KH, Hastings JW (1979) Microbiol. Rev. 43: 496

1378. Donoghue NA, Norris DB, Trudgill PW (1976) Eur. J. Biochem. 63: 175
1379. Britton LN, Markavetz AJ (1977) J. Biol. Chem. 252: 8561
1380. Trower MK, Buckland RM, Griffin M (1989) Eur. J. Biochem. 181: 199
1381. Mihovilovic MD, Müller B, Stanetty P (2002) Eur. J. Org. Chem. 3711
1382. Kayser M, Chen G, Stewart J (1999) Synlett. 153
1383. Stewart JD, Reed KW, Kayser MM (1996) J. Chem. Soc., Perkin Trans. 1, 755
1384. Kayser MM, Chen G, Stewart JD (1998) J. Org. Chem. 63: 7103
1385. Stewart JD, Reed KW, Martinez CA, Zhu J, Chen G, Kayser MM (1998) J. Am. Chem. Soc. 120: 3541
1386. Stewart JD (1998) Curr. Org. Chem. 2: 211
1387. Grogan G, Roberts SM, Willetts AJ (1993) J. Chem. Soc., Chem. Commun. 699
1388. Torres Pazmino DE, Snajdrova R, Baas BJ, Ghobrial M, Mihovilovic MD, Fraaije MW (2008) Angew. Chem. Int. Ed. 47: 2275
1389. Willetts AJ, Knowles CJ, Levitt MS, Roberts SM, Sandey H, Shipston NF (1991) J. Chem. Soc., Perkin Trans. 1, 1608
1390. Kyte BC, Rouviere P, Cheng Q, Stewart JD (2004) J. Org. Chem. 69: 12
1391. Hamberg M (1996) Acta Chem. Scand. 50: 219
1392. Yagi K (ed) (1982) Lipid Peroxides in Biology and Medicine. Academic Press, New York
1393. Subramanian V, Sugumaran M, Vaidyanathan CS (1978) J. Ind. Inst. Sci. 60: 143
1394. Jeffrey H, Yeh HJC, Jerina DM, Patel TR, Davey JF, Gibson DT (1975) Biochemistry 14: 575
1395. Axelrod B (1974) ACS Adv. Chem. Ser. 136: 324
1396. Vick BA, Zimmerman DC (1984) Plant. Physiol. 75: 458
1397. Corey EJ, Nagata R (1987) J. Am. Chem. Soc. 109: 8107
1398. Theorell H, Holman RT, Akeson A (1947) Acta Chem. Scand. 1: 571
1399. Finnazzi-Agro A, Avigliano L, Veldink GA, Vliegenhart JFG, Boldingh J (1973) Biochim. Biophys. Acta 326: 462
1400. Axelrod B, Cheesbrough TM, Laakso TM (1981) Methods Enzymol. 71: 441
1401. Funk Jr MO, Andre JC, Otsuki T (1987) Biochemistry 26: 6880
1402. Van Os CPA, Vente M, Vliegenhart JFG (1979) Biochim. Biophys. Acta 547: 103
1403. Corey EJ, Albright JO, Burton AE, Hashimoto S (1980) J. Am. Chem. Soc. 102: 1435
1404. Iacazio G, Langrand G, Baratti J, Buono G, Triantaphylides C (1990) J. Org. Chem. 55: 1690
1405. Gunstone FD (1979) In: Barton DHR, Ollis WD, Haslam E (eds) Comprehensive Organic Chemistry. Pergamon Press, New York, p 587
1406. Datcheva VK, Kiss K, Solomon L, Kyler KS (1991) J. Am. Chem. Soc. 113: 270
1407. Corey EJ, Nagata R (1987) Tetrahedron Lett. 28: 5391
1408. Novak MJ (1999) Bioorg. Med. Chem. Lett. 9: 31
1409. Martini D, Buono G, Iacazio G (1996) J. Org. Chem. 61: 9062
1410. Zhang P, Kyler KS (1989) J. Am. Chem. Soc. 111: 9241
1411. Smith MR, Ratledge C (1989) Appl. Microbiol. Biotechnol. 32: 68
1412. Zylstra GJ, Gibson DT (1989) J. Biol. Chem. 264: 14940
1413. Gibson DT, Koch JR, Kallio RE (1968) Biochemistry 7: 2653.
1414. Gibson DT, Koch JR, Kallio RE (1968) Biochemistry 7: 3795
1415. Brazier AJ, Lilly MD, Herbert AB (1990) Enzyme Microb. Technol. 12: 90
1416. Parales RE, Resnick SM (2007) Application of aromatic hydrocarbon dioxygenases. In: Patel R N (ed) Biocatalysis in the Pharmaceutical and Biotechnological Industries. CRC Press, Boca Raton, p 299
1417. Kauppi B, Li K, Carredano E, Parales RE, Gibson DT, Eklund H, Ramaswamy S (1998) Structure 6: 571.
1418. Karlsson A, Parales JV, Parales RE, Gibson DT, Eklund H, Ramaswamy S (2003) Science 299: 1039.
1419. Gibson DT, Parales RE (2000) Curr. Opin. Biotechnol. 11: 236

1420. Butler CS, Mason JR (1997) Adv. Microb. Physiol. 38: 47
1421. Gibson DT, Parales RE (2000) Curr. Opin. Biotechnol. 11: 236.
1422. Ribbons DW, Evans CT, Rossiter JT, Taylor SCJ, Thomas SD, Widdowson DA, Williams
      DJ (1990) Biotechnology and biodegradation. In: Kamely D, Chakrabarty A, Omenn GS
      (eds) Advances in Applied Biotechnology Series, vol 4. Gulf Publ. Co., Houston, p 213
1423. Crosby J (1991) Tetrahedron 47: 4789
1424. Widdowson DA, Ribbons DW (1990) Janssen Chim. Acta 8 (3): 3
1425. Ballard DHG, Courtis A, Shirley IM, Taylor SC (1988) Macromolecules 21: 294
1426. Sheldrake GN (1992) Biologically derived arene cis-dihydrodiols as synthetic building
      blocks. In: Collins AN, Sheldrake GN, Crosby J (eds) Chirality in Industry. Wiley,
      New York, p 127
1427. Ribbons DW, Cass AEG, Rossiter JT, Taylor SJC, Woodland MP, Widdowson DA,
      Williams SR, Baker PB, Martin RE (1987) J. Fluorine Chem. 37: 299
1428. Taylor SC, Ribbons DW, Slawin AMZ, Widdowson DA, Williams DJ (1987) Tetrahedron
      Lett. 28: 6391
1429. Rossiter JT, Williams SR, Cass AEG, Ribbons DW (1987) Tetrahedron Lett. 28: 5173
1430. Boyd DR, Sharma ND, Hand MV, Groocock MR, Kerley NA, Dalton H, Chima J, Sheldrake
      GN (1993) J. Chem. Soc., Chem. Commun. 974
1431. Geary PJ, Pryce RJ, Roberts SM, Ryback G, Winders JA (1990) J. Chem. Soc., Chem.
      Commun. 204
1432. Deluca ME, Hudlicky T (1990) Tetrahedron Lett. 31: 13
1433. Wackett LP, Kwart LD, Gibson DT (1988) Biochemistry 27: 1360
1434. Boyd DR, Sharma ND, Boyle R, Malone JF, Chima J, Dalton H (1993) Tetrahedron
      Asymmetry 4: 1307
1435. Allen CCR, Boyd DR, Dalton H, Sharma ND, Brannigan I, Kerley NA, Sheldrake GN,
      Taylor SS (1995) J. Chem. Soc., Chem. Commun. 117
1436. Boyd DR, Sharma ND, Boyle R, McMurry BT, Evans TA, Malone JF, Dalton H, Chima J,
      Sheldrake GN (1993) J. Chem. Soc., Chem. Commun. 49
1437. Lakshman MK, Chaturvedi S, Zaijc B, Gibson DT, Resnick SM (1998) Synthesis 1352
1438. Hudlicky T, Boros EE, Boros CH (1993) Tetrahedron Asymmetry 4: 1365
1439. Boyd DR, Sharma ND, Bowers NI, Brannigan IN, Groocock MR, Malone JF,
      McConville G, Allen CCR (2005) Adv. Synth. Catal. 347: 1081
1440. Gally C, Nestl BM, Hauer B (2015) Angew. Chem. Int. Ed. 54: 12952.
1441. Ley SV, Sternfeld F (1989) Tetrahedron 45: 3463
1442. Ley SV, Parra M, Redgrave AJ, Sternfeld F (1990) Tetrahedron 46: 4995
1443. Hudlicky T, Luna H, Barbieri G, Kwart LD (1988) J. Am. Chem. Soc. 110: 4735
1444. Hudlicky T, Luna H, Price JD, Rulin F (1989) Tetrahedron Lett. 30: 4053
1445. Hudlicky T, Price JD (1990) Synlett. 159
1446. Hudlicky T, Luna H, Price JD, Rulin F (1990) J. Org. Chem. 55: 4683
1447. Mermod N, Harayamas S, Timmis KN (1986) Bio/Technology 4: 321
1448. Reddy J, Lee C, Neeper M, Greasham R, Zhang J (1999) Appl. Microbiol. Biotechnol. 51:
      614
1449. Boyd DR, Sharma ND, Malone JF, Allen CCR (2009) Chem. Commun. 3633.
1450. Strukul G (ed) (1992) Catalytic Oxidations with Hydrogen Peroxide as Oxidant. Kluwer,
      Dordrecht
1451. Butler A, Walker JV (1993) Chem. Rev. 93: 1937
1452. Kutney JP (1991) Synlett. 11
1453. Nakayama T, Amachi T (1999) J. Mol. Catal. B 6: 185
1454. Flohe L (1979) CIBA Found. Symp. 65: 95
1455. de Boer E, Y van Kooyk, MGM Tromp, Plat H, Wever R (1986) Biochim. Biophys. Acta
      869: 48
1456. Butler A (1998) Curr. Opinion Chem. Biol. 2: 279
1457. Kuwahara M, Glenn JK, Morgan MA, Gold MH (1984) FEBS Lett. 169: 247

1458. Dolin MI (1957) J. Biol. Chem. 225: 557
1459. Dunford HB (1991) Horseradish peroxidase: structure and kinetic properties. In: Everse J, Everse KE, Grisham MB (eds) Peroxidases in Chemistry and Biology. CRC Press, Boca Raton, p 1
1460. Anni H, Yonetani T (1992) Mechanism of action of peroxidases. In: Sigel H, Sigel A (eds) Metal Ions in Biological Systems: Degradation of Environmental Pollutants by Microorganisms and their Related Metalloenzymes. Marcel Dekker, New York, p 219
1461. Ortiz de Montellano PR (1992) Annu. Rev. Pharm. Toxicol. 32: 89
1462. Berglund GI, Carlsson GH, Smith A T, Szöke H, Hensiksen A, Hajdu J (2002) Nature 417:463
1463. Colonna S, Gaggero N, Richelmi C, Pasta P (1999) Trends Biotechnol. 17: 163
1464. Adam W, Lazarus M, Saha-Möller CR, Weichold O, Hoch U, Häring D, Schreier P (1999) Adv. Biochem. Eng. Biotechnol. 63: 73
1465. van Deurzen MPJ, van Rantwijk F, Sheldon RA (1997) Tetrahedron 53: 13183
1466. Dordick JS (1992) Trends Biotechnol. 10: 287
1467. Uyama H, Kurioka H, Sugihara J, Kobayashi S (1996) Bull. Chem. Soc. Jpn. 69: 189
1468. Kobayashi S, Shoda S, Uyama H (1995) Adv. Polym. Sci. 121: 1
1469. Schmitt MM, Schüler E, Braun M, Häring D, Schreier P (1998) Tetrahedron Lett. 39: 2945
1470. Fukunishi K, Kitada K, Naito I (1991) Synthesis 237
1471. Littlechild J (1999) Curr. Opinion Chem. Biol. 3: 28
1472. Franssen MCR (1994) Biocatalysis 10: 87
1473. Shaw PD, Hager LP (1961) J. Biol. Chem. 236: 1626
1474. Blanke SR, Hager LP (1989) Biotechnol. Lett. 11: 769
1475. Sundaramoothy M, Terner J, Poulos TL (1995) Structure 3: 1367
1476. McCarthy MB, White RE (1983) J. Biol. Chem. 258: 9153
1477. Miller VP, Tschirret-Guth RA, Ortiz de Montellano PG(1995) Arch. Biochem. Biophys. 319: 333
1478. Zaks A, Dodds DR (1995) J. Am. Chem. Soc. 117: 10419
1479. Hu S, Hager LP (1999) J. Am. Chem. Soc. 121: 872
1480. Seelbach K, van Deurzen MPJ, van Rantwijk F, Sheldon RA, Kragl U (1997) Biotechnol. Bioeng. 55: 283
1481. van Deurzen MPJ, van Rantwijk F, Sheldon RA (1996) J. Mol. Catal. B 2: 33
1482. Hager LP, Lakner FJ, Basavapathruni A (1998) J. Mol. Catal. B 5: 95
1483. Hu S, Hager LP (1999) Tetrahedron Lett. 40: 1641
1484. Lakner FJ, Cain KP, Hager LP (1997) J. Am. Chem. Soc. 119: 443
1485. Allain EJ, Hager LP, Deng L, Jacobsen EN (1993) J. Am. Chem. Soc. 115: 4415
1486. Dexter AF, Lakner FJ, Campbell RA, Hager LP (1995) J. Am. Chem. Soc. 117: 6412
1487. Lakner FJ, Hager LP (1996) J. Org. Chem. 61: 3923
1488. Colonna S, Gaggero N, Casella L, Carrea G, Pasta P (1993) Tetrahedron Asymmetry 4: 1325
1489. Kobayashi S, Nakano M, Kimura T, Schaap AP (1987) Biochemistry 26: 5019
1490. Colonna S, Gaggero N, Casella L, Carrea G, Pasta P (1992) Tetrahedron Asymmetry 3: 95
1491. Colonna S, Gaggero N, Manfredi A, Casella L, Gullotti M, Carrea G, Pasta P (1990) Biochemistry 29: 10465
1492. Andersson M, Willetts A, Allenmark S (1997) J. Org. Chem. 62: 8455
1493. Allenmark SG, Andersson MA (1996) Tetrahedron Asymmetry 7: 1089
1494. Andersson MA, Allenmark SG (1998) Tetrahedron 54: 15293
1495. van Deurzen MPJ, Seelbach K, van Rantwijk F, Kragl U, Sheldon RA (1997) Biocatalysis 15: 1

References to Sect. 2.4

1496. Evans DA, Nelson JV, Taber TR (1982) Topics Stereochem. 13: 1
1497. Heathcock CH (1984) Asymm. Synthesis 3: 111
1498. Mukaiyama T (1982) Org. React. 28: 203
1499. Paterson I, Goodman JM, Lister MA, Schumann RC, McClure CK, Norcross RD (1990) Tetrahedron 46: 4663
1500. Masamune S, Choy W, Peterson J, Sita LR (1986) Angew. Chem. Int. Ed. 24: 1
1501. Kazmeier U (2005) Angew. Chem. Int. Ed. 44: 2186
1502. Fessner W-D (1992) Kontakte (Merck) (3) 3
1503. Fessner W-D (1993) Kontakte (Merck) (1) 23
1504. Toone EJ, Simon ES, Bednarski MD, Whitesides, GM (1989) Tetrahedron 45: 5365
1505. Gijsen HJM, Qiao L, Fitz W, Wong CH (1996) Chem. Rev. 96: 443
1506. Wong CH (1993) Chimia 47: 127
1507. Drueckhammer DG, Hennen WJ, Pederson RL, Barbas CF, Gautheron CM, Krach T, Wong CH (1991) Synthesis 499
1508. Look GC, Fotsch CH, Wong CH (1993) Acc. Chem. Res. 26: 182
1509. Wong CH (1990) Aldolases in organic synthesis. In: Abramowicz DA (ed) Biocatalysis. Van Nostrand Reinhold, New York, p 319
1510. Dean SM, Greenberg WA, Wong CH (2007) Adv. Synth. Catal. 349: 1308
1511. Fessner WD, Helaine V (2001) Curr. Opinion Biotechnol. 12: 574
1512. Seoane G (2000) Curr. Org. Chem. 4: 283
1513. Machajewski TD, Wong CH (2000) Angew. Chem. Int. Ed. 39: 1352
1514. Fessner WD, Jennewein S (2007) Biotechnological applications of aldolases. In: Patel R N (ed) Biocatalysis in the Pharmaceutical and Biotechnology Industries. CRC Press, Boca Raton, p 363
1515. Clapes P, Fessner WD, Sprenger GA, Samland AK (2010) Curr. Opin. Chem. Biol. 14: 154
1516. Clapes P, Garrabou X (2011) Adv. Synth. Catal. 353: 2263.
1517. Mueller M (2012) Adv. Synth. Catal. 354: 3161.
1518. Windle CL, Muller M, Nelson A, Berry A (2014) Curr. Opin. Chem. Biol. 19: 25.
1519. Busto E (2016) ChemCatChem 8: 2589.
1520. Meyerhof O, Lohmann K (1934) Biochem. Z. 271: 89
1521. Horecker L, Tsolas O, Lai CY (1972) In: Boyer PD (ed) The Enzymes, vol 7. Academic Press, New York, p 213
1522. Brockamp HP, Kula MR (1990) Appl. Microbiol. Biotechnol. 34: 287
1523. von der Osten CH, Sinskey AJ, Barbas III CF, Pederson RL, Wang YF, Wong C-H (1989) J. Am. Chem. Soc. 111: 3924
1524. Fessner WD, Schneider A, Held H, Sinerius G, Walter C, Hixon M, Schloss JV (1996) Angew. Chem., Int. Ed. Engl. 35: 2219
1525. Dreyer MK, Schulz GE (1993) J. Mol. Biol. 231: 549
1526. Shelton MC, Cotterill IC, Novak STA, Poonawala RM, Sudarshan S, Toone EJ (1996) J. Am. Chem. Soc. 118: 2117
1527. Schoevaart R, van Rantwijk F, Sheldon RA (1999) Tetrahedron Asymmetry 10: 705
1528. Jones JKN, Sephton HH (1960) Can. J. Chem. 38: 753
1529. Jones JKN, Kelly RB (1956) Can. J. Chem. 34: 95
1530. Horecker BL, Smyrniotis PZ (1952) J. Am. Chem. Soc. 74: 2123
1531. Huang PC, Miller ON (1958) J. Biol. Chem. 330: 805
1532. Kajimoto T, Chen L, Liu KKC, Wong C-H (1991) J. Am. Chem. Soc. 113: 6678
1533. Effenberger F, Straub A, Null V (1992) Liebigs Ann Chem. 1297
1534. Wong C-H, Mazenod FP, Whitesides GM (1983) J. Org. Chem. 3493
1535. Bednarski MD, Waldmann HJ, Whitesides GM (1986) Tetrahedron Lett. 27: 5807
1536. Jones JKN, Matheson NK (1959) Can. J. Chem. 37: 1754
1537. Gorin PAJ, Hough L, Jones JKN (1953) J. Chem. Soc. 2140

1538. Lehninger AL, Sice J (1955) J. Am. Chem. Soc. 77: 5343
1539. Charalampous FC (1954) J. Biol. Chem. 211: 249
1540. Duncan R, Drueckhammer DG (1996) J. Org. Chem. 61: 438
1541. Fessner WD, Sinerius G (1994) Angew. Chem. Int. Ed. 33: 209
1542. Simon ES, Plante R, Whitesides GM (1989) Appl. Biochem. Biotechnol. 22: 169
1543. Bischofberger N, Waldmann H, Saito T, Simon ES, Lees W, Bednarski MD, Whitesides GM (1988) J. Org. Chem. 53: 3457
1544. Arth HL, Fessner WD (1997) Carbohydr. Res. 305: 131
1545. Liu KKC, Kajimoto T, Chen L, Zhong Z, Ichikawa Y, Wong CH (1991) J. Org. Chem. 56: 6280
1546. Ziegler T, Straub A, Effenberger F (1988) Angew. Chem. Int. Ed. 27: 716
1547. Schultz M, Waldmann H, Kunz H, Vogt W (1990) Liebigs Ann. Chem. 1010
1548. Lees WJ, Whitesides GM (1993) J. Org. Chem. 58: 1887
1549. Durrwachter JR, Drueckhammer DG, Nozaki K, Sweers HM, Wong CH (1986) J. Am. Chem. Soc. 108: 7812
1550. Fessner WD (1992) A building block strategy for asymmetric synthesis: the DHAP-aldolases. In: Servi S (ed) Microbial Reagents in Organic Synthesis. Kluwer, Dordrecht, p 43
1551. Ozaki A, Toone EJ, von der Osten CH, Sinskey AJ, Whitesides GM (1990) J. Am. Chem. Soc. 112: 4970
1552. Pederson RL, Esker J, Wong CH (1991) Tetrahedron 47: 2643
1553. Effenberger F, Straub A (1987) Tetrahedron Lett. 28: 1641
1554. Gefflaut T, Lemaire M, Valentin ML, Bolte J (1997) J. Org. Chem. 62: 5920
1555. Fessner WD, Walter C (1992) Angew. Chem. Int. Ed. 31: 614
1556. Suhiyama M, Hong Z, Whalen LJ, Greenberg WA, Wong CH (2006) Adv. Synth. Catal. 348: 2555
1557. Garrabou X, Calveras J, Joglar J, Parella T, Bujons J, Clapes P (2011) Org. Biomol. Chem. 9: 8430.
1558. Schürmann M, Sprenger GA (2001) J. Biol. Chem. 276: 11055
1559. Schürmann M, Schürmann M, Sprenger GA (2002) J. Mol. Catal. B: Enzym. 19–20: 247
1560. Sugiyama M, Hong Z, Liang PH, Dean SM, Whalen LJ, Greenberg WA, Wong CH (2007) J. Am. Chem. Soc. 129: 14811
1561. Castillo JA, Guerard-Helaine C, Gutierrez M, Garrabou X, Sancelme M, Schürmann M, Inoue T, Helaine V, Charmantray F, Gefflaut T, Hecquet L, Joglar J, Clapes P, Sprenger GA, Lemaire M (2010) Adv. Synth. Catal. 352: 1039
1562. Rale M, Schneider S, Sprenger GA, Samland AK, Fessner W-D (2011) Chem. Eur. J. 17: 2623.
1563. Schneider S, Gutierrez M, Sandalova T, Schneider G, Clapes P, Sprenger GA, Samland AK (2010) ChemBioChem 11: 681.
1564. Castillo JA, Calveras J, Casas J, Mitjans M, Vinardell MP, Parella T, Inoue T, Sprenger GA, Joglar J, Clapes P (2006) Org. Lett. 8: 6067
1565. Simon ES, Bednarski MD, Whitesides GM (1988) J. Am. Chem. Soc. 110: 7159
1566. Uchida Y, Tsukada Y, Sugimori T (1985) Agric. Biol. Chem. 49: 181
1567. Kragl U, Kittelmann M, Ghisalba O, Wandrey C (1995) Ann. NY Acad. Sci. 750: 300
1568. Kragl U, Gygax D, Ghisalba O, Wandrey C (1991) Angew. Chem. Int. Ed. 30: 827
1569. Maru I, Ohnishi J. Ohta H, Tsukuda Y (1998) Carbohydr. Res. 306: 575
1570. Mahmoudian M, Noble D, Drake CS, Middleton RF, Montgomery DS, Piercey JE, Ramlakhan D, Todd M, Dawson MJ (1997) Enzyme Microb. Technol. 20: 393
1571. Schauer R (1985) Trends Biochem. Sci. 10: 357
1572. Ota Y, Shimosaka M, Murata K, Tsudaka Y, Kimura A (1986) Appl. Microbiol. Biotechnol. 24: 386
1573. Brunetti P, Jourdian GW, Roseman S (1962) J. Biol. Chem. 237: 2447

1574. Brossmer R, Rose U, Kasper D, Smith TL, Grasmuk H, Unger FM (1980) Biochem. Biophys. Res. Commun. 96: 1282
1575. Augé C, David S, Gautheron C, Malleron A, Cavayre B (1988) New J. Chem. 12: 733
1576. Kim MJ, Hennen WJ, Sweers HM, Wong CH (1988) J. Am. Chem. Soc. 110: 6481
1577. Lin CH, Sugai T, Halcomb RL, Ichikawa Y, Wong CH (1992) J. Am. Chem. Soc. 114: 10138
1578. Augé C, Gautheron C, David S, Malleron A, Cavayé B, Bouxom B (1990) Tetrahedron 46: 201
1579. Bednarski MD, Crans DC, DiCosmio R, Simon ES, Stein PD, Whitesides GM, Schneider M (1988) Tetrahedron Lett. 29: 427
1580. Sugai T, Shen GJ, Ichikawa Y, Wong CH (1993) J. Am. Chem. Soc. 115: 413
1581. Gillingham DG, Stallforth P, Adibekian A, Seeberger PH, Hilvert D (2010) Nat. Chem. 2: 102.
1582. Henderson DP, Cotterill IC, Shelton MC, Toone EJ (1998) J. Org. Chem. 63: 906
1583. Barbas III CF, Wang YF, Wong CH (1990) J. Am. Chem. Soc. 112: 2013
1584. Wong CH, Garcia-Junceda E, Chen L, Blanco O, Gijsen HJM, Stennsma DH (1995) J. Am. Chem. Soc. 117: 3333
1585. Greenberg WA, Varvak A, Hanson SR, Wong K, Huang H, Chen P, Burk MJ (2004) Proc. Nat. Acad. Sci. USA 101: 5788
1586. DeSantis G, Liu J, Clark DP, Heine A, Wilson JA, Wong CH (2003) Bioorg. Med. Chem. 11: 43
1587. Müller M (2005) Angew. Chem. Int. Ed. 44: 362
1588. Gijsen HJM, Wong CH (1995) J. Am. Chem. Soc. 117: 2947
1589. Gijsen HJM, Wong CH (1995) J. Am. Chem. Soc. 117: 7585
1590. Lotz BT, Gasparski CM, Peterson K, Miller MJ (1990) J. Chem. Soc., Chem. Commun. 1107
1591. Steinreiber J, Fesko K, Reisinger C, Schürmann M, van Assema F, Wolberg M, Mink D, Griengl H (2007) Tetrahedron 63: 918
1592. Fesko K, Reisinger C, Steinreiber J, Weber H, Schürmann M, Griengl H (2008) J. Mol. Catal. B Enzym. 52-53: 19.
1593. Liu LQ, Dairi T, Itoh N, Kataoka M, Shimizu S, Yamada H (2000) J. Mol. Catal. B Enzym. 10: 107.
1594. Baik S-H, Yoshioka H (2009) Biotechnol. Lett. 31: 443.
1595. Liu JQ, Dairi T, Itoh N, Kataoka M, Shimizu S, Yamada H (2000) J. Mol. Catal. B 10: 107
1596. Saeed A, Young DW (1992) Tetrahedron 48: 2507
1597. Vassilev VP, Uchiyama T, Kajimoto T, Wong CH (1995) Tetrahedron Lett. 36: 4081
1598. Nozaki H, Kuroda S, Watanabe K, Yokozeki K (2008) Appl. Environ. Microbiol. 74: 7596.
1599. Fesko K, Uhl M, Steinreiber J, Gruber K, Griengl H (2010) Angew. Chem. Int. Ed. 49: 121.
1600. Steinreiber J, Schürmann M, van Assema F, Wolberg M, Fesko K, Reisinger C, Mink D, Griengl H (2007) Adv. Synth. Catal. 349: 1379
1601. Lohmann W, Schuster G (1937) Biochem. Z. 294: 188
1602. Müller M, Gocke D, Pohl M (2009) FEBS J. 276: 2894
1603. Pohl M, Lingen B, Müller M (2002) Chem. Eur. J. 8: 5289
1604. Pohl M, Sprenger GA, Müller M (2004) Curr. Opinion Biotechnol. 15: 335
1605. Frank RAW, Leeper FJ (2007) Cell. Mol. Life Sci. 64: 892
1606. Breslow R (1957) J. Am. Chem. Soc. 79: 1762
1607. Kern D, Kern G, Neef H, Tittmann K, Killenberg-Jabs M, Wikner C, Schneider G, Hübner G (1997) Science 275: 67
1608. Kluger R, Tittmann K (2008) Chem. Rev. 108: 1797.
1609. Lintner CJ, Liebig HJ (1913) Z. physiol. Chem. 88: 109
1610. Neuberg C, Hirsch J (1921) Biochem. Z. 115: 282
1611. Goetz G, Ivan P, Hauer B, Breuer M, Pohl M (2001) Biotechnol. Bioeng. 74: 317
1612. Pohl M (1997) Adv. Biochem. Eng. Biotechnol. 58: 16

1613. Fuganti C, Grasselli P (1989) Baker's yeast-mediated synthesis on natural products. In: Whitaker JR, Sonnet PE (eds) Biocatalysis in Agricultural Biotechnology. ACS Symp. Ser., ACS, Washington DC, p 359
1614. Fuganti C, Grasselli P (1977) Chem. Ind. 983
1615. Crout DHG, Dalton H, Hutchinson DW, Miyagoshi M (1991) J. Chem. Soc., Perkin Trans. 1, 1329
1616. Pohl M (1999) Protein design on pyruvate decarboxylase (PDC) by site-directed mutagenesis. In: Scheper T (ed) New Enzymes for Organic Synthesis. Springer, Berlin Heidelberg New York, p 15
1617. Sprenger GA, Pohl M (1999) J. Mol. Catal. B 6: 145
1618. Fuganti C, Grasselli P, Poli G, Servi S, Zorzella A (1988) J. Chem. Soc., Chem. Commun. 1619
1619. Fuganti C, Grasselli P, Servi S, Spreafico F, Zirotti C (1984) J. Org. Chem. 49: 4087
1620. Suomalainen H, Linnahalme T (1966) Arch. Biochem. Biophys. 114: 502
1621. Neuberg C, Liebermann L (1921) Biochem. Z. 121: 311
1622. Behrens M, Ivanoff N (1926) Biochem. Z. 169: 478
1623. Fronza G, Fuganti C, Majori L, Pedrocchi-Fantoni G, Spreafico F (1982) J. Org. Chem. 47: 3289
1624. Fuganti C, Grasselli P, Spreafico F, Zirotti C (1984) J. Org. Chem. 49: 543
1625. Pohl M, Lingen B, Müller M (2002) Chem. Eur. J. 8: 5288
1626. Kren V, Crout DHG, Dalton H, Hutchinson DW, König W, Turner MM, Dean G, Thomson N (1993) J. Chem. Soc., Chem. Commun. 341
1627. Bornemann S, Crout DHG, Dalton H, Hutchinson DW, Dean G, Thomson N, Turner MM (1993) J. Chem. Soc., Perkin Trans. 1, 309
1628. Dünkelmann P, Kolter-Jung D, Nitsche A, Demir AS, Siegert P, Lingen B, Baumann M, Pohl M, Müller M (2002) J. Am. Chem. Soc. 124: 12084
1629. Pohl M, Gocke D, Müller M (2009) Thiamine-based enzymes for biotransformations. In: Crabtree RH (ed) Handbook of Green Chemistry, vol. 3: Biocatalysis. Wiley-VCH, Weinheim, p 75
1630. Müller M, Sprenger GA, Pohl M (2013) Curr. Opin. Chem. Biol. 17: 261.
1631. Hailes HC, Rother D, Müller M, Westphal R, Ward JM, Pleiss J, Vogel C, Pohl M (2013) FEBS J. 280: 6374.
1632. Iding H, Siegert P, Mesch K, Pohl M (1998) Biochim. Biophys. Acta 1385: 307
1633. Baykal A, Chakraborty S, Dodoo A, Jordan F (2006) Bioorg. Chem. 34: 380
1634. Bringer-Meyer S, Sahm H (1988) Biocatalysis 1: 321
1635. Neuser F, Zorn H, Berger RG (2000) Z. Naturforsch. 55: 560
1636. Neuser F, Zorn H, Berger RG (2000) J. Agric. Food Chem. 48: 6191
1637. Kurniadi T, Bel-Rhlid R, Fay LB, Juillerat MA, Berger RG (2003) J. Agric. Food Chem. 51: 3103
1638. Andrews FH, McLeish MJ (2012) Bioorg. Chem. 43: 26.
1639. Siegert P, McLeish MJ, Baumann M, Iding H, Kneen MM, Kenyon GL, Pohl M (2005) Prot. Eng. Des. Sel. 18: 345
1640. Iding H, Dünnwald T, Greiner L, Liese A, Müller M, Siegert P, Grötzinger J, Demir AS, Pohl M (2000) Chem. Eur. J. 6: 1483
1641. Demir AS, Dünnwald T, Iding H, Pohl M, Müller M (1999) Tetrahedron Asymmetry 10: 4769
1642. Demir AS, Ayhan P, Sopaci SB (2007) Clean 35: 406
1643. Dünkelmann P, Pohl M, Müller M (2004) Chimica Oggi/Chem. Today 22: 24
1644. Demir AS, Ayan P, Igdir A C, Guygu AN (2004) Tetrahedron 60: 6509
1645. Demir AS, Pohl M, Janzen E, Müller M (2001) J. Chem. Soc. Perkin Trans. 1, 633
1646. Demir AS, Sesenoglu O, Eren E, Hosrik B, Pohl M, Janzen E, Kolter D, Feldmann R, Dünkelmann P, Müller M (2002) Adv. Synth. Catal. 344: 96
1647. Cosp A, Dresen C, Pohl M, Walter L, Röhr C, Müller M (2008) Adv. Synth. Catal. 350: 759

1648. Beigi M, Waltzer S, Fries A, Eggeling L, Sprenger GA, Müller M (2013) Org. Lett. 15: 452.
1649. Kurutsch A, Richter M, Brecht V, Sprenger GA, Müller M (2009) J. Mol. Catal. B Enzym. 61: 56.
1650. Thunberg L, Backstrom G, Lindahl U (1982) Carbohydr. Res. 100: 393
1651. Morris KG, Smith MEB, Turner NJ, Lilly MD, Mitra RK, Woodley JM (1996) Tetrahedron Asymmetry 7: 2185
1652. Demuynck C, Bolte J, Hecquet L, Dalmas V (1991) Tetrahedron Lett. 32: 5085
1653. Humphrey AJ, Turner NJ, McCague R, Taylor SCJ (1995) J. Chem. Soc., Chem. Commun. 2475
1654. Bolte J, Demuynck C, Samaki H (1987) Tetrahedron Lett. 28: 5525
1655. Kobori Y, Myles DC, Whitesides GM (1992) J. Org. Chem. 57: 5899
1656. Mocali A, Aldinucci D, Paoletti F (1985) Carbohydr. Res. 143: 288
1657. Ranoux A, Hanefeld U (2013) Top. Catal. 56: 750.
1658. Datta A, Racker E (1961) J. Biol. Chem. 236: 617
1659. Villafranca J, Axelrod B (1971) J. Biol. Chem. 246: 3126
1660. Hobbs GR, Lilly MD, Turner NJ, Ward JM, Willetts AJ, Woodley JM (1993) J. Chem. Soc., Perkin Trans. 1, 165
1661. Hobbs GR, Mitra RK, Chauhan RP, Woodley JM, Lilly MD (1996) J. Biotechnol. 45: 173
1662. Turner NJ (2000) Curr. Opinion Biotechnol. 11: 527
1663. Effenberger F, Null V, Ziegler T (1992) Tetrahedron Lett. 33: 5157

References to Sect. 2.5

1664. Gocke D, Nguyen C L, Pohl M, Stillger T, Walter L, Müller M (2007) Adv. Synth. Catal. 349: 1425
1665. Sheldon RA (2000) Pure Appl. Chem. 72: 1233
1666. Sheldon RA (1997) J. Chem. Technol. Biotechnol. 68: 381
1667. Resch V, Hanefeld U (2015) Catal. Sci. Technol. 5: 1385.
1668. Hiseni A, Arends IWCE, Otten LG (2015) ChemCatChem 7: 29
1669. Mohrig JR (2013) Acc. Chem. Res. 46: 1407.
1670. Holden HM, Benning MM, Haller T, Gerlt JA (2001) Acc. Chem. Res. 34: 145.
1671. Kim B-N, Joo Y-J, Kim Y-S, Kim K-R, Oh D-K (2012) Appl. Microbiol. Biotechnol. 95: 929.
1672. Bicas JL, Fontanille P, Pastore GM, Larroche C (2010) Proc. Biochem. 45: 481.
1673. Hill RL, Teipel JW (1971) Fumarase and Crotonase. In: Boyer P (ed) The Enzymes, vol 5, Academic Press, New York, p 539
1674. Botting NP, Akhtar M, Cohen MA, Gani D (1987) J. Chem. Soc., Chem. Commun. 1371
1675. Nuiry II, Hermes JD, Weiss PM, Chen C, Cook PF (1984) Biochemistry 23: 5168
1676. Mattey M (1992) Crit. Rev. Biotechnol. 12: 87
1677. Tosa T, Shibatani T (1995) Ann. NY Acad. Sci. 750: 364
1678. Findeis MA, Whitesides GM (1987) J. Org. Chem. 52: 2838
1679. Michielsen MJF, Frielink C, Wijffels RH, Tramper J, Beeftink HH (2000) J. Biotechnol. 79: 13
1680. van der Werf M, van den Tweel W, Hermans S (1992) Appl. Environ. Microbiol. 58: 2854
1681. Subramanian SS, Rao MRR (1968) J. Biol. Chem. 243: 2367
1682. Kieslich K (1991) Acta Biotechnol. 11: 559
1683. Hasegawa J, Ogura M, Kanema H, Noda N, Kawaharada H, Watanabe K (1982) J. Ferment. Technol. 60: 501
1684. Obon JM, Maiquez JR, Canovas M, Kleber HP, Iborra JL (1997) Enzyme Microb. Technol. 21: 531

1685. Meyer HP (1993) Chimia 47: 123
1686. Hoeks FWJMM, Muehle J, Boehlen L, Psenicka I (1996) Chem. Eng. J. 61: 53
1687. Abraham WR, Arfmann HA (1989) Appl. Microbiol. Biotechnol. 32: 295
1688. Holland HL, Gu JX (1998) Biotechnol. Lett. 20: 1125
1689. Poppe L, Paizs C, Kovacs K, Irimie F-D, Vertessy B (2012) Meth. Mol. Biol. 794: 3.
1690. Hanson KR, Havir EA (1972) The enzymic elimination of ammonia. In: Boyer P (ed) The
      Enzymes, vol. 7. Academic Press, New York, p 75
1691. Poppe L, Retey J (2005) Angew. Chem. Int. Ed. 44: 3668.
1692. Chibata I, Tosa T, Sato T (1976) Methods Enzymol. 44: 739
1693. Kumagai H (2000) Adv. Chem. Eng. Biotechnol. 69: 71
1694. Terasawa M, Yukawa H, Takayama Y (1985) Proc. Biochem. 20: 124
1695. Yamagata H, Terasawa M, Yukawa H (1994) Catal. Today 22: 621
1696. Shi W, Dunbar J, Jayasekera MMK, Viola RE, Farber GK (1997) Biochemistry 36: 9136
1697. Viola RE (2000) Adv. Enzymol. Relat. Areas Mol. Biol. 74: 295
1698. Fujii T, Sakai H, Kawata Y, Hata Y (2003) J. Mol. Biol. 328: 635
1699. Weiner B, Poelarends GJ, Janssen DB, Feringa BL (2008) Chem. Eur. J. 14: 10094
1700. Gulzar MS, Akhtar M, Gani D (1997) J. Chem. Soc., Perkin Trans. 1, 649
1701. Emery TF (1963) Biochemistry 2: 1041
1702. Akhtar M, Botting NP, Cohen MA, Gani D (1987) Tetrahedron 43: 5899
1703. Akhtar M, Cohen MA, Gani D (1986) J. Chem. Soc., Chem. Commun. 1290
1704. Breuer M, Hauer B (2003) Curr. Opinion Biotechnol. 14: 570
1705. Paizs C, Tosa MI, Bencze LC, Brem J, Irimie FD, Retey J (2011) Heterocycles 82: 1217.
1706. Steele CL, Chen Y, Dougherty BA, Li W, Hofstead S, Lam KS, Xing Z, Chiang SJ (2005)
      Arch. Biochem. Biophys. 438: 1
1707. Christianson CV, Montavon TJ, Festin GM, Cooke HA, Shen B, Bruner SD (2007) J. Am.
      Chem. Soc. 129: 15744
1708. Szymanski W, Wu B, Weiner B, de Wildeman S, Feringa BL, Janssen DB (2009) J. Org.
      Chem. 74: 9152
1709. Wu B, Szymanski W, Wietzes P, de Wildeman S, Poelarends GJ, Feringa BL, Janssen DB
      (2009) ChemBioChem 10: 338
1710. Smitskamp-Wilms E, Brussee J, van der Gen A, van Scharrenburg GJM, Sloothaak JB
      (1991) Rec. Trav. Chim. Pays-Bas 110: 209
1711. Johnson DV, Griengl H (1999) Adv. Biochem. Eng. Biotechnol. 63: 31
1712. Effenberger FX (1992) (R)- and (S)-cyanohydrins – their enzymatic synthesis and their
      reactions. In: Servi S (ed) Microbial Reagents in Organic Synthesis. Kluwer, Dordrecht,
      p 25
1713. Kruse CG (1992) Chiral cyanohydrins – their manufacture and utility as chiral building
      blocks. In: Collins AN, Sheldrake GN, Crosby J (eds) Chirality in Industry. Wiley,
      New York, p 279
1714. van Scharrenburg GJM, Sloothaak JB, Kruse CG, Smitskamp-Wilms E (1993) Ind. J. Chem.
      32B: 16
1715. Bracco P, Busch H, von Langermann J, Hanefeld U (2016) Org. Biomol. Chem. 14: 6375.
1716. Rosenthaler I (1908) Biochem. Z. 14: 238
1717. Pichersky E, Lewinsohn E (2011) Ann. Rev. Plant. Biol. 62: 549.
1718. Effenberger F (1994) Angew. Chem., Int. Ed. 33: 1555
1719. Matsuo T, Nishioka T, Hirano M, Suzuki Y, Tsushima K, Itaya N, Yoshioka H (1980)
      Pestic. Sci. 202
1720. Nahrstedt A (1985) Pl. Syst. Evol. 150: 35
1721. Fechter MH, Griengl H (2004) Food Technol. Biotechnol. 42: 287
1722. Sharma M, Sharma NN, Bhalla TC (2005) Enzyme Microb. Technol. 37: 279
1723. Effenberger F, Förster S, Wajant H, (2000) Curr. Opinion Biotechnol. 11: 532
1724. Griengl H, Schwab H, Fechter M (2000) Trends Biotechnol. 18: 252
1725. Hochuli E (1983) Helv. Chim. Acta 66: 489

1726. Jorns MS, Ballenger C, Kinney G, Pokora A, Vargo D (1983) J. Biol. Chem. 258: 8561
1727. Kiljunen E, Kanerva LT (1997) Tetrahedron Asymmetry 8: 1225
1728. Becker W, Pfeil E (1964) Naturwissensch. 51: 193
1729. Niedermeyer U, Kula MR (1990) Angew. Chem. Int. Ed. 29: 386
1730. Klempier N, Griengl H, Hayn M (1993) Tetrahedron Lett. 34: 4769
1731. Förster S, Roos J, Effenberger F, Wajant H, Sprauer A (1996) Angew. Chem. Int. Ed. 35: 437
1732. Effenberger F, Hörsch B, Förster S, Ziegler T (1990) Tetrahedron Lett. 31: 1249
1733. Seely MK, Criddle RS, Conn EE (1966) J. Biol. Chem. 241: 4457
1734. Selmar D, Lieberei R, Biehl B, Conn EE (1989) Physiol. Plantarum 75: 97
1735. Klempier N, Pichler U, Griengl H (1995) Tetrahedron Asymmetry 6: 845
1736. Kuroki GW, Conn EE (1989) Proc. Natl. Acad. Sci. USA 86: 6978
1737. Bourquelot E, Danjou E (1905) J. Pharm. Chim. 22: 219
1738. Hughes J, De Carvalho FJP, Hughes MA (1994) Arch. Biochem. Biophys. 311: 496
1739. Wagner UG, Hasslacher M, Griengl H, Schwab H, Kratky C (1996) Structure 4: 811
1740. Gruber K, Kratky C (2004) J. Polym. Sci [A] 42: 479
1741. Zuegg J, Gruber K, Gugganig M, Wagner UG, Kratky C (1999) Protein Sci. 8: 1990
1742. Brussee J, Loos WT, Kruse CG, van der Gen A (1990) Tetrahedron 46: 979
1743. Effenberger F, Ziegler T, Förster S (1987) Angew. Chem. Int. Ed. 26: 458
1744. Ziegler T, Hörsch B, Effenberger F (1990) Synthesis 575
1745. Becker W, Pfeil E (1966) J. Am. Chem. Soc. 88: 4299
1746. Becker W, Pfeil E (1966) Biochem. Z. 346: 301
1747. Effenberger F, Hörsch B, Weingart F, Ziegler T, Kühner S (1991) Tetrahedron Lett. 32: 2605
1748. Effenberger F, Heid S (1995) Tetrahedron Asymmetry 6: 2945
1749. Weis R, Gaisberger R, Skranc W, Gruber K, Glieder A (2005) Angew. Chem. Int. Ed. 44: 4700
1750. Glieder A, Weis R, Skranc W, Poechlauer P, Dreveny I, Majer S, Wubbolts M, Schwab H, Gruber K (2003) Angew. Chem. Int. Ed. 42, 4815
1751. Griengl H, Hickel A, Johnson DV, Kratky C, Schmidt M, Schwab H (1997) Chem. Commun. 1933
1752. Sheldon RA, Schoemaker HE, Kamphuis J, Boesten WHJ, Meijer EM (1988) Enzymatic methods for the industrial synthesis of optically active compounds. In: Ariens EJ, van Rensen JJS, Welling W (eds) Stereoselectivity of Pestcides. Elsevier, Amsterdam, p 409
1753. Wehtje E, Adlercreutz P, Mattiasson B (1990) Biotechnol. Bioeng. 36: 39
1754. Ognyanov VI, Datcheva VK, Kyler KS (1991) J. Am. Chem. Soc. 113: 6992
1755. Menendez E, Brieva R, Rebolledo F, Gotor V (1995) J. Chem. Soc., Chem. Commun. 989
1756. Gruber-Khadjawi M, Purkarthofer T, Scranc W, Griengl H (2007) Adv. Synth. Catal. 349: 1445.
1757. Purkarthofer T, Gruber K, Gruber-Khadjawi M, Waich K, Skranc W, Mink D, Griengl H (2006) Angew. Chem. Int. Ed. 45: 3454
1758. Gruber-Khadjawi M, Purkarthofer T, Skranc W, Griengl H (2007) Adv. Synth. Catal. 349: 1445
1759. Michael A (1887) Am. Chem. J. 9: 112.
1760. Harutyunyan SR, den Hartog T, Geurts K, Minnaard AJ, Feringa BL (2008) Chem. Rev. 108: 2824.
1761. Alexakis A, Bäckvall J-E, Krause N, Pamies O, Dieguez M (2008) Chem. Rev. 108: 2796.
1762. Kusebauch B, Busch B, Scherlach K, Roth M, Hertweck C (2009) Angew. Chem. Int. Ed. 48: 5001.
1763. Müller M, Sprenger GA, Pohl M (2013) Curr. Opin. Chem. Biol. 17: 261.
1764. Emmons GT, Campbell IM, Bentley R (1985) Biochem. Biophys. Res. Commun. 131: 956.
1765. Xu J-M, Zhang F, Liu B-K, Wu Q, Lin X-F (2007) Chem. Commun 2078.
1766. Xie B-H, Guan Z, He Y-H (2012) J. Chem. Technol. Biotechnol. 87: 1709.

1767. Svedendahl M, Hult K, Berglund P (2005) J. Am. Chem. Soc. 127: 17988.
1768. Strohmeier GA, Sovic T, Steinkellner G, Hartner FS, Andryushkova A, Purkarthofer T, Glieder A, Gruber K, Griengl H (2009) Tetrahedron 65: 5663.
1769. Svedendahl M, Jovanovic B, Fransson L, Berglund P (2009) ChemCatChem 1: 252
1770. Strohmeier GA, Sovic T, Steinkellner G, Hartner FS, Andryushkova A, Purkarthofer T, Glieder A, Gruber K, Griengl H (2009) Tetrahedron 65: 5663
1771. Svedendahl M, Hult K, Berglund P (2005) J. Am. Chem. Soc. 127: 17988
1772. Carlqvist P, Svedendahl M, Branneby C, Hult K, Brinck T, Berglund P (2005) ChemBio-Chem 6: 331
1773. Kitazume T, Ikeya T, Murata K (1986) J. Chem. Soc., Chem. Commun. 1331
1774. Zandvoort E, Geertsema EM, Baas B-J, Quax WJ, Poelarends GJ (2012) Angew. Chem. Int. Ed. 51: 1240.
1775. Miao Y, Geertsema EM, Tepper PG, Zandvoort E, Poelarends GJ (2013) ChemBioChem 14: 191.
1776. Geertsema EM, Miao Y, Tepper PG, de Haan P, Zandvoort E, Poelarends GJ (2013) Chem. Eur. J. 19: 14407.

References to Sect. 2.6

1777. Stern R, Jedrzejas MJ (2008) Chem. Rev. 108: 5061.
1778. Kennedy JF, White CA (1983) Bioactive Carbohydrates. Ellis Horwood, West Sussex
1779. Kadokawa J (2011) Chem. Rev. 111: 4308.
1780. Ginsburg V, Robbins PW (eds) (1984) Biology of Carbohydrates. Wiley, New York
1781. Karlsson KA (1989) Ann. Rev. Biochem. 58: 309
1782. Hakomori S (1984) Ann. Rev. Immunol. 2: 103
1783. Laine RA (1994) Glycobiology 4: 759
1784. Boons GJ (1996) Tetrahedron 52: 1095
1785. Nilsson KGI (1988) Trends Biotechnol. 6: 256
1786. Wong CH (1996) Acta Chem. Scand. 50: 211
1787. Okamoto K, Goto T (1990) Tetrahedron 46: 5835
1788. Gigg J, Gigg R (1990) Topics Curr. Chem. 154: 77
1789. Cote GL, Tao BY (1990) Glycoconjugate J. 7: 145
1790. Wong CH, Halcomb RL, Ichikawa Y, Kajimoto T (1995) Angew. Chem. Int. Ed. 34: 521
1791. Edelman J (1956) Adv. Enzymol. 17: 189
1792. Desmet T, Soetaert W, Bojarova P, Kren V, Dijkhuizen L, Eastwick-Field V, Schiller A (2012) Chem. Eur. J. 18: 10786.
1793. Desmet T, Soetaert W (2011) Biocatal. Biotrans. 29: 1.
1794. Lairson LL, Withers SG (2004) Chem. Commun. 20: 2243.
1795. Leloir LF (1971) Science 172: 1299
1796. Sharon N (1975) Complex Carbohydrates. Addison-Wesley, Reading, MA
1797. Andre I, Potocki-Veronese G, Morel S, Monsan P, Remaud-Simeon M (2010) Top. Curr. Chem. 294: 25.
1798. Schachter H, Roseman S (1980) In: Lennarz WJ (ed) The Biochemistry of Glycoproteins and Proteoglycans. Plenum Press, New York, p 85
1799. Leloir LF (1971) Science 172: 1299.
1800. Beyer TA, Sadler JE, Rearick JI, Paulson JC, Hill RL (1981) Adv. Enzymol. 52: 23
1801. Öhrlein R (1999) Topics Curr. Chem. 200: 227
1802. Elling L (1999) Glycobiotechnologhy: enzymes for the synthesis of nucleotide sugars. In: Scheper T (ed) New Enzymes for Organic Synthesis. Springer, Berlin Heidelberg New York, p 89

1803. Elling L (1997) Adv. Biochem. Eng. Biotechnol. 58: 89
1804. Utagawa T (1999) J. Mol. Catal. B 6: 215
1805. Wong CH (1996) Practical synthesis of oligosaccharides based on glycosyl transferases and glycosylphosphites. In: Khan SH, O'Neill RA (eds) Modern Methods in Carbohydrate Synthesis, vol 19. Harwood, Amsterdam, p 467
1806. Bojarova P, Rosencrantz RR, Elling L, Kren V (2013) Chem. Soc. Rev. 42: 4774.
1807. Sadler JE, Beyer TA, Oppenheimer CL, Paulson JC, Prieels JP, Rearick JI, Hill RL (1982) Methods Enzymol. 83: 458
1808. Aoki D, Appert HE, Johnson D, Wong SS, Fukuda MN (1990) EMBO J. 9: 3171
1809. Ginsburg V (1964) Adv. Enzymol. 26: 35
1810. Palcic MM, Srivastava OP, Hindsgaul O (1987) Carbohydr. Res. 159: 315
1811. Berliner LJ, Robinson RD (1982) Biochemistry 21: 6340
1812. Lambright DG, Lee TK, Wong SS (1985) Biochemistry 24: 910
1813. Augé C, David S, Mathieu C, Gautheron C (1984) Tetrahedron Lett. 25: 1467
1814. Srivastava OP, Hindsgaul O, Shoreibah M, Pierce M (1988) Carbohydr. Res. 179: 137
1815. Palcic MM, Venot AP, Ratcliffe RM, Hindsgaul O (1989) Carbohydr. Res. 190: 1
1816. Nunez HA, Barker R (1980) Biochemistry 19: 489
1817. David S, Augé C (1987) Pure Appl. Chem. 59: 1501
1818. Wong CH, Haynie SL, Whitesides GM (1982) J. Org. Chem. 47: 5416
1819. Schmölzer K, Lemmerer M, Gutmann A, Nidetzky B (2017) Biotechnol. Bioeng. 114: 924.
1820. Goedl C, Schwarz A, Mueller M, Brecker L, Nidetzky B (2008) Carbohydr. Res. 343: 2032
1821. Kitaoka M, Hayashi K (2002) Trends Glycosci. Glycotechnol. 14: 35
1822. Luley-Goedl C, Nidetzky B (2010) Biotechnol. J. 5: 1324.
1823. Haynie SL, Whitesides GM (1990) Appl. Biochem. Biotechnol. 23: 155
1824. Monsan P, Remaud-Simeon M, Andre I (2010) Curr. Opin. Microbiol. 13: 293.
1825. Goedl C, Sawangwan T, Wildberger P, Nidetzky B (2010) Biocatal. Biotransform. 28: 10
1826. Sawangwan T, Goedl C, Nidetzky B (2009) Org. Biomol. Chem. 7: 4267
1827. Goedl C, Sawangwan T, Mueller M, Schwarz A, Nidetzky B (2008) Angew. Chem. Int. Ed. 47: 10086
1828. Wolfenden R, Lu X, Young G (1998) J. Am. Chem. Soc. 1220: 6814.
1829. Henrissat B, Davies G (1997) Curr. Opinion Struct. Biol. 7: 637
1830. Wallenfels K, Weil R (1972) In: Boyer PD (ed) The Enzymes, vol VII. Academic Press, New York, p 618
1831. Pan SC (1970) Biochemistry 9: 1833
1832. Scigelova M, Singh S, Crout DHG (1999) J. Mol. Catal. B 6: 483
1833. Crout DHG, Vic G (1998) Curr. Opinion Chem. Biol. 2: 98
1834. Nilsson KGI (1996) Synthesis with glycosidases. In: Khan SH, O'Neill RA (eds) Modern Methods in Carbohydrate Synthesis, vol 21. Harwood, Amsterdam, p 518
1835. van Rantwijk F, Woudenberg-van Oosterom M, Sheldon RA (1999) J. Mol. Catal. B 6: 511
1836. Bourquelot E, Bridel M (1913) Ann. Chim. Phys. 29: 145
1837. Wang Q, Graham RW, Trimbur D, Warren RAJ, Withers SG (1994) J. Am. Chem. Soc. 116: 11594
1838. Post CB, Karplus M (1986) J. Am. Chem. Soc. 108: 1317
1839. Huang X, Tanaka KSE, Bennet AJ (1997) J. Am. Chem. Soc. 119: 11147
1840. Chiba S (1997) Biosci. Biotechnol. Biochem. 61: 1233
1841. Withers SG, Street IP (1988) J. Am. Chem. Soc. 110: 8551
1842. Kengen SWM, Luesink EJ, Stams AJM, Zehnder AJB (1993) Eur. J. Biochem. 213: 305
1843. Koshland DE (1953) Biol. Rev. 28: 416
1844. Heightman TD, Vasella AT (1999) Angew. Chem. Int. Ed. 38: 750
1845. Sinnot ML (1990) Chem. Rev. 90: 1171
1846. Wang Q, Withers SG (1995) J. Am. Chem. Soc. 117: 10137
1847. Veibel S (1936) Enzymologia 1: 124
1848. Li YT (1967) J. Biol. Chem. 242: 5474

1849. Johansson E, Hedbys L, Mosbach K, Larsson PO, Gunnarson A, Svensson S (1989) Enzyme Microb. Technol. 11: 347
1850. Likolov ZL, Meagher MM, Reilly PJ (1989) Biotechnol. Bioeng. 34: 694
1851. Rastall RA, Bartlett TJ, Adlard MW, Bucke C (1990) The production of hetero-oligosaccharides using glycosidases. In: Copping LG, Martin RE, Pickett JA, Bucke C, Bunch AW (eds) Opportunities in Biotransformations. Elsevier, London, p 47
1852. Vulfson EN, Patel R, Beecher JE, Andrews AT, Law BA (1990) Enzyme Microb. Technol. 12: 950
1853. Beecher JE, Andrews AT, Vulfson EN (1990) Enzyme Microb. Technol. 12: 955
1854. Fujimoto H, Nishida H, Ajisaka K (1988) Agric. Biol. Chem. 52: 1345
1855. Kren V, Thiem J (1997) Chem. Soc. Rev. 26: 463
1856. Vocadlo DJ, Withers SG (2000) In: Ernst B, Hart GW, Sinay P (eds) Carbohydrates in Chemistry and Biology. Wiley-VCH, Weinheim 2: 723
1857. Husakova L, Riva S, Casali M, Nicotra S, Kuzma M, Hunkova Z, Kren V (2001) Carbohydr. Res. 331: 143
1858. Fialova P, Weignerova L, Rauvolfova J, Prikrylova V, Pisvejcova A, Ettrich R, Kuzma M, Sedmera P, Kren V (2004) Tetrahedron 60: 693
1859. Gold AM, Osber MP (1971) Biochem. Biophys. Res. Commun. 42: 469
1860. Williams SJ, Withers SG (2000) Carbohydr. Res. 327: 27
1861. Fialova P, Carmona AT, Robina I, Ettrich R, Sedmera P, Prikrylova V, Petraskova-Husakova L, Kren V (2005) Tetrahedron Lett. 46: 8715
1862. Day AG, Withers SG (1986) Biochem. Cell Biol. 64: 914
1863. Yasukochi T, Fukase K, Kusumoto S (1999) Tetrahedron Lett. 40: 6591
1864. Chiffoleau-Giraud V, Spangenberg P, Rabiller C (1997) Tetrahedron Asymmetry 8: 2017
1865. Kobayashi S, Kiyosada T, Shoda S (1997) Tetrahedron Lett. 38: 2111
1866. Nilsson KGI (1990) Asymmetric synthesis of complex oligosaccharides. In: Copping LG, Martin RE, Pickett JA, Bucke C, Bunch AW (eds) Opportunities in Biotransformations. Elsevier, London, p 131
1867. Nilsson KGI (1990) Carbohydr. Res. 204: 79
1868. Nilsson KGI (1988) Ann. NY Acad. Sci. USA 542: 383
1869. Nilsson KGI (1987) A comparison of the enzyme-catalysed formation of peptides and oligosaccharides in various hydroorganic solutions using the nonequilibrium approach. In: Laane C, Tramper J, Lilly MD (eds) Biocatalysis in Organic Media. Elsevier, Amsterdam, p 369
1870. Nilsson KGI (1987) Carbohydr. Res. 167: 95
1871. Nilsson KGI (1989) Carbohydr. Res. 188: 9
1872. Nilsson KGI (1988) Carbohydr. Res. 180: 53
1873. Crout DHG, Howarth OW, Singh S, Swoboda BEP, Critchley P, Gibson WT (1991) J. Chem. Soc., Chem. Commun. 1550
1874. Huber RE, Gaunt MT, Hurlburt KL (1984) Arch. Biochem. Biophys. 234: 151
1875. Ooi Y, Mitsuo N, Satoh T (1985) Chem. Pharm. Bull. 33: 5547
1876. Mitsuo N, Takeichi H, Satoh T (1984) Chem. Pharm. Bull. 32: 1183
1877. Trincone A, Improta R, Nucci R, Rossi M, Giambacorta A (1994) Biocatalysis 10: 195
1878. Björkling F, Godtfredsen SE (1988) Tetrahedron Lett. 44: 2957
1879. Gais HJ, Zeissler A, Maidonis P (1988) Tetrahedron Lett. 29: 5743
1880. Crout DHG, MacManus DA, Critchley P (1991) J. Chem. Soc., Chem. Commun. 376
1881. Matsumura S, Yamazaki H, Toshima K (1997) Biotechnol. Lett. 19: 583
1882. Perugino G, Trincone A, Rossi M, Moracci M (2004) Trends Biotechnol. 22: 31
1883. Williams SJ, Withers SG (2002) Austr. J. Chem. 55: 3
1884. Jahn M, Chen H, Muellegger J, Marles J, Warren RAJ, Withers SG (2004) Chem. Commun. 274
1885. Perugino G, Cobucci-Ponzano B, Rossi M, Moracci M (2005) Adv. Synth. Catal. 347: 941
1886. Jahn M, Marles J, Warren RAJ, Withers SG (2003) Angew. Chem. Int. Ed. 42: 352

1887. Kim YW, Fox DT, Hekmat O, Kantner T, McIntosh LP, Warren RAJ, Withers SG (2006) Org. Biomol. Chem. 4: 2025
1888. Turner N J, Truppo M D (2010) Biocatalytic routes to non-racemic chiral amines, In: Nugent TC (ed) Chital Amine Synthesis. Wiley-VCH, Weinheim, p 431
1889. Koszelewski D, Tauber K, Faber K, Kroutil W (2010) Trends Biotechnol. 28: 324
1890. Zhu D, Hua L (2009) Biotechnol. J. 4: 1420
1891. Höhne M, Bornscheuer UT (2009) ChemCatChem 1: 42
1892. Stewart JD (2001) Curr. Opinion Chem. Biol. 5: 120
1893. Ager DJ, Li T, Pantaleone DP, Senkpeil RF, Taylor PP, Fotheringham IG (2001) J. Mol. Catal. B: Enzym. 11: 199
1894. Cassimjee KE, Humble MS, Miceli V, Colomina CG, Berglund P (2011) ACS Catal. 1: 1051.
1895. Cassimjee KE, Manta B, Himo F (2015) Org. Biomol. Chem. 13: 8453.
1896. Noe FF, Nickerson WJ (1958) J. Bacteriol. 75: 674
1897. Kim K H (1964) J. Biol. Chem. 239: 783
1898. Matcham GW, Bowen ARS (1996) Chim. Oggi 14: 20
1899. Cho BK, Park HY, Seo JH, Kinnera K, Lee BS, Kim BG (2004) Biotechnol. Bioeng. 88: 512
1900. Fuchs M, Farnberger JE, Kroutil W (2015) Eur. J. Org. Chem. 6965.
1901. Kohls H, Steffen-Munsberg F, Höhne M (2014) Curr. Opin. Chem. Biol. 19: 180.
1902. Mehta PK, Hale TI, Christen P (1993) Eur. J. Biochem. 214: 549
1903. Hwang BY, Cho BK, Yun H, Koteshwar K, Kim BG (2005) J. Mol. Catal. B: Enzym. 37: 47
1904. Steffen-Munsberg F, Vickers C, Kohls H, Land H, Mallin H, Nobili A, Skalden L, van den Bergh T, Joosten HJ, Berglund P, Höhne M, Bornscheuer UT (2015) Biotechnol. Adv. 33: 566.
1905. Yamada H, Kimura T, Tanaka A, Ogata K (1964) Agric. Biol. Chem. 28: 443.
1906. Shin JS, Kim BG (2001) Biosci. Biotechnol. Biochem. 65: 1782
1907. Iwasaki A, Yamada Y, Ikenaka Y, Hasegawa J (2003) Biotechnol. Lett. 25: 1843
1908. Koszelewski D, Lavandera I, Clay D, Rozzell D, Kroutil W (2008) Adv. Synth. Catal. 350: 2761
1909. Hwang BY, Kim BG (2004) Enzyme Microb. Technol. 34: 429
1910. Yun H, Cho BK, Kim BG, (2004) Biotechnol. Bioeng. 87: 772
1911. Smithies K, Smith MEB, Kaulmann U, Galman JL, Ward JM, Hailes HC (2009) Tetrahedron Asymmetry 20: 570
1912. Kaulmann U, Smithies K, Smith MEB, Hailes H C, Ward JM (2007) Enzyme Microb. Technol. 41: 628
1913. Ingram CU, Bommer M, Smith MEB, Dalby PA, Ward JM, Hailes HC, Lye GJ (2007) Biotechnol. Bioeng. 96: 559
1914. Hanson RL, Davis BL, Chen Y, Goldberg SL, Parker WL, Tully TP, Montana MA, Patel RN (2008) Adv. Synth. Catal. 350: 1367
1915. Yun H, Lim S, Cho BK, Kim BG, (2004) Appl. Environ. Microbiol. 70: 2529
1916. Koszelewski D, Göritzer M, Clay D, Seisser B, Kroutil W (2010) ChemCatChem 2: 73
1917. Nakamichi K (1990) Appl. Microbiol. 33: 637.
1918. Tufvesson P, Lima-Ramos J, Jensen JS, Al-Haque N, Neto W, Woodley JM (2011) 108: 1479.
1919. Shin JS. Kim BG (1998) Biotechnol. Bioeng. 60: 534
1920. Matcham G, Bhatia M, Lang W, Lewis C, Nelson R, Wang A, Wu W (1999) Chimia 53: 584
1921. Hoene M, Kuehl S, Robins K, Bornscheuer UT (2008) ChemBioChem 9: 363
1922. Hwang JY, Park J, Seo JH, Cha M, Cho BK, Kim J, BG Kim (2009) Biotechnol. Bioeng. 102: 1323
1923. Truppo MD, Rozzell JD, Moore JC, Turner NJ (2009) Org. Biomol. Chem. 7: 395
1924. Koszelewski D, Lavandera I, Clay D, Guebitz GM, Rozzell D, Kroutil W (2008) Angew. Chem. Int. Ed. 47: 9337
1925. Wang B, Land H, Berglund P (2013) Chem. Commun. 161.

1926. Li T, Kootstra AB, Fotheringham IG (2002) Org. Proc. Res. Dev. 6: 533
1927. Lo HH, Hsu SK, Lin WD, Chan NL, Hsu WH (2005) Biotechnol. Progr. 21: 411
1928. Green AP, Turner NJ, O'Reilly E (2014) Angew. Chem. Int. Ed. 53: 10714.
1929. Pavlidis I, Weiß MS, Genz M, Spurr P, Hanlon SP, Wirz B, Iding H, Bornscheuer UT (2016) Nat. Chem. 8: 1076.
1930. Fuchs M, Koszelewski D, Tauber K, Sattler J, Banko W, Holzer AK, Pickl M, Kroutil W, Faber K (2012) Tetrahedron 68: 7691.
1931. Savile CK, Janey JM, Mundorff EC, Moore JC, Tam S, Jarvis WR, Colbeck JC, Krebber A, Fleitz FJ, Brands J, Devine PN, Huisman GW, Hughes GJ (2010) Science 329: 305.
1932. Limanto J, Ashley ER, Yin JJ, Beutner GL, Grau BT, Kassim AM, Kim MM, Klapars A, Liu ZJ, Strotman HR, Truppo MD (2014) Org. Lett. 16: 2716.
1933. Chung CK, Bulger PG, Kosjek B, Belyk KM, Rivera N, Scott ME, Humphrey GR, Limanto J, Bachert DC, Emerson KM (2014) Org. Proc. Res. Dev. 18: 215.

References to Sect. 2.7

1934. Petty MA (1961) Bacteriol. Rev. 25: 111
1935. Field JA, Verhagen FJM, de Jong E (1995) Trends Biotechnol. 13: 451
1936. Gschwend PM, MacFarlane JK, Newman KA (1985) Science 227: 1033
1937. Siuda JF, De Barnardis JF (1973) Lloydia 36: 107
1938. Paul C, Pohnert G (2011) Nat. Prod. Rep. 28: 186.
1939. Fowden L (1968) Proc. Royal Soc. London B 171: 5
1940. Fenical W (1979) Recent Adv. Phytochem. 13: 219
1941. Wever R, Krenn BE (1990) Vanadium Haloperoxidases. In: Chasteen ND (ed) Vanadium in Biological Systems. Kluwer, Dordrecht, p 81
1942. Krenn BE, Tromp MGM, Wever R (1989) J. Biol. Chem. 264: 19287
1943. Faulkner DJ (1984) Nat. Prod. Rep. 1: 251
1944. Morris DR, Hager LP (1966) J. Biol. Chem. 241: 1763
1945. van Pée KH (1990) Kontakte (Merck) 41
1946. Neidleman SL, Geigert J (1986) Biohalogenation. Ellis Horwood, Chichester
1947. Franssen MCR, van der Plas HC (1992) Adv. Appl. Microbiol. 37: 41
1948. Neidleman SL (1980) Hydrocarbon Proc. 60: 135
1949. van Pee KH (2001) Arch. Microbiol. 175: 250
1950. Hager LP, Morris DR, Brown FS, Eberwein H (1966) J. Biol. Chem. 241: 1769.
1951. Fujimori D G, Walsh C T (2007) Curr. Opinion Chem. Biol. 11: 553
1952. Anderson JLR, Chapman SK (2006) Mol. BioSyst. 2: 350
1953. Murphy C D (2006) Nat. Prod. Rep. 23: 147
1954. van Pee KH, Patallo EP (2006) Appl. Microbiol. Biotechnol. 70: 631
1955. van Pee KH, Flecks S, Patallo EP (2007) Chimica Oggi 25: 22
1956. Krebs K, Fujimori D G, Walsh CT, Bollinger Jr JM, (2007) Acc. Chem. Res. 40: 484
1957. van Pee KH (1990) Biocatalysis 4: 1
1958. Neidleman SL, Geigert J (1987) Endeavour 11: 5
1959. Hofrichter M, Ullrich R (2006) Appl. Microbiol. Biotechnol. 71: 276
1960. Smith DRM, Grüschow S, Goss RJM (2013) Curr. Opin. Chem. Biol.17: 276.
1961. O'Hagan D, Schaffrath C, Cobb SL, Hamilton JTG, Murphy CD (2002) Nature 416: 279.
1962. Sundaramoorthy M, Terner J, Poulos TL (1998) Chem. Biol. 5: 461.
1963. Poulos TL (2014) Chem. Rev. 114: 3919.
1964. Hemrika W, Renirie R, Macedo-Ribeiro S, Messerschmidt A, Wever R (1999) J. Biol. Chem. 274: 23820.
1965. Winter JM, Moore BS (2009) J. Biol. Chem. 284: 18577.

1966. Wagenknecht HA, Woggon WD (1997) Chem. Biol. 4: 367
1967. Itoh N, Izumi Y, Yamada H (1985) J. Biol. Chem. 261: 5194
1968. van Pée KH, Lingens F (1985) J. Bacteriol. 161: 1171
1969. Wiesner W, van Pée KH, Lingens F (1985) Hoppe-Seylers Z. Physiol. Chem. 366: 1085
1970. van Pée KH, Lingens F (1985) J. Gen. Microbiol. 131: 1911
1971. Wagner AP, Psarrou E, Wagner LP (1983) Anal. Biochem. 129: 326
1972. Thomas JA, Morris DR, Hager LP (1970) J. Biol. Chem. 245: 3135
1973. Yamada H, Itoh N, Izumi Y (1985) J. Biol. Chem. 260: 11962
1974. Libby RD, Shedd AL, Phipps AK, Beachy TM, Gerstberger SM (1992) J. Biol. Chem. 267: 1769
1975. Turk J, Henderson WR, Klebanoff SJ, Hubbard WC (1983) Biochim. Biophys. Acta 751: 189
1976. Boeynaems JM, Watson JT, Oates JA, Hubbard WC (1981) Lipids 16: 323
1977. Geigert J, Neidleman SL, Dalietos DJ, DeWitt SK (1983) Appl. Environ. Microbiol. 45: 1575
1978. Lee TD, Geigert J, Dalietos DJ, Hirano DS (1983) Biochem. Biophys. Res. Commun. 110: 880
1979. Neidleman SL, Oberc MA (1968) J. Bacteriol. 95: 2424
1980. Levine SD, Neidleman SL, Oberc MA (1968) Tetrahedron 24: 2979
1981. Geigert J, Neidleman SL, Dalietos DJ, DeWitt SK (1983) Appl. Environ. Microbiol. 45: 366
1982. Neidleman SL, Levin SD (1968) Tetrahedron Lett. 9: 4057
1983. Ramakrishnan K, Oppenhuizen ME, Saunders S, Fisher J (1983) Biochemistry 22: 3271
1984. Carter-Franklin JN, Parrish JD, Tschirret-Guth RA, Little RD, Butler A (2003) J. Am. Chem. Soc. 125: 3688.
1985. Geigert J, Neidleman SL, Dalietos DJ (1983) J. Biol. Chem. 258: 2273
1986. Neidleman SL, Cohen AI, Dean L (1969) Biotechnol. Bioeng. 2: 1227
1987. van Pée KH, Lingens F (1984) FEBS Lett. 173: 5
1988. Itoh N, Izumi Y, Yamada H (1987) Biochemistry 26: 282
1989. Jerina D, Guroff G, Daly J (1968) Arch. Biochem. Biophys. 124: 612
1990. Matkovics B, Rakonczay Z, Rajki SE, Balaspiri L (1971) Steroidologia 2: 77
1991. Corbett MD, Chipko BR, Batchelor AO (1980) Biochem. J. 187: 893
1992. Wischang D, Brücher O, Hartung J (2011) Coord. Chem. Rev. 255: 2204.
1993. Wischang D, Hartung J (2012) Tetrahedron 68: 9456.
1994. Loo TL, Burger JW, Adamson RH (1964) Proc. Soc. Exp. Biol. Med. 114: 60
1995. Beissner RS, Guilford WJ, Coates RM, Hager LP (1981) Biochemistry 20: 3724
1996. Libby RD, Thomas JA, Kaiser LW, Hager LP (1982) J. Biol. Chem. 257: 5030
1997. Franssen MCR, van der Plas HC (1984) Recl. Trav. Chim. Pays-Bas 103: 99
1998. Neidleman SL, Diassi PA, Junta B, Palmere RM, Pan SC (1966) Tetrahedron Lett. 7: 5337
1999. Theiler R, Cook JC, Hager LP, Siuda JF (1978) Science 202: 1094
2000. Zaks A, Yabannavar AV, Dodds DR, Evans CA, Das PR, Malchow R (1996) J. Org. Chem. 61: 8692
2001. Grisham MB, Jefferson MM, Metton DF, Thomas EL (1984) J. Biol. Chem. 259: 10404
2002. Silverstein RM, Hager LP (1974) Biochemistry 13: 5069
2003. Tsan M-F (1982) J. Cell. Physiol. 111: 49
2004. Lal R, Saxena DM (1982) Microbiol. Rev. 46: 95
2005. Alexander M (1977) Introduction to Soil Microbiology. Wiley, Chichester, p. 438
2006. Ghisalba O (1983) Experientia 39: 1247
2007. Rothmel RK, Chakrabarty AM (1990) Pure Appl. Chem. 62: 769
2008. Müller R, Lingens F (1986) Angew. Chem. Int. Ed. 25: 778
2009. Vogel TM, Criddle CS, McCarthy PL (1987) Environ. Sci. Technol. 21: 722
2010. Castro CE, Wade RS, Belser NO (1985) Biochemistry 24: 204
2011. Chacko CI, Lockwood JL, Zabik M (1966) Science 154: 893
2012. Markus A, Klages V, Krauss S, Lingens F (1984) J. Bacteriol. 160: 618

2013. Yoshida M, Fujita T, Kurihara N, Nakajima M (1985) Pest. Biochem. Biophysiol. 23: 1
2014. Leisinger T, Bader R (1993) Chimia 47: 116
2015. Pavlova M, Klvana M, Prokop Z, Chaloupkova R, Banas P, Otyepka M, Wade RC, Tsuda M, Nagata Y, Damborsky J (2009) Nat. Chem. Biol. 5: 727.
2016. Janssen DB (2004) Curr. Opin. Chem. Biol. 8: 150.
2017. Janssen DB (2007) Adv. Appl. Microbiol. 61: 233.
2018. Koudelakova T, Bidmanova S, Dvorak P, Pavelka A, Chaloupkova R, Prokop Z, Damborsky J (2013) Biotechnol. J. 8: 32.
2019. Pieters RJ, Lutje Spelberg JH, Kellogg RM, Janssen DB (2001) 42: 469.
2020. Prokop Z, Sato Y, Brezovsky J, Mozga T, Chaloupkova R, Koudelakova T, Jerabek P, Stepankova V, Natsume R, van Leeuwen JGE, Janssen DB, Florian J, Nagata Y, Senda T, Damborsky J (2010) Angew. Chem. Int. Ed. 49: 6111.
2021. Westerbeek A, Szymanski W, Wijma HJ, Marrink SJ, Feringa BL, Janssen DB (2011) Adv. Synth. Catal. 353: 931.
2022. Motosugi K, Esaki N, Soda K (1982) Agric. Biol. Chem. 46: 837
2023. Allison N, Skinner AJ, Cooper RA (1983) J. Gen. Microbiol. 129: 1283
2024. Kawasaki H, Tone N, Tonomura K (1981) Agric. Biol. Chem. 45: 35
2025. Onda M, Motosugi K, Nakajima H (1990) Agric. Biol. Chem. 54: 3031
2026. Tsang JSH, Sallis PJ, Bull AT, Hardman DJ (1988) Arch. Microbiol. 150: 441
2027. Little M, Williams PA (1971) Eur. J. Biochem. 21: 99
2028. Vyazmensky M, Geresh S (1998) Enzyme Microb. Technol. 22: 323
2029. Cambou B, Klibanov AM (1984) Appl. Biochem. Biotechnol. 9: 255
2030. Taylor SC (1990) (S)-2-Chloropropanoic acid by biotransformation. In: Copping LG, Martin RE, Pickett JA, Bucke C, Bunch AW (eds) Opportunities in Biotransformations. Elsevier, London, p 170
2031. Smith JM, Harrison K, Colby J (1990) J. Gen. Microbiol. 136: 881
2032. Barth PT, Bolton L, Thomson JC (1992) J. Bacteriol. 174: 2612
2033. Taylor SC (1997) (S)-2-Chloropropanoic Acid: Developments in Its Industrial Manufacture. In: Collins AN, Sheldrake GN, Crosby J (eds) Chirality in Industry II. Wiley, Chichester, p 207
2034. Hasan AKMQ, Takata H, Esaki N, Soda K (1991) Biotechnol. Bioeng. 38: 1114
2035. Castro CE, Bartnicki EW (1968) Biochemistry 7: 3213
2036. Geigert J, Neidleman SL, Liu TN, DeWitt SK, Panschar BM, Dalietos DJ, Siegel ER (1983) Appl. Environ. Microbiol. 45: 1148
2037. Nagasawa T, Nakamura T, Yu F, Watanabe I, Yamada H (1992) Appl. Microbiol. Biotechnol. 36: 478
2038. Assis HMS, Sallis PJ, Bull AT, Hardman DJ (1998) Enzyme Microb. Technol. 22: 568
2039. Assis HMS, Bull AT, Hardman DJ (1998) Enzyme Microb. Technol. 22: 545
2040. Kasai N, Tsujimura K, Unoura K, Suzuki T (1990) Agric. Biol. Chem. 54: 3185
2041. Nakamura T, Yu F, Mizunashi W, Watanabe I (1991) Agric. Biol. Chem. 55: 1931
2042. Nakamura T, Nagasawa T, Yu F, Watanabe I, Yamada H (1992) J. Bacteriol. 174: 7613
2043. Kasai N, Tsujimura K, Unoura K, Suzuki T (1992) J. Ind. Microbiol. 9: 97
2044. Kasai N, Suzuki T, Furukawa Y (1998) J. Mol. Catal. B 4: 237
2045. Suzuki T, Kasai N (1991) Bioorg. Med. Chem. Lett. 1: 343
2046. Suzuki T, Kasai N, Minamiura N (1994) Tetrahedron Asymmetry 5: 239
2047. Kasai N, Suzuki T (2002) Adv. Synth. Catal. 345: 437
2048. de Vries EJ, Janssen DB (2003) Curr. Opinion Biotechnol. 14: 414
2049. Tang L, Lutje Spelberg JH, Fraaije MW, Janssen DB (2003) Biochemistry 42: 5378
2050. De Jong RM, Dijkstra BW (2003) Curr. Opinion Struct. Biol. 13: 722
2051. Haak RM, Tarabiono C, Janssen DB, Minnaard AJ, de Vries J G, Feringa BL (2007) Org. Biomol. Chem. 5: 318
2052. van Hylckama Vlieg JET, Tang L, Lutje Spelberg JH, Smilda T, Poelarends GJ, Bosma T, van Merode AEJ, Fraaije MW, Janssen DB (2001) J. Bacteriol. 183: 5058.

2053. van den Wijngaard AJ, Reuvekamp PT, Janssen DB (1991) J. Bacteriol. 173: 124.
2054. Elenkov MM, Hoeffken HW, Tang L, Hauer B, Janssen DB (2007) Adv. Synth. Catal. 349: 2279
2055. Hasnaoui G, Lutje Spelberg JH, de Vries E, Tang L, Hauer B, Janssen DB (2005) Tetrahedron Asymmetry 16: 1685
2056. Elenkov MM, Hauer B, Janssen DB (2006) Adv. Synth. Catal. 348: 579
2057. Hasnaoui-Dijoux G, Elenkov MM, Lutje Spelberg JH, Hauer B, Janssen DB (2008) ChemBioChem 9: 1048
2058. Fuchs M, Simeo Y, Ueberbacher BT, Mautner B, Netscher T, Faber K (2009) Eur. J. Org. Chem. 833
2059. Fox RJ, Davis SC, Mundorff EC, Newman LM, Gavrilovic V, Ma SK, Chung LM, Ching C, Tam S, Muley S, Grate J, Gruber J, Whitman JC, Sheldon RA, Huisman GW (2007) Nature Biotechnol. 25: 338
2060. Pattabiraman TN, Lawson WB (1972) Biochem. J. 126: 645 and 659

Chapter 3
Special Techniques

Most biocatalysts can be used in a straightforward manner by regarding them as chiral catalysts and by applying standard methodology, i.e., in buffered aqueous solution. In order to broaden the applicability of enzymes, some special techniques have been developed. In particular, using biocatalysts in *nonaqueous media* rather than in water can lead to the gain of some significant advantages as long as some specific guidelines are followed [1]. Furthermore, 'fixation' of the enzyme by immobilization may be necessary, and the use of membrane technology may be advantageous as well. For both of the latter topics only the most simple techniques which can be adopted in an average organic chemistry laboratory are discussed (Sect. 3.3).

3.1 Enzymes in Organic Solvents

Water is a poor solvent for nearly all reactions in preparative organic chemistry because most organic compounds are poorly soluble in this medium. Furthermore, the removal of water is tedious and expensive due to its high boiling point and high heat of vaporization. Side-reactions such as hydrolysis, racemization, polymerization, and decomposition are often facilitated in the presence of water. These limitations were circumvented by the introduction of organic solvents for the majority of organic chemical processes. On the other hand, conventional biocatalysis has mainly been performed in aqueous solutions due to the perceived notion that enzymes are most active in water and it has been tacitly assumed that organic solvents only serve to destroy their catalytic power. However, this commonly held opinion is certainly too simplistic, bearing in mind that in nature many enzymes or multienzyme complexes function in hydrophobic environments, for instance, in the presence of, or bound onto, a membrane [2]. Therefore it should not be surprising that enzymes can be catalytically active in the presence of organic solvents [3–14]. The role of water in biocatalytic systems is contradictory: On the

© Springer International Publishing AG 2018
K. Faber, *Biotransformations in Organic Chemistry*,
DOI 10.1007/978-3-319-61590-5_3

one hand, the enzyme depends on water for the majority of the noncovalent interactions – salt bridges and hydrogen bonding – that help to maintain its catalytically active conformation [15] but water also participates in most of the reactions which lead to denaturation. As a consequence, it may be anticipated that replacing *some* (but not *all*) of the water with an organic solvent would retain enzymatic activity. Hence, it is conceivable that completely anhydrous solvents are incapable of supporting enzymatic activity because *some* water is always necessary for catalysis. The crucial answer to the question concerning *how much* water is required to retain catalytic activity is enzyme-dependent [16]. For example, α-chymotrypsin needs only 50 molecules of water per enzyme molecule to remain catalytically active [17], which is much less than is needed to form a monolayer of water around the enzyme. Other enzymes, like subtilisin and various lipases are similar in their need for trace quantities of water [18]. In other cases, however, much more water is required. Polyphenol oxidase, for instance, prefers a rather 'wet' environment and requires the presence of about 3.5×10^7 molecules of water [19].

The water present in a biological system can be separated into two physically distinct categories [20–22]. Whereas the majority of the water (>98%) serves as a true solvent ('bulk water'), a small fraction of it is tightly bound to the enzyme's surface ('bound water'). The physical state of bound water – as monitored by differences in melting point, heat capacity, EPR- and IR-spectroscopical properties – is clearly distinct from the bulk water and it should be regarded as a crucial integral part of the enzyme's structure rather than as adventitious residual solvent. Thus, bound water is also often referred to as 'structural water'. For a picture displaying the crystal structure of *Candida antarctica* lipase B with (and without) structural water molecules see Sect. 1.4.1, Fig. 1.1. If one extends this concept to enzymatic catalysis in organic media, it should be possible to replace the *bulk water* by an organic solvent without significant alteration of the enzyme's environment, as long as the *structural water* remains unaffected.

Biocatalytic transformations performed in organic media offer the following advantages:

- The overall yields of processes performed in organic media are usually better due to the omission of an extractive step during work-up. Thus, the loss-causing formation of emulsions can be avoided and the recovery of product(s) is facilitated by the use of low-boiling organic solvents.
- Nonpolar substrates are transformed at better rates due to their increased solubility [23].
- Since an organic medium is a hostile environment for living cells, microbial contamination is negligible. This is particularly important for reactions on an industrial scale, where maintaining sterility may be a serious problem.
- Deactivation and/or inhibition of the enzyme caused by lipophilic substrates and/or products is minimized since their enhanced solubility in the organic medium leads to a reduced local concentration at the enzyme's surface.
- Many side-reactions such as hydrolysis of labile groups (e.g., epoxides, acid anhydrides [24]), polymerization of quinones [19], racemization of

cyanohydrins [25], or acyl migration [26] are water-dependent and are therefore largely suppressed in an organic medium.

- Immobilization of enzymes is not necessary because they may be recovered by simple filtration after the reaction due to their insolubility in organic solvents. Nevertheless, if it is desired, experimentally simple adsorption onto the surface of a cheap macroscopic carrier such as diatomaceous earth (Celite), silica, or glass beads is possible. Desorption from the carrier into the medium – 'leaking' – is largely impeded in a lipophilic environment.

- Since many of the reactions which are responsible for the denaturation of enzymes (see Sect. 1.4.1) are hydrolytic reactions and therefore require water, it can be expected that enzymes should be more stable in an environment of low water content [27, 28]. For instance, porcine pancreatic lipase is active for many hours at 100 °C in a 99% organic medium but it is rapidly denaturated at this temperature when placed in pure water [29].

- Due to the conformational change (i.e., a partial unfolding and refolding) of the enzyme during the formation of the enzyme – substrate complex (the 'induced-fit'), numerous hydrogen bonds are reversibly broken and reformed. This process is greatly facilitated in an aqueous medium, where the broken bonds are rapidly replaced by hydrogen bonds to the surrounding water. Thus, it serves as a 'molecular lubricant' [22]. In an organic solvent, this process is impeded and, as a consequence, enzymes appear to be there more 'rigid' [30]. Thus, it is often possible to control some of the enzyme's catalytic properties such as the substrate specificity [17, 31–33], the chemo- [34], regio- [35] and enantioselectivity [36–39] by variation of the solvent (Sect. 3.1.7).

- The most important advantage, however, is the possibility of shifting thermodynamic equilibria to favor *synthesis* over *hydrolysis*. Thus, by using hydrolase enzymes (mainly lipases and proteases), esters [40–42], polyesters [43, 44], lactones [45, 46], amides [37, 47], and peptides [48] can be *synthesized* in a chemo-, regio-, and enantioselective manner.

The solvent systems which have commonly been used for enzyme-catalyzed reactions containing organic media can be classified into three different categories.

Enzyme Dissolved in a Monophasic Aqueous-Organic Solution
The enzyme, the substrate and/or product are dissolved in a monophasic solution consisting of water and a *water-miscible* organic cosolvent, such as dimethyl sulfoxide, dimethyl formamide, tetrahydrofuran, dioxane, acetone or one of the lower alcohols, e.g., *iso*-propanol or *tert*-butanol. Systems of this type are mainly used for the transformation of lipophilic substrates, which are sparingly soluble in an aqueous system alone and which would therefore be impeded by low reaction rates. In some cases, selectivities of esterases and proteases may be enhanced by using water-miscible organic cosolvents (see Scheme 2.42). As a rule of thumb, most water-miscible solvents can be applied in concentrations up to ~10–20% of the total volume, in rare enzyme/solvent combinations even 50–70% of cosolvent are tolerated. If the proportion of the organic solvent exceeds a certain threshold, the essential structural water is stripped from the enzyme's surface leading to deactivation. Only rarely do

enzymes remain catalytically active in water-miscible organic solvents with an extremely low water content; these cases are limited to unusually stable enzymes such as subtilisin and some lipases, e.g., from *Candida antarctica* [49–51]. Water-miscible organic solvents have also been successfully used to decrease the freezing temperature of aqueous systems when biocatalytic reactions were conducted at temperatures below 0 °C ('cryoenzymology') [52–55].

Enzyme Dissolved in a Biphasic Aqueous-Organic Solution

Reaction systems consisting of two discrete macroscopic phases, namely the aqueous phase containing the dissolved enzyme, and a second phase of a nonpolar organic solvent (preferably lipophilic and of high molecular weight) such as (chlorinated) hydrocarbons, aromatics or ethers, may be advantageous to achieve a spatial separation of the biocatalyst from the organic phase [56–58]. Thus, the biocatalyst is in a favorable aqueous environment and not in direct contact with the organic solvent, where most of the substrate/product is located. Therefore, the limited concentrations of organic material in the aqueous phase may circumvent inhibition phenomena. Furthermore, the removal of product from the enzyme surface drives the reaction towards completion. Due to the fact that in such biphasic systems the enzymatic reaction proceeds only in the aqueous phase, a sufficient mass transfer of the reactant(s) to and product(s) from the catalyst and between the two phases is necessary [59]. It is obvious that shaking or stirring represents a crucial parameter in such systems.

The number of phase distributions, measured as the partition coefficient, in a given reaction depends on the number of reactants and products (A, B, C, D) which are involved in the transformation (Table 3.1). Each distribution is dependent on the solubilities of substrate(s) and product(s) in the two phases and represents a potential rate-limiting factor.

Table 3.1 Partition coefficients involved in biphasic reactions

Type of reaction	Number of partition coefficients
A → B	3
A + B → C	4
A → B + C	4
A + B → C + D	7
Any type[a]	1

[a]for monophasic systems

Therefore, in biphasic systems the partition coefficient (a *thermodynamic* dimension) and the mass-transfer coefficient (a *kinetic* dimension) will dominate the k_{cat} of the enzyme. As a consequence, the overall reaction rate is mainly determined by the physical properties of the system (such as solubilities and stirring) and only to a lesser extent by the enzyme's catalytic power. In other words, the enzyme could work faster, but is unable to get enough substrate. Enhanced agitation (stirring, shaking) would improve the mass transfer but, on the other hand, it increasingly leads to deactivation of the enzyme due to mechanical shear and chemical stress.

Despite these problems, water-organic solvent two-phase systems have been successfully used to transform highly lipophilic substrates such as steroids [60], fats [61], and alkenes [62]. In addition, the use of biphasic solvent systems was beneficial for the asymmetric epoxidation of alkenes (Sect. 2.3.3.3) to minimize toxic effects of the epoxide produced by *Nocardia corallina* cells [63].

Enzyme Suspended in a Monophasic Organic Solution

Replacing all of the bulk water (which accounts for >98%) by a *water-immiscible* organic solvent leads to a suspension of the solid enzyme in a monophasic organic solution [64, 65]. Since enzymes are insoluble in organic media, such reactions resemble a heterogeneous catalytic system. Although the biocatalyst seems to be 'dry' on a macroscopic level, it must maintain the necessary residual structural water to remain catalytically active. Most of the research on such systems (which have proved to be extremely reliable, versatile and easy to use) has been performed during the 1980s, but it is striking that the first biotransformation of this kind was already reported in 1900! [66]. Due to the importance of this technique and its simplicity, all of the examples discussed below have been performed using solid 'dry' enzymes in organic solvents having a water content of <2%. However, it should be kept in mind that the catalytic activity of enzymes in nonconventional solvent systems is significantly reduced (often by one order of magnitude) compared to the activity in water [67].

The remarkable catalytic activity of solid proteins in organic solvents can be explained by their special properties [68]. In contrast to densely packed crystals of organic compounds of comparatively low molecular weight, which form rather dense and impenetrable structures, solid proteins represent soft and delicate aggregates. Since the average (monomeric) protein used in biotransformations has a diameter of ~5 nm (50 Å), there is limited contact between the single protein molecules thereby allowing minor conformational changes consonant with formation of the enzyme – substrate complex [69]. The total surface of solid enzymes is within the range of 1–3×10^6 m^2/kg, which is close to that of silica or activated carbon. About one to two thirds of the total volume is hollow, with large solvent-filled cavities and channels running through a 'sponge-like' macroscopic aggregate. Thus, if sufficient agitation is provided, the substrate is not only transformed by the active sites exposed to the surface of the crystal but also at those buried inside. In order to tune a biocatalytic reaction in a monophasic *water-immiscible* organic solvent system, the following parameters should be considered [70, 71]:

pH-Memory One particularly important aspect is the effect of the pH of the reaction medium, which cannot be measured or controlled easily in organic solutions that lack a distinct aqueous phase [72]. However, the pH determines the ionization state of the enzyme and hence its conformation and its catalytic properties, such as activity and selectivity. Since the ionization state of a protein does not change when placed in an organic solvent, but remains 'frozen', it is important to employ solid enzymes that have been recovered by lyophilization or precipitation from a buffer at their pH optimum [73]. The latter fact has vividly been described as the 'pH-memory' of enzymes.

Enzyme State The physical state of the enzyme may be crystalline, lyophilized or precipitated. Adsorption of enzymes onto the surface of a macroscopic (inorganic or organic) carrier material generates a better distribution of the biocatalyst and generally gives significantly enhanced reaction rates, in some cases up to one order of magnitude [74]. Any inorganic material such as diatomaceous earth (Celite), silica gel or an organic nonionic support (e.g., XAD-8, Accurel [75]) may be used as the carrier.

Biocompatibility of Organic Solvent In order to provide a measure for the 'compatibility' of an organic solvent in a monophasic system with high enzyme activity, several parameters describing the hydrophobicity of the solvent, such as the Hildebrandt solubility parameter (δ), the Reichardt–Dimroth polarity parameter (ET), the dielectric constant (ε), and the dipole moment (μ), have been proposed [76, 77]. However, the most reliable results were obtained by using the logarithm of the partition coefficient (log P) according to the Nernst distribution law of a given solvent between 1-octanol and water (Table 3.2) [78]. Although the effects of organic solvents on enzyme *stability* can be predicted with reasonably accuracy, the effects on enzyme (*stereo*)*selectivity* are only poorly understood and reliable predictions are very difficult to make [79, 80].

Table 3.2 Biocompatibility of organic solvents determined by partition coefficients (log P values)

log P	Water-miscibility	Solvent effects on enzyme activity
−2.5 to 0	Completely miscible	May be used to solubilize lipophilic substrates in aqueous systems in concentrations of ~10–20% v/v without deactivating the enzyme
0–1.5	Partially miscible	Causes serious enzyme distortion at elevated concentrations, may be used with unusually stable enzymes[a] but deactivation is common for average proteins
1.5–2.0	Low miscibility	Causes some enzyme distortion, may be used with many enzymes but activities are often unpredictable
>2.0	Immiscible	Causes negligible enzyme distortion and ensures high retention of activity for almost all enzymes

[a]For instance, subtilisin and *Candida antarctica* lipase B

If the log P value is not available in the literature, it can be calculated from hydrophobic fragmental constants [81]. As may be deduced from the log P values of some selected common organic solvents (Table 3.3), *water-miscible* hydrophilic solvents such as DMF, DMSO, acetone, and lower alcohols cannot be used as 'neat' organic solvents, whereas *water-immiscible* lipophilic solvents such as (halo) alkanes, ethers and aromatics retain an enzyme's high catalytic activity. Only in certain cases, in which polar substrates such as polyhydroxy compounds and amino acid derivatives have to be dissolved, should water-miscible solvents such as dioxane, tetrahydrofuran, 3-methyl-3-pentanol, or DMSO be considered for monophasic systems. However, in these solvents, most enzymes are deactivated and only exceptionally stable enzymes (for instance, subtilisin and *Candida antarctica* lipase B) can be used. It is an empirical phenomenon, that ball-shaped (round) organic solvents (e.g., *di-iso*propyl ether) often display higher biocompatibility than the corresponding straight-chain analog (e.g., *di-n*-propyl ether).

Table 3.3 log P values for common organic solvents

Solvent	log P
Dimethylsulfoxide	−1.3
Dioxane	−1.1
N,N-Dimethylformamide	−1.0
Methanol	−0.76
Acetonitrile	−0.33
Ethanol	−0.24
Acetone	−0.23
Tetrahydrofuran	0.49
Ethyl acetate	0.68
Pyridine	0.71
Butanol[a]	0.80
Diethyl ether	0.85
Propyl acetate[a]	1.2
Butyl acetate[a]	1.7
Dipropylether[a]	1.9
Chloroform	2.0
Benzene	2.0
Pentyl acetate[a]	2.2
Toluene	2.5
Octanol[a]	2.9
Dibutyl ether[a]	2.9
Pentane[a]	3.0
Carbon tetrachloride	3.0
Cyclohexane	3.2
Hexane[a]	3.5
Octane[a]	4.5
Decane[a]	5.6
Dodecane[a]	6.6

[a]Since the specific place of a molecular fragment is not significant for the log P value, only one solvent for every structural isomer is listed. Thus, the log P values of *n-/iso*propanol, *n-/tert*-butanol and di-*n*-/di-*iso*propyl ether are identical

Water Content (Water Activity) The ability of the solvent to strip off the bound water from the enzyme's surface depends not only on its polarity but also on its water content. Thus, the water content – more precisely the water activity (a_W) [82] – of an organic solvent has to be adjusted to the enzyme's requirements in order to ensure optimum activity [83–85]. The minimum amount of water required to maintain enzyme activity also depends on the enzyme type: Whereas lipases are able to operate at extremely low water activities of a_W 0.0–0.2, oxidoreductases and glycosidases require a_W of 0.1–0.7 and 0.5–0.8, respectively [86].

As a rule of thumb, acceptable activities are obtained in water-saturated organic solvents by using a buffer of low ionic strength. For large-scale applications, however, careful adjustment and maintenance of the water activity of the system is highly recommended to ensure optimal results. This can be conveniently achieved by a pair of salt hydrates added to the solvent by functioning as a 'water-buffer' (Table 3.4) [87–89]. Alternatively, a saturated salt solution (being in equilibrium with a sufficient amount of undissolved salt) is circulated through the reaction compartment via a silicone tubing that is submerged in the reaction medium. Any water produced (or consumed) during the reaction is removed (or added) by diffusion through the tube walls, thus maintaining an equilibrium a_W set by the salt solution used [90].

Table 3.4 Water activity (a_W) of saturated salt solutions and pairs of salt hydrates

Salt[a]	a_W	Salt-hydrate pair	a_W[b]
LiBr	0.06	$CaCl_2 \cdot H_2O/2\ H_2O$	0.037
LiCl	0.11	NaI anh./2 H_2O	0.12
$MgCl_2$	0.33	Na_2HPO_4 anh./2 H_2O	0.16
K_2CO_3	0.43	NaOAc anh./3 H_2O	0.28
$Mg(NO_3)_2$	0.54	NaBr anh./2 H_2O	0.33
NaBr	0.58	$Na_4P_2O_7$ anh./7 H_2O	0.46
NaCl	0.75	$Na_2HPO_4 \cdot 2\ H_2O/7\ H_2O$	0.57
KCl	0.84	Na_2SO_4 anh./10 H_2O	0.76
K_2SO_4	0.97	$Na_2HPO_4 \cdot 7\ H_2O/12\ H_2O$	0.80

[a]In equilibrium with a saturated salt solution
[b]At 20 °C

Effects of Additives   The addition of enzyme-stabilizing agents – often denoted as 'activators' or 'enhancers' – at low concentration may be beneficial [91–93]. Although the effects of the stabilizers on the protein are only poorly understood on a molecular level making this technique therefore rather empirical, several groups of additives can be recommended [94, 95]:

- Polyalcohols such as carbohydrates, sugar alcohols, or glycerol are well known to stabilize proteins [96, 97] as well as inactive proteins (bovine serum albumin) and polymers which have a certain structural resemblance to that of water (e.g., polyethylene glycol, polyvinyl alcohol, and derivatives thereof).
- Small polar organic molecules (e.g., N,N-dimethyl formamide and formamide) are known to enhance reaction rates by acting as 'molecular lubricants' [98–100].
- The addition of salts (LiCl, NaCl, KCl [101]) or weak organic bases (e.g., triethylamine, pyridine [102, 103]) may improve reaction rates and selectivities via formation of salt-pairs of substrate and/or product, which shift equilibria into the desired direction.

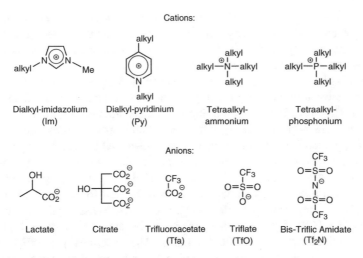

Fig. 3.1 Ionic components for the generation of ionic liquids

Ionic Liquids and Deep Eutectic Solvents Salts that do not crystallize at (or close to) room temperature are called 'ionic liquids' (IL). Due to their exceptional properties, such as outstanding thermal stability (up to 300 °C [104]), near-zero vapor pressure [105] and unconventional miscibility properties they are heralded as safe 'green' solvents and they are expected to replace some volatile and flammable organic solvents in the future [106, 107]. The most widely used components for the generation of ILs consist of peralkylated imidazolium, pyridinium, ammonium, and phosphonium cations and carboxylate, triflate, and triflic amidate anions (Fig. 3.1). Heavily fluorinated anions, such as BF_4^- and PF_6^- were recently replaced by more inocuous carboxylates for environmental reasons. Although tests using Reichardt's dye indicate that the polarity of ILs is similar to that of methanol, N-methylformamide, and 2-chloroethanol – which rapidly inactivate enzymes – the ILs surprisingly don't [108, 109] and it appears that enzymes that work in lipophilic organic solvents will also act in more polar ILs.

The polar nature of ILs increases the solubility of polar substrates, such as carbohydrates, which ensures enhanced reaction rates. Other potential advantages are increased enzyme stability [110] or stereoselectivity [111]. Furthermore, the properties of ILs can be easily tailored by simply choosing another combination of ions. On the down-side, ILs are considerably more expensive than organic solvents and are more viscous, which complicates their handling. Some components of ILs are quite toxic and are not easily biodegradable [112]; in addition, the commonly used anion PF_6^- (and to some extent also BF_4^-) is hydrolytically unstable and releases HPO_2F_2, H_2PO_3F and H_3PO_4 together with highly corrosive HF in water [113].

Biocatalysis in ILs dates back to the year 2000 [114–116] and more recent studies indicate that (almost) all types of enzymes may act in such systems. Successful examples were demonstrated for transesterification, perhydrolysis and ammonolysis (using lipases [117], esterases, and proteases), amide/peptide synthesis (using proteases), epoxide hydrolysis (using epoxide hydrolases), and glycoside synthesis (using glycosidases). Even redox-transformations, such as carbonyl reduction and sulfoxidation are possible [118–124].

Many disadvantages of ILs, such as toxicity, limited biodegradability and high cost, can be circumvented by a subgroup of ionic liquids termed 'deep eutectic solvents'. They are formed by mixing quaternary ammonium salts, e.g. choline citrate or acetate, with an uncharged hydrogen-bond donating component, e.g. urea, isosorbide, glycerol or ethylene glycol. These advanced ILs are biodegradable and sustain enzyme activity remarkably well despite the presence of high concentrations of denaturing agents, such as urea [125–127]. The first proof-of-concept concerning the use of deep eutectic solvents in biotransformations only appeared in 2008 [128]. The reactions studied so far encompass transesterification by lipases and proteases [129, 130], epoxide hydrolysis by epoxide hydrolases [131] and protease-catalysed peptide synthesis [132]. The reactions generally exhibited rates and (enantio)selectivities comparable to or higher than those reported for conventional organic solvents.

Alternatively, enzyme-catalyzed reactions may be performed in nonconventional media composed of microemulsions and liquid crystals [133]. The use of these systems, however, requires a great deal of knowledge of bioprocess engineering for the separation of the surfactant from substrate(s) and/or product(s).

Supercritical Gases Instead of a lipophilic organic solvent, supercritical gases such as carbon dioxide,[1] freons (e.g., CHF_3), hydrocarbons (ethane, ethene, propane), or inorganic compounds (SF_6, N_2O) which exhibit solubility properties similar to that of a hydrocarbon such as hexane, can be used as solvent or cosolvent for the enzymatic transformation of lipophilic organic compounds [134–137]. Enzymes are as stable in these media as in lipophilic organic solvents. The use of supercritical gases is not restricted to a particular class of enzyme but, not surprisingly, the use of hydrolases is dominant. For instance, esterification [138], transesterification [139, 140], alcoholysis [141], and hydrolysis [142] are known as well as hydroxylation [143] and dehydrogenation reactions [144]. The most striking advantages of this type of solvent are a lack of toxicity, easy removal and the low viscosity, which is intermediate between those of gases and those of liquids. This latter property ensures high diffusion, being about one to two orders of magnitude higher than in common solvents. Furthermore, small variations in temperature or pressure may result in large solubility changes near the critical point, which allows to control an enzyme's catalytic properties such as reaction rate or stereoselectivity [145]. However, some disadvantages should be mentioned. The high-pressure equipment, that must withstand several hundred atmospheres pressure, requires a considerable initial investment and the depressurization step may cause enzyme denaturation due to mechanical stress [146, 147]. In addition, some supercritical

---

[1]T_{crit} 31 °C and p_{crit} 73 bar.

fluids may react with sensitive groups located at the enzyme's surface causing a loss of activity. For instance, carbon dioxide is known to reversibly react with ε-amino groups of lysine residues by forming carbamates, going in hand with the removal/ formation of a positive charge at the enzyme's surface [148, 149]. The main use of supercritical gases as solvents is the production of 'natural' compounds used in cosmetics and food.

The following basic rules should be considered for the application of solid ('dry') enzymes in organic media having a low water content:

- Hydrophobic solvents are more compatible than hydrophilic ones (log P of the organic solvent should be greater than ~1.5).
- The water layer bound to the enzyme must be maintained; this is accomplished by using water-saturated organic solvents or, alternatively, via control of the water activity.
- The 'micro-pH' must be that of the pH-optimum of the enzyme in water, a prerequisite that is fulfilled if the protein was isolated from an aqueous solution at the pH-optimum.
- Stirring, shaking, or sonication is necessary in order to maximize diffusion of substrate to the catalyst's surface.
- The addition of enzyme-stabilizing agents may improve the stability of the solid enzyme preparation significantly.

3.1.1   Ester Synthesis

Esterification
In every synthetic reaction where a net amount of water is formed (such as an ester synthesis from an alcohol and a carboxylic acid [40–42]) physicochemical problems arise. Due to the fact that the lipophilic solvent (log $P > 1.5$) is unable to accommodate the water which is gradually produced during the course of the reaction, it is collected at the hydrophilic enzyme surface. As a consequence, the water forms a discrete aqueous phase which entraps the enzyme, finally separating substrate and enzyme from each other by a polar interface, which is difficult to penetrate for lipophilic substrate/product molecules. Thus, the rate slows down and the reaction may cease before reaching the desired extent of conversion. Furthermore, at elevated water activity the reverse hydrolysis reaction prevails setting a low ceiling for the conversion. In order to solve this problem, two techniques have been developed.

- Removal of water from the system [150] (e.g., by evaporation [151], azeotropic distillation [152]), or via chemical drying [153, 154] via addition of molecular sieves or water-scavenging inorganic salts [155].
- Alternatively, the formation of water may be avoided by employing an *acyl-transfer* step rather than an esterification reaction.

Some of the methods for the removal of water have inherent disadvantages and are therefore not trivial. Evaporation of water from the reaction mixture can only be efficient if the alcohol and acid reactants have a low volatility (high boiling point). On the other hand, recovery of (solid) enzymes from organic solvents in the presence of solid inorganic water-scavengers such as salts or molecular sieves may be troublesome.

Direct enzymatic esterification on industrial scale is commonly used for the production of cosmetic ingredients, where purity and appearance (odor, color) are a critical issue. Furthermore, EU-legislation allows to classify products obtained from natural sources via physical or biocatalytic processes as 'natural', which translates into a significantly higher price. Enzymatic esterification is usually performed in solvent-free conditions at elevated temperatures (60–70 °C) using a thermostable lipase. In order to drive the reaction towards completion, the water produced during the reaction is evaporated at reduced pressure (0.01 bar). The key parameters of the *Candida antarctica* lipase B catalyzed production of the emmolient ester myristyl myristate are compared to those of the state-of-the-art conventional process based on Sn^{+2} oxalate (Scheme 3.1, top) [156, 157]. It is obvious, that the mild reaction conditions require less energy input and the absence

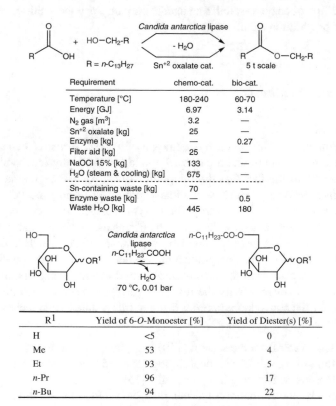

Requirement	chemo-cat.	bio-cat.
Temperature [°C]	180-240	60-70
Energy [GJ]	6.97	3.14
N_2 gas [m³]	3.2	—
Sn^{+2} oxalate [kg]	25	—
Enzyme [kg]	—	0.27
Filter aid [kg]	25	—
NaOCl 15% [kg]	133	—
H_2O (steam & cooling) [kg]	675	—
Sn-containing waste [kg]	70	—
Enzyme waste [kg]	—	0.5
Waste H_2O [kg]	445	180

R^1	Yield of 6-*O*-Monoester [%]	Yield of Diester(s) [%]
H	<5	0
Me	53	4
Et	93	5
n-Pr	96	17
n-Bu	94	22

Scheme 3.1 Biocatalytic synthesis of long-chain fatty acid esters used in cosmetics

of side reactions make four steps of downstream processing (deodorization, bleaching, drying, filtration) redundant.

Using the same technology, 6-*O*-acyl derivatives of alkyl glucopyranosides, which are used in cosmetics as fully biodegradable nonionic surfactants [158], were synthesized from fatty acids and the corresponding 1-*O*-alkyl glucopyranosides under catalysis of thermostable *Candida antarctica* lipase B in the absence of solvents [151] (Scheme 3.1, bottom).

Acyl Transfer

Trans- or interesterifications, which do not form water during the course of the reaction, are usually easier to perform (Schemes 2.1 and 3.2) [159–161]. Furthermore, the water content of the reaction medium (more accurately the 'water activity', a_W), which is a crucial parameter for retaining the enzyme's activity, remains constant. As a consequence, it has only to be adjusted at the beginning of the reaction, but not constantly monitored. Any trace of chemically available 'bulk' water, which may be present in the reaction medium, is quickly consumed at the expense of acyl donor, which is usually used in excess. The structural water, which is required to retain the enzyme's activity is chemically 'not available' because it is too tightly bound onto the enzyme's surface to be removed.

In contrast to hydrolytic reactions, where the nucleophile (water) is always in excess (55 mol/L), the concentration of the 'foreign' nucleophile in acyl transfer reactions (such as another alcohol) is always limited. As a result, trans- and interesterification reactions involving non-activated esters as acyl donors are generally *reversible* in contrast to the *irreversible* nature of a hydrolytic reaction. This leads to a slow reaction rate and can cause a severe depletion of the selectivity of the reaction for kinetic reasons (Fig. 2.6).

In order to avoid the undesired depletion of the optical purity of (predominantly) the remaining substrate during an enzymatic resolution under *reversible* reaction conditions, two tricks can be applied to shift the equilibrium of the reaction.

- Use of a large excess of acyl donor may impede enzyme activity.
- A better solution, however, is the use of special acyl donors which ensure a more or less *irreversible* type of reaction.

The reversibility of transesterification reactions is caused by the comparable nucleophilicity of the incoming nucleophile (Nu^1) and the leaving group of the acyl donor (Nu^2), both of which compete for the acyl-enzyme intermediate in the forward and the reverse reaction (Scheme 3.2). If the nucleophilicity of the leaving group Nu^2 is decreased by the introduction of electron-withdrawing substituents, the reaction is shifted to the right, i.e., towards completion. This concept has been verified by the introduction of 'activated' esters [162], such as 2-haloethyl, cyanomethyl and oxime esters (Scheme 3.3). Although acyl transfer using activated esters is still reversible in principle, the equilibrium of the reaction is shifted so far to the product side that for preparative purposes it can be regarded as quasi-irreversible [163].

As shown in Scheme 3.2, the relative rate of the enantioselective acylation of (±)-2-octanol, catalyzed by porcine pancreatic lipase (PPL), was one to two orders of magnitude faster when 'activated' esters were used as acyl donors instead of 'nonactivated' methyl or ethyl alkanoates.

The following parameters should be considered before the 'activated' ester is chosen. Cyanomethyl esters have been used only rarely due to toxicity problems arising from formaldehyde cyanohydrin, which is liberated. 2-Haloethyl esters have been applied more widely. Among them, trifluoroethyl esters are the acyl donors of choice when the reactions are performed on a laboratory scale because their degree of activation is high and trifluoroethanol can be evaporated easily during workup procedures. For larger batches, trichloroethyl esters are more economic but the removal of trichloroethanol during work-up can be troublesome due to its high boiling point (151 °C). 2-Chloroethyl esters are cheap and the resulting 2-chloroethanol is easier to remove (bp 130 °C), but their degree of activation is limited.

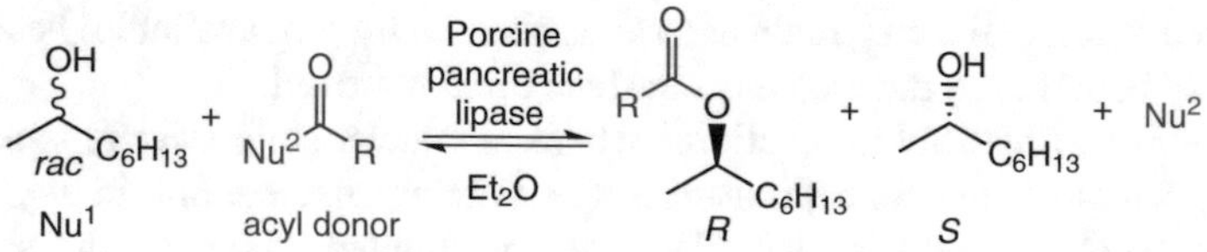

Acyl Donor	R	Leaving Group Nu^2	Initial Rate [%]
Ethyl acetate	Me	EtOH	0.3
2-Chloroethyl acetate	Me	$ClCH_2$-CH_2OH	1
Methyl butanoate	n-Pr	MeOH	5
Ethyl cyanoacetate	$N{\equiv}C$-CH_2-	EtOH	6
Trichloroethyl trichloroacetate	Cl_3C-	Cl_3C-CH_2OH	7
Methyl bromoacetate	$BrCH_2$-	MeOH	14
Tributyrin	n-Pr	dibutyrin	34
Trichloroethyl butanoate	n-Pr	Cl_3C-CH_2OH	58
Trichloroethyl heptanoate	n-C_6H_{13}-	Cl_3C-CH_2OH	100

Scheme 3.2   Reaction rate of enzymatic acylation of (±)-2-octanol depending on acyl donor

As an alternative, oxime esters have been proposed as acyl donors for acyl-transfer reactions [164]. During the reaction a weakly nucleophilic oxime is liberated which is unable to compete with the substrate alcohol for the acyl-enzyme intermediate. However, cosubstrate inhibition and problems in separating the nonvolatile oxime from the substrate alcohol during work-up may be encountered. Alternatively, thioesters have been used [165, 166]. The thiols liberated as byproducts are highly volatile and are easily removed by evaporation, thus driving the equilibrium. However, excellent ventilation is recommended in order to avoid a noxious laboratory atmosphere and complaints of labmates.

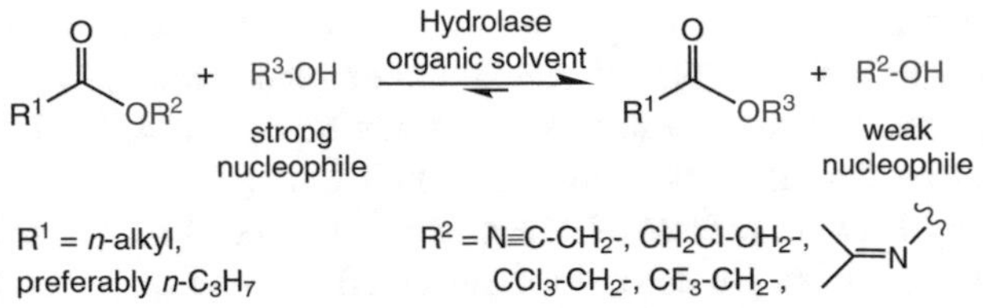

Scheme 3.3 Quasi-irreversible enzymatic acyl transfer using activated esters

In contrast to the above-mentioned acyl donors which shift the equilibrium of the reaction to the product side by liberating a weakly nucleophilic co-product alcohol species, several concepts have been proposed for making the reaction completely irreversible (Scheme 3.4).

Enol esters such as vinyl or isopropenyl esters liberate unstable enols as coproducts, which tautomerize to give the corresponding aldehydes or ketones [167, 168] (Scheme 3.4). Thus, the reaction becomes *completely irreversible* and this ensures that all the benefits with regard to a rapid reaction rate and a high selectivity are accrued. Acyl transfer using enol esters has been shown to be about only ten times slower than hydrolysis (in aqueous solution) and about 10–100 times faster than acyl-transfer reactions using activated esters. In contrast, when nonactivated esters such as ethyl acetate were used, reaction rates of about 10^{-3}–10^{-4} of that of the hydrolytic reaction are observed (Table 3.5) [169].

Table 3.5 Relative rates of reactions catalyzed by hydrolases

Reaction	Acyl donor	Relative rate
Ester hydrolysis	–	10,000
Acyl transfer	Enol esters	1000
Acyl transfer	Acid anhydrides	1000
Acyl transfer	Activated esters	10–100
Acyl transfer	Nonactivated esters	1–10

Due to steric reasons, vinyl esters give better reaction rates than isopropenyl esters and the former are therefore used most widely, but their use is not without drawbacks. Acetaldehyde, which is liberated during the reaction, is known to act as an alkylating agent by forming Schiff bases with the terminal amino group of lysine residues [170]. Thus, a positive charge is removed from the enzyme's surface during the course of this reaction, which may cause enzyme deactivation. The extent of this depends on the nature of the enzyme [171, 172]. Whereas the majority of the more widely employed lipases seem to be quite stable, *Candida rugosa* (CRL) and *Geotrichum candidum* lipase are very sensitive.

Covalent immobilization of CRL onto an epoxy-activated macroscopic carrier leads to selective monoalkylation of the lysine amino residues which are involved in the deactivation reaction with retention of the positive charge. In contrast to the native enzyme, the immobilized enzyme is inert towards the formation of Schiff bases, which results not only in a greatly stabilized activity but also in a significant enhancement in selectivity [173]. The addition of a molecular sieve to the medium in order to trap acetaldehyde seems to have some benefit [174, 175]. Alternatively, the lipase may be stabilized by adsorption onto Celite [176].

The possibly harmful effects of acetaldehyde can be avoided by employing *i*-propenyl acetate, which yields (more innocuous) acetone as byproduct. Alternatively, ethoxyvinyl acetate can be used as acyl donor (Scheme 3.4) [177–179]. The latter renders ethyl acetate as byproduct, which is generally regarded as innocuous to ester-hydrolyzing enzymes. Unfortunately, ethoxyvinyl esters are rather expensive.

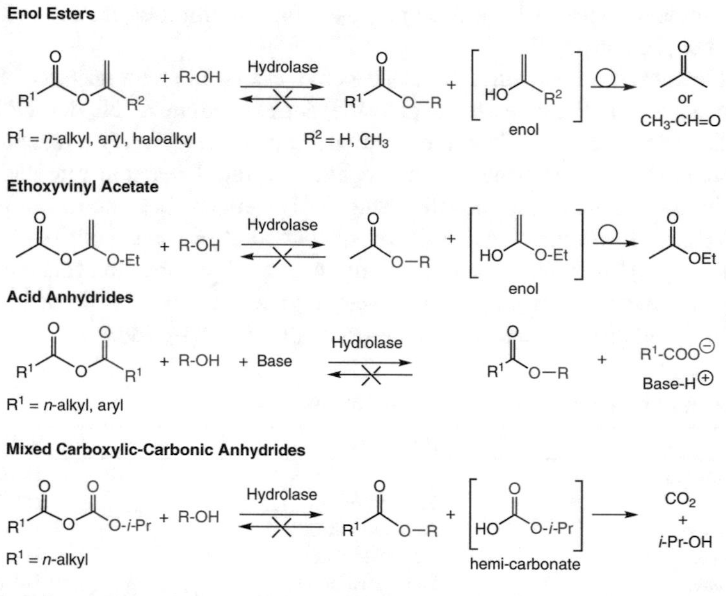

Enol Esters

R^1 = n-alkyl, aryl, haloalkyl   R^2 = H, CH$_3$

Ethoxyvinyl Acetate

Acid Anhydrides

R^1 = n-alkyl, aryl

Mixed Carboxylic-Carbonic Anhydrides

R^1 = n-alkyl   hemi-carbonate

Scheme 3.4   Irreversible enzymatic acylation using enol esters and acid anhydrides

Acid Anhydrides Another useful method of achieving completely irreversible acyl-transfer reactions is the use of *acid anhydrides* (Scheme 3.4) [180]. The selectivities achieved are usually high and the reaction rates are about the same as with enol esters. One of the advantages of this technique is that no aldehydic byproducts are formed and the enzyme is not acylated under the conditions employed, making its reuse possible.

However, the carboxylic acid formed as by-product may lead to a decrease of the pH in the micro-environment of the enzyme, thus leading to a depletion of activity and selectivity. The CRL-catalyzed resolution of the bicyclic tetrachloroalcohol shown in Scheme 3.5, using acetic anhydride as acyl donor, initially proceeded with only moderate selectivity ($E = 18$). Addition of a weak inorganic or (preferably) organic base such as 2,6-lutidine which functions as an acid scavenger, led to a greater than tenfold increase in selectivity [181]. A similar acid-quenching effect could be observed by immobilization of CRL onto diatomaceous earth (Celite).

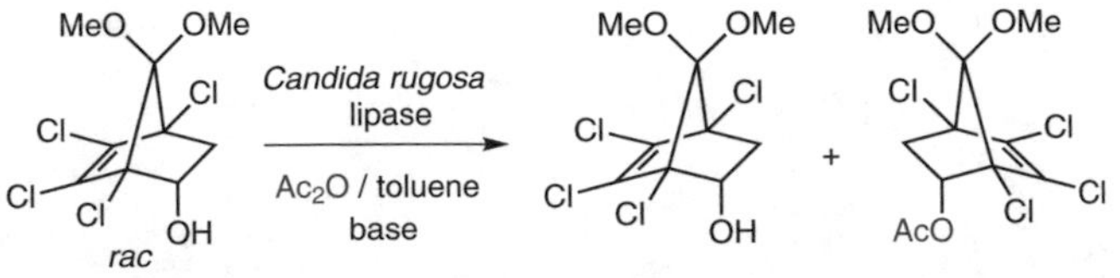

MeO OMe   *Candida rugosa* lipase   MeO OMe   MeO OMe

Cl Cl   Ac₂O / toluene base   Cl Cl   +   Cl Cl

rac

Base	Reaction Rate	Selectivity (E)
none	good	18
KHCO₃	low	>200
2,6-lutidine	good	>200

Scheme 3.5 Selectivity enhancement of acyl transfer using acetic anhydride via addition of base

In contrast to symmetric acid anhydrides, which liberate one equivalent of carboxylic acid as byproduct, mixed anhydrides composed of a straight-chain carboxylic acid (R^1-CO_2H) and a carbonic ester bearing a branched secondary alcohol group (e.g., isopropyl) can be used instead. The hydrolase takes off the straight-chain carboxylic acid moiety from the acyl donor by liberating an unstable hemi-carbonate ester, which undergoes rapid decarboxylation, forming innocuous carbon dioxide and a *sec*-alcohol, thereby rendering the reaction completely irreversible (Scheme 3.4) [182]. Cyclic acid anhydrides, such as succinic and glutaric acid anhydrides lead to the formation of a hemiester [183–185]. Due to the presence of the carboxylic acid moiety, separation of the formed hemiester product from the nonreacted alcohol enantiomer is particularly easy using a (basic) aqueous-organic solvent system. Consequently, cyclic acid anhydrides are advantageous in large-scale applications.

Besides the more often-used acyl donors mentioned above, others which would also ensure an irreversible type of reaction have been investigated [186]. Bearing in mind that most of the problems of irreversible enzymatic acyl transfer arise from the formation of unavoidable byproducts, emphasis has been put on finding acyl donors that possess cyclic structures, which would not liberate any byproducts at all. However, with candidates such as lactones, lactams, enol lactones (e.g., diketene [187, 188]), and oxazolin-5-one derivatives [189], the drawbacks often outweighed their merits.

Enzyme-catalyzed acyl transfer can be applied to a number of different synthetic problems. The majority of applications that have been reported involve the desymmetrization of prochiral and *meso*-diols or the kinetic resolution of racemic primary and secondary alcohols. Since, as a rule, an enzyme's preference for a specific enantiomer remains constant when water is replaced by an organic solvent, it is always the *same enantiomer* which is preferably accepted in hydrolysis and ester synthesis. Taking into consideration that hydrolysis and esterification represent reactions in *opposite directions*, products of *opposite configuration* are obtained (Scheme 3.6). In other words, if the (R)-enantiomer of an ester is *hydrolyzed* at a faster rate than its (S)-counterpart [yielding an (R)-alcohol and an (S)-ester], *esterification* of the racemic alcohol will lead to the formation of an (S)-alcohol and an (R)-ester.

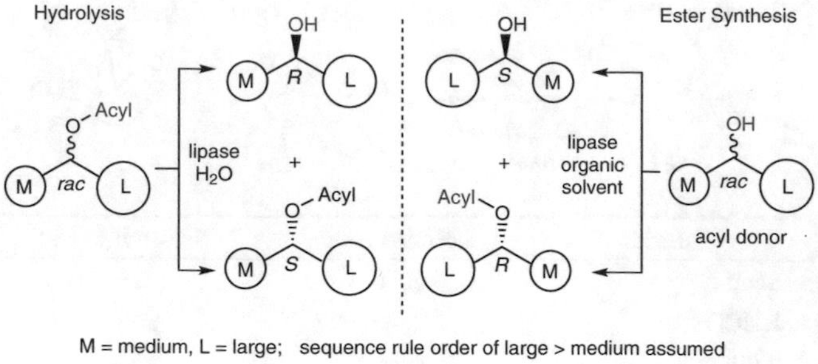

M = medium, L = large;   sequence rule order of large > medium assumed

Scheme 3.6 Symmetry in hydrolysis and ester synthesis reactions

Separation of E/Z-Stereoisomers Stereoisomeric mixtures of the allylic terpene alcohols, geraniol and nerol, which are used in flavor and fragrance formulations, were separated by selective acylation with an acid anhydride using porcine pancreatic lipase (PPL) as catalyst (Scheme 3.7) [190]. Depending on the acyl donor employed, the slightly less hindered geraniol was more quickly acylated to give geranyl acetate leaving nerol unreacted. Acetic anhydride proved to be unsuitable, giving a low yield and poor selectivity, but longer-chain acid anhydrides were used successfully.

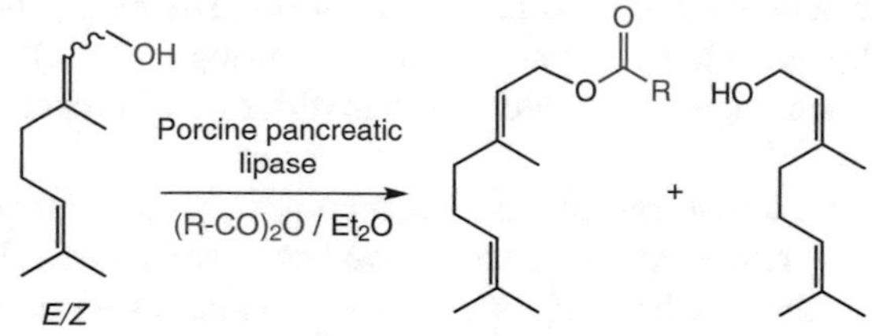

R	E-Geranyl Ester [%]	Z-Nerol [%]	Selectivity [k_E/k_Z]
n-C$_3$H$_7$-	85	16	11
n-C$_5$H$_{11}$-	66	7	13
n-C$_7$H$_{15}$-	72	7	15

Scheme 3.7 E/Z-Stereoselective enzymatic acylation of terpene alcohols

Desymmetrization of Prochiral and *meso*-Diols Chiral 1,3-propanediol derivatives are useful building blocks for the preparation of enantiomerically pure bio-active compounds such as phospholipids [191], platelet activating factor antagonists [192], and renin inhibitors [193]. A simple access to these synthons starts from 2-substituted 1,3-propanediols (Scheme 3.8, top). Depending on the substituent R, (R)- or (S)-monoesters were obtained in excellent optical purities using *Pseudomonas* sp. lipase (PSL) [194–197]. The last three entries demonstrate an enhancement in selectivity upon lowering the reaction temperature [198].

The desymmetrization of a prochiral 2-substituted 1,3-propanediol building block using *Candida antarctica* lipase B allowed the efficient synthesis of multiton quantities of the antifungal agent posaconazol (Scheme 3.8, bottom) [199]. Moderate chemical and optical yields of monoacetate obtained using vinyl acetate as acyl donor were overcome by using the sterically more demanding *i*-butyric anhydride in presence of $NaHCO_3$ at low temperatures to suppress undesired background acylation and acyl migration.

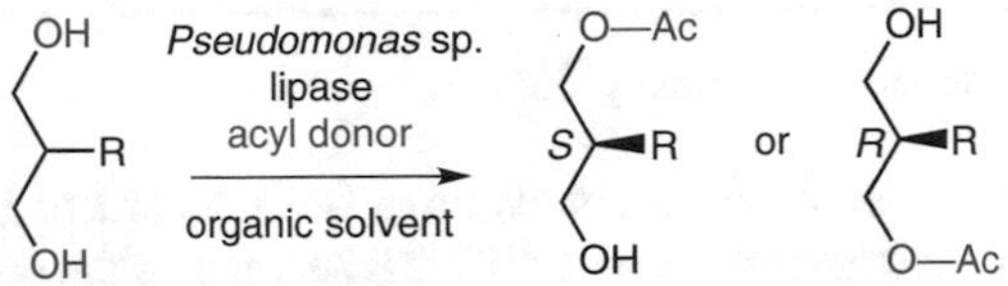

R	Acyl Donor	Solvent	Temperature [°C]	Configuration	e.e. [%]
Me	vinyl acetate	$CHCl_3$	r.t.	S	>98
CH_2-Ph	vinyl acetate[a]	none	r.t.	R	>94
O-CH_2-Ph	*i*-propenyl acetate	$CHCl_3$	r.t.	S	96
O-CH_2-Ph	vinyl acetate	none	25	S	90
O-CH_2-Ph	vinyl acetate	none	17	S	92
O-CH_2-Ph	vinyl acetate	none	8	S	94

[a] used as acyl donor and as solvent.

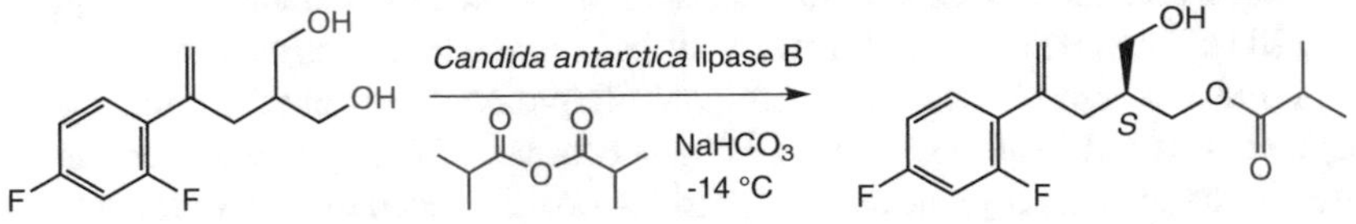

Scheme 3.8 Desymmetrization of 2-substituted propane-1,3-diols

Cyclic *meso-cis*-diols were asymmetrically acylated quite efficiently to give the respective chiral monoester by a PSL [200]. Whereas a slow reaction rate was observed in a reversible reaction using ethyl acetate as acyl donor, the reaction was about ten times faster when vinyl acetate was employed.

Kinetic Resolution of Alcohols Primary alcohols are difficult to resolve because their chirality center is more distant from the reacting alcohol group compared to *sec*-alcohols. However, *Pseudomonas* sp. lipase showed moderate to good selectivities using vinyl acetate [201] or acetic anhydride as the acyl donor (Scheme 3.9). It is apparent that the difference in size of substituents (R^1, R^2) has a strong impact on the enantioselectivity (compare Scheme 2.45). Whereas the selectivities achieved were moderate with alkyl and aryl substituents, substrate modification via introduction of a bulky sulfur atom in R^2 helped considerably. In this way, chiral isoprenoid synthons having a C_5-backbone were obtained in >98% enantiomeric excess.

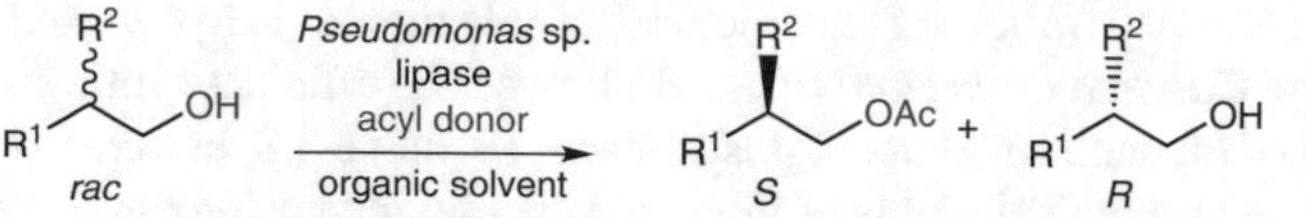

Scheme 3.9 Kinetic resolution of primary alcohols

Acyl Donor	Solvent	R^1	R^2	Selectivity (E)
Ac$_2$O	benzene	Et	n-Bu	2
Ac$_2$O	benzene	Me	Ph	12
vinyl acetate	CHCl$_3$	Me	(CH$_2$)$_2$SPh	>100

Numerous acyclic secondary alcohols have been separated into their enantiomers using lipase-catalyzed acyl-transfer [202]. As long as the difference in size of the substituents is substantial, excellent selectivities were obtained with lipases from *Candida antarctica* (CAL) and *Pseudomonas* sp. (PSL) [203]. *sec*-Alcohols bearing unsaturated functional groups, such as olefins, alkynes or allenes, which serve as handles for further functionalization, are given as an illustrative example for this methodology (Scheme 3.10, entries 1–7) [204, 205].

A generally applicable method for the preparation of optically active epoxides makes use of a lipase-catalyzed resolution of halohydrins bearing the halogen in the terminal position (Scheme 3.10, entries 8–10). *Pseudomonas* sp. lipase-catalyzed acylation of racemic halohydrins affords a readily separable mixture of (R)-halohydrin and the corresponding (S)-ester in good to excellent optical purities [206, 207]. Treatment of the former with base leads to the formation of epoxides with no loss of optical purity. A semiquantitative comparison of the reaction rate obtained with different acyl donors using substrates of this type revealed that they were in the order ethyl acetate << trichloroethyl acetate < isopropenyl acetate < vinyl butanoate ~ vinyl octanoate ~ vinyl acetate [208].

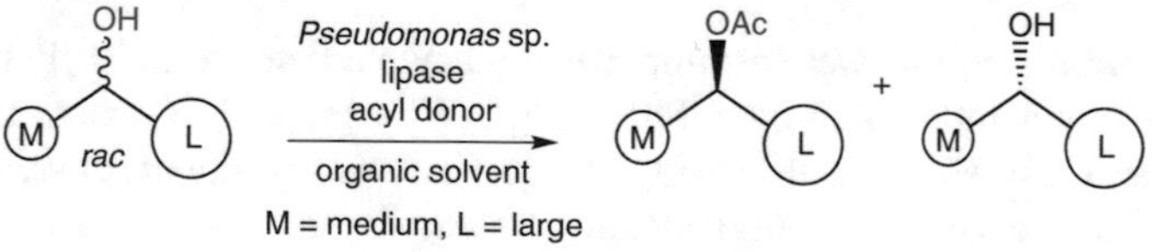

Medium	Large	Acyl Donor	Selectivity (E)
Me	Ph-C(=CH$_2$)-	vinyl actate	>20
CH$_2$=CH-	(E)-Ph-CH=CH-	vinyl actate	>20
CH$_2$=CH-	Ph-C≡C-	vinyl actate	>20
HC≡C-	(E)-Ph-CH=CH-	vinyl actate	>20
Me	n-Bu-C≡C-	vinyl actate	>20
CH$_2$=C=CH-	Ph-CH$_2$-	vinyl actate	>20
Me	Me$_3$Si-C≡C-	vinyl actate	>20
CH$_2$-Cl	Ph-	i-propenyl acetate	100
CH$_2$-Br	2-Naphthyl-	i-propenyl acetate	95
CH$_2$-Cl	p-Tos-O-CH$_2$-	i-propenyl acetate	>100

Scheme 3.10 Kinetic resolution of unsaturated *sec*-alcohols and halohydrins

Enantiomerically pure *trans*-cycloalkane-1,2-diols are of interest for the synthesis of optically active crown-ethers [209] or as chiral auxiliaries for the preparation of bidentate ligands [210]. A convenient method for their preparation consists in PSL-catalyzed enantioselective acylation (following a sequential resolution pattern, see Fig. 2.7), which yields varying amounts of diester, monoester, and remaining nonreacted diol in excellent optical purities [211]. The advantage of acyl-transfer in organic solvents lies in the suppression of undesired acyl migration, which plagues the hydrolysis of the corresponding diesters [26, 212, 213].

Along the same lines, the remarkable synthetic potential of enzyme-catalyzed irreversible acyl transfer in nearly anhydrous organic solvents can be demonstrated particularly well by the transformation of alcoholic substrates (such as organometallics or cyanohydrins) which are prone to decomposition reactions in an aqueous medium and thus cannot be transformed via enzyme-catalyzed hydrolysis reactions.

For instance, organometallic compounds such as hydrolytically labile chromiumtricarbonyl complexes, which are of interest as chiral auxiliary reagents for asymmetric synthesis due to their axial chirality [214], were easily resolved by PSL (Scheme 3.11) [215, 216]. A remarkable enhancement in selectivity was obtained when the acyl moiety of the vinyl ester used as acyl donor was varied. This concept was successfully employed for the resolution of 1-ferrocenylethanol, which cannot be well resolved via enzymatic hydrolysis due to the lability of 1-ferrocenyl acetate in aqueous systems [217, 218].

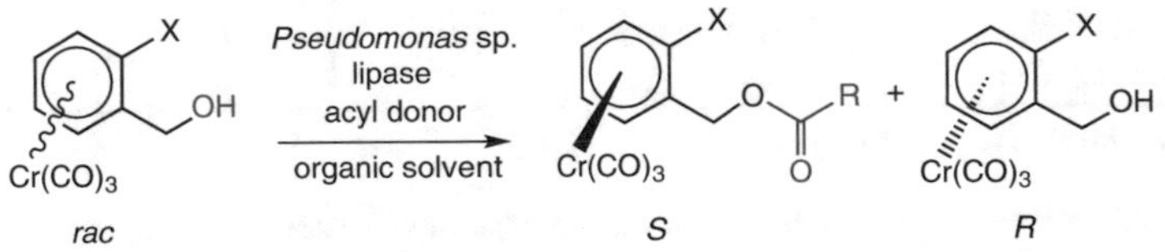

X	Acyl Donor	R	Solvent	Selectivity (E)
SiMe3	*i*-propenyl acetate	Me	none	30
Me	*i*-propenyl acetate	Me	none	>200
Me	vinyl acetate	Me	toluene	39
Me	vinyl octanoate	n-C7H15	toluene	67
Me	vinyl palmitate	n-C15H31	toluene	>200

Scheme 3.11 Kinetic resolution of organo-metallic hydroxy compounds

Chiral hydroxyesters would be accessible via enzymatic hydrolysis of their acyloxy esters, but a commonly encountered disadvantage in such resolutions is the undesired hydrolysis of the carboxyl ester moiety which leads to the formation of hydroxyacids as byproducts [219]. In contrast to hydrolysis, the acyl transfer mode is highly selective for *O*-acylation, because the hydroxyl functionality is the only nucleophile in the substrate molecule which can be acylated (Scheme 3.12), and no hydrolysis of the carboxylic acid ester can take place due to the absence of water [220]. This concept was successfully applied to the resolution of γ-hydroxy-α,β-unsaturated esters which are used for the synthesis of statin analogs [221, 222]. A reversal of the stereopreference of PSL was observed when the size of the side-chain substituent R was gradually increased.

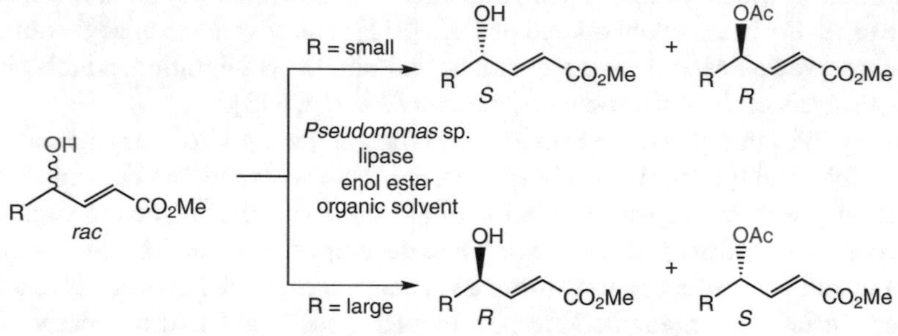

R	Acyl Donor	Configuration		Selectivity (E)
		Acetoxy Ester	Hydroxy Ester	
Me	*i*-propenyl acetate	R	S	>30
Et	*i*-propenyl acetate	R	S	>150
n-Pr	*i*-propenyl acetate	R	S	>20
i-Pr	vinyl acetate	S	R	1.6
i-Pr-CH$_2$-	vinyl acetate	S	R	2.5
C$_6$H$_{11}$-CH$_2$-	vinyl acetate	S	R	13
Me$_2$ThexSiO-(CH$_2$)$_2$-	vinyl acetate	S	R	>150

Thex = thexyl (1,1,2-trimethylpropyl).

Scheme 3.12   Kinetic resolution of γ-hydroxy-α,β-unsaturated esters

This strategy has also been successfully applied to the preparation of optically active α-methylene-β-hydroxy esters and -ketones [223], which cannot be resolved using the Sharpless epoxidation technique because of the deactivating influence of the electron-withdrawing carbonyl group on the alkene unit [224]. Similarly, optically active cyclopentanoids carrying a terminal carboxylate group useful for prostaglandin synthesis were obtained without the occurrence of undesired side reactions [225].

Racemic hydroperoxides may be resolved in organic solvents via lipase-catalyzed acyl transfer (Scheme 3.13). Although the so-formed acetylated (R)-peroxy-species is unstable and spontaneously decomposes to form the corresponding ketone via elimination of acetic acid, the non-reacted (S)-hydroperoxide was isolated in varying optical purity [226]. This concept was also applied to the resolution of a hydroperoxy derivative of an unsaturated fatty acid ester [227].

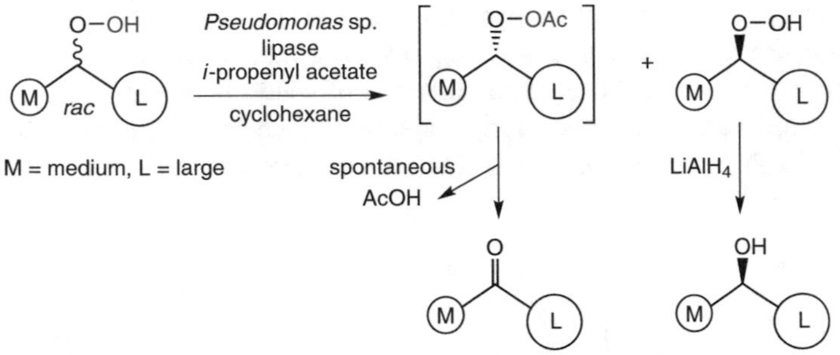

Medium	Large	E.e. Hydroperoxide [%]	Selectivity (E)
Me	n-Pr	10	1.2
Me	2-naphthyl	58	2.3
Et	Ph	62	3.7
Me	Ph	100	>20

Scheme 3.13 Kinetic resolution of hydroperoxides

Dynamic Resolution Lipase-catalyzed acyl transfer has become a well-established and popular method for the kinetic resolution of primary and secondary alcohols on industrial scale. In order to circumvent the limitations of kinetic resolution (i.e., a 50% theoretical yield of both enantiomers), several strategies have been developed, which achieve a more economic *dynamic* resolution process and allow the formation of a single stereoisomer as the sole product (for the theoretical background see Sect. 2.1.1). In contrast to compounds bearing a chiral center adjacent to an electron-withdrawing group (e.g., carboxylic acid esters, Scheme 2.39), which facilitates in-situ racemization via an achiral enolate, *sec*-alcohols are more difficult to racemize.

Two techniques of general applicability are worth considering (Scheme 3.14):

- Several types of *sec*-alcohols bearing a leaving group (Nu) attached to the carbinol moiety are chemically unstable and therefore prone to decomposition via a reversible elimination-addition process of a nucleophile (HNu) onto an aldehyde or ketone, respectively. This applies to cyanohydrins (Nu=C≡N), and hemi(thio)acetals (Nu=OR^2, SR^2) or hemiaminals (Nu=NHR^2), respectively. It is obvious that the corresponding cyclic structures – (thio)lactols, etc. – behave in the same way [228–230].

Optically pure cyanohydrins are required for the synthesis of synthetic pyrethroids, which are more environmentally acceptable agents for agricultural pest control than the classic highly chlorinated phenol derivatives (Scheme 2.208) [231]. They are important intermediates for the synthesis of chiral α-hydroxyacids, α-hydroxyaldehydes [232], and aminoalcohols [233, 234]. By asymmetric hydrolysis of their respective acetates using microbial lipases [235], only the remaining nontransformed substrate enantiomer can be obtained in high optical purity because the cyanohydrin product is spontaneously racemized as it is in equilibrium with the corresponding aldehyde and hydrocyanic acid at values above pH ~ 4 (Scheme 2.65). In the absence of water, however, the cyanohydrins are stable and can be isolated in high optical and chemical yields (Scheme 3.15) [25, 236]. In this manner, *both* enantiomers are accessible.

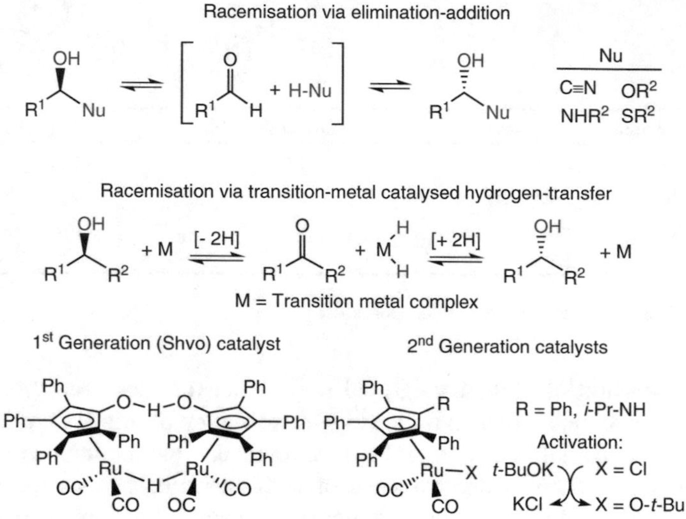

Scheme 3.14 Strategies for in-situ racemization of *sec*-alcohols in organic solvents

The kinetic resolution of cyanohydrins via enantioselective acylation may be converted into a dynamic process by making use of the chemical instability of cyanohydrins (Scheme 3.15) [237]. Thus, racemic cyanohydrins were generated from an aldehyde and acetone cyanohydrin (as a relatively safe source of hydrogen cyanide) under catalysis by an anion exchange resin. The latter also served as catalytic base for the in-situ racemization. Enantioselective acylation using PSL and *i*-propenyl acetate led to the exclusive formation of the corresponding (*S*)-cyanohydrin acetates in 47–91% optical purity.

An α-acetoxysulfide shown in Scheme 3.16 was used as central chiral building block for the synthesis of Lamivudine, a drug candidate for the treatment of HIV and HBV infections. Due to the different toxicities of the two enantiomers, an enantioselective route was required. Furthermore, applicability to large-scale synthesis and absence of any 'unwanted' enantiomers were important issues. The solution was found by using an approach which is closely related to the dynamic resolution of cyanohydrins, i.e., a lipase-catalyzed enantioselective esterification employing vinyl acetate as acyl donor [238]. Dynamic resolution was attempted by making use of the inherent instability of the racemic hemithioacetal substrate, which is in equilibrium with the corresponding aldehyde and thiol. Initially, the *Pseudomonas* sp. lipase catalyzed acyl transfer reaction spontaneously stopped at 50% conversion, indicating an insufficient in-situ racemization of the substrate. The

latter problem was circumvented by adding silica gel to the mixture, which catalyzed the reversible dissociation of the thioacetal into thiol and aldehyde. With the
latter modification, a range of (S)-hemithioacetal esters were obtained in excellent
optical purities with yields being considerably beyond the usual 50% limitation for
classical kinetic resolutions.

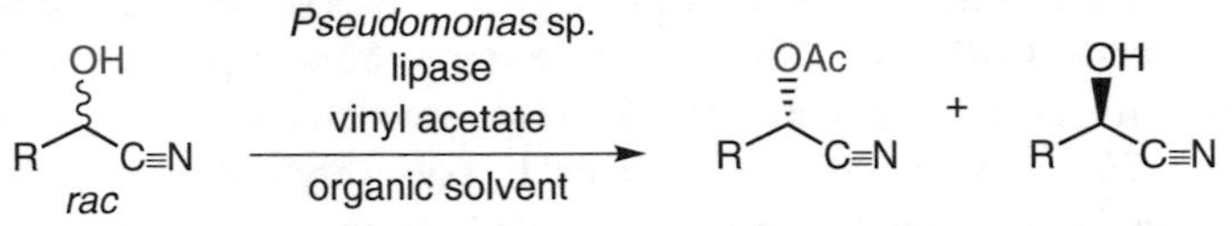

R	Solvent	e.e. [%] of		Selectivity
		Acetate	Cyanohydrin	(E)
n-Pr	CH2Cl2	55	33	5
Ph-CH2-O-CH2-	none	55	95	12
Ph-	CH2Cl2	68	80	13
4-HO-C6H4-	CH2Cl2	79	50	14
Ph-(CH2)2-	CH2Cl2	95	86	100

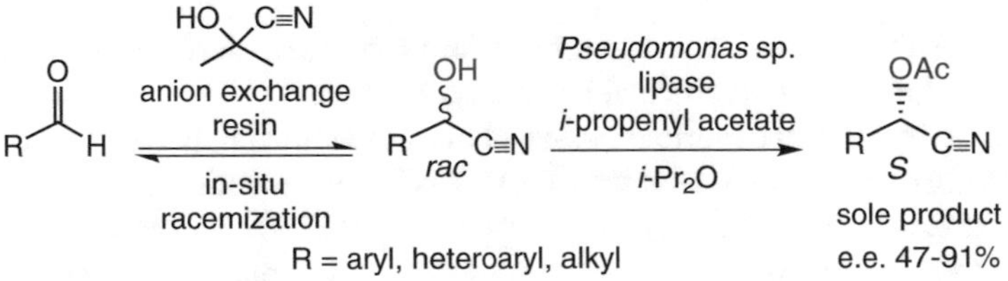

Scheme 3.15 Kinetic and dynamic resolution of cyanohydrins

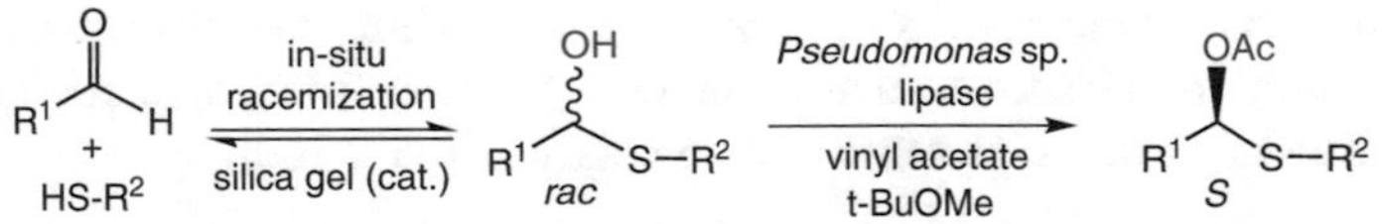

R^1	R^2	Product	
		yield [%]	e.e. [%]
-CO2Me	n-Bu-	63	>95
-CO2Me	Et3Si-O-(CH2)2-	87	90
Ac-O-CH2-	n-Bu-	87	87
Ac-O-CH2-	i-Pr-	65	>95
Ac-O-CH2-	n-Octyl-	85	>95

Scheme 3.16 Dynamic resolution of hemithioacetals

- Stereochemically stable sec-alcohols can be racemized via an oxidation-
  reduction sequence catalyzed by (transition) metal complexes based on Al, Ru,
  Rh, and Ir [239, 240]. More labile allylic alcohols could be racemized using a
  vanadium-catalyst [241]. However, both racemization techniques have their

potential pitfalls, since transition metal complexes may cause enzyme deactivation, whose biochemical mechanism is only poorly understood. Furthermore,
some transition metal complexes are incompatible with the commonly used enol
esters serving as acyl donors.

The first dynamic resolution making use of transition-metal catalyzed substrate-
racemization in presence of an enzyme was reported in 1996 [242]. Since then,
rapid progress was made and this technology is nowadays used on industrial-scale
[243] (Scheme 3.14) [244, 245]. First generation (Shvo-type) racemization catalysts were impeded by slow racemization rates, which required elevated temperatures (ca. 70 °C), which could be tolerated by only very few thermostable lipases.
In addition, popular enol-ester-type acyl donors, such as vinyl or *iso*propenyl
acetate were incompatible with the transition metal complex, which required the
use of *p*-chlorophenyl acetate liberating *p*-chlorophenol as toxic byproduct
[246]. During recent years, most of these initial drawbacks were circumvented
by the development of second-generation racemization catalysts, which do not
react with enol esters and show high racemization rates already at room temperature. The pre-catalysts have to be activated by the displacement of a Cl atom by *t*-
BuOK to render the catalytically active species [247–251].

Examples for the successful dynamic resolution of *sec*-alcohols using transition-
metal-lipase/protease combo-catalysis are shown below.

Dynamic resolution of various *sec*-alcohols was achieved by coupling a *Candida
antarctica* lipase-catalyzed acyl transfer to in-situ racemization based on a second-
generation transition metal complex (Scheme 3.17) [252]. In accordance with the
Kazlauskas rule (Scheme 2.45) (*R*)-acetate esters were obtained in excellent optical
purity and chemical yields were far beyond the 50% limit set for classical kinetic
resolution. This strategy is highly flexible and is also applicable to mixtures of
functional *sec*-alcohols [253–256] and *rac*- and *meso*-diols [257, 258]. In order to
access products of opposite configuration, the protease subtilisin, which shows
opposite enantiopreference to that of lipases (Scheme 2.46), was employed in a
dynamic transition-metal-protease combo-catalysis [259, 260].

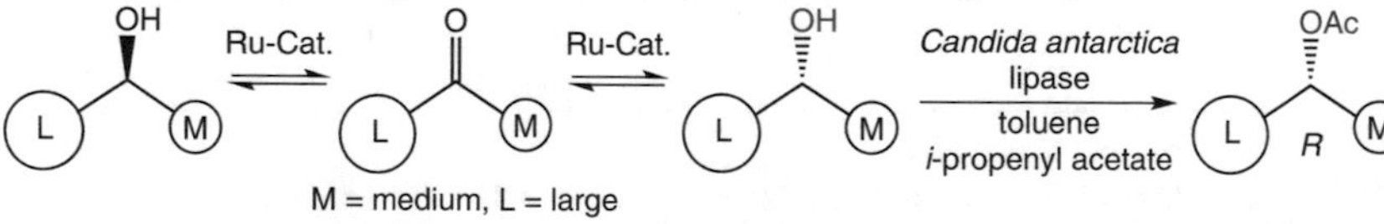

Large	Medium	E.e. [%]	Yield [%]
Ph	Me	>99	95
c-C$_6$H$_{11}$	Me	>99	86
c-C$_6$H$_{11}$	CH=CH$_2$	>99	90
Ph-CH$_2$	Me	>99	90
n-Hexyl	Me	91	89
(*E*)-Ph-CH=CH	Me	98	93
Ph	CH=CH$_2$	81	62
t-Bu-O-CH$_2$	Me	99	97

Scheme 3.17 Dynamic resolution of *sec*-alcohols via Ru-catalyzed in-situ racemization

β-Vinyl arylpropionic acids, which are key building blocks for the synthesis of γ-aminobutyric acid analogs for the treatment of neurodegenerative disorders, were subjected to dynamic kinetic resolution via enantioselective esterification catalyzed by *Candida antarctica* lipase B (Scheme 3.18). In order to avoid the formation of water and to maintain a constant water activity during the reaction, an *ortho*-ester (triethyl orthobenzoate) was employed as ethanol donor [261]. Racemization of the substrate, which cannot be achieved via redox or acid-base catalysis, was accomplished by $Rh(OAc)_2$ catalysis in toluene at 60 °C. Several (*S*)-esters were obtained in up to >99% e.e. and excellent isolated yields.

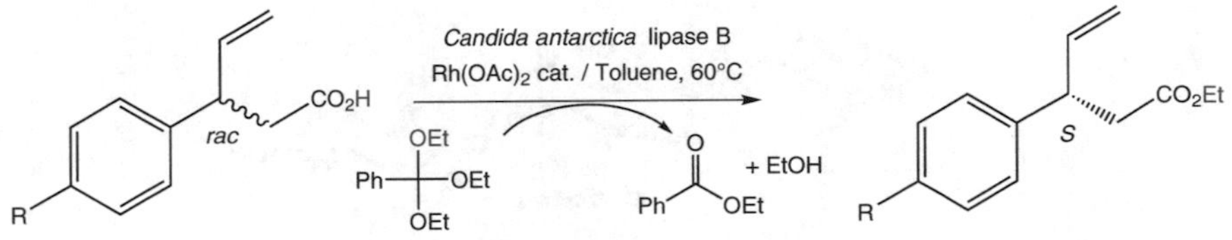

R	Yield [%]	E.e. [%]
NO_2	98	10
MeO	37	43
Cl	78	90
F	92	95
H	98	>99

Scheme 3.18 Dynamic resolution of β-vinyl arylpropionic esters via enantioselective esterification with in-situ substrate racemization

Regioselective Protection of Polyhydroxy Compounds Selective protection and deprotection of compounds containing multiple hydroxyl groups such as carbohydrates and steroids is a current problem in organic synthesis [262]. By using standard methodology, a series of multiple steps is usually required to achieve the desired combination of protected and free hydroxyl functionalities. By contrast, enzymatic acyl transfer reactions in organic solvents have proven to be extremely powerful for such transformations by making use of the *regioselectivity* of hydrolytic enzymes [263–266]. Whereas the acylation of steroids in lipophilic organic solvents having a desired log *P* of greater than 2 is comparatively facile, carbohydrate derivatives are only scarcely soluble in these media. Thus, more polar solvents such as dioxane, THF, 3-methyl-3-pentanol, DMF, or even pyridine have to be used. Thus, only the most stable enzymes such as PPL, *Candida antarctica* lipase, or subtilisin can be used.

Paralleling the difference in reactivity between primary and secondary hydroxyl groups, the former can be selectively acylated by using PPL in THF [267] or pyridine [49] as the solvent (Scheme 3.19). Unlike most of the other hydrolases, the protease subtilisin is stable enough to remain active even in anhydrous DMF [51]. In general, activated esters such as trihaloethyl esters have been used as acyl donors.

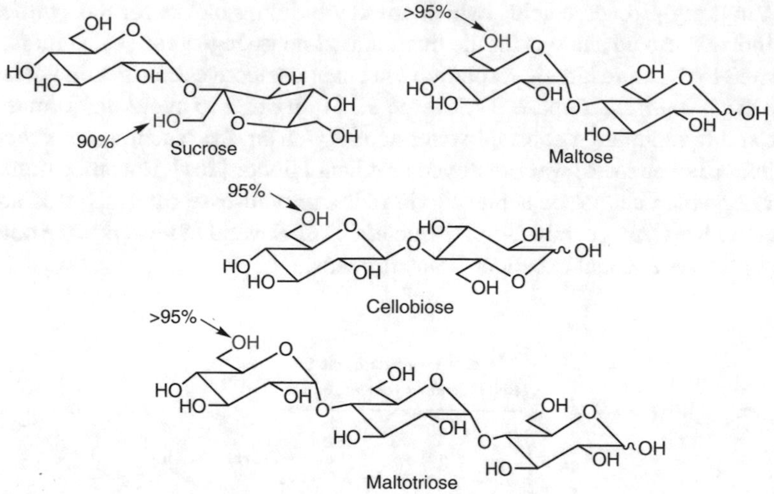

Scheme 3.19 Regioselective protection of primary hydroxy groups of carbohydrates (→ acylation site)

A greater challenge, however, is the regioselective discrimination of secondary hydroxyl groups due to the close similarity of their reactivity. This has been accomplished with steroids using subtilisin and several lipases [50, 268]. For the regioselective acylation of secondary hydroxyl groups of more polar sugar derivatives with blocked primary hydroxyl groups, pyridine [269] or mixed solvent systems (CH₂Cl₂/THF/acetone) have been used [270].

The plant alkaloid castanospermine is a potent glucosidase inhibitor and is being considered as antiviral agent [271]. Some of the corresponding *O*-acyl derivatives, which have been reported as being more active than castanospermine itself, were obtained by an enzyme-catalyzed regioselective acylation reaction (Scheme 3.20) [272]. Thus, the OH group at C1 was selectively acylated using subtilisin and *Chromobacterium viscosum* lipase was employed to esterify the hydroxyl group in position 7. Only minor amounts of the 1,6-isomer were detected.

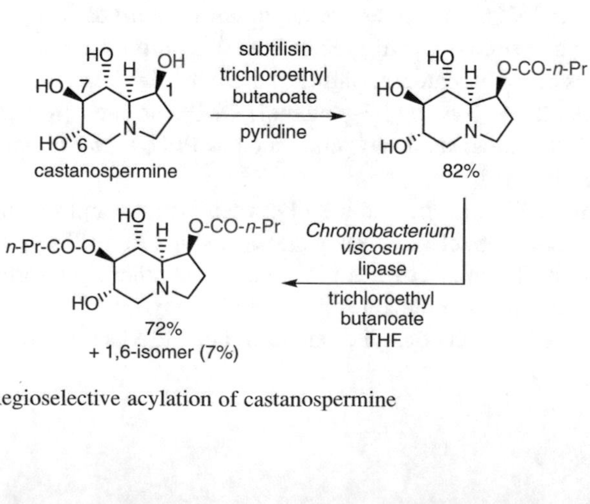

Scheme 3.20 Regioselective acylation of castanospermine

3.1.2   *Lactone and Lactame Synthesis*

Bearing in mind the enzyme-catalyzed esterification and transesterification described in the foregoing chapter, it is not surprising that lactones may be obtained from hydroxy acids or esters by cyclization via *intramolecular* esterification or acyl transfer reactions [273]. Under chemical catalysis, the course of the reaction is relatively simple and the formation of either lactones or open-chain oligomers mainly depends on the ring size of the product: Lactones with less than five or more than seven atoms in the ring are not favored and, thus, linear condensation products are formed predominantly. In contrast, five-membered lactones are easily formed, whereas the formation of six-membered structures is often accompanied by the formation of straight-chain oligomers. Usually the corresponding cyclic dimers – diolides – are not obtained by chemical methods.

In contrast, lipase-catalyzed lactonization often leads to a product pattern that is different from that obtained by chemical catalysis (Scheme 3.21). The outcome depends on several parameters, i.e., the length of the hydroxy acid, the type of lipase, the solvent, the dilution, and even the temperature [274, 275]. In addition, when racemic or prochiral hydroxyacids are employed as substrates, a kinetic resolution [276, 277] or desymmetrization may be accomplished with high selectivities [45]. It is obvious that enzymatic lactone formation is particularly easy with γ-hydroxy derivatives, which lead to the formation of (favored) five-membered ring lactones [278]. The most important synthetic aspect of enzymatic lactone formation, however, is the possibility of directing the condensation reaction towards the formation of macrocyclic lactones and dilactones – i.e., macrolides and macrodiolides, respectively – which are difficult to obtain by chemical catalysis (Scheme 3.21) [46, 279]. This strategy was employed as the key step in the synthesis of the naturally occurring antifungal agent (—)-pyrenophorin, a 16-membered ring macrocyclic diolide [280].

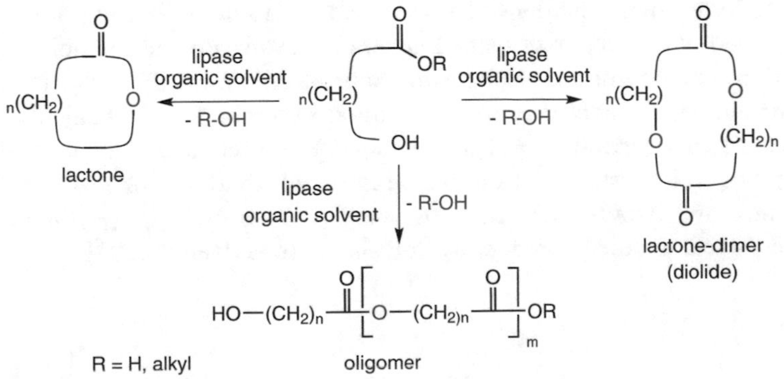

Scheme 3.21 Lipase-catalyzed formation of lactones, diolides, and oligomers

To ensure a highly desirable irreversible lactonization reaction, the use of hydroxy-substituted vinyl carboxylates has been suggested in analogy to the use of enol esters as acyl donors [281].

In analogy to intramolecular cyclization of hydroxy-esters, lactams can be obtained from amino-esters using PLE [282] and PPL [283].

3.1.3 Amide Synthesis

When N-nucleophiles such as ammonia, amines or hydrazine are subjected to acyl-transfer reactions, the corresponding N-acyl derivatives – amides or hydrazides – are formed through interception of the acyl-enzyme intermediate by the N-nucle-ophile (Schemes 2.1 and 3.23) [284]. Due to the pronounced difference in nucle-ophilicity of the amine (or hydrazine) as compared to the leaving alcohol (R^2–OH), *aminolysis* reactions can be regarded as quasi-irreversible. Any type of serine hydrolase which forms an acyl-enzyme intermediate (esterases, lipases, and most proteases) is able to catalyze these reactions. Among them, proteases such as subtilisin and penicillin acylase and lipases from *Candida antarctica* and *Pseudo-monas* sp. have been used most often.

Chemoselective Amide Synthesis Enzyme-catalyzed chemoselective *ammonolysis* [285, 286] or *aminolysis* of esters [287] may be advantageous for the synthesis of carboxamides bearing an additional functional group, which is susceptible to nucleophilic attack, for instance, β-keto-, α,β-unsaturated, or propargylic amides [288]. The latter compounds cannot be obtained by using chemical catalysis due to competing side reactions which lead to enaminoesters and Michael adducts, respectively. The analogous *hydrazinolysis* of esters leading to the formation of hydrazides under mild reaction conditions, may be performed using enzymatic catalysis in a similar manner [289–291]. As an alternative to ammonia, urea may serve as NH_3-source [292].

The mild reaction conditions of enzymatic amide synthesis have been exploited in the industrial production of second-generation semisynthetic antibiotics, such as ampicillin, amoxicillin and cephalexin (Scheme 3.22). D-α-Phenylglycin methyl ester (or amide) is coupled to the free amino group of 6-aminopenicillanic acid under catalysis of penicillin G acylase in aqueous buffer. Due to the low solubility of the product, it crystallizes from the mixture, which drives the reaction towards completion and avoids side reactions, such as hydrolytic ring opening of the reactive β-lactam moiety, product hydrolysis or epimerization [293].

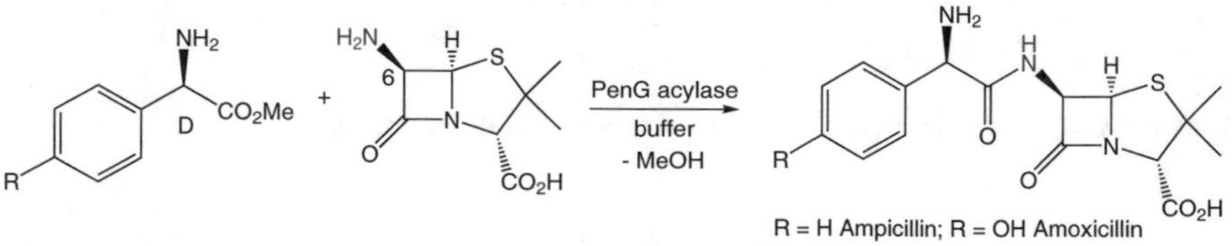

R = H Ampicillin; R = OH Amoxicillin

Scheme 3.22 Industrial synthesis of amoxicillin via enzymatic amide bond formation

Enantioselective Amide Synthesis More important, however, are transformations where chirality is involved. As may be deduced from Scheme 2.23, three different types of chiral recognition are possible, depending on the location of the chiral center in either R^1, R^2, or R^3.

- Esters of chiral acids (chiral R^1) can be resolved via acyl transfer using *N*-nucleophiles [47, 287].
- Chiral alcohols (center in R^2) may be separated via their esters through ammonolysis or aminolysis in a similar fashion [294].
- The most intriguing aspect, however, lies in the enantioselective *formation* of amides, where the center of chirality is located on the amine (chiral R^3) [295]. Thus, kinetic resolution of amines may be achieved [296–298].
- If both the ester and the amine are chiral, *diastereomeric* amides are formed, going in hand with recognition of both chiral entities [299].

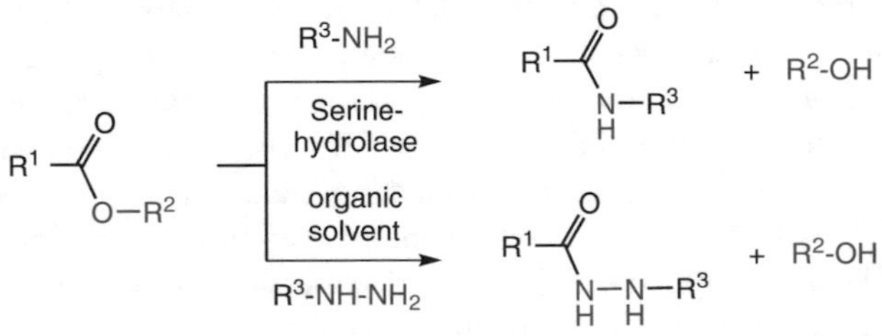

Scheme 3.23 Ammonolysis, aminolysis, and hydrazinolysis of esters

Because the nucleophilicity of an amine is significantly larger than that of an alcohol, the choice of the acyl donor is of crucial importance [300]. Many activated esters,[2] such as ethyl trifluoroacetate and trifluoroethyl butanoate used as acyl donors for the acylation of alcohols are too reactive and lead to a certain amount of spontaneous (nonselective) background reaction, which causes a depletion of selectivity. Less reactive acyl donors, such as benzyl *iso*-propenyl carbonate might be used in a solvent system (e.g., 3-methylpentan-3-ol), which suppresses the background reaction. Simple carboxylic acid esters, such as ethyl acetate, may be used, but the (chemical) cleavage of the resulting carboxamides is rather difficult and requires harsh reaction conditions, which preclude the presence of other sensitive functional groups in the molecule. The acyl donors of choice for aminolysis reactions are as follows:

- Ethyl methoxyacetate yields a fast reaction rate in lipase-catalyzed aminolysis (i.e. about 100 times faster than ethyl butanoate) and the *N*-methoxyacetamides thus formed can be hydrolyzed under reasonably mild conditions using aqueous base. This technique is used for the resolution of amines on a multi-ton industrial scale [301] (Scheme 3.24).

---

[2]In principle, ethyl fluoroacetate would also fall into this category. However, its use is not recommended since fluoroacetic acid is a severe toxin by acting as inhibitor of the Krebs-cycle.

• Diallyl carbonate leads to the formation of allyl carbamates (Scheme 3.24). The
  free amines can be selectively deprotected by using mild Pd(0)-catalysis [302].

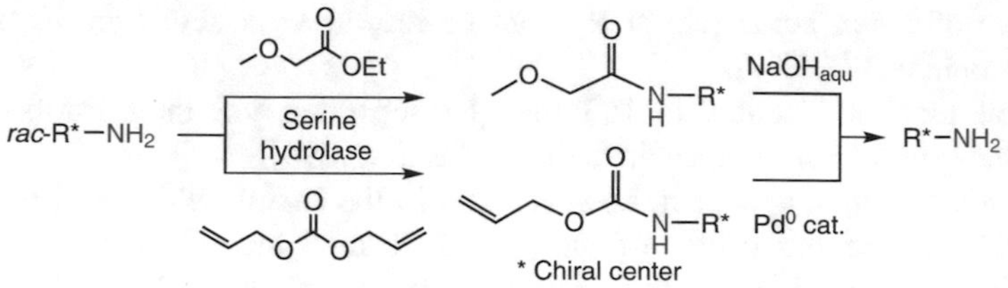

Scheme 3.24 Special acyl donors for the resolution of amines via aminolysis reactions

Representative examples for the resolution of amines via lipase-catalyzed
aminolysis are given in Scheme 3.25. Among numerous enzymes, lipases from
Candida antarctica and *Pseudomonas* sp. have been proven to be most useful
[303]. Since primary amines of the type $R^1R^2CH-NH_2$ are isosteric with secondary
alcohols, the rule predicting the faster reacting enantiomer in a lipase-catalyzed
reaction (for *sec*-alcohols this rule is commonly referred to as 'Kazlauskas rule',
Scheme 2.45) can be applied [304]. Thus, (*R*)-amines are preferentially acylated if
the CIP-sequence priority of the substituents is large > medium. This process has
been scaled up to a capacity of >1000 t/year using ethyl methoxyacetate as acyl
donor and produces a wide variety of α-chiral primary amines for pharma- and
agro-applications, among them (*S*)-methoxy-isopropylamine, which represents the
key building block for the herbicide Outlook™. The separation of formed amide
from nonreacted amine can be achieved via extraction and undesired amine enantiomers are recycled via ex-situ racemization using Raney-Ni as catalyst [296].

In contrast to the facile in-situ racemization of *sec*-alcohols via Ru-catalysts
(Schemes 3.14 and 3.17), which allows dynamic resolution, the isomerization of
α-chiral amines requires more drastic conditions [305, 306]. Hydrogen transfer
catalyzed by Pd [307, 308], Ru [309, 310] Ni, or Co [311] is slow and requires
elevated temperatures close to 100 °C, which still requires the spatial separation of
(metal-catalyzed) racemization from the lipase aminolysis [312]. More recently,
Pd-nanoparticles [313, 314], Ru- [315] or Ir-based catalysts [316] were developed.
Interestingly, also free thiol radicals generated by AIBN were applicable [317].

Although lipase-catalysed acyl transfer appears to be applicable also to cyclic
sec-amines [318], acyl transfer onto SH-groups (corresponding to 'ester thiolysis'
in analogy to aminolysis) does not take place [319]. As a result, the resolution of
sec-thiols is not feasible by this method and has to be performed via hydrolysis or
alcoholysis of the corresponding thioesters [320–322].

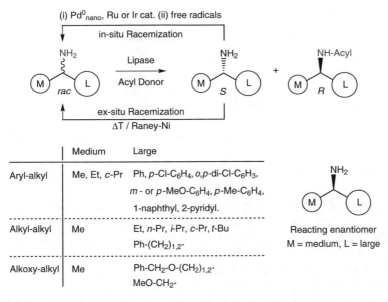

	Medium	Large
Aryl-alkyl	Me, Et, *c*-Pr	Ph, *p*-Cl-C$_6$H$_4$, *o,p*-di-Cl-C$_6$H$_3$, *m*- or *p*-MeO-C$_6$H$_4$, *p*-Me-C$_6$H$_4$, 1-naphthyl, 2-pyridyl.
Alkyl-alkyl	Me	Et, *n*-Pr, *i*-Pr, *c*-Pr, *t*-Bu Ph-(CH$_2$)$_{1,2}$-
Alkoxy-alkyl	Me	Ph-CH$_2$-O-(CH$_2$)$_{1,2}$- MeO-CH$_2$-

Reacting enantiomer
M = medium, L = large

Scheme 3.25 Lipase-catalyzed kinetic and dynamic resolution of amines via ester aminolysis

3.1.4 Peptide Synthesis

Peptides display a diverse range of biological activity. They may be used as sweeteners and toxins, antibiotics and chemotactic agents, as well as growth factors. They play an important role in hormone release either as stimulators or as inhibitors. The most recent application is their use as immunogens for the generation of specific antisera. At present, the most frequently used methods of peptide synthesis are predominantly chemical in nature, generally proceeding through a sequence of four steps. (a) First, all the functionalities of the educts, which are not to participate in the reaction, must be selectively protected. (b) Then, the carboxyl group must be activated in a second step to enable (c) the formation of the peptide bond. (d) Finally, the protective groups have to be removed in toto, if the synthesis is complete, or the amino- or carboxy-terminus must be selectively liberated if the synthesis is to be continued. Thus, an extensive protection and deprotection methodology is required. Two of the major problems associated with chemical peptide synthesis are a danger of racemization – particularly during the activation step – and the tedious (and sometimes impossible) purification of the final product from (diastereo)isomeric peptides with a closely related sequence. To circumvent these problems, peptide synthesis is increasingly carried out by making use of the specificities of enzymes, in particular proteases [48, 323–329]. The pros and cons of conventional versus enzymatic peptide synthesis are summarized in Table 3.6.

Table 3.6 Pros and cons of chemical and enzymatic peptide synthesis

	Chemical	Enzymatic
Stereoselectivity	Low	High
Regioselectivity	Low	High
Amino acid range	Broad	Limited
Protective group requirements	High	Low
Purity requirements of starting materials	High	Moderate
Byproducts	Some	Negligible
Danger of racemization	Some	None

Non-ribosomal enzymatic peptide synthesis occurs via two major pathways (Scheme 3.26) [330]:

Activation of the carboxyl terminus to be coupled with an amine (at the expense of ATP) furnishes either an acyl-adenylate- or acyl-phosphate intermediate, catalysed by acyl-adenylate-forming- or 'ATP-grasp enzymes'. This high-energy mixed phosphoric carboxylic acid anhydride intermediate reacts with an amine nucleophile in an irreversible reaction.[3] Although this is a clear advantage for synthesis, these enzymes are difficult to handle and their dependency on ATP requires its recycling. Consequently, these systems are currently not employed for large-scale peptide synthesis.

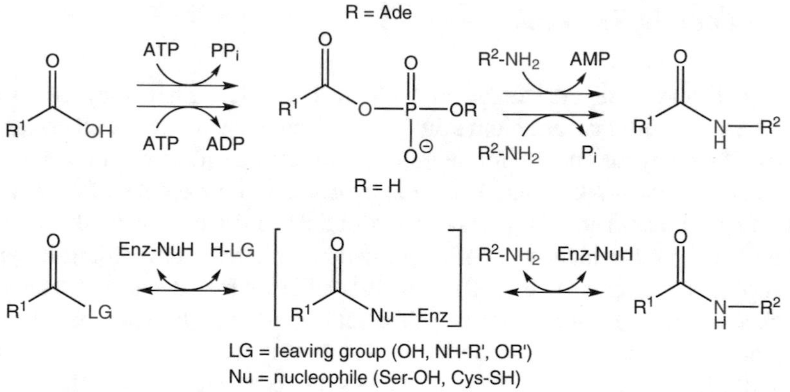

Scheme 3.26 Principles of enzymatic peptide synthesis

Alternatively, peptide-*cleaving* proteases can be used to catalyse the reverse (condensation) reaction, provided that the mechanism proceeds via a covalent acyl-enzyme intermediate, usually involving a Ser-OH or Cys-SH nucleophile [331].[4] Since many carboxyl ester hydrolases (esterases, lipases) also form an acyl-enzyme

---

[3]Sometimes, this proceeds via thioester intermediates.

[4]Metallo- or carboxy-proteases, which do not form a covalent acyl enzyme intermediate are usually unsuitable.

intermediate, they may be likewise be employed. This methodology has been intensely investigated during the last decades, although the first report dates back to 1938 [332]. Occasionally, the proteases used for peptide synthesis are also misleadingly called 'peptide ligases' [333, 334], they are, however, simple hydrolases [EC 3.x.x.x] and have nothing in common with peptide ligases [EC 6.x.x.x] (Scheme 3.27).

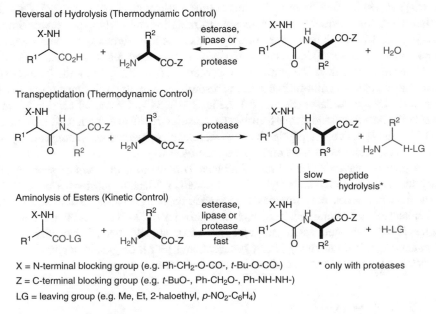

X = N-terminal blocking group (e.g. Ph-CH₂-O-CO-, t-Bu-O-CO-)   * only with proteases
Z = C-terminal blocking group (e.g. t-BuO-, Ph-CH₂O-, Ph-NH-NH-)
LG = leaving group (e.g. Me, Et, 2-haloethyl, p-NO₂-C₆H₄)

Scheme 3.27 Enzymatic peptide synthesis using proteases and carboxyl ester hydrolases

Thermodynamic Approach Reversed hydrolysis and transpeptidation are reversible and are therefore thermodynamically controlled. Under physiological conditions, the equilibrium position in protease-catalyzed reactions is far over on the side of proteolysis. In order to create a driving force in the reverse direction towards peptide *synthesis*, the following constraints may be applied.

- One of the reactants is used in excess.
- Removal of product via formation of an insoluble derivative [335], by specific complex formation [336], or by extraction of the product into an organic phase by using a water-immiscible organic cosolvent.
- Lowering the water-activity (concentration) of the system by addition of water-miscible organic cosolvents. In this respect, polyhydroxy compounds such as glycerol or 1,4-butanediol have been shown to conserve enzyme activity better than the solvents which are more commonly employed, such as DMF, DMSO, ethanol, acetone, or acetonitrile [337]. The use of water immiscible neat organic solvents is limited by the low solubility of reactants in these lipophilic systems. Alternatively, peptide synthesis may also be performed with neat reactants – i.e., in the absence of solvents [338].

Kinetic Approach The third method – aminolysis of esters – involves a kinetically controlled irreversible reaction, in which a weak and a strong nucleophile (water and an amine) are competing for the acyl-enzyme intermediate [339]. As mentioned above this reaction can be regarded as irreversible. Thus, it is not surprising that besides proteases, other serine hydrolases which are capable of forming an acyl-enzyme intermediate (mainly esterases or lipases, such as PPL [340], PLE, and CRL [341, 342]) may be used for this reaction (Schemes 3.23 and 3.28). On the other hand, this method is not applicable to metallo- and carboxyproteases. Since the peptide formed during ester aminolysis may be hydrolytically cleaved by proteases in a subsequent (slow) reaction in the presence of water, these reactions have to be terminated before the equilibrium is reached (Fig. 3.2). Thus, the kinetics of enzymatic peptide synthesis has a strong resemblance to glycosyl transfer reactions mediated by glycosidases (Sect. 2.6.1.3, Fig. 2.19) [343]. Undesired peptide hydrolysis may be suppressed by performing the reaction in a frozen aqueous medium at −15 °C [344]. Of course peptide hydrolysis may be neglected when nonproteolytic hydrolases, such as esterases and lipases, are used.

In general, naturally occurring (*N*-protected) L-amino acid esters of short-chain alcohols, such as methyl and ethyl esters are usually sufficiently reactive as 'acceptors' to achieve reasonable reaction rates in enzymatic peptide synthesis via aminolysis. For less reactive (nonnatural) analogs, such as α-substituted [345] or D-configured amino acids [346], activated esters are recommended, among them, 2-haloethyl (e.g., 2-chloroethyl, trifluoroethyl) [347], *p*-nitrophenyl [348], or guanidinophenyl esters [349]. For the use of 'cyclic activated esters' [5(4*H*)-oxazolones] see below.

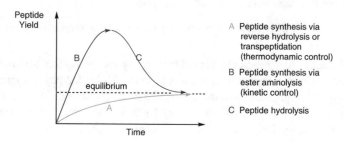

Fig. 3.2 Enzymatic peptide synthesis under thermodynamic and kinetic control

Because all hydrolases exhibit a certain substrate selectivity, the availability of a library of enzymes which can cover all possible types of peptide bonds is of crucial importance in order to make enzymatic peptide synthesis applicable to all possible amino acid combinations. Although the existing range of proteases is far from complete, it provides a reasonable coverage (Table 3.7). The most striking 'shortage' involves proline derivatives. The most commonly used proteases and their approximate selectivities are listed below, where X stands for an unspecified amino acid or peptide residue.

Table 3.7 Selectivities of proteases

Protease	Type	Specificity
Achromobacter protease	Serine protease	–Lys–X
α-Chymotrypsin, subtilisins	Serine protease	–Trp(Tyr,Phe,Leu,Met)–X
Carboxypeptidase Y	Serine protease	Nonspecific
Elastase	Serine protease	–Ala(Ser,Met,Phe)–X
Trypsin, *Streptomyces griseus* protease	Serine protease	–Arg(Lys)–X
Staphylococcus aureus V8 protease	Serine protease	–Glu(Asp)–X
Papain, ficin	Thiol protease	–Phe(Val,Leu)–X
Clostripain, cathepsin B	Thiol protease	–Arg–X
Cathepsin C	Thiol protease	H–X–Phe(Tyr,Arg)–X
Thermolysin, *B. subtilis* protease	Metalloprotease	–Phe(Gly,Leu)–Leu(Phe)
Myxobacter protease II	Metalloprotease	X–Lys
Pepsin, cathepsin D	Carboxyl protease	–Phe(Tyr,Leu)–Trp(Phe,Tyr)

X = unspecified amino acid or peptide residue

The dipeptide ester L-Asp-L-Phe-OMe (Aspartame) is used in large amounts as a low-calorie sweetener. One of the most economical strategies for its synthesis involves an enzymatic step, which is run at a capacity of 2000 t/year (Scheme 3.28). Benzyloxycarbonyl-(Z)-protected L-aspartic acid is linked with L-Phe–OMe in a thermodynamically controlled condensation reaction catalyzed by the protease thermolysin without formation of the undesired (bitter) β-isomer. Removal of the product via formation of an insoluble salt was used as driving force to shift the equilibrium of the reaction in the synthetic direction [350, 351].

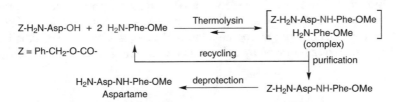

Scheme 3.28 Enzymatic synthesis of Aspartame

α-Chymotrypsin has been used for the kg-scale synthesis of the dipeptide *N*-benzoyl-L-Tyr-L-Arg-NH$_2$, a precursor to the powerful analgesic kyotorphin [352]. The wide potential of enzymatic peptide synthesis is illustrated with the chemo-enzymatic synthesis of an enkephalin derivative – (D-Ala$_2$, D-Leu$_5$)-enkephalin amide (Scheme 3.29) [353]. Enkephalins are pain regulators and commonly undergo rapid enzymatic degradation in vivo, and therefore exert only limited biological activities. However, by substitution of a D-amino acid into the sequence, 'chirally muted' derivatives are obtained, which can elicit long-lasting pharmacological effects [354]. The Tyr-D-Ala and Gly-Phe subunits were obtained via an ester aminolysis and a condensation reaction using α-chymotrypsin and thermolysin, respectively. The latter dipeptide was extended by a D-Leu unit using the same methodology. After converting the C-terminal hydrazide protective group into an azide moiety and removal of the *N*-terminal Z-group by hydrogenolysis, the fragments were coupled by conventional methodology.

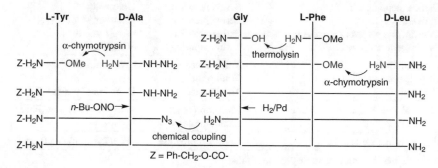

Scheme 3.29 Chemoenzymatic synthesis of an enkephalin derivative

Probably the most prominent example of an enzymatic peptide synthesis is the transformation of porcine insulin into its human counterpart (Scheme 3.30). As millions of diabetics suffer from insulin deficiency, and due to the fact that the demands cannot be satisfied by exploiting natural sources for obvious reasons, numerous attempts have been undertaken to convert porcine insulin into human insulin via exchange of a terminal alanine by a threonine residue [355–357]. A protease from *Achromobacter lyticus* which is completely specific for peptide bonds formed by a lysine residue [358] is used to selectively hydrolyze the terminal Ala^{30} residue from porcine insulin. The same enzyme is then used to catalyze the condensation reaction with threonine, protected as its *tert*-butyl ester. Finally, the *tert*-butyl group was removed by acid treatment to yield human insulin [359].

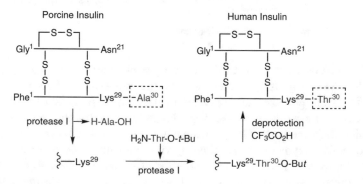

Scheme 3.30 Enzymatic conversion of porcine into human insulin

Enzymatic peptide synthesis using proteases is nowadays sufficiently solid to enable the coupling of large proteins, such as the biologically active 493–515 sequence of human thyroid PKA anchoring protein Ht31 [360], cyclic somatostatin [361] and oxytocin [362]. Due to the absence of isomerization and reduction of chemical protection-deprotection steps, total yields of 30–50% are common.

A special type of enzymatic peptide synthesis employs activated heterocyclic amino acid/peptide derivatives as acyl donors (Scheme 3.31) [363]. In a kinetically controlled approach, 5(4H)-oxazolones (which may be regarded as 'cyclic activated esters') are cleaved by α-chymotrypsin thereby generating an acyl enzyme intermediate. Then the amino acid/peptide segment is coupled onto the N-terminus of the acyl acceptor by forming a new peptide bond.

Some features of this technique are worthy of attention:

- Racemization of the activated acyl donor involving its α-center is largely suppressed.
- When racemic acyl donors are employed (with respect to the α-center), kinetic resolution proceeds with incomplete selectivity and with moderate preference for the L-enantiomer.
- Protection of the C-terminal carboxyl moiety (R^4) is not required since the oxazolone is a much better acyl donor than the (competing) carboxyl group on the acyl acceptor.
- The chemical yields may be diminished by undesired enzymatic hydrolysis of the acyl donor, the extent of which depends on the nature of the acyl donor.

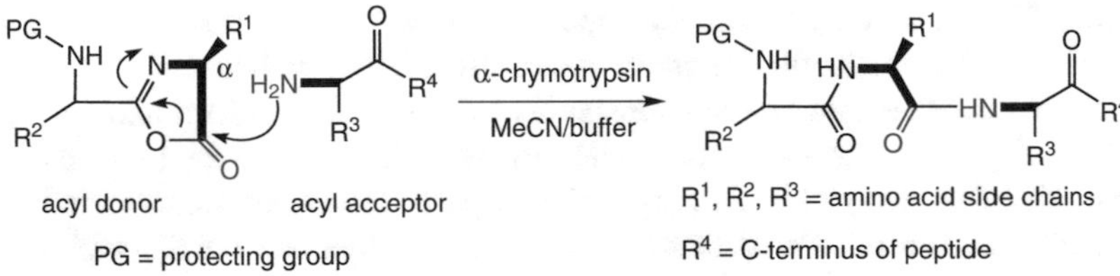

Scheme 3.31 Peptide synthesis using 5(4H)-oxazolones as acyl donors

3.1.5 Peracid Synthesis

In contrast to the acidic conditions usually applied for in situ generation of peroxycarboxylic acids, they may be generated under virtually neutral conditions in a suitable organic solvent directly from the parent carboxylic acid and hydrogen peroxide via lipase catalysis (Scheme 3.32).

The mechanism involves a perhydrolysis of the acyl-enzyme intermediate by the nucleophile hydrogen peroxide (Scheme 2.1) [364]. The peroxy acids thus formed can be used in situ for the epoxidation of alkenes [365], the Baeyer-Villiger oxidation of carbonyl compounds [366, 367] and the sulfoxidation of thioethers, while the liberated fatty acid re-enters the cyclic process [368, 369]. It should be noted that the oxidation reaction itself takes place without involvement of the enzyme; therefore no significant enzyme-induced selectivities were detected. If the fatty acid contains an olefin, it can epoxidize itself via the peroxy fatty acid intermediate [370].

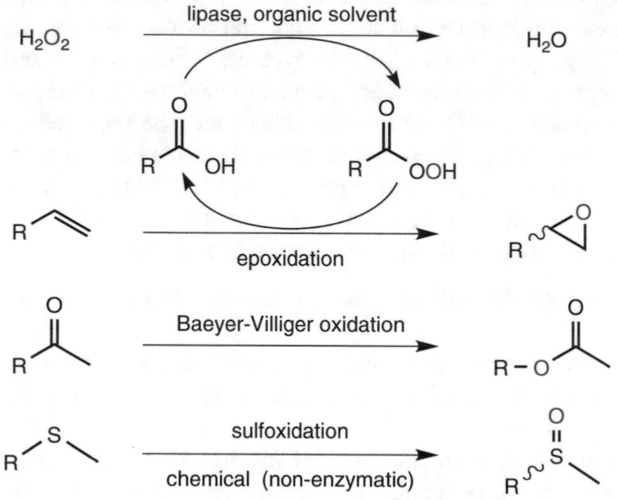

Scheme 3.32 Lipase-catalyzed peracid formation and catalytic epoxidation

The main advantages of this method are the mild conditions employed and a higher safety margin due to the fact that only catalytic concentrations of peracid are involved. This aspect is particularly important for oxidation reactions on an industrial scale, such as the sulfur-oxidation of penicillin G into its 1-(s)-oxide, which is a key intermediate en route to cephalosporins [371]. Medium-chain alkanoic acids (C_8–C_{16}) and a biphasic aqueous-organic solvent system containing toluene or *tert*-butanol give the best yields. Among various lipases tested, an immobilized lipase from *Candida antarctica* was shown to be superior to lipases from *Candida rugosa* and *Pseudomonas* sp.

3.1.6   Redox Reactions

In contrast to hydrolases, redox enzymes such as dehydrogenases and oxygenases have been used less often in organic solvents because they require cofactors, e.g., nicotin-amide adenine dinucleotide species. The latter are highly polar (charged) compounds and are therefore completely insoluble in a lipophilic medium. As a consequence, the cofactor is irreversibly bound to a protein molecule and cannot freely be exchanged between enzymes, which is necessary for its recycling. These limitations can be circumvented to some extent by co-precipitation of enzyme and cofactor onto a macroscopic carrier provided that a minimum amount of water is present [372, 373]. Thus, the cofactor is able to freely enter and exit the active site of the enzyme but it cannot disaggregate from the carrier into the medium because it is trapped together with the enzyme in the hydration layer on the carrier surface. As a result, acceptable turnover numbers are achieved [374–376].

Redox reactions catalyzed by alcohol dehydrogenases (e.g., from horse liver, HLADH) may be performed in organic solvents in both the reduction and oxidation mode using the coupled substrate method for cofactor recycling, because in this case nicotinamide always remains bound to a single dehydrogenase and is not exchanged to a recycling enzyme (Sect. 2.2.1, Scheme 2.108). Reduction of aldehydes/ketones and oxidation of alcohols is effected by NADH- or NAD$^+$-recycling, using ethanol or *iso*butyraldehyde, respectively.

An alternative way of NAD$^+$ recycling makes use of a three-enzyme cascade with molecular oxygen as the ultimate oxidant (Scheme 3.33) [377]. As in the methods described above, all the enzymes and cofactors have to be precipitated together. Thus, NADH which is produced by HLADH-catalyzed oxidation of a secondary alcohol is re-oxidized by diaphorase at the expense of pyrroloquinoline quinone (PQQ) [378]. The reduced form of the latter (PQQH$_2$) is spontaneously oxidized by molecular oxygen producing hydrogen peroxide, which, in turn, is destroyed by catalase.

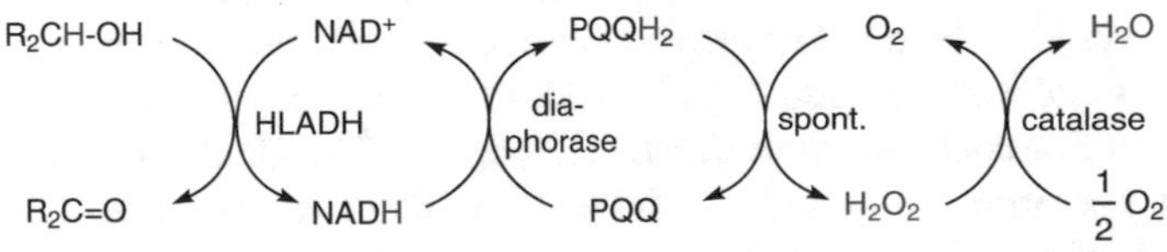

Scheme 3.33 NAD$^+$-Recycling via a diaphorase-catalase system

Polyphenol oxidase catalyzes the hydroxylation of phenols to catechols and subsequent dehydrogenation to *o*-quinones (Sect. 2.3.3.2, Scheme 2.154) [379]. The preparative use of this enzyme for the regioselective hydroxylation of phenols is impeded by the instability of *o*-quinones in aqueous media, which rapidly polymerize to form polyaromatic pigments leading to enzyme deactivation [380]. Since water is an essential component of the polymerization reaction, the *o*-quinones formed are stable when the enzymatic reaction is performed in an organic solvent (Scheme 3.34). Subsequent nonenzymatic chemical reduction of the *o*-quinones (e.g., by ascorbic acid) to form stable catechols leads to a net regioselective hydroxylation of phenols [19]. Depending on the substituent R in the *p*-position, cresols were obtained in good yields. Electron-withdrawing and bulky substituents decreased the reactivity, and *o*-, and *m*-cresols were unreactive. The preparative use of this method was demonstrated by the conversion of *N*-acetyl-L-tyrosine ethyl ester into the corresponding L-DOPA-derivative. An incidental observation that the related enzyme horseradish peroxidase remains active in nearly anhydrous organic solvent was already reported in the late 1960s [381].

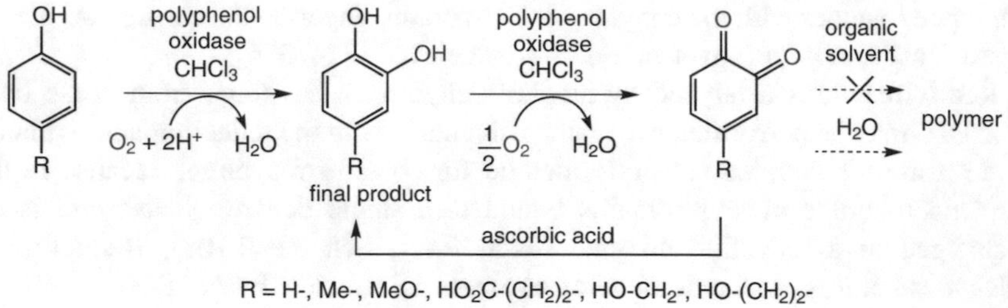

R = H-, Me-, MeO-, HO₂C-(CH₂)₂-, HO-CH₂-, HO-(CH₂)₂-

Scheme 3.34 Polyphenol-oxidase-catalyzed regioselective hydroxylation of phenols

3.1.7 Medium Engineering

As may be deduced from the introductory chapter, any solvent exerts a significant influence on the conformation of an enzyme, which in turn governs its catalytic efficiency and its chemo-, regio- and stereoselectivity. Thus, it is reasonable to expect that an enzyme's specificity may be controlled by varying the solvent's properties. For reactions performed in water, however, this is hardly possible, because its physicochemical properties are fixed by Nature and can only be altered within a very narrow margin, e.g., by addition of water-miscible (polar) organic cosolvents at low concentrations (Sect. 2.1.3, pp. 74–75). On the other hand, when a reaction is performed in an organic solvent, the latter can be chosen from a large repertoire having different physicochemical parameters, such as dipole moment, polarity, solubility, boiling point, straight-chain or cyclic structure, etc. Therefore, the outcome of an enzyme-catalyzed reaction may be controlled by choosing the appropriate organic solvent [382]. The modulation of enzyme specificity by variation of the solvent properties has been commonly denoted as 'medium engineering' [79].

For instance, the almost exclusive specificity of proteases for L-configurated amino acid derivatives may be 'destroyed' when reactions are carried out in organic solvents [383]. This makes them useful for the synthesis of peptides containing nonnatural D-amino acids, which are usually not substrates for proteases.

The influence of organic solvents on enzyme enantioselectivity is not limited to the group of proteases, but has also been observed with lipases, and is a general phenomenon [384–388]. As a rule of thumb, the stereochemical preference of an enzyme for one specific enantiomer usually remains the same, although its selectivity may vary significantly depending on the solvent. In rare cases, however, it was possible to even invert an enzyme's enantioselectivity [389–391].

As shown in Scheme 3.35, resolution of the mycolytic drug *trans*-sobrerol was achieved by acyl transfer using vinyl acetate and PSL as the catalyst. The selectivity of the reaction markedly depended on the solvent used, with *tert*-amyl alcohol being best. As may be deduced from the physicochemical data given, any attempts

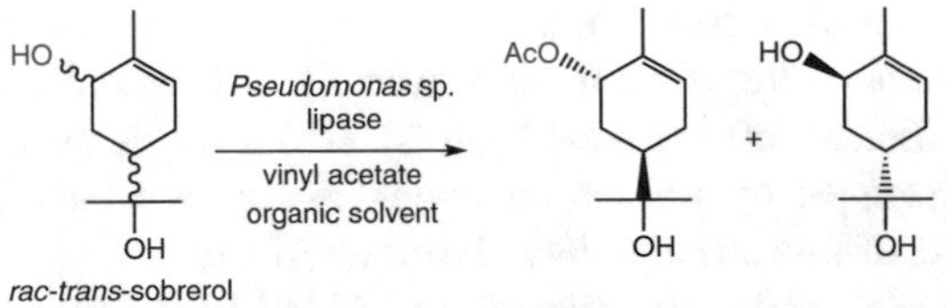

Scheme 3.35 Optimization of selectivity by solvent variation

Solvent	logP	Dielectric Constant (ε)	Selectivity (E)
vinyl acetate	0.31	--	89
THF	0.49	7.6	69
acetone	-0.23	20.6	142
dioxane	-1.14	2.2	178
3-pentanone	0.80	17.0	212
t-amyl alcohol	1.45	5.8	518

to link the observed change in selectivity to the lipophilicity (expressed as the log P) or the dielectric constant (ε) were unsuccessful.

Besides the nature of the solvent itself, it is also the intrinsic water content (more precisely, the water activity, a_W), which has an influence on the enzyme selectivity (Scheme 3.36). For instance, the resolution of 2-bromopropionic acid by esterification using *Candida rugosa* lipase proceeded with a significantly enhanced selectivity when the water content of the solvent was gradually increased [392].

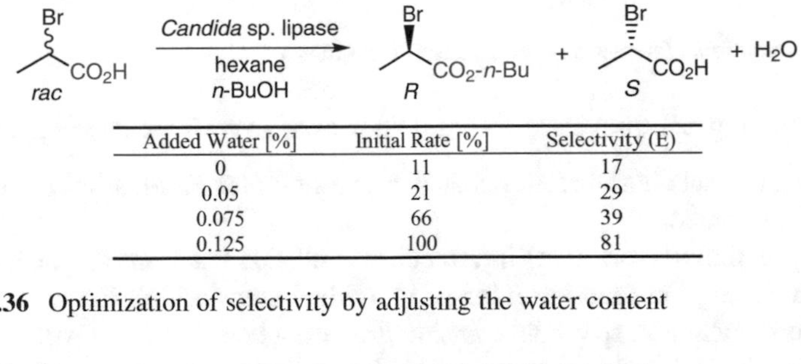

Added Water [%]	Initial Rate [%]	Selectivity (E)
0	11	17
0.05	21	29
0.075	66	39
0.125	100	81

Scheme 3.36 Optimization of selectivity by adjusting the water content

The data available so far demonstrate that the nature of the organic solvent and its water content exerts a strong influence on the catalytic properties of an enzyme. However, no general rationale, which would allow the prediction of the selectivity enhancement mediated by medium engineering has been presented so far [393].

3.2 Cascade-Reactions

Traditional chemical synthesis proceeds in a stepwise fashion to convert starting material **A** into a final product **D** and involves isolation of intermediates **B** and **C** beween the individual steps during the whole sequence (Fig. 3.3). This is not only time-consuming, but also causes unavoidable loss of material due to incomplete recovery and generates large amounts of waste [394]. In order to render organic synthesis more efficient, efforts have been directed towards the design of cascade-like reaction sequences by linking individual steps onto each other [395]. The following variants of this strategy have been described:

In a 'domino-reaction',[5] all reagents are added at the start and one catalyst triggers the formation of a reactive intermediate, which undergoes spontaneous subsequent transformation(s) [396, 397]. In contrast, 'telescoping' reactions implies performing each step individually in a one-pot fashion, where the catalyst/reagent(s) of a subsequent step is added after completion of the previous reaction [398]. A 'cascade-' or or 'tandem-reaction' implies the presence of all catalysts/reagents from the start during the whole sequence, which enable all reactions to proceed in a concurrent fashion [399–403].

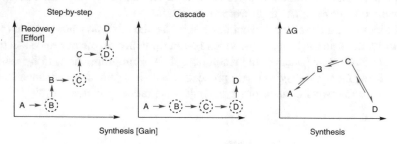

Fig. 3.3 Conventional (step-wise) versus cascade-synthesis

The following advantages resulting in improved overall yields are accrued:

- Loss of materials is reduced, because separation and purification of intermediates is eliminated.
- Decomposition of (sensitive) intermediates and side reactions can be minimized by maintaining the reaction rates of individual steps within the same order of magnitude, which keeps their concentration on a constant low level.
- Unfavorable equilibria can be overcome by pulling out the final product from the sequence using a 'downhill' reaction as last step.
- Redox-neutral sequences can be designed by linking an oxidation to a reduction step, whereby the redox equivalents are shuffled inbetween, for instance, by 'borrowing' hydrogen [404].

The most challenging requirement, however, is the compatibility of catalysts and reaction conditions of the individual steps, which – for most chemical transformations – is very difficult to meet. In contrast, enzymes share very similar reaction

---

[5]Domino-reactions are also called 'zipper-reactions'.

parameters – ambient temperature, (near) neutral pH, aqueous solvent – and are nicely compatible with each other.[6] After all, 6144 proteins work nicely side-by-side to manage the whole metabolism of a yeast cell [405]. These properties make enzymes ideal for the construction of biocatalytic cascades [406]. Two different philosophies are currently persued:

Metabolic engineering employs a top-down approach, whereby the metabolism of a living cell is redirected by minimizing (but maintaining) its vital pathways while directing the bulk of its metabolic flux from a cheap carbon source (usually a carbohydrate) into a desired metabolite. This powerful technology enables the large-scale production of fermentation products, not only simple carboxylic acids (citric, succinic, fumaric, itaconic, lactic acid) and alcohols (ethanol, 1,2- and 1,3-propanediol, 1-butanol, 1,4-butanediol), but also complex natural products (vitamins, antibiotics, terpenoids, steroids, etc.). Overall, this technology relies on a (minimized) living system – a 'cell factory' – and is only applicable to *natural* compounds, which are non-toxic and hence do not disrupt vital pathways. Problems to overcome are material erosion by competing metabolic pathways, kinetic restrictions by physical (membrane) barriers and the re-wiring of regulating circuits.

Cascade biocatalysis, also denoted as *systems biocatalysis* [407] or *artificial metabolism* [408], is a bottom-up approach, where individual enzymes are combined to an artificial biocatalytic network in analogy to the assembly of Lego-blocks by imitating the metabolism occurring in living cells. Since this is a 'dead' system, toxicity is not a major problem and the adjustment of the overall kinetics is comparatively simple. The most striking difference to metabolic engineering is the possibility to convert *non-natural* 'foreign' compounds.

During the past years, the impressive potential of combining enzymes into biocatalytic cascades has been recognized and several systems have been proposed (Fig. 3.4) [409].

• Linear cascades aim at the straightforward synthesis of a target compound through sequential combination of n steps.

• Parallel cascades are the most common type and are applied in cofactor-recycling using the 'coupled-substrate' or the 'coupled enzyme' approach (Sect. 2.2.1); thus, product formation by reduction (or oxidation) is coupled with a second parallel reaction proceeding in the opposite oxidation (or reduction) direction to provide redox equivalents from a sacrificial co-substrate. Alternatively, two reactions are joined in a concomitant fashion to yield two products, e.g. in transamination using a sacrificial amine donor (Sect. 2.6.2).

• Closely related to parallel cascades are orthogonal cascades, where a sacrificial cosubstrate yields a co-product, which is further utilized in the cascade or decomposes into a final waste product, for instance, by decarboxylation. This circumvents co-product inhibition or enables to shift equilibria.

---

[6]With the rare exception of proteases.

- Convergent cascades are typical for deracemization processes, where two enantiomeric substrates are converted simultaneously to yield a single stereoisomeric product (Sect. 2.1.1) [410].
- The main characteristics of cyclic cascades is that one substrate out of a mixture is converted into an intermediate, which is then recycled back to one (or both) starting materials, which may be either two different compounds or two enantiomers. Repetition of this cycle leads to the accumulation of the compound that is not converted during the first transformation. The cyclic deracemization of α-amino acids, α-hydroxy acids and amines is a typical example for such a system (Sect. 2.3.2).

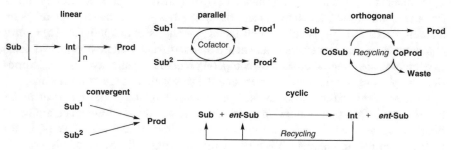

Fig. 3.4 Types of biocatalytic cascade-reactions

In the following section, various types of multienzyme cascades are described with their particular merits.

The power of multienzyme catalysis was demonstrated already in 1992 in a landmark-study directed to facilitate enzymatic C–C bond formation employing aldolases for the synthesis of ketose 1-phosphates by making an essential substrate – dihydroxyacetone phosphate – efficiently available from an economical source [411]. The overall strategy resembles in essential parts the glycolysis pathway in a test tube and combines seven enzymes (including ATP-regeneration in three of the transformations) in one pot (Scheme 3.37). All enzymes are commercially available and are biochemically well characterized. First, sucrose is hydrolysed by invertase (EC 3.2.1.26) to yield equimolar amounts of glucose and fructose. Both of the latter are independently phosphorylated by hexokinase (EC 2.7.1.1) in presence of ATP recycling using phosphoenol pyruvate and pyruvate kinase (EC 2.7.1.40) to yield the corresponding 6-phosphates. In order to make use of both hexose 6-phosphate intermediates, glucose-6-phosphate is isomerised to the fructose isomer by glucose 6-phosphate isomerase (EC 5.3.1.9). Further phosphorylation by fructose 5-phosphate kinase (EC 2.7.1.11) yields the central metabolite fructose 1,6-bisphosphate. The latter is cleaved by fructose 1,6-bisphosphate aldolase (EC 4.1.2.13) via a retro-aldol reaction to furnish two C-3 fragments –

glyceraldehyde-3-phosphate and dihydroxyacetone phosphate.[7] Since only the latter functions as donor in aldol reactions, glyceraldehyde 3-phosphate is isomerized by triosephosphate isomerase (EC 5.3.1.1). Finally, fructose 1,6-bisphosphate aldolase catalysed the asymmetric C–C bond formation between dihydroxyacetone phosphate and various non-natural α- and β-hydroxyaldehydes in excellent yields. Due to the double convergence of this sequence, the starting C12-disaccharide is completely converted into four C3-equivalents without loss of carbon.

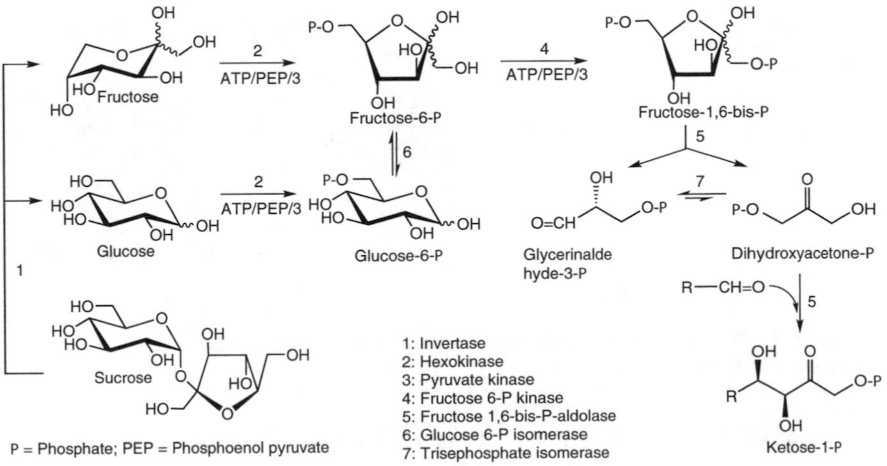

Scheme 3.37 Multienzymatic cascade-synthesis of ketose 1-phosphates from sucrose

Cascade-synthesis can also be beneficial if toxic or sensitive intermediates are involved, which are cumbersome to handle in isolated form.

For instance, the asymmetric epoxidation of alkenes catalysed by P450 monooxygenases produces epoxides, which are toxic at elevated concentrations, which severely limits the efficiency of these systems (Sect. 2.3.3.3). This problem was circumvented by combining epoxidation with a sequential hydrolytic step using an epoxide hydrolase (Sect. 2.1.5) to yield an innocuous *vic*-diol as final product (Scheme 3.38) [412]. The first step was catalysed by a styrene monooxygenase from *E. coli* JM 101 pSPZ10, whereas the hydrolytic step was mediated by an epoxide hydrolase from *Sphingomonas* sp. HXN-200. A two-phase liquid system was used to overcome competing unspecific epoxide hydrolysis. The validity of this cascade was demonstrated with several styrene derivatives, which were transformed in up to 95% yield and >99% e.e. in a formal asymmetric

---

[7]The enzyme from rabbit muscle (RAMA) is commonly used to catalyse the reverse aldol reaction (Sect. 2.4.1).

dihydroxylation procedure. The glycols could be further converted into the corresponding α-hydroxy acids or amino alcohols by extending the cascade with an alcohol-aldehyde dehydrogenase or an alcohol dehydrogenase-transaminase module.

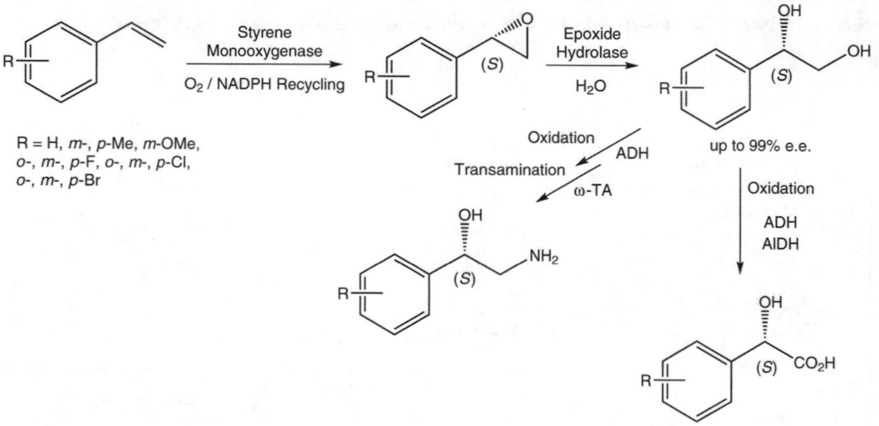

Scheme 3.38 Formal asymmetric dihydroxylation of styrenes through combination of enzymatic epoxidation with epoxide hydrolysis

Hydroxynitriles are notoriously unstable compounds, which give rise to the formation of HCN at pH values ≥ 5 going in hand with spontaneous racemization. This can be circumvented by employing cyanohydrin formation – using a stereoselective hydroxynitrile lyase (Sect. 2.5.3) – with a nitrile-hydrolysing enzyme (Sect. 2.1.6) in a cascade (Scheme 3.39), which results in the formation of stable α-hydroxy acids or -amides, respectively. Although this appears to be a straightforward task, several problems had to solved: Cyanohydrin synthesis is usually performed at acidic pH (≤ 5) in aqueous-organic solvents, where nitrile-hydrolysing enzymes are rapidly inactivated. (S)-Mandelic acid was prepared from benzaldehyde and HCN by chosing (S)-HNL from cassava (*Manihot esculenta*) and a non-stereoselective nitrilase from *Pseudomonas fluorescens* EBC191, which is more active at acidic pH than other nitrilases. Minor amounts of undesired carboxamide were converted to the acid by addition of an amidase. All three enzymes were co-immobilised and gave almost pure (S)-mandelic acid [413]. Enantiomeric (R)-α-hydroxy acids [R=o-Cl-C$_6$H$_4$, Ph-(CH$_2$)$_2$-] were obtained by using an (R)-HNL from almond (*Prunus amygdalus*) [414]. A bi-enzymatic cascade leading to (S)-α-hydroxy amides was realized by employing (S)-HNL (*Manihot esculenta*) and a relatively stable nitrile hydratase from the halophilic extremophile *Nitriliruptor alkaliphilus*. Careful adjustment of the reaction conditions with portionwise feed of HCN had to be employed to compensate for the difference in pH optima of both enzymes (pH 4.5 versus 8) and the sensitivity of the nitrile hydratase [415]. Here, the low enantioselectivity of nitrile-hydrolysing enzymes is turned into an advantage, if combined with a stereoselective hydroxynitrile lyase.

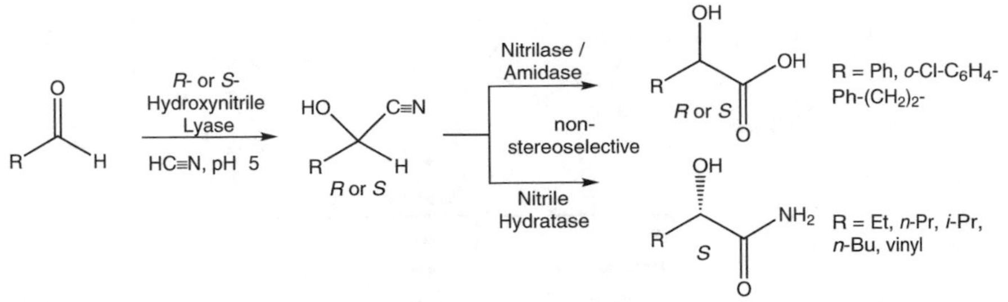

Scheme 3.39 Cascade-synthesis of α-hydroxy acids or -amides employing a stereoselective hydroxynitrile lyase in combination with a non-selective nitrilase or nitrile hydratase, respectively

Several enzymatic transformations, in particular C–C bond formation catalysed by aldolases or transketolases, are plagued by unfavorable equilibria. This can be overcome by linking an irreversible step to the synthetic transformation. For instance, L-threonine aldolases catalyse the (reversible) formation of α-amino-β-hydroxy acids from aldehydes and the donor glycine (Sect. 2.4.1). The reaction equilibrium can be pulled towards product formation via (irreversible) decarboxylation of the aldol product, which yields (R)-2-amino-1-phenylethanol derivatives, which are a common structural motif in natural and synthetic bioactive compounds (Scheme 3.40, top). This was realized by a DYKAT process combining L-threonine aldolase with L-tyrosine decarboxylase by matching the D/L and *syn*/*anti* selectivity of the respective enzymes [416].

Closely related phenylpropanolamines from the amphetamine family, such as norephedrine and norpseudoephedrine, are important pharmaceutically active molecules. As a shortcut to traditional multistep synthesis, an elegant cascade-reaction was designed (Scheme 3.40, bottom) [417, 418]. In a first step, benzaldehyde and pyruvate were ligated by thiamine diphosphate-dependent acetohydroxyacid synthase I to yield (R)-phenylacetyl carbinol (PAC) with >98% e.e. The second chiral center bearing the amino group was introduced by diastereoselective transamination of the ketone using either (R)-ω-transaminase from *Aspergillus terreus* or (S)-ω-transaminase from *Chromobacterium violaceum* at the expense of L-alanine as amino donor to furnish norpseudoephedrine or norephedrine, respectively. The latter reaction produces pyruvate as by-product, which has to be removed to shift the equilibrium of transamination. This may be achieved via decarboxylation, reduction to lactate or by reductive amination (Sect. 2.6.2). In the cascade-reaction, pyruvate is cycled back and used as C-donor in the carboligation step, which significantly enhances the atom economy of the whole process. Overall, benzaldehyde and alanine yield the target amino alcohol with CO_2 as the only by-product. The main obstacle in this cascade was the dominant amination of benzaldehyde (yielding benzylamine) owing to its higher carbonyl activity compared to the ketone moiety of PAC, which was managed by running the sequence in a one-pot two-step fashion, i.e. the transaminase was only added after benzaldehyde was completely consumed in the carboligation. This one-pot two step sequence provided norpseudoephedrine (conv. 96%) and norephedrine (conv. 80%) in >99% e.e. and >98% d.e.

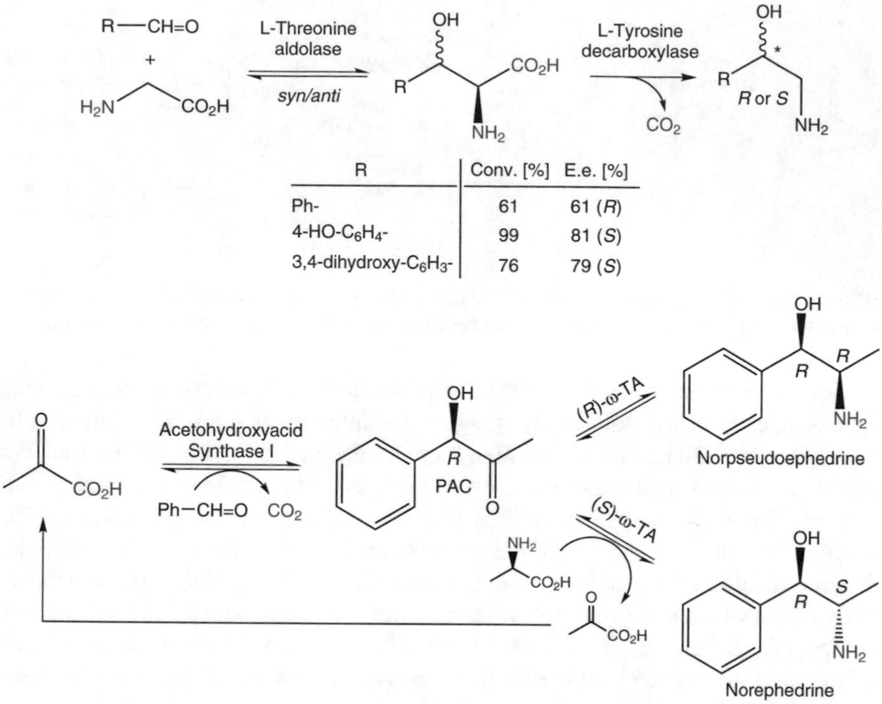

R	Conv. [%]	E.e. [%]
Ph-	61	61 (*R*)
4-HO-C_6H_4-	99	81 (*S*)
3,4-dihydroxy-C_6H_3-	76	79 (*S*)

Scheme 3.40 Linear and cyclic cascades in carboligation. Pulling the equilibium of Thr-aldolase towards C–C bond formation by irreversible decarboxylation using Tyr-decarboxylase (*top*), recycling of pyruvate (co-product from transamination) as donor in benzoin condensation (*bottom*)

In large-scale synthesis, redox reactions are avoided wherever possible, because they require costly oxidants or reductants in stoichiometric amounts. In biocatalysis, these redox mediators are cofactors, which are recycled. Although the recycling of nicotinamide cofactor in its oxidised or reduced form is nowadays feasible with high efficiency (Sect. 2.2.1), it requires usually a second enzyme and a sacrificial co-substrate in stoichiometric amounts. In cascade-reactions, redox-transformations can be interconnected such that their overall demand for redox-equivalents, e.g. reduced and oxidised nicotinamide, cancels out and an overall redox-neutral balance is achieved, which omits the neccessity for sacrificial co-substrates. This concept is widely persued in traditional chemical synthesis and has been termed 'borrowing hydrogen'[8] [419, 420]. In the context of biotransformations, internal hydrogen-transfer between two concurring redox reactions was first demonstrated in 1984 by the conversion of lactate to alanine via the intermediate pyruvate

[8]Hydrogen-transfer reactions are known since 1925: Meerwein H, Schmidt H (1925) Liebigs Ann. Chem. 444: 221; Verley A (1925) Bull. Soc. Fr. 37: 537.

(Scheme 3.41, top) [421]. In the first step, racemic lactate was oxidised by D- and L-lactate dehydrogenase to yield pyruvate with concomitant formation of NADH, which is employed for the reductive amination step catalysed by L-alanine dehydrogenase to form L-Ala. Overall, this redox-neutral functional group interconversion of a hydroxy to an amino moiety requires only NH_3 and produces H_2O. This strategy has been extensively applied to the conversion of *prim-* and *sec-*(di) alcohols to (di)amines by coupling an alcohol dehydrogenase (from *Bacillus stearothermophilus*) with transaminases (from *Arthrobacter citreus* or *Chromobacterium violaceum*) in the reductive amination mode using alanine dehydrogenase (from *Bacillus subtilis*) for the recycling of Ala [422]. In order to avoid side-reactions involving intermediate aldehydes, the concentration of the latter was kept on a low level by keeping the relative rate of alcohol oxidation at a slower rate than amine formation. Selected amines produced via this cascade are shown in Scheme 3.41 (bottom). An alternative system consists in the combination of an alcohol dehydrogenase with an amine dehydrogenase [423].

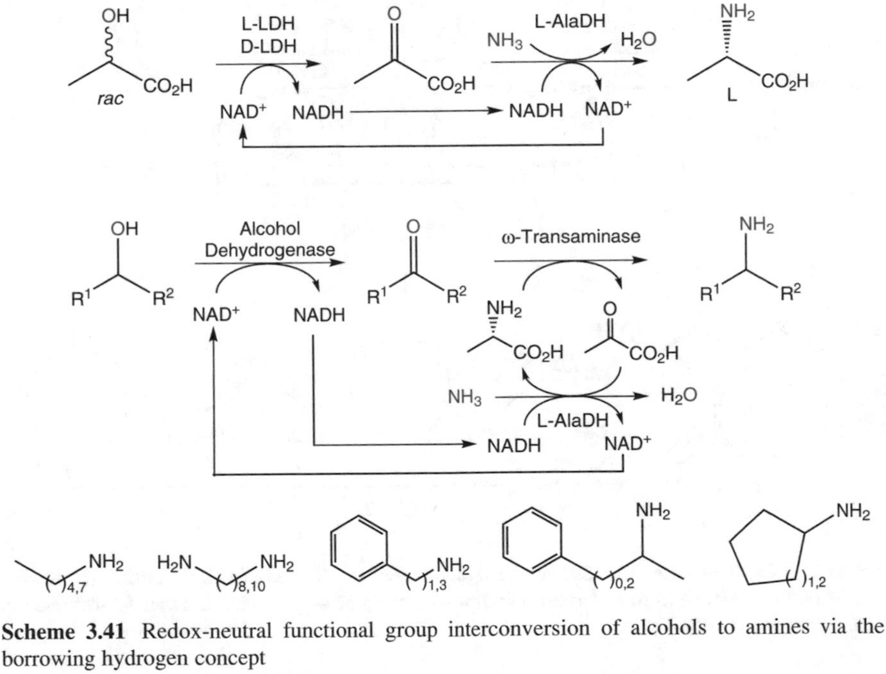

Scheme 3.41 Redox-neutral functional group interconversion of alcohols to amines via the borrowing hydrogen concept

The bi-enzymatic hydrogen borrowing concept can also be applied to run NAD(P)H-dependent monooxygenase oxidations using O_2 (Sect. 2.3.3), e.g. the Baeyer-Villiger oxidation, in a redox-neutral mode by starting from an alcohol precursor of the ketone substrate, which is oxidised by an alcohol dehydrogenase, generating NAD(P)H, which is used to balance the redox-balance of the monooxygenase reaction producing H_2O

(Scheme 3.42, top). The driving force of this sequence is imposed by the irreversible oxygenation step [424]. More recently, this concept was successfully applied to the synthesis of ε-caprolactone in >94% conversion from cyclohexanol [425].

A clever combination of an alcohol dehydrogenase and a Baeyer-Villiger monooxygenase was used for the redox-neutral synthesis of ε-caprolactone from 1,6-hexanediol and 2 equivalents of cyclohexanone (Scheme 3.42, bottom). Oxidation of the latter by the BVMO required two equivalents of NADPH, which were generated via step-wise oxidation of 1,6-hexanediol by *Thermoanaerobacter ethanolicus* ADH. In the first step, the intermediate 6-hydroxyhexanal spontaneously underwent cyclization to yield the corresponding lactol, which was further oxidised to ε-caprolactone in a second step. In practice, however, the formation of three equivalents of ε-caprolactone was never observed due to undesired hydrolysis and oligomerization of the seven-membered lactone. Nevertheless, an analytical yield of 61% was achieved [426].

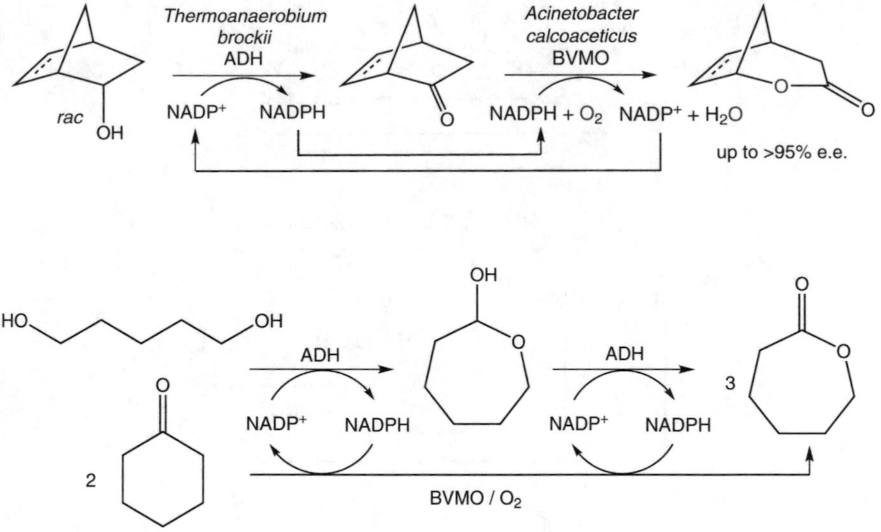

Scheme 3.42 Redox-neutral Baeyer-Villiger oxidation of ketones by starting from the corresponding alcohol (*top*) and redox-neutral formation of ε-caprolactone from 1,6-hexanediol cyclohexanone (*bottom*)

At present, there is almost no limit set for the combination of individual enzymes in cascades, which enable the multistep transformation of renewable carbon sources, e.g. fatty acids, into valuable products with a higher degree of functionalization [427], or to design completely new reactions, which are neither feasible with conventional chemical catalysts nor have a biological counterpart, such as the direct vinylation of unprotected phenols (Scheme 3.43). This cascade was designed based on a combination of three enzymatic reactions: (i) Coupling of a phenol and ammonium pyruvate mediated by tyrosine phenol lyase (variant M379V from *Citrobacter freundii*) goes in hand with simultaneous C–C and C–N bond formation and leads to the formation of L-Tyr derivatives in a regio- and stereoselective fashion. (ii) The second concurring step is catalysed by tyrosine ammonia lyase (from *Rhodobacter sphaeroides*) and consists of elimination of NH_3, which is re-used in the first carboligation step. (iii) The *p*-hydroxycinnamates thus formed are finally decarboxylated by a ferulic acid decarboxylase (from *Enterobacter* sp.) in an irreversible fashion to furnish *p*-vinyl phenols in $\geq 99\%$ conversion.

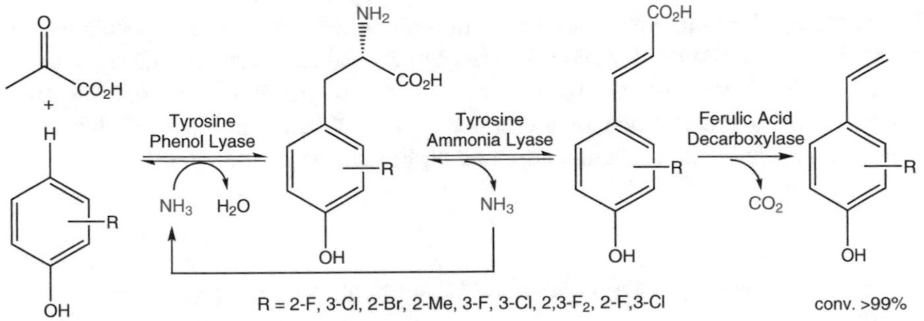

Scheme 3.43 Formal *p*-vinylation of phenols via a three-enzyme cascade at the expense of pyruvate producing CO_2 and H_2O as by-products

In order to optimize the flux in metabolic pathways, nature has evolved sophisticated strategies to optimize the cooperative action of enzymes in cascades. Thus proteins directly transfer reactants from one active site to another without diffusing them into the bulk solvent by mimicking a molecular assembly line. This reduces decomposition of sensitive intermediates and minimizes the detrimental influence of competing enzymes. The molecular mechanisms of this 'substrate channelling'[428] along metabolic pathways is brought about by nanoscale spatial organization or electrostatic guidance [429, 430]. Toxic intermediates (e.g. carbon monoxide) [431] are passed through intramolecular tunnels, and substrates requiring chemical activation can be temporarily mounted onto swing arms by covalent binding [432]. In order to boost the performance of cascade-reactions, the effect of substrate channeling is currently imitated by the clustering of enzymes into agglomerates, but it remains to be seen whether this technology will be applicable to preparative-scale reactions [433].

An interesting 'processive' enzyme with synthetic potential acting via a spacer arm is carboxylate reductase (CAR) [434]. The chemoselective reduction of carboxylic acids to the corresponding aldehydes by conventional methods is plagued by over-reduction yielding *prim*-alcohols, but carboxylate reductases selectively stop at the aldehyde stage. In order to make a carboxylic acid amenable to bioreduction at the expense of nicotinamide, it has to be activated as thio-ester involving a nucleophilic SH-residue in the active site of the enzyme, which requires a sophisticated mechanism (Scheme 3.44). First, apo-CAR is post-translationally activated by attachment of a phosphopantetheinyl residue (P-pant-SH) from Coenzyme A (Scheme 2.203) onto a conserved Ser-OH moiety in its active site under catalysis of phosphopantetheinyl transferase [435]. The holo-CAR thus obtained is now equipped with a flexible linker bearing a terminal SH residue, which serves as robotic swing arm to safely transfer the substrate between the activating (adenylating) and the reducing domain of CAR. In the first step of the catalytic cycle, the carboxylate substrate is transformed into a mixed carboxylic-phosphoric anhydride at the expense of ATP in the activating domain. Next, the activated carboxylate is bound onto the SH-residue on the spacer arm, which swings over to the reducing domain, where the acyl-thioester moiety is reduced by NADPH to yield the corresponding aldehyde. The SH-linker is liberated and re-enters the cycle again. CARs from various sources have shown to possess a relaxed substrate spectrum for a broad range of aromatic, O- and N-heteroaromatic, arylaliphatic and straight-chain and branched aliphatic carboxylic acids [436–438].

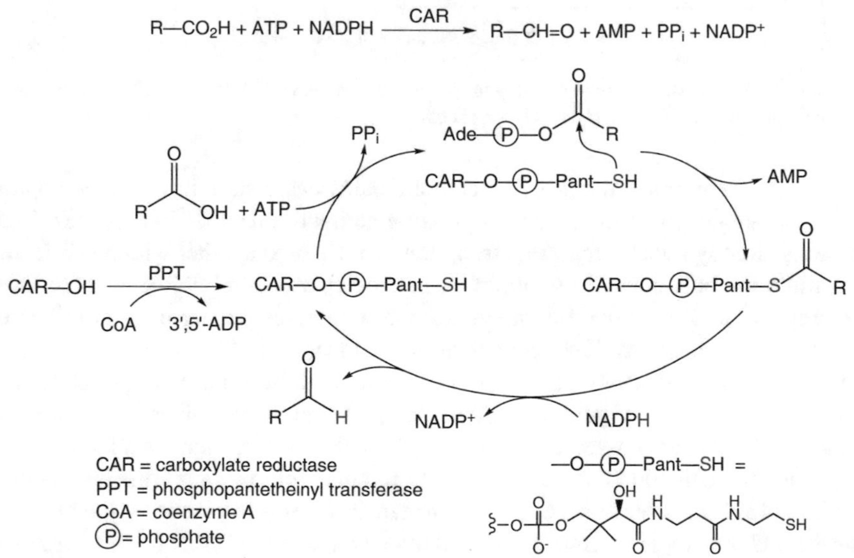

Scheme 3.44 Chemoselective reduction of carboxylic acids by carboxylate reductase

3.3 Immobilization

In practice, three significant drawbacks are often encountered in enzyme-catalyzed reactions.

- Many enzymes are not sufficiently stable under the operational conditions and they may lose catalytic activity due to auto-oxidation, self-digestion and/or denaturation by the solvent, the solutes or due to mechanical shear forces.
- Since enzymes are water-soluble molecules, their repeated use, which is important to ensure their economic application [439], is problematic because their separation from substrates and products and their recovery from aqueous systems is difficult.
- The productivity of industrial processes, measured as the space-time yield is often low due to the limited tolerance of enzymes to high concentrations of substrate(s) and product(s).

These problems may be overcome by 'immobilization' of the enzyme [440–452]; two main strategies exist: (i) The enzyme is attached onto a solid support via coupling onto a carrier or the enzyme molecules are linked onto each other via cross-linking. (ii) Alternatively, the biocatalyst may be spatially confined to a restricted area from which it cannot leave but where it remains catalytically active by entrapment into a solid matrix or a membrane-restricted compartment. Thus, *homogeneous* catalysis using a native enzyme turns into *heterogeneous* catalysis when the biocatalyst is immobilized. Depending on the immobilization technique, enzyme properties, such as stability, selectivity [453, 454], k_{cat} and K_M value, pH and temperature characteristics, may be significantly altered [455] – sometimes for the better, sometimes for the worse. At present, predictions about the effects of immobilization on the catalytic performance of proteins are very difficult to make. The following immobilization strategies are most widely employed (Fig. 3.5) [456, 457].

Adsorption
Adsorption of a biocatalyst onto a water-insoluble macroscopic carrier is the easiest and oldest method of immobilization. It may be equally well applied to isolated enzymes as well as to whole viable cells. For example, adsorption of whole cells of *Acetobacter* onto wood chips for the fermentation of vinegar from ethanol was first used in 1815! Adsorbing forces are of different types, such as van der Waals (London) forces, ionic interactions, and hydrogen bonding, and are all relatively weak. The appealing feature of immobilization by adsorption is the simplicity of the procedure. As a result of the weak binding forces, losses in enzyme activity are usually low, but desorption (leakage) from the carrier may be caused by even minor changes in the reaction parameters, such as a variation of substrate concentration, the solvent, temperature, or pH.

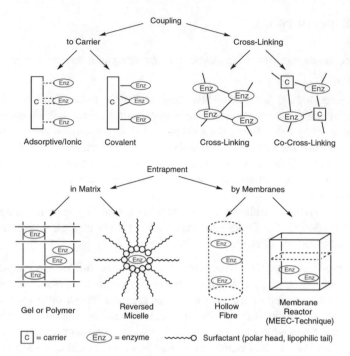

Fig. 3.5 Principles of immobilization techniques

Numerous inorganic and organic materials have been used as carriers: activated charcoal [458], alumina [459], silica [460, 461], diatomaceous earth (Celite) [180, 462], cellulose [463], controlled-pore glass, and synthetic resins [464]. In contrast to the majority of enzymes, which preferably adsorb to materials having a polar surface, lipases are better adsorbed onto lipophilic carriers due to their peculiar physicochemical character (Sect. 2.1.3.2) [74, 465–468]. Adsorption is the method of choice when enzymes are used in lipophilic organic solvents, where desorption cannot occur due to their insolubility in these media.

Ionic Binding

Due to their polar surfaces, ion exchange resins readily adsorb proteins. Thus, they have been widely employed for enzyme immobilization. Both cation exchange resins such as carboxymethyl cellulose or Amberlite IRA [469], and anion exchange resins, e.g., N,N-diethyl-aminoethylcellulose (DEAE cellulose) [470] or sephadex [471],[9] are used industrially. Although the binding forces are stronger than the forces involved in simple physical adsorption, ionic binding is particularly susceptible to the presence of other ions. As a consequence, proper maintenance of ion concentrations and pH is important for continued immobilization by ionic binding and for prevention of desorption of the enzyme. When the biocatalytic

---

[9]DEAE cellulose and sephadex are popular stationary phases for protein chromatography.

activity is exhausted, the carrier may easily be reused by reloading it with fresh biocatalyst.

Covalent Attachment

Covalent binding of an enzyme onto a macroscopic carrier leads to the irreversible formation of stable chemical bonds, which prevents leakage completely. A disadvantage of this method is that rather harsh conditions are required since the protein must undergo a chemical reaction. Consequently, some loss of activity is always observed [472, 473]. As a rule of thumb, each bond attached to an enzyme decreases its native activity by about one fifth. Consequently, residual activities generally do not exceed 60–80% of the activity of the native enzyme, and values of around 50% are normal. The functional groups of the enzyme which are commonly involved in covalent binding are nucleophilic, i.e., mainly N-terminal and ε-amino groups of lysine, but also carboxy-, sulfhydryl-, hydroxyl-, and phenolic functions. In general, covalent immobilization involves two steps, i.e., (i) activation of the carrier with a reactive 'spacer' group and (ii) enzyme attachment. Since viable cells usually do not survive the drastic reaction conditions required for the formation of covalent bonds, this type of immobilization is only recommended for isolated enzymes.

Porous glass is a popular inorganic carrier for covalent immobilization [474]. Activation is achieved by silylation of the hydroxy groups using aminoalkylethoxy- or aminoalkyl chlorosilanes as shown in Scheme 3.45. In a subsequent step, the aminoalkyl groups attached to the glass surface are either transformed into reactive isothiocyanates or into Schiff bases by treatment with thiophosgene or glutardialdehyde, respectively. Both of the latter species are able to covalently bind an enzyme through its amino groups.

Carriers based on natural polymers of the polysaccharide type (such as cellulose [475], dextran [476], starch [477], chitin, or agarose [478]) can be useful alternatives to inorganic material due to their well-defined pore size. Activation is achieved by reaction of adjacent hydroxyl groups with cyanogen bromide leading to the formation of reactive imidocarbonates (Scheme 3.46). Again, coupling of the enzyme involves its amino groups. To avoid the use of hazardous cyanogen bromide, pre-activated polysaccharide-type carriers are commercially available.

During recent years, synthetic copolymers (e.g., based on polyvinyl acetate) have become popular (Scheme 3.47) [479–481]. Partial hydrolysis of some of the acetate groups liberates a hydroxyl functionality, which is activated by epichlorohydrin [482]. A number of such epoxy-preactivated resins are commercially available. Among them, Eupergit C™ is widely used. It is a macroporous copolymer of N,N'-methylene-bi-(methacrylamide), glycidyl methacrylate, allyl glycidyl ether and methacrylamide with an average particle size of 170 µm and a pore diameter of 25 nm [483]. Enzyme attachment occurs with the formation of a stable C–N bond via nucleophilic opening of the epoxide groups by amino groups of the enzyme under mild conditions. In contrast to the majority of the above-mentioned covalent immobilization reactions involving ε-amino groups of Lys, which remove a positive (ammonium) charge from the enzyme's surface due to Schiff base formation or N-acylation, this method preserves the charge distribution of the enzyme since it constitutes an N-alkylation process.

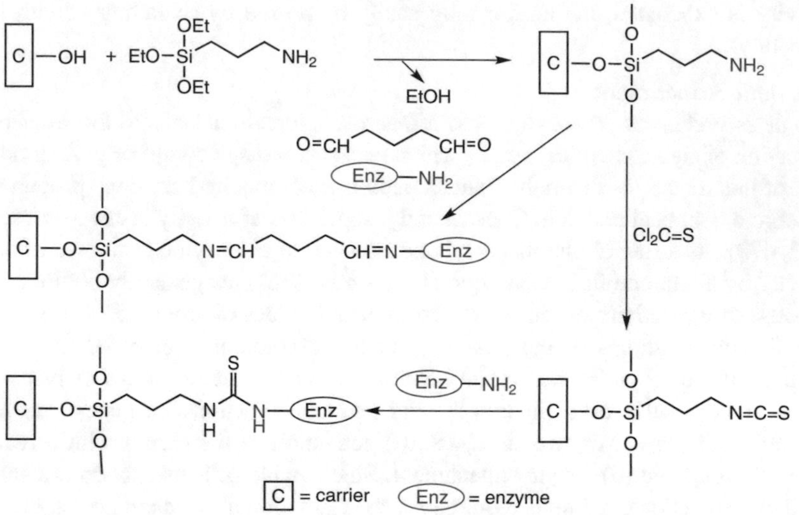

Scheme 3.45 Covalent immobilization of enzymes onto inorganic carriers

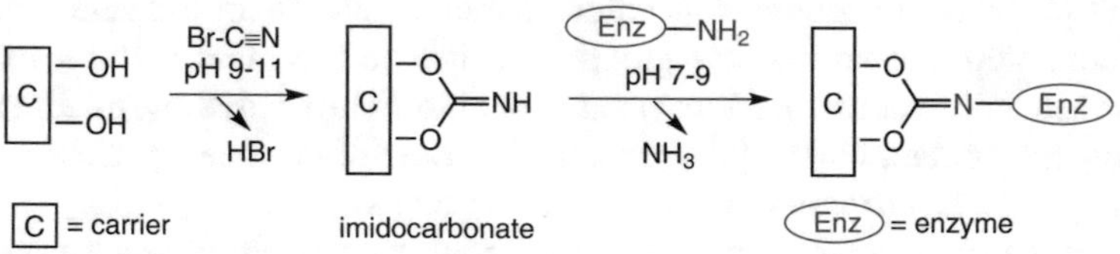

Scheme 3.46 Covalent immobilization of enzymes onto natural polymers

Alternatively, cation exchange resins can be activated by transforming their carboxyl groups into acid chlorides, which then form stable amide bonds with the amino groups of an enzyme.

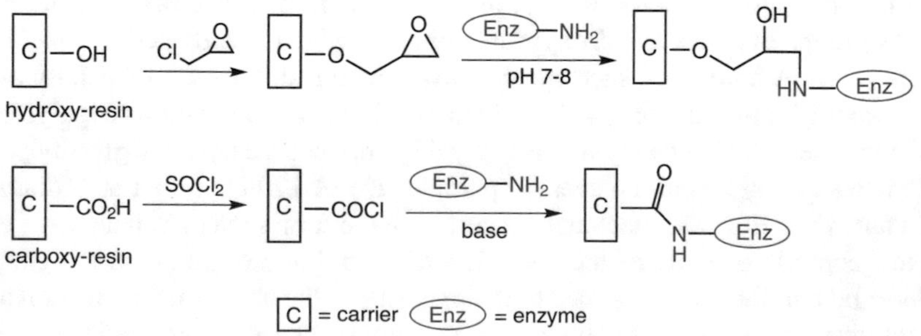

Scheme 3.47 Covalent immobilization of enzymes onto synthetic polymers

A novel approach to immobilization of enzymes via covalent attachment is the use of stimulus-responsive 'smart' polymers, which undergo dramatic conformational changes in response to small alterations in the environment, such as temperature, pH, and ionic strength [484–486]. The most prominent example is a thermo-responsive and biocompatible polymer (poly-*N*-isopropyl-acrylamide), which exhibits a critical solution temperature around 32 °C, below which it readily dissolves in water, while it precipitates at elevated temperatures due to the expulsion of water molecules from its polymeric matrix. Hence, the biotransformation is performed under conditions, where the enzyme is soluble. Raising the temperature leads to precipitation of the immobilized protein, which allows its recovery and reuse. In addition, runaway reactions are avoided because in case the reaction temperature exceeds the critical solution temperature, the catalyst precipitates and the reaction shuts down.

Cross-Linking
The use of a macroscopic (polymeric) carrier inevitably leads to 'dilution' of catalytic activity owing to the introduction of a large proportion of inactive ballast, ranging from 90 to >99%, which leads to reduced productivities. This can be avoided via attachment of enzymes *onto each other* via 'cross-linking' through covalent bonds [487, 488]. By this means, insoluble high-molecular aggregates are obtained. The enzyme molecules may be crosslinked either with themselves or may be co-crosslinked with other inactive 'filler' proteins such as albumins. The most widely used bifunctional reagents used for this type of immobilization are α,ω-glutardialdehyde (Scheme 3.48) [489, 490], dextran polyaldehyde [491], dimethyl adipimidate, dimethyl suberimidate, and hexamethylenediisocyanate or -isothiocyanate. The advantage of this method is its simplicity, but it is not without drawbacks. The soft aggregates are often of a gelatine-like nature, which prevents their use in packed-bed reactors. Furthermore, the activities achieved are often limited due to diffusional problems, since many of the biocatalyst molecules are buried inside the complex structure which impedes their access by the substrate. The reactive groups involved in the crosslinking of an enzyme are not only free amino functions but also sulfhydryl- and hydroxyl groups.

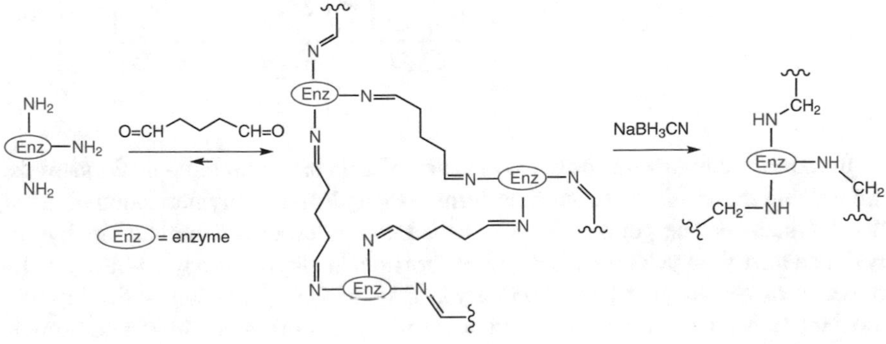

Scheme 3.48 Crosslinking of enzymes by glutaraldehyde

Crosslinking can be performed with dissolved (monomeric) proteins, and also by using them in microcrystalline or amorphous form. Thus, either 'cross-linked enzyme crystals' (CLECs) [492–497] or 'cross-linked enzyme aggregates' (CLEAs) are obtained [491, 498, 499]. The advantage of the solid aggregates thus obtained are increased stability against chemical and mechanical stress. Since they constitute a very firm solid matrix, they cannot be attacked by proteases [500] and are therefore inert towards (self)digestion. Due to the close vicinity of the individual enzymes molecules, cross-linking is particularly effective for oligomeric proteins.

However, some experimentation is usually necessary to find out the optimum conditions to obtain optimal results.

Entrapment into Gels

In case an enzyme does not tolerate direct binding, it may be physically 'encaged' in a macroscopic matrix. To ensure catalytic activity, it is necessary that substrate and product molecules can freely pass into and out of the macroscopic structure. Due to the lack of covalent binding, entrapment is a mild immobilization method which is also applicable to the immobilization of viable cells.

Entrapment into a biological matrix such as agar gels [501], alginate gels [502], or κ-carrageenan [503] is frequently used for viable cells. As depicted in Fig. 3.6, the gel-formation may be initiated either by variation of the temperature or by changing the ionotropic environment of the system. For instance, an agar gel is easily obtained by dropping a mixture of cells suspended in a warm (40 °C) solution of agar into well-stirred ice-cold aqueous buffer. Alternatively, calcium alginate or κ-carrageenan gels are prepared by adding a sodium alginate solution to a $CaCl_2$ or KCl solution, respectively. The main drawbacks of such biological matrices are their instability towards changes in temperature or the ionic environment and their low mechanical stability.

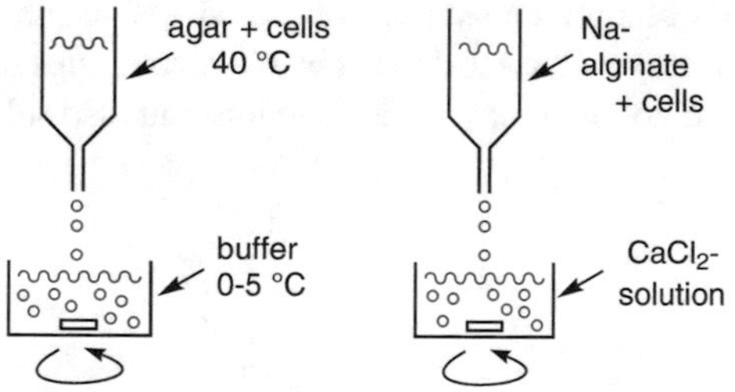

Fig. 3.6 Gel-entrapment of viable cells

In order to circumvent the disadvantages of gels based on biological materials, more stable *inorganic* silica matrices formed by hydrolytic polymerization of metal Si-alkoxides became popular. This so-called sol-gel process is initiated by hydrolysis of a tetraalkoxysilane $Si(OR)_4$ (R = short-chain alkyl, *n*-propyl, *n*-butyl) in the presence of the enzyme [504, 505]. Hydrolysis and condensation of the $Si(OR)_4$ monomers, catalyzed by a weak acid or base, triggers the cross-linking and simultaneous formation of amorphous $(SiO_2)_n$ (Fig. 3.7).

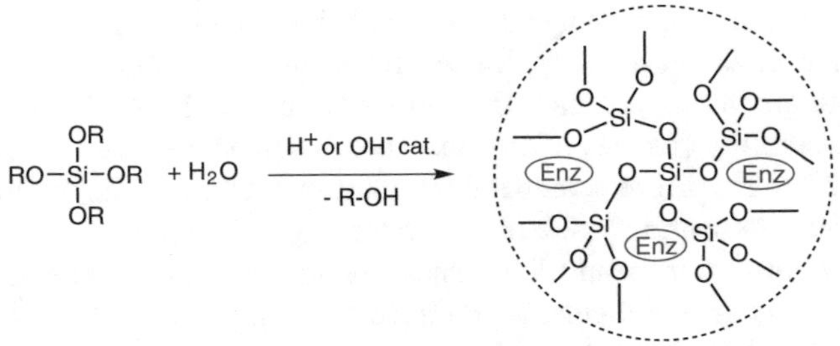

$$RO-\underset{\underset{OR}{|}}{\overset{\overset{OR}{|}}{Si}}-OR \ + H_2O \quad \xrightarrow[- R-OH]{H^+ \ or \ OH^- \ cat.}$$

Fig. 3.7 Entrapment of enzymes in silica sol-gels

The latter constitutes a highly stable porous inorganic polymeric matrix that grows around the enzyme in a three-dimensional fashion. The sol-gel aggregates thus obtained ensure high enzyme activity and long operational stability. In addition, this process is well suited for large-scale applications since the materials are comparatively inexpensive [506–509].

A tight network which is able to contain isolated enzyme molecules (which are obviously much smaller than whole cells) may be obtained by polymerization of synthetic monomers such as polyacrylamide in the presence of the enzyme [510–512]. It is obvious that the harsh conditions required for the polymerization makes this method inapplicable to whole cells.

Entrapment into Membrane Compartments
Enzymes may be enclosed in a restricted compartment bordered by a membrane. Although this does not lead to 'immobilization' per se, it provides a restricted space for the enzyme, which is separated from the rest of the reaction vessel. Small substrate and/or product molecules can freely diffuse through the pores of the membrane, but the large enzyme cannot. The separation of a reaction volume into compartments by membranes is a close imitation of 'biological immobilization' within a living cell, since many enzymes are membrane-bound in order to provide a safe micro-environment. Two general methods exist for the entrapment of enzymes into membrane-restricted compartments.

Micelles and Vesicles Mixtures of certain compositions containing water, an organic solvent and a detergent (Fig. 3.8, Scheme 3.49) give transparent solutions in which the organic solvent is the continuous phase [513]. The water is present in microscopic droplets which have a diameter of 6–40 nm [514] and are surrounded by the surfactant. The whole structure is embedded in the organic solvent and represents a micelle which is turned inside out. It is therefore termed a 'reverse micelle' (see Fig. 3.8). The latter are mimics for the micro-environment of the cell and can be regarded as artificial micro-cells. Thus, they provide high enzyme activity. On the other hand, when *water* constitutes the bulk phase, micelles may be formed by a symmetrical double layer of surfactant. The latter structure constitutes a 'vesicle' (liposome). The water trapped inside these micro-environments has several chemical and physical properties that deviate from 'normal' water, such as restricted molecular motion, decreased hydrogen bonding, increased viscosity, and a lower freezing point [515, 516]. Enzymes can be accommodated in these water-pools and can stay catalytically active [517–519]. The exchange of material from one micelle to another occurs by means of collisions and is a very fast process. Like in other immobilization methods, the catalytic activity of enzymes may be altered through entrapment in micelles. For instance, the activity [520–522] and the temperature stability [523] is often enhanced; in some cases also the specificity is changed [524, 525]. From a preparative point of view, it is important that compounds which are sparingly soluble in water (e.g., steroids) can be converted at much higher rates than would have been possible in aqueous media [526]. The disadvantages encountered when using micelles on a preparative scale are considerable operational problems during (extractive) workup, caused by the presence of a considerable amount of surfactant, which behaves like a soap.

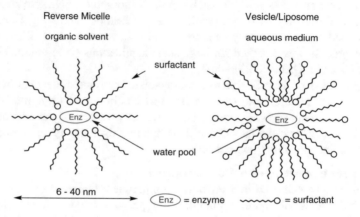

Fig. 3.8 Entrapment of enzymes in reversed micelles and vesicles

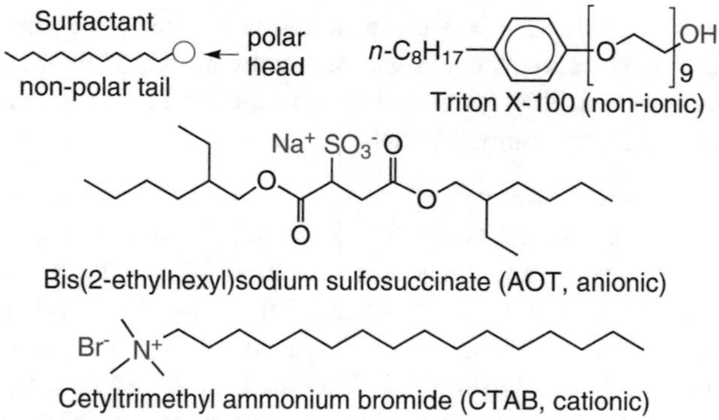

Scheme 3.49 Commonly employed surfactants

Synthetic Membranes A practical alternative to the use of sensitive biological matrices is the use of synthetic membranes [527, 528] based on polyamide or polyethersulfone (Fig. 3.9).[10] They have long been employed for the purification

---

[10]A popular membrane from daily life is Gore-Tex™.

of enzymes by ultrafiltration, which makes use of the large difference in size between high-molecular biocatalysts and small substrate/products molecules. Synthetic membranes of defined pore size, covering the range between 500 and 300,000 Da, are commercially available at reasonable cost. The biocatalyst is detained in the reaction compartment by the membrane, but small substrate/product molecules can freely diffuse through the pores of the barrier. This principle allows biocatalytic reactions to be performed in highly desirable *continuous* processes. Furthermore, disadvantages caused by heterogeneous catalysis, such as mass-transfer limitations and alteration of catalytic properties, are largely avoided [529]. A variety of synthetic membranes are available in various shapes such as flat or cylindrical foils or hollow fibers.

A simplified form of a membrane reactor which does not require any special equipment may be obtained by using an enzyme solution enclosed in dialysis tubing like a tea bag (Fig. 3.9). This simple technique termed 'membrane-enclosed enzymatic catalysis' (MEEC) seems to be applicable to most types of enzymes except lipases [530–532]. It consists of a dialysis bag containing the enzyme solution, mounted on a gently rotating magnetic stirring bar.

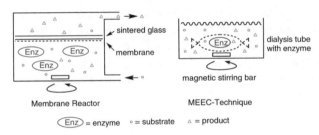

Membrane Reactor   MEEC-Technique

Enz = enzyme   ○ = substrate   △ = product

Fig. 3.9 Principle of a membrane reactor and the membrane-enclosed enzyme catalysis (MEEC)-technique

In some enzyme-catalyzed processes it is of an advantage to couple an additional (chemical) reaction onto the process in a cascade-like fashion in order to drive an unfavorable equilibrium in the desired direction. Very often, however, the harsh reaction conditions required for the auxiliary step are incompatible with the enzyme(s). In such cases, a membrane may be used to separate the enzyme-catalyzed reaction from the auxiliary process, while the chemical intermediates can pass freely through the barrier.

This concept has been applied to improve the cofactor recycling in a stereoselective reduction catalyzed by lactate dehydrogenase (LDH) (Scheme 3.50) [533]. Thus, pyruvate was reduced to D-lactate at the expense of NADH using D-LDH. The cofactor was recycled via the yeast alcohol dehydrogenase (YADH)/ethanol system. In order to avoid enzyme deactivation by acetaldehyde and also to drive the reaction towards completion, acetaldehyde was reduced by sodium borohydride to yield ethanol, which in turn re-enters the process. To save the enzymes from being deactivated by borohydride and a strongly alkaline pH, the process was carried out in a reactor consisting of two compartments, which are separated by a supported gas-membrane, comparable to Gore-Tex. All volatile species (acetaldehyde and ethanol) can freely pass through the air-filled micropores, but the nonvolatiles (enzyme, cofactor, substrate, product, salts) cannot. Thus, a cycle number of >10,000 was achieved.

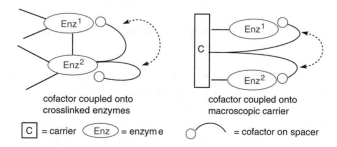

Scheme 3.50 Cofactor recycling using a gas-membrane reactor

Immobilization of Cofactors

All of the above-mentioned immobilization techniques can readily be used for enzymes which are independent of cofactors and for those in which the cofactors are tightly bound (e.g., flavins, Sect. 1.4.4). For enzymes, which depend on charged cofactors, such as NAD(P)H or ATP, which readily dissociate into the medium, coimmobilization of the cofactor is often required to ensure a proper functioning of the overall system (Sects. 2.1.4 and 2.2.1).

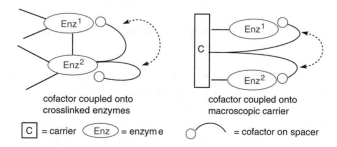

Fig. 3.10 Co-immobilization of cofactors

Two solutions to this problem have been put forward (Fig. 3.10).

- The cofactor may be bound onto the surface of a crosslinked enzyme or it may be attached to a macroscopic carrier. In either case it is essential that the spacer arm is long enough so that the cofactor can freely swing back and forth between both enzymes from Enz^1 to perform the reaction to Enz^2 to be regenerated. However, these requirements are very difficult to meet in practice.
- In membrane reactors a more promising approach has been developed in which the molecular weight of the cofactor is artificially increased by covalent attachment of large groups such as polyethylene glycol (MW 20,000) [534–536],

polyethylenimine [537], or dextran moieties [538]. Although the cofactor is freely dissolved, it cannot pass the membrane barrier due to its high molecular weight.

It has to be emphasized that it is difficult to directly compare the different methodologies of enzyme immobilization since enzymes differ in their properties and details of the immobilization of industrial biocatalysts are often not disclosed.

3.4   Artificial and Modified Enzymes

Synthetic chemists have always admired the unparalleled catalytic efficiency and specificity of enzymes with some envy and have attempted to copy the catalytic principles which were developed by nature during evolution and to adopt them to the needs of chemical synthesis. The following strategies can be distinguished.

3.4.1   Non-Proteinogenic Enzyme Mimics

In a *biomimetic approach*, enzyme models are created de novo from a nonproteinogenic artificial scaffold, such a (natural) cyclodextrin, or a (synthetic) dendrimer or polymer, which is endowded with a rationally designed catalytic site [539]. The catalytic machinery represents a simplified chemical model of the chemical operators found in enzymes and is often derived from a cofactor (e.g., cytochrome, pyridoxal or pyridoxamine phosphate) or a catalytic metal acting as Lewis acid (e.g., Zn) or as central mediator in a redox reaction (e.g., Cu, Co, Mn, Fe). Although many of these so-called *synzymes* show astonishing catalytic activites, the catalytic efficiencies (expressed as rate accelerations and turnovers) and (stereo)selectivities are way too low to use them for synthetic transformations. What we gain from these studies, is a better understanding (and appreciation) of the numerous subtle contributions which add up in a synergistic fashion in enzymes and thereby make them as efficient as they are. From these studies we have learned that the structural complexity of the whole three-dimensional structure of a protein made up of several hundred amino acids is not out of mere biological luxury, but due to mechanistic necessity.

3.4.2   Modified Enzymes

Modified enzymes can be created by leaving their protein scaffold basically intact but altering some of their properties by chemical or genetic methods, which yields altered and/or improved enzymes for synthetic purposes. The following general strategies are discussed below [540]:

3.4.2.1 Chemically Modified Enzymes

Historically, site-directed chemical modification of enzymes using group-specific reagents was established mainly during the 1960s aiming at the elucidation of enzyme structures and mechanisms [541, 542] rather than for the creation of biocatalysts with a better performance. In other words, enzyme modification has been developed more as an *analytical* rather than a *synthetic* tool.

Surface-Modified Enzymes Enzymes acting in nearly anhydrous organic solvents always give rise to heterogeneous systems (Sect. 3.1). In order to turn them into homogeneous systems, which can be controlled more easily, proteins can be modified in order to make them soluble in lipophilic organic solvents. This can be readily achieved by covalent attachment of the amphipathic polymer polyethylene glycol (PEG) onto the surface of enzymes [543]. The pros and cons of PEG-modified enzymes are as follows [544, 545]:

- They dissolve in organic solvents such as toluene or chlorinated hydrocarbons, such as chloroform, 1,1,1-trichloroethane, or trichloroethylene [546].
- Their properties such as stability, activity [547, 548], and specificity [549, 550] may be altered.
- Due to their solubility in various organic solvents, spectroscopic studies on the conformation of enzymes are simplified [551].

The most widely used modifier is monomethyl PEG having a molecular weight of about 5000 Da. Linkage of the polymer chains onto the enzyme's surface may be achieved by several methods, all of which involve reaction at the ε-amino groups of lysine residues. The latter are preferably located on the surface of the enzyme. A typical 'linker' is cyanuric chloride (Scheme 3.51). Nucleophilic displacement of two of the chlorine atoms of cyanuric chloride by monomethyl PEG yields a popular modifier, which is attached to the enzyme via an alkylation reaction involving the remaining chlorine atom.

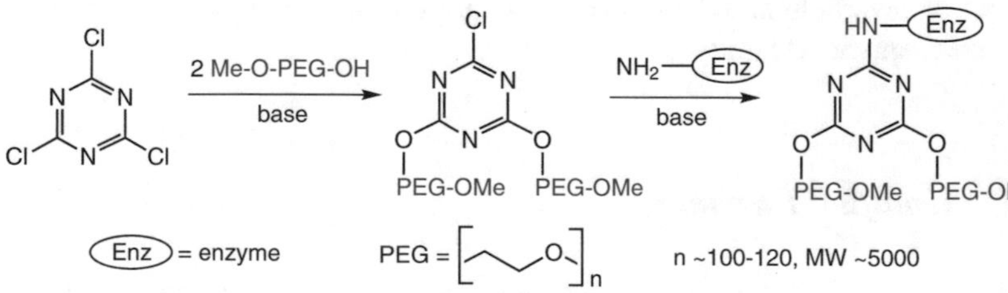

Scheme 3.51 Surface modification of enzymes by polyethylene glycol

The cyanuric chloride/PEG method seems to work for all classes of enzymes, including hydrolases (lipases [552], proteases, glucosidases [553]) and redox enzymes (dehydrogenases, oxidases [554]). The residual activities are usually high (50–80%), and for most enzymes about five to ten PEG chains per enzyme molecule are sufficient to render them soluble in organic solvents. Care has to be taken to avoid extensive modification which leads to deactivation. PEG-modified enzymes may be recovered from a toluene solution by precipitation upon the addition of a hydrocarbon such as petroleum ether or hexane [555].

A special type of PEG modification, which allows the simple recovery of the enzyme from solution by using magnetic forces, is shown below (Scheme 3.52) [556]. When α,ω-dicarboxyl-PEG is exposed to a mixture of ferric ions and hydrogen peroxide, ferromagnetic magnetite particles (Fe_3O_4) are formed which are tightly adsorbed to the carboxylate. The remaining 'free end' of the magnetic modifier is then covalently linked to the enzyme via succinimide activation. When a magnetic field of moderate strength (5000–6000 Oe≅53–75 A/m) is applied to the reaction vessel, the modified enzyme is removed from the solvent by being pulled to the walls of the container. When the magnet is switched off, the biocatalyst is 'released'.

The majority of applications using PEG-modified enzymes have involved the synthesis of esters [162, 557–561], polyesters [562], amides [563], and peptides [564, 565].

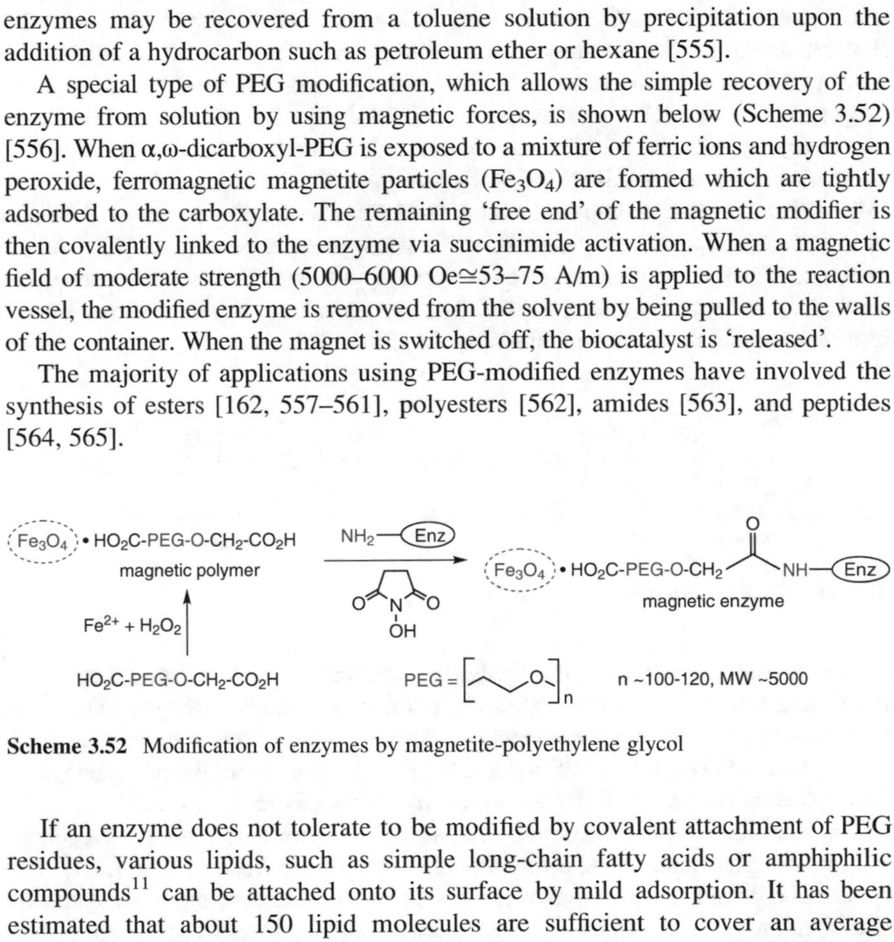

Scheme 3.52 Modification of enzymes by magnetite-polyethylene glycol

If an enzyme does not tolerate to be modified by covalent attachment of PEG residues, various lipids, such as simple long-chain fatty acids or amphiphilic compounds[11] can be attached onto its surface by mild adsorption. It has been estimated that about 150 lipid molecules are sufficient to cover an average protein with a lipophilic layer, which makes it soluble in organic solvents [64, 566]. This so-called 'lipid-coating' seems to be applicable to various types of enzymes, such as lipases [567], phospholipases [521], glycosidases [568], and catalase [569]. Whereas the (enantio)selectivity was not significantly altered in most cases, lipid-coated enzymes showed significantly enhanced reaction rates in organic solvents (up >100-fold).

---

[11]For instance, glutamic acid dioleyl ester ribitol amide, didodecyl N-D-glucono-L-glutamate or Brij35.

Chemically Modified Enzymes Besides varying the physicochemical properties of enzymes (such as their solubility) through modification of its surface, the catalytic properties of an enzyme can be fundamentally altered by chemical modification of the chemical operator in the active site. This leads to 'semisynthetic' enzymes, which often do not have much in common with their natural ancestors [570–572]. Early efforts focussed on the modification of nucleophilic hydroxy- or thiol-residues in Ser- or Cys-hydrolases, such as subtilisin or papain, respectively. For example, the Ser-hydroxy group within the active site of subtilisin was converted to its selenium analog via chemical activation followed by nucleophilic displacement with HSe^- (Scheme 3.53) [573–575]. The seleno-subtilisin thus obtained showed a 700-fold enhanced rate for aminolysis versus hydrolysis in comparison to native subtilisin [576]. Furthermore, the Se–H group may be reversibly oxidized to its seleninic acid analog (subtilisin-Se–OH) by hydrogen peroxide or organic hydroperoxides giving rise to a semisynthetic peroxidase [577]. In an analogous fashion, a flavin-type redox cofactor was attached to the nucleophilic thiol group in the active site of papain. By this means, a hydrolase was transformed into the artificial redox enzyme flavopapain [578–583].

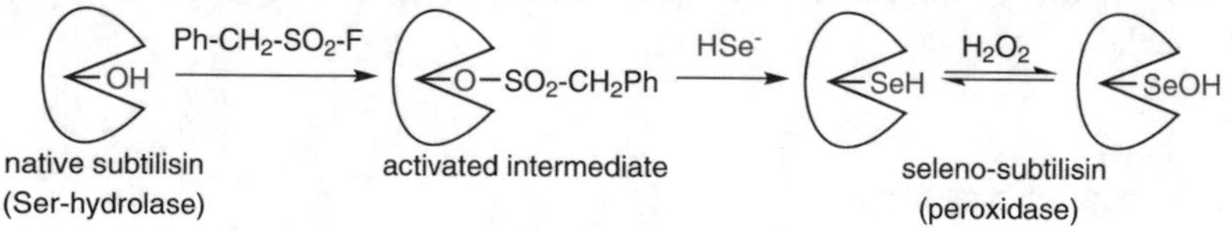

Scheme 3.53 Synthesis of seleno-subtilisin

During the past decade, the design of artificial metalloenzymes became a flourishing sector of enzymology which combines a synthetic catalytic machinery (usually a base metal or precious metal complex) within a natural protein scaffold. These semisynthetic hybrid catalysts combine the first coordination sphere of a homogeneous catalyst with the second coordination sphere of a protein.

An example for the transformation of a non-redox protein into a peroxygenase by simple replacement of its catalytic metal is shown in Scheme 3.54 [584]. The lyase carbonic anhydrase is one of the fastest enzymes known and it catalyzes the equilibration of bicarbonate (derived from oxidative metabolism) into carbon dioxide to facilitate respiration. Removal of the central (redox-neutral) Zn^{2+}-ion in the active site by a strong chelating agent led to the metal-free apoenzyme, which was reconstituted by addition of Mn^{2+} to yield an artificial peroxygenase, which was able to catalyze the asymmetric epoxidation of styrene-type alkenes at the expense of hydrogen peroxide as oxidant. Although its reaction rate and stereoselectivity was within the range of (heme-containing) native horseradish peroxidase, its stability was about 100 times lower and the artificial enzyme lost activity after only 20 cycles due to autooxidation.

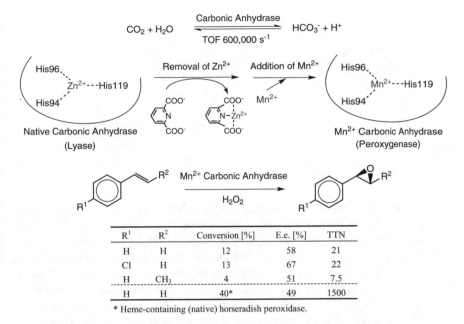

Scheme 3.54 Conversion of carbonic anhydrase (a lyase) into a peroxygenase

The incorporation of base metals (Mn, Co, Fe, Cu) into the active site of an enzyme is comparatively simple due to its Lewis-acid properties, which allows to use ligands (usually His-residues) in the active site for anchoring. In contrast, precious metals are generally encaged in a (synthetic) ligand, which is bound in the active site. The most popular strategy employs the biotin-(strept)avidin technology [585]. Avidin and streptavidin are proteins, which serve to bind biotin (vitamin B_7[12]) and are found in eggwhite or are produced by *Streptomyces* sp., with exceptionally low dissociation constants of $K_D \sim 10^{-15}$ M, making it one of the strongest non-covalent bonds known. This affinity is used to firmly anchor a biotin moiety, which carries a transition metal complex (containing Rh, Ru, Ir, Pt or Pd bound via phospine ligands) on its spacer into the enzyme's active site (Fig. 3.11). Such artificial metalloenzymes show two remarkable features.

- Although the metal-containing ligand representing the first coordination sphere is achiral, remarkable stereoselectivities can be induced by the surrounding protein scaffold (the second coordination sphere).
- Incorporation of precious transition metals, not encountered in catalytic proteins, enable to catalyze reactions, which are not found in nature.

---

[12]Also denoted as vitamin H.

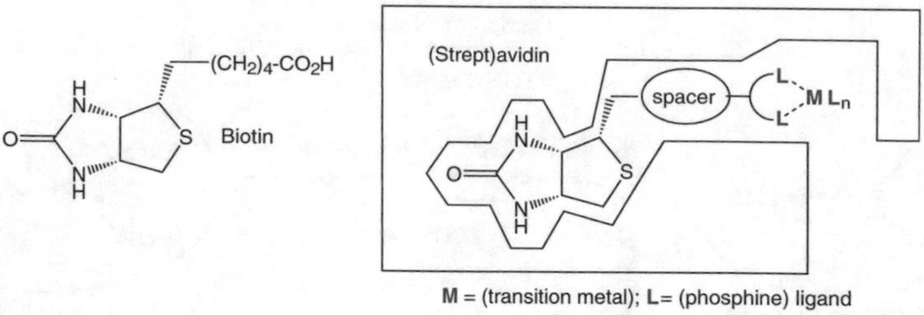

M = (transition metal); L= (phosphine) ligand

Fig. 3.11 Schematic structure of a chimeric metalloprotein based on (strept)avidin technology: A catalytically active metal (M) bound via ligands (L) onto biotin via a spacer is encaged in the biotin-binding site of the protein (stept)avidin

Since the first visionary communication [586], chimeric transition metal containing enzymes were constructed for a broad range of reactions [587, 588], including asymmetric hydrogenation, allylic alkylation, sulfur oxidation, C=C hydration [589], etc. Among them, a 'Suzukiase'[590] and a 'metathase'[591] are most remarkable, because for the Suzuki-coupling and the metathesis-reactions, no biological counterparts are imaginable (Scheme 3.55). Suzuki-coupling of iodonaphthalene with a boronic acid partner gave the atropisomeric (R)-biaryl product in 90% e.e. after careful mutational optimization of the streptavidin cavity serving as active site under Pd-catalysis. In a related setup, ring closing metathesis of an open-chain terminal 1,6-diene furnished the corresponding cyclopropene product with expulsion of ethylene mediated by Ru. In both cases, the catalytic efficiency of the chimeric transition metal protein hybrids was remarkable, but still far from what naturally evolved enzymes are able to achieve.

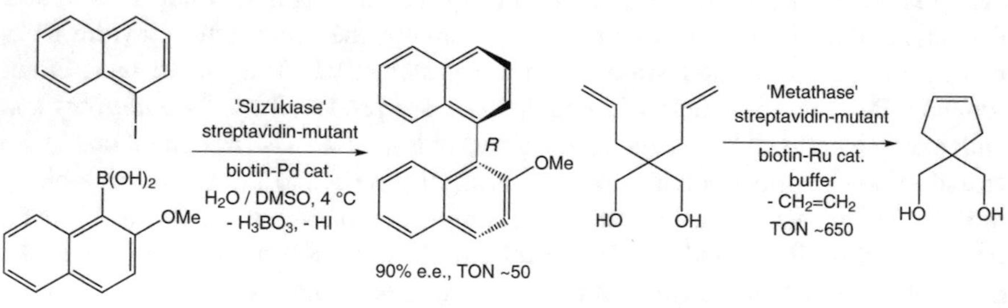

Scheme 3.55 Asymmetric C–C coupling and meathesis reactions catalysed by chimeric metalloproteins

3.4.2.2   Genetically Modified Enzymes

On the other hand, enzyme modification by 'natural' methods via site-directed mutagenesis using the power of molecular biology became a powerful and well-established technique for the generation of proteins possessing altered properties [592–599]. The most important strategies are briefly discussed with increasing order of difficulty.

Changing Substrate Specificity The re-design of the three-dimensional structure of the active site of an enzyme by rational site-directed mutagenesis or via directed evolution (see pp. 76–79, Scheme 2.43) is a state-of-the-art method to generate enzymes with altered substrate specificities. Thus, mutants possessing a more spacious active site can accommodate bulky substrates and enhanced stereoselectivities can be accrued by tightening the enzyme-substrate fit. This technology has been demonstrated to work for virtually all enzyme classes, including hydrolases [600–604], dehydrogenases [605, 606], lyases and transferases [607, 608].

Altering Cofactor Requirements Although enzymes are generally very faithful concerning their cofactors, their specificites can be altered within a narrow frame. Replacement of NADPH by the cheaper and more easily recyclable NADH is possible using dehydrogenase mutants lacking the specific (often Arg-containing) phosphate binding site for NADPH [609–611].

Inverting Stereochemistry Altering the stereochemical outcome of an enzyme reaction in order to access both stereoisomers is more challenging, because the main scaffold of the protein, which is invariably constructed from L-amino acids, remains the same. Nevertheless, successful examples were reported using rational [612–616] or directed evolution methods [617–619].

Engineering Catalysis The 'holy grail' of enzyme redesign is the engineering of entirely new catalytic activities, a property which is often denoted as catalytic promiscuity [620–623]. The latter has been driven by the rapidly increasing number of crystal structures of proteins, which allow to understand the molecular details of their catalytic mechanism. In this context, it was possible to re-engineer the catalytic activities of well studied proteins to furnish switched activities or even completely novel functions, which are rarely found in Nature (Table 3.8).

Table 3.8 Examples for the catalytic promiscuity of rationally designed enzymes

Parent enzyme	Designer enzyme	References
Protease	Nitrile hydratase	[624]
Esterase	Organophosphorous hydrolase	[625, 626]
Carbonyl reductase	Ene-reductase	[627]
Esterase	Epoxide hydrolase	[628]
Lipase	Aldolase	[629]
Lipase	Michael lyase	[630, 631]
Racemase	Amino transferase	[632]
Transaminase	β-Decarboxylase	[633]

Although the exchange of active site residues furnished novel catalytic functions, the overall efficiencies concerning catalytic rates or stereoselectivities were almost invariably disappointing, indicating that the whole environment within the active site exerting numerous (energetically incremental) interactions with the substrate plays an important role in the catalytic mechanism, and not just the chemical operator.

3.4.3   Catalytic Antibodies and Computational De-Novo Design of Enzymes

One of the most striking drawbacks of enzyme catalysis is that proteins cannot be designed and synthesized from scratch due to the boundaries of ribosomal protein synthesis [634]. As a consequence, some synthetically useful transformations, which are not (or only rudimentary) found in nature, are impossible by enzymatic catalysis, in particular cycloadditions or rearrangement reactions. Furthermore, a search for biocatalysts possessing opposite stereochemical properties is usually a tedious empirical undertaking which is often unsuccessful. This gap may be filled by the development of synthetic enzymes [635–637].

Like all catalysts, enzymes accelerate a reaction by stabilizing its transition state. If the latter is known for a given organic transformation, an idealized cavity can be designed possessing matching properties to that of the transition state, such as H-bond donor/acceptor, δ^+/δ^- pairs, π-donor/acceptor, lipophilic binding sites, etc.[13] This entity will bind the transition state, thereby lower its ΔG and diminish the activation energy (E_a). Consequently, the reaction will be accelerated (Fig. 1.7). For the de novo creation of transition state stabilizing proteins, two strategies were evolved: (i) catalytic antibodies [638–642] and (ii) computational design of existing protein scaffolds [643].

Catalytic Antibodies
The immune system produces a vast repertoire – in the range of 10^8–10^{10} [644] – of exquisitely specific proteins (antibodies) that protect vertebrates from 'foreign' invaders such as pathogenic bacteria and viruses, parasites and cancer cells. Antibodies have become invaluable tools in the detection, isolation and analysis of biological materials and are the key elements in many diagnostic procedures. Their potential, however, is not limited to biology and medicine, because they can also be elicited to a large array of synthetic molecules. In a very simplified version they can be regarded as *enzymes possessing an active site, but no chemical operator*. To indicate their derivation, catalytic antibodies are also called 'abzymes' – antibodies as enzymes.

Antibodies are large proteins consisting of four peptide chains which have a molecular weight of about 150,000 Da (Fig. 3.12). There are two identical heavy (H, 50,000 Da) and light chains (L, 25,000 Da) which are crosslinked by disulfide bonds. The light chains are divided into variable (V) and constant (C) regions (V_L, C_L), while the heavy chains consist of V_H, C_H1, C_H2, and C_H3 domains. The variable V_H and V_L domains (located within the first ~110 amino acids of the heavy and light chains) are highly polymorphic and are adapted to the antigen (hapten) they are supposed to trap. Consequently, binding occurs in this region. On the other hand, the constant regions represent the basic framework of the antibody and are relatively invariant. Antibodies bind molecules ranging in size from about 6 to 34 Å

---

[13]The transition state is related to the binding cavity like the 'negative' to a photographic picture.

with association constants of 10^4–10^{14} M^{-1}, in other words, binding is extremely strong. As with enzymes, binding occurs by Van der Waals (London), hydrophobic, electrostatic and hydrogen-bonding interactions.

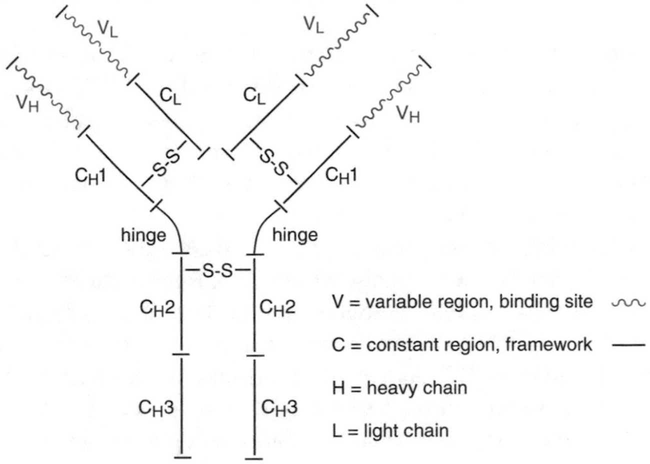

Fig. 3.12   Schematic structure of a catalytic antibody

Antibodies for the stabilization of the transition state of the reaction to be catalyzed may be generated as follows (Scheme 3.56). First, a chemically stable *transition state mimic* is synthesized. This 'model' is then used as an antigen to elicit antibodies. Then, the template is removed from the antibody, which acts as catalyst. If required, the latter can then be genetically or chemically modified by attachment of a reactive group [645], or a cofactor, in order to catalyze redox reactions [646].

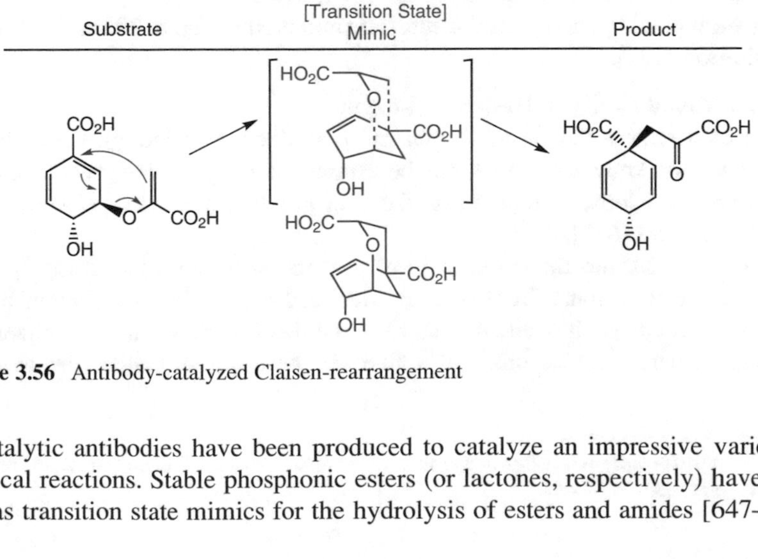

Scheme 3.56   Antibody-catalyzed Claisen-rearrangement

Catalytic antibodies have been produced to catalyze an impressive variety of chemical reactions. Stable phosphonic esters (or lactones, respectively) have been used as transition state mimics for the hydrolysis of esters and amides [647–649], acyl transfer [650] and lactonization [651] reactions, which all proceed via a tetrahedral carbanionic intermediate. In some instances, the reactions proved to be enantioselective [652]. Elimination reactions [653, 654], reductions [655], formation and breakage of C–C bonds [656, 657], ether cleavage [658], *cis-trans* isomerization of alkenes [659] and even photochemical reactions [660] can be catalyzed.

More importantly, pericyclic reactions, such as the Claisen rearrangement [661, 662] (Scheme 3.56) or the Diels-Alder reaction [663, 664], have been catalyzed by antibodies, which were raised against bicyclic and tricyclic transition-state mimics. The latter reactions normally cannot be catalyzed by enzymes. Also cationic cyclization reactions [665, 666] and an enantioselective Robinson annulation were achieved [667].

The catalytic power of abzymes is most conveniently measured as the rate acceleration compared to the uncatalyzed reaction. Recent studies have produced artificial enzymes with catalytic activities approaching those of natural enzymes [668], i.e. rate accelerations of abzymes are in the range of 10^3–10^5, and only in rare cases they reach values of 10^8. As yet, most enzymes are far better (~10^7–10^{17}).

The data availble so far indicate that catalytic antibodies will certainly help us to gain more insight into enzyme mechanisms but won't playing an important role in preparative biotransformations for the following reasons:

- They offer a unique access to artificial tailor-made enzyme-like catalysts and are able to catalyze reactions which have no equivalent counterpart within the diverse natural enzymes.
- The construction of abzymes having an opposite stereochemical preference is possible.
- The catalytic efficiency is low in comparison to that of natural enzymes, because abzymes were elicited against a *model* of the true transition state.
- The production of abzymes on scale is tedious [669] and 'preparative-scale' reactions are in the micromolar range [670, 671].
- The high binding energy for the substrate and product (up to 20 kcal/mol) causes inhibition [672].

Computational De-Novo Design of Enzymes

Instead of a protein devoid of a chemical operator (an antibody), the active site architecture of known proteins can be computationally re-designed to catalyze non-natural reactions, by tapping on the ever growing number of high resolution crystal structures [673].[14]

In a first step, the transition state of a reaction is described using quantum mechanical calculations. From this, an idealized active site of a protein bearing functional groups positioned such as to maximize transition state stabilization is computer-generated. This minimal active site description, called 'theozyme' or

---

[14]Today, approximately 100,000 structure files are available from the Protein Data Bank and about a dozen are added every day.

'catalophor', serves as template for a search in protein crystal structures. The hit structures are analyzed and the active site is redesigned to maximize its affinity for the transition state and its conformational stability. The obtained (variant) proteins are then screened for activity.

The power of this method was demonstrated by generating an enzyme, which catalyzes the Kemp elimination, which does not have a counterpart in nature (Scheme 3.57). The mechanism proceeds via a single transition state, which can be stabilized by an acid-base pair for the deprotonation of the carbon and protonation of the phenolic oxygen. The Rosetta algorithm identified two native proteins as candidates (indole-3-glycerolphosphate synthase from *Sulfolobus solfataricus* and deoxyribose-phosphate aldolase from *E. coli*), which were further optimized to furnish 'Kemp-eliminase' constructs, which achieved remarkable rate accelerations of k_{cat}/k_{uncat} of 10^5–10^6 [674]. Analogous studies led to the construction of a stereoselective Diels-Alderase [675] and the identification of ene-reductases possessing opposite stereoselectivity [676].

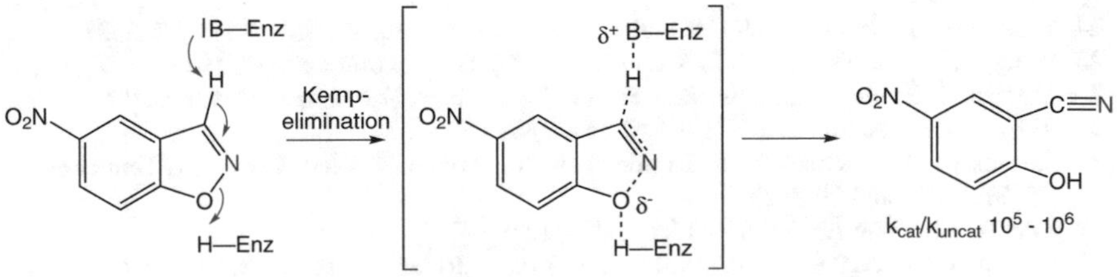

Scheme 3.57 Kemp-elimination catalysed by de-novo designed proteins

By reviewing the catalytic efficiencies and (stereo)selectivities of artificial enzyme constructs it becomes evident that these studies have greatly enhanced our understanding of the catalytic mechanism of enzymes but their use for the synthesis of organic compounds on preparative scale is still far away.

References

1. Carrea G, Riva S (eds) (2008) Organic Synthesis with Enzymes in Non-aqueous Media. Wiley-VCH, Weinheim
2. Borgström B, Brockman HL (eds) (1984) Lipases. Elsevier, Amsterdam
3. Dordick JS (1989) Enzyme Microb. Technol. 11: 194
4. Tramper J, Vermue MH, Beeftink HH, von Stockar U (eds) (1992) Biocatalysis in Non-Conventional Media. Elsevier, Amsterdam
5. Brink LES, Tramper J, Luyben KCAM, Vant't Riet K (1988) Enzyme Microb. Technol. 10: 736
6. Klibanov AM (1990) Acc. Chem. Res. 23: 114
7. Klibanov AM (1986) ChemTech 354
8. Klibanov AM (1989) Trends Biochem. Sci. 14: 141

9. Laane C, Tramper J, Lilly MD (eds) (1987) Biocatalysis in Organic Media. Studies in Organic Chemistry, vol 29. Elsevier, Amsterdam

10. Gutman AL, Shapira M (1995) Adv. Biochem. Eng. Biotechnol. 52: 87

11. Koskinen AMP, Klibanov AM (eds) (1996) Enzymatic Reactions in Organic Media. Blackie, New York

12. Khmelnitsky YL, Levashov AV, Klyachko NL, Martinek K (1988) Enzyme Microb. Technol. 10: 710

13. Klibanov AM (2001) Nature 409: 241.

14. Cantone S, Hanefeld U, Basso A (2007) Green Chem. 9: 954.

15. Schultz GE, Schirmer RH (1979) Principles of Protein Structure. Springer, Berlin, Heidelberg, New York

16. Bell G, Halling PJ, Moore BD, Partidge J, Rees DG (1995) Trends Biotechnol. 13: 468

17. Zaks A, Klibanov AM (1986) J. Am. Chem. Soc. 108: 2767

18. Zaks A, Klibanov AM (1988) J. Biol. Chem. 263: 3194

19. Kazandjian RZ, Klibanov AM (1985) J. Am. Chem. Soc. 107: 5448

20. Cooke R, Kuntz ID (1974) Ann. Rev. Biophys. Bioeng. 3: 95

21. Bone S (1987) Biochim. Biophys. Acta 916: 128

22. Rupley JA, Gratton E, Careri G (1983) Trends Biochem. Sci. 8: 18

23. Cotterill IC, Sutherland AG, Roberts SM, Grobbauer R, Spreitz J, Faber K (1991) J. Chem. Soc. Perkin Trans. 1: 1365

24. Yamamoto Y, Yamamoto K, Nishioka T, Oda J (1989) Agric. Biol. Chem. 52: 3087

25. Wang YF, Chen ST, Liu KKC, Wong CH (1989) Tetrahedron Lett. 30: 1917

26. Laumen K, Seemayer R, Schneider MP (1990) J. Chem. Soc. Chem. Commun. 49

27. Aldercreutz P, Mattiasson B (1987) Biocatalysis 1: 99

28. Stepankova V, Bidmanova S, Koudelakova T, Prokop Z, Chaloupkova R, Damborsky J (2013) ACS Catal. 3: 2823.

29. Zaks A, Klibanov AM (1984) Science 224: 1249

30. Broos J, Visser AJWG, Engbersen JFJ, van Hoek A, Reinhoudt DN (1995) J. Am. Chem. Soc. 117: 12657

31. Russell AJ, Klibanov AM (1988) J. Biol. Chem. 263: 11624

32. Gaertner H, Puigserver A (1989) Eur. J. Biochem. 181: 207

33. Ferjancic A, Puigserver A, Gaertner A (1990) Appl. Microbiol. Biotechnol. 32: 651

34. Tawaki S, Klibanov AM (1993) Biocatalysis 8: 3

35. Ottolina G, Carrea G, Riva S (1991) Biocatalysis 5: 131

36. Sakurai T, Margolin AL, Russell AJ, Klibanov AM (1988) J. Am. Chem. Soc. 110: 7236

37. Kitaguchi H, Fitzpatrick PA, Huber JE, Klibanov AM (1989) J. Am. Chem. Soc. 111: 3094

38. Fitzpatrick PA, Klibanov AM (1991) J. Am. Chem. Soc. 113: 3166

39. Kise H, Hayakawa A, Noritomi H (1990) J. Biotechnol. 14: 239

40. Langrand G, Baratti J, Buono G, Triantaphylides C (1986) Tetrahedron Lett. 27: 29

41. Koshiro S, Sonomoto K, Tanaka A, Fukui S (1985) J. Biotechnol. 2: 47

42. Inagaki T, Ueda H (1987) Agric. Biol. Chem. 51: 1345

43. Morrow CJ, Wallace JS (1990) Synthesis of polyesters by lipase-catalysed polycondensation in organic media. In: Abramowicz DA (ed) Biocatalysis. Van Nostrand Reinhold, New York, p 25

44. Margolin AL, Fitzpatrick PA, Klibanov AM (1991) J. Am. Chem. Soc. 113: 4693

45. Gutman AL, Bravdo T (1989) J. Org. Chem. 54: 4263

46. Makita A, Nihira T, Yamada Y (1987) Tetrahedron Lett. 28: 805

47. Gotor V, Brieva R, Rebolledo F (1988) Tetrahedron Lett. 29: 6973

48. Kullmann W (1987) Enzymatic Peptide Synthesis. CRC, Boca Raton

49. Therisod M, Klibanov AM (1986) J. Am. Chem. Soc. 108: 5638

50. Riva S, Klibanov AM (1988) J. Am. Chem. Soc. 110: 3291

51. Riva S, Chopineau J, Kieboom APG, Klibanov AM (1988) J. Am. Chem. Soc. 110: 584

52. Douzou P (1977) Cryobiochemistry. Academic, London

53. Fink AL, Cartwright SJ (1981) CRC Crit. Rev. Biochem. 11: 145
54. Sakai T, Mitsutomi H, Korenaga T, Ema T (2005) Tetrahedron Asymmetry 16: 1535
55. Sakai T (2004) Tetrahedron Asymmetry 15: 2749
56. Carrea G (1984) Trends Biotechnol. 2: 102
57. Halling PJ (1989) Trends Biotechnol. 7: 50
58. Anderson E, Hahn-Hägerdal B (1990) Enzyme Microb. Technol. 12: 242
59. Lilly MD (1982) J. Chem. Technol. Biotechnol. 32: 162
60. Antonini E, Carrea G, Cremonesi P (1981) Enzyme Microb. Technol. 3: 291
61. Kim KH, Kwon DY, Rhee JS (1984) Lipids 19: 975
62. Brink LES, Tramper J (1985) Biotechnol. Bioeng. 27: 1258
63. Furuhashi K (1992) Biological routes to optically active epoxides. In: Crosby J, Collins AN, Sheldrake GN (eds) Chirality in Industry. Wiley, Chichester, p 167
64. Khmelnitsky YL, Rich JO (1999) Curr. Opinion Chem. Biol. 3: 47
65. Zaks A, Klibanov AM (1988) J. Biol. Chem. 263: 8017
66. Kastle JH, Loevenhart AS (1900) Am. Chem. J. 24: 491; see: (1901) J. Chem. Soc. Abstr. Sect. 80: 178
67. Klibanov AM (1997) Trends Biotechnol. 15: 97
68. Faber K (1991) J. Mol. Catalysis 65: L49
69. Johnson LN (1984) Incl. Compds. 3: 509
70. Hudson EP, Eppler RK, Clark DS (2005) Curr. Opinion Biotechnol. 16: 637
71. Gupta MN, Roy I (2004) Eur. J. Biochem. 271: 2575
72. Valivety RH, Brown L, Halling PJ, Johnston GA, Suckling CJ (1990) Enzyme reactions in predominantly organic media: Measurement and changes of pH. In: Copping LG, Martin RE, Pickett JA, Bucke C, Bunch AW (eds) Opportunities in Biotransformations. Elsevier, London, p 81
73. Zaks A, Klibanov AM (1985) Proc. Natl. Acad. Sci. USA 82: 3192
74. Hsu SH, Wu SS, Wang YF, Wong CH (1990) Tetrahedron Lett. 31: 6403
75. Clark DS (1994) Trends Biotechnol. 12: 439
76. Laane C, Boeren S, Hilhorst R, Veeger C (1987) Optimization of biocatalysts in organic media. In: Laane C, Tramper J, Lilly MD (eds) Biocatalysis in Organic Media. Elsevier, Amsterdam, p 65
77. Carlson R (1992) Design and Optimisation in Organic Synthesis, vol 8. Elsevier, Amsterdam
78. Laane C, Boeren S, Vos K, Veeger C (1987) Biotechnol. Bioeng. 30: 81
79. Jongejan JA (2008) Effects of organic solvents on enzyme selectivity. In: Carrea G, Riva S (eds) Organic Synthesis with Enzymes in Non-aqueous Media. Wiley-VCH, Weinheim, p 25
80. Kulchewski T, Pleiss J (2013) Protein Engineering Handbook 3: 407.
81. Rekker RF, de Kort HM (1979) Eur. J. Med. Chim. Ther. 14: 479
82. Bell G, Janssen AEM, Halling PJ (1997) Enzyme Microb. Technol. 20: 471
83. Halling PJ (1994) Enzyme Microb. Technol. 16: 178
84. Wehtje E, Costes D, Adlercreutz P (1997) J. Mol. Catal. B 3: 221
85. Hutcheon GA, Halling PJ, Moore BD (1997) Methods Enzymol. 286: 465
86. Adlercreutz P (2000) Fundamentals of biocatalysis in neat organic solvents. In: Carrea G, Riva S (eds) Organic Synthesis with Enzymes in Non-aqueous Media. Wiley-VCH, Weinheim, p 3
87. Kuhl P, Posselt S, Jakubke HD (1981) Pharmazie 36: 436
88. Wehtje E, Kaur J, Adlercreutz P, Chand S, Mattiasson B (1997) Enzyme Microb. Technol. 21: 502
89. Robb DA, Yang Z, Halling PJ (1994) Biocatalysis 9: 277
90. Wehtje E, Svensson I, Adlercreutz P, Mattiasson B (1993) Biotechnol. Tech. 7: 873
91. Tomazic SJ (1991) Protein stabilization. In: Dordick JS (ed) Biocatalysts for Industry. Plenum Press, New York, p 241
92. Yamane T, Ichiryu T, Nagata M, Ueno A, Shimizu S (1990) Biotechnol. Bioeng. 36: 1063
93. Freeman A (1984) Trends Biotechnol. 2: 147

94. Theil F (2000) Tetrahedron 56: 2905
95. O'Fagain C (2003) Enzyme Microb. Technol. 33: 137
96. Mejri M, Pauthe E, Larreta-Garde V, Mathlouthi M (1998) Enzyme Microb. Technol. 23: 392
97. Colaco CALS, Collett M, Roser BJ (1996) Chimica Oggi, July/August, 32
98. Yamamoto Y, Kise H (1994) Bull. Chem. Soc. Jpn. 67: 1367
99. Reslow M, Adlercreutz P, Mattiasson B (1992) Biocatalysis 6: 307
100. Kitaguchi H, Klibanov AM (1989) J. Am. Chem. Soc. 111: 9272
101. Okamoto T, Ueji S (1999) Chem. Commun. 939
102. Ke T, Klibanov AM (1999) J. Am. Chem. Soc. 121: 3334
103. Parker MC, Brown SA, Robertson L, Turner NJ (1998) Chem. Commun. 2247
104. Kosmulski M, Gustafsson J, Rosenholm JB (2004) Thermochim. Acta 412: 47
105. Earle MJ, Esperanca JMSS, Gilea MA, Lopes JNC, Rebelo LPN, Magee JW, Seddon KR, Widegren JA (2006) Nature 439: 831
106. Wasserscheidt P, Welton T (eds) (2003) Ionic Liquids in Synthesis. Wiley-VCH, Weinheim
107. Welton T (1999) Chem. Rev. 99: 2071
108. Chin JT, Wheeler SL, Klibanov AM (1994) Biotechnol. Bioeng. 44: 140
109. Park S, Kazlauskas RJ (2003) Curr. Opin. Biotechnol. 14: 432
110. Lozano P, Diego TD, Carrie D, Vaultier M, Iborra JL (2001) Biotechnol. Lett. 23: 1529
111. Kim KW, Song B, Choi MY, Kim MJ (2001) Org. Lett. 3: 1507
112. Wells AS, Coombie VT (2006) Org. Proc. Res. Dev. 10: 794
113. Huddleston JG, Visser AE, Reichert WM, Willauer HD, Broker GA, Rogers RD (2001) Green Chem 3: 156.
114. Cull SG, Holbrey JD, Vargas-Mora V, Seddon KR, Lye GJ (2000) Biotechnol. Bioeng. 69: 227
115. Lau RM, van Rantwijk F, Seddon KR, Sheldon RA (2000) Org. Lett. 2: 4189
116. Erbeldinger M, Mesiano AJ, Russell AJ (2000) Biotechnol. Progr. 16: 1129
117. Salihu A, Alam MZ (2015) Proc. Biochem. 50: 86.
118. Jain N, Kumar A, Chauhan S, Chauhan SMS (2005) Tetrahedron 61: 1015
119. Yang Z, Pan W (2005) Enzyme Microb. Technol. 37: 19
120. Song CE (2004) Chem. Commun. 1033
121. Sheldon R A (2001) Chem. Commun. 2399
122. Kragl U, Eckstein M, Kaftzik N (2002) Curr. Opinion Biotechnol. 13: 565
123. van Rantwijk F, Lau RM, Sheldon RA (2003) Trends Biotechnol. 21: 131
124. Sheldon RA, Lau RM, Sorgedrager MJ, van Rantwijk F, Seddon KR (2002) Green Chem. 4: 147
125. Monhemi H, Housaindokht MR, Moosavi-Movahedi AA, Bozorgmehr MR (2014) ChemPhysChem 16: 14882.
126. Gorke J, Srienc F, Kazlauskas R (2010) Biotechnol. Bioproc. Eng. 15: 40.
127. Guajardo N, Mueller CR, Schrebler R, Carlesi C, Dominguez de Maria P (2016) ChemCatChem 8: 1020.
128. Gorke JT, Srienc F, Kazlauskas RJ (2008) Chem. Commun. 1235.
129. Zhao H, Baker GA, Holmes, S (2011) J. Mol. Catal. B: Enzym. 72: 163.
130. Petrenz A, Dominguez de Maria P, Ramanathan A, Hanefeld U, Ansorge-Schumacher MB, Kara S (2015) J. Mol. Catal. B: Enzym. 114: 42.
131. Lindberg D, de la Fuente Revenga M, Widersten M (2010) J. Biotechnol. 147: 169.
132. Maugeri Z, Leitner W, Dominguez de Maria P (2013) Eur. J. Org. Chem. 4223.
133. Ballesteros A, Bornscheuer U, Capewell A, Combes D, Condoret J-S, Koenig K, Kolisis FN, Marty A, Menge U, Scheper T, Stamatis H, Xenakis A (1995) Biocatalysis Biotrans. 13: 1
134. Nakamura K (1990) Trends Biotechnol. 8: 288
135. Mesiano AJ, Beckman EJ, Russell AJ (1999) Chem. Rev. 99: 623
136. Russell AJ, Beckman EJ (1991) Appl. Biochem. Biotechnol. 31: 197
137. Aaltonen O, Rantakylä M (1991) Chemtech 240

138. Marty A, Chulalaksananukul W, Condoret JS, Willemot RM, Durand G (1990) Biotechnol. Lett. 12: 11
139. Chi YM, Nakamura K, Yano T (1988) Agric. Biol. Chem. 52: 1541
140. Pasta P, Mazzola G, Carrea G, Riva S (1989) Biotechnol. Lett. 11: 643
141. van Eijs AMM, de Jong PJP (1989) Procestechniek 8: 50
142. Randolph TW, Blanch HW, Prausnitz JM, Wilke CR (1985) Biotechnol. Lett. 7: 325
143. Hammond DA, Karel M, Klibanov AM, Krukonis V (1985) J. Appl. Biochem. Biotechnol. 11: 393
144. Randolph TW, Clark DS, Blanch HW, Prausnitz JM (1988) Science 238: 387
145. Beckman EJ, Russell AJ (1993) J. Am. Chem. Soc. 115: 8845
146. Kasche V, Schlothauer R, Brunner G (1988) Biotechnol. Lett. 10: 569
147. Chulalaksananukul W, Condoret JS, Combes D (1993) Enzyme Microb. Technol. 15: 691
148. Loriner GH, Miziorko HM (1980) Biochemistry 19: 5321
149. Liu Y, Chen D, Xu L, Yan Y (2012) Enzyme Microb. Technol. 51: 354.
150. Bornscheuer UT (1995) Enzyme Microb. Technol. 17: 578
151. Björkling F, Godtfredsen SE, Kirk O (1989) J. Chem. Soc. Chem. Commun. 934
152. Bloomer S, Adlercreutz P, Mattiasson B (1992) Enzyme Microb. Technol. 14: 546
153. Bell G, Blain JA, Paterson JDE, Shaw CEL, Todd RJ (1978) FEMS Microbiol. Lett. 3: 223
154. Paterson JDE, Blain JA, Shaw CEL, Todd RJ (1979) Biotechnol. Lett. 1: 211
155. Kvittingen L, Sjursnes B, Anthonsen T, Halling P (1992) Tetrahedron 48: 2793
156. Thum O, Oxenboll KM (2008) SOFW J. 134: 44.
157. Thum O (2004) Tenside Surf. Det. 41: 287.
158. Sarney DB, Vulfson EN (1995) Trends Biotechnol. 13: 164
159. Riva S, Faber K (1992) Synthesis. 895
160. Santaniello E, Ferraboschi P, Grisenti P (1993) Enzyme Microb. Technol. 15: 367
161. Andersch P, Berger M, Hermann J, Laumen K, Lobell M, Seemayer R, Waldinger C, Schneider MP (1997) Methods Enzymol. 286 B: 406
162. Kirchner G, Scollar MP, Klibanov AM (1985) J. Am. Chem. Soc. 107: 7072
163. Mischitz M, Pöschl U, Faber K (1991) Biotechnol. Lett. 13: 653
164. Ghogare A, Kumar GS (1989) J. Chem. Soc. Chem. Commun. 1533
165. Orrenius C, Öhrner N, Rotticci D, Mattson A, Hult K, Norin T (1995) Tetrahedron Asymmetry 6: 1217
166. Öhrner N, Martinelle M, Mattson A, Norin T, Hult K (1994) Biocatalysis 9: 105
167. Degueil-Castaing M, De Jeso B, Drouillard S, Maillard B (1987) Tetrahedron Lett. 28: 953
168. Wang YF, Wong CH (1988) J. Org. Chem. 53: 3127
169. Wang YF, Lalonde JJ, Momongan M, Bergbreiter DE, Wong CH (1988) J. Am. Chem. Soc. 110: 7200
170. Donohue TM, Tuma DJ, Sorrell MF (1983) Arch. Biochem. Biophys. 220: 239
171. Weber HK, Stecher H, Faber K (1995) Biotechnol. Lett. 17: 803
172. Weber HK, Faber K (1997) Methods Enzymol. 286 B: 509
173. Berger B, Faber K (1991) J. Chem. Soc. Chem. Commun. 1198
174. Sugai T, Ohta H (1989) Agric. Biol. Chem. 53: 2009
175. Holla EW (1989) Angew. Chem. Int. Ed. 28: 220
176. Kaga H, Siegmund B, Neufellner E, Faber K, Paltauf F (1994) Biotechnol. Tech. 8: 369
177. Schudok M, Kretzschmar G (1997) Tetrahedron Lett. 38: 387
178. Akai S, Naka T, Takebe Y, Kita Y (1997) Tetrahedron Lett. 38: 4243
179. Kita Y, Takebe Y, Murata K, Naka T, Akai S (1996) Tetrahedron Lett. 37: 7369
180. Bianchi D, Cesti P, Battistel E (1988) J. Org. Chem. 53: 5531
181. Berger B, Rabiller CG, Königsberger K, Faber K, Griengl H (1990) Tetrahedron Asymmetry 1: 541
182. Guibe-Jampel E, Chalecki Z, Bassir M, Gelo-Pujic M (1996) Tetrahedron 52: 4397
183. de Gonzalo G, Brieva R, Sanchez VM, Bayod M, Gotor V (2003) J. Org. Chem. 68: 3333
184. Tokuyama S, Yamano T, Aoki I, Takanohashi K, Nakahama K (1993) Chem. Lett. 741

185. Yamamoto K, Nishioka T, Oda J (1989) Tetrahedron Lett. 30: 1717
186. Keumi T, Hiraoka Y, Ban T, Takahashi I, Kitajima H (1991) Chem. Lett. 1989
187. Nicotra F, Riva S, Secundo F, Zucchelli L (1990) Synth. Commun. 20: 679
188. Suginaka K, Hayashi Y, Yamamoto Y (1996) Tetrahedron Asymmetry 7: 1153
189. Bevinakatti HS, Newadkar RV (1993) Tetrahedron Asymmetry 4: 773
190. Fourneron JD, Chiche M, Pieroni G (1990) Tetrahedron Lett. 31: 4875
191. Caer E, Kindler A (1962) Biochemistry 1: 518
192. Suemune H, Mizuhara Y, Akita H, Sakai K (1986) Chem. Pharm. Bull. 34: 3440
193. Morishima H, Koike Y, Nakano M, Atsuumi S, Tanaka S, Funabashi H, Hashimoto J,
     Sawasaki Y, Mino N, Nakano K, Matsushima K, Nakamichi K, Yano M (1989) Biochem.
     Biophys. Res. Commun. 159: 999
194. Santaniello E, Ferraboschi P, Grisenti P (1990) Tetrahedron Lett. 31: 5657
195. Banfi L, Guanti G (1993) Synthesis. 1029
196. Atsuumi S, Nakano M, Koike Y, Tanaka S, Ohkubo M (1990) Tetrahedron Lett. 31: 1601
197. Baba N, Yoneda K, Tahara S, Iwase J, Kaneko T, Matsuo M (1990) J. Chem. Soc. Chem.
     Commun. 1281
198. Terao Y, Murata M, Achiwa K (1988) Tetrahedron Lett. 29: 5173
199. Morgan B, Dodds DR, Homann MJ, Zaks A, Vail R (2001) Methods Biotechnol. 15: 423
200. Ader U, Breitgoff D, Laumen KE, Schneider MP (1989) Tetrahedron Lett. 30: 1793
201. Ferraboschi P, Grisenti P, Manzocchi A, Santaniello E (1990) J. Org. Chem. 55: 6214
202. Laumen K, Breitgoff D, Schneider MP (1988) J. Chem. Soc. Chem. Commun. 1459
203. Anderson EM, Larson KM, Kirk O (1998) Biocatalysis Biotrans. 16: 181
204. Burgess K, Jennings LD (1990) J. Am. Chem. Soc. 112: 7434
205. Burgess K, Jennings LD (1991) J. Am. Chem. Soc. 113: 6129
206. Chen CS, Liu YC, Marsella M (1990) J. Chem. Soc. Perkin Trans. 1: 2559
207. Chen CS, Liu YC (1989) Tetrahedron Lett. 30: 7165
208. Hiratake J, Inagaki M, Nishioka T, Oda J (1988) J. Org. Chem. 53: 6130
209. Hayward RC, Overton CH, Witham GH (1976) J. Chem. Soc. Perkin Trans. 1: 2413
210. Cunningham AF, Kündig EP (1988) J. Org. Chem. 53: 1823
211. Seemayer R, Schneider MP (1991) J. Chem. Soc. Chem. Commun. 49
212. Hemmerle H, Gais HJ (1987) Tetrahedron Lett. 28: 3471
213. Xie ZF, Nakamura I, Suemune H, Sakai K (1988) J. Chem. Soc. Chem. Commun. 966
214. Solladié-Cavallo A (1989) In: Liebeskind LS (ed) Advances in Metal-Organic Chemistry, vol
     1. JAI Press, Greenwich, pp 99–131
215. Nakamura K, Ishihara K, Ohno A, Uemura M, Nishimura H, Hayashi Y (1990) Tetrahedron
     Lett. 31: 3603
216. Yamazaki Y, Hosono K (1990) Tetrahedron Lett. 31: 3895
217. Gokel GW, Marquarding D, Ugi IK (1972) J. Org. Chem. 37: 3052
218. Boaz NW (1989) Tetrahedron Lett. 30: 2061
219. Feichter C, Faber K, Griengl H (1989) Tetrahedron Lett. 30: 551
220. Feichter C, Faber K, Griengl H (1990) Biocatalysis 3: 145
221. Burgess K, Henderson I (1990) Tetrahedron Asymmetry 1: 57
222. Burgess K, Cassidy J, Henderson I (1991) J. Org. Chem. 56: 2050
223. Burgess K, Jennings LD (1990) J. Org. Chem. 55: 1138
224. Pfenninger A (1986) Synthesis. 89
225. Babiak KA, Ng JS, Dygos JH, Weyker CL, Wang YF, Wong CH (1990) J. Org. Chem. 55:
     3377
226. Baba N, Mimura M, Hiratake J, Uchida K, Oda J (1988) Agric. Biol. Chem. 52: 2685
227. Baba N, Tateno K, Iwasa J, Oda J (1990) Agric. Biol. Chem. 54: 3349
228. Thuring JWJF, Klunder AJH, Nefkens GHL, Wegman MA, Zwanenburg B (1996) Tetrahe-
     dron Lett. 37: 4759
229. van den Heuvel M, Cuiper AD, van der Deen H, Kellogg RM, Feringa BL (1997) Tetrahedron
     Lett. 38: 1655

230. van der Deen H, Cuiper AD, Hof RP, van Oeveren A, Feringa BL, Kellogg RM (1996) J. Am. Chem. Soc. 118: 3801
231. Mitsuda S, Nabeshima S, Hirohara H (1989) Appl. Microbiol. Biotechnol. 31: 334
232. Tinapp P (1971) Chem. Ber. 104: 2266
233. Kruse CG (1992) Chiral Cyanohydrins. In: Collins AN, Sheldrake GN, Crosby J (eds) Chirality in Industry. Wiley, New York, p 279
234. Satoh T, Suzuki S, Suzuki Y, Miyaji Y, Imai Z (1969) Tetrahedron Lett. 10: 4555
235. Mitsuda S, Yamamoto H, Umemura T, Hirohara H, Nabeshima S (1990) Agric. Biol. Chem. 54: 2907
236. Effenberger F, Gutterer B, Ziegler T, Eckhardt E, Aichholz R (1991) Liebigs Ann. Chem. 47
237. Inagaki M, Hiratake J, Nishioka T, Oda J (1992) J. Org. Chem. 57: 5643
238. Brand S, Jones MF, Rayner CM (1995) Tetrahedron Lett. 36: 8493
239. Larsson ALE, Persson BA, Bäckvall JE (1997) Angew. Chem. Int. Ed. 36: 1211
240. Dinh PM, Howarth JA, Hudnott AR, Williams JMJ, Harris W (1996) Tetrahedron Lett. 37: 7623
241. Akai S, Tanimoto K, Kanao Y, Egi M, Yamamoto T, Kita Y (2006) Angew. Chem. Int. Ed. 45: 2592
242. Allen JV, Williams JMJ (1996) Tetrahedron Lett. 37: 1859
243. Verzijl GKM, de Vries JG, Broxterman QB (2005) Tetrahedron Asymmetry 16: 1603
244. de Miranda AS, Miranda LSM, de Souza ROMA (2015) Biotechnol. Adv. 33: 372.
245. Verho O, Baeckvall J-E (2015) J. Am. Chem. Soc. 137: 3996.
246. Persson BA, Larsson ALE, Le Ray M, Bäckvall JE (1999) J. Am. Chem. Soc. 121: 1645
247. Martin-Matute B, Bäckvall JE (2007) Curr. Opinion Chem. Biol. 11: 226
248. Huerta FF, Minidis ABE, Bäckvall JE (2001) Chem. Soc. Rev. 30: 321
249. Pamies O, Bäckvall JE (2003) Chem. Rev. 103: 3247
250. Kim MJ, Ahn Y, Park J (2002) Curr. Opinion Biotechnol. 13: 578
251. Lee JH, Han K, Kim MJ, Park J (2010) Eur. J. Org. Chem. 987
252. Choi JH, Kim YH, Nam SH, Shin ST, Kim MJ, Park J (2002) Angew. Chem. Int. Ed. 41: 2373
253. Pamies O, Bäckvall JE (2001) J. Org. Chem. 66: 4022
254. Pamies O, Bäckvall JE (2002) Adv. Synth. Catal. 344: 947
255. Pamies O, Bäckvall JE (2001) Adv. Synth. Catal. 343: 726
256. Pamies O, Bäckvall JE (2002) J. Org. Chem. 67: 9006
257. Fransson ABL, Xu Y, Leijondahl K, Bäckvall JE (2006) J. Org. Chem. 71: 6309
258. Martin-Matute B, Bäckvall JE (2004) J. Org. Chem. 69: 9191
259. Boren L, Martin-Matute B, Xu Y, Cordova A, Bäckvall JE (2006) Chem. Eur. J. 12: 225
260. Kim MJ, Chung YI, Choi Y C, Lee HK, Kim D, Park J (2003) J. Am. Chem. Soc. 125: 11494
261. Koszelewski D, Brodzka A, Zadlo A, Paprocki D, Trepizur D, Zysk M, Ostaszewski R (2016) ACS Catal. 6: 3287.
262. Greene TW (1981) Protective groups in Organic Chemistry. Wiley, New York
263. Bashir NB, Phythian SJ, Reason AJ, Roberts SM (1995) J. Chem. Soc. Perkin Trans. 1: 2203
264. Park HG, Do JH, Chang HN (2003) Biotechnol. Bioproc. Eng. 8: 1.
265. Danieli B, Riva S (1994) Pure Appl. Chem. 66: 2215.
266. Chebil L, Humeau C, Falcimaigne A, Engasser J-M, Ghoul M (2006) Proc. Biochem. 41: 2237.
267. Hennen WJ, Sweers HM, Wang YF, Wong CH (1988) J. Org. Chem. 53: 4939
268. Riva S, Bovara R, Ottolina G, Secundo F, Carrea G (1989) J. Org. Chem. 54: 3161
269. Colombo D, Ronchetti F, Toma L (1991) Tetrahedron 47: 103
270. Therisod M, Klibanov AM (1987) J. Am. Chem. Soc. 109: 3977
271. Gruters RA, Neefjes JJ, Tersmette M, De Goede REJ, Tulp A, Huisman HG, Miedema F, Ploegh HL (1987) Nature 330: 74
272. Margolin AL, Delinck DL, Whalon MR (1990) J. Am. Chem. Soc. 112: 2849
273. Gatfield IL (1984) Ann. N. Y. Acad. Sci. 434: 569
274. Gutman AL, Oren D, Boltanski A, Bravdo T (1987) Tetrahedron Lett. 28: 5367

275. Guo Z, Ngooi TK, Scilimati A, Fülling G, Sih CJ (1988) Tetrahedron Lett. 29: 5583
276. Gutman AL, Zuobi K, Boltansky A (1987) Tetrahedron Lett. 28: 3861
277. Henkel B, Kunath A, Schick H (1993) Tetrahedron Asymmetry 4: 153
278. Zhang Y, Schaufelberger F, Sakulsombat M, Liu C, Ramström O (2014) Tetrahedron 70: 3826.
279. Guo Z, Sih CJ (1988) J. Am. Chem. Soc. 110: 1999
280. Ngooi TK, Scilimati A, Guo Z-W, Sih CJ (1989) J. Org. Chem. 54: 911
281. Lobell M, Schneider MP (1993) Tetrahedron Asymmetry 4: 1027
282. Barker CV, Page MI, Korn SR, Monteith M (1999) Chem. Commun. 721.
283. Gutman AL, Meyer E, Yue X, Abell C (1992) Tetrahedron Lett. 33: 3943.
284. Gotor V (1992) Enzymatic aminolysis, hydrazinolysis and oximolysis reactions. In: Servi S (ed) Microbial Reagents in Organic Synthesis. Kluwer, Dordrecht, p 199
285. Garcia MJ, Rebolledo F, Gotor V (1993) Tetrahedron Lett. 34: 6141
286. Chen ST, Jang MK, Wang KT (1993) Synthesis. 858
287. Gotor V, Brieva R, Gonzalez C, Rebolledo F (1991) Tetrahedron 47: 9207
288. Rebolledo F, Brieva R, Gotor V (1989) Tetrahedron Lett. 30: 5345
289. Fastrez J, Fersht AR (1973) Biochemistry 12: 2025
290. Yagisawa S (1981) J. Biochem. (Tokyo) 89: 491
291. Astorga C, Rebolledo F, Gotor V (1991) Synthesis. 350
292. Al-Mulla EAJ, Yunus WMZW, Ibrahim NAB, Rahman MZA (2010) J. Oleo. Sci. 59: 59.
293. Bruggink A, Roos EC, De Vroom E (1998) Org. Proc. Res. Dev. 2: 128.
294. de Zoete MC, Kock-van Dalen AC, van Rantwijk F, Sheldon RA (1993) J. Chem. Soc. Chem. Commun. 1831
295. Gotor V, Brieva R, Rebolledo F (1988) J. Chem. Soc. Chem. Commun. 957
296. Balkenhohl F, Dietrich K, Hauer B, Ladner W (1997) J. Prakt. Chem. 339: 381
297. Messina F, Botta M, Corelli F, Schneider MP, Fazio F (1999) J. Org. Chem. 64: 3767
298. Roche D, Prasad K, Repic O (1999) Tetrahedron Lett. 40: 3665
299. Brieva R, Rebolledo F, Gotor V (1990) J. Chem. Soc. Chem. Commun. 1386
300. Takayama S, Lee ST, Chung SC, Wong CH (1999) Chem. Commun. 127
301. Ladner WE, Ditrich K (1999) Chimica Oggi, July/August, 51
302. Orsat B, Alper PB, Moree W, Mak CP, Wong CH (1996) J. Am. Chem. Soc. 118: 712
303. van Rantwijk F, Sheldon RA (2004) Tetrahedron 60: 501
304. Smidt H, Fischer A, Fischer P, Schmidt RD (1996) Biotechnol. Tech. 10: 335
305. Ahn Y, Ko S-B, Kim M-J, Park J, (2008) Coord. Chem. Rev. 252: 647.
306. Parvulescu A, Janssens J, Vanderleyden J, Vos D (2010) Top. Catal. 53: 931.
307. Reetz MT, Schimossek K (1996) Chimia 50: 668
308. Parvulescu A, De Vos D, Jacobs P (2005) Chem. Commun. 5307
309. Pamies O, Ell AH, Samec JSM, Hermanns N, Bäckvall JE (2002) Tetrahedron Lett. 43: 4699
310. Paetzold J, Bäckvall JE (2005) J. Am. Chem. Soc. 127: 17620
311. Parvulescu AN, Jacobs PA, De Vos DE (2008) Adv. Synth. Catal. 350: 113
312. Livingston AG, Roengpithya C, Patterson D A, Irwin JL, Parrett MR, Taylor PC (2007) Chem. Commun. 3462
313. Shakeri M, Tai C, Goethelid E, Oscarsson S, Baeckvall, Jan-E (2011) Chem. Eur. J. 17: 13269.
314. Engström K, Johnston EV. Verho O, Gustafson KPJ, Shakeri M, Tai C-W, Bäckvall J-E (2016) Angew. Chem. Int. Ed. 52: 14006.
315. Pamies Oscar, Ell AH, Samec JSM, Hermanns N, Backvall, Jan-E (2002) Tetrahedron Lett. 43: 4699.
316. Stirling MJ, Mwansa JM, Sweeney G, Blacker AJ, Page MI (2016) Org. Biomol. Chem. 14: 7092.
317. Gastaldi S, Escoubet S, Vanthuyne N, Gil G, Bertrand MP (2007) Org. Lett. 9: 837.
318. Ding W, Li M, Deng Y (2012) Tetrahedron: Asymm. 23: 1376.
319. Öhrner N, Orrenius C, Mattson A, Norin T, Hult K (1996) Enzyme Microb. Technol. 19: 328

320. Baba N, Mimura M, Oda J, Iwasa J (1990) Bull. Inst. Chem. Res. Kyoto Univ. 68: 208
321. Bianchi D, Cesti P (1990) J. Org. Chem. 55: 5657
322. Kiefer M, Vogel R, Helmchen G, Nuber B (1994) Tetrahedron 50: 7109
323. Fruton JS (1982) Adv. Enzymol. 53: 239
324. Jakubke HD, Kuhl P, Könnecke A (1985) Angew. Chem. Int. Ed. 24: 85
325. Jakubke HD (1987) The Peptides 9: 103
326. Morihara K (1987) Trends Biotechnol. 5: 164
327. Glass JD (1981) Enzyme Microb. Technol. 3: 2
328. Gill I, Lopez-Fandino R, Jorba X, Vulfson EN (1996) Enzyme Microb. Technol. 18: 162
329. Chaiken IM, Komoriya A, Ojno M, Widmer F (1982) Appl. Biochem. Biotechnol. 7: 385
330. Goswami A, Van Lanen SG (2015) Mol. Biosyst. 11: 338.
331. Yazawa K, Numata K (2014) Molecules 19: 13755.
332. Bergmann M, Fraenkel-Conrat H (1938) J. Biol. Chem. 124: 1
333. Jakubke HD (1995) Angew. Chem. Int. Ed. 34: 175
334. Jackson DY, Burnier JP, Wells JA (1995) J. Am. Chem. Soc. 117: 819
335. Kuhl P, Wilsdorf A, Jakubke HD (1983) Monatsh. Chem. 114: 571
336. Homandberg GA, Komoriya A, Chaiken IM (1982) Biochemistry 21: 3385
337. Inouye K, Watanabe K, Tochino Y, Kobayashi M, Shigeta Y (1981) Biopolymers 20: 1845
338. Gill I, Vulfson EN (1993) J. Am. Chem. Soc. 115: 3348
339. Schellenberger V, Jakubke HD (1991) Angew. Chem. Int. Ed. 30: 1437
340. Margolin AL, Klibanov AM (1987) J. Am. Chem. Soc. 109: 3802
341. West JB, Wong CH (1987) Tetrahedron Lett. 28: 1629
342. Matos JR, West JB, Wong CH (1987) Biotechnol. Lett. 9: 233
343. Crout DHG, MacManus DA, Ricca JM, Singh S (1993) Indian J. Chem. 32B: 195
344. Gerisch S, Jakubke H-D, Kreuzfeld H-J (1995) Tetrahedron Asymmetry 6: 3039
345. Sekizaki H, Itoh K, Toyota E, Tanizawa K (1997) Tetrahedron Lett. 38: 1777
346. Sekizaki H, Itoh K, Toyota E, Tanizawa K (1996) Chem. Pharm. Bull. 44: 1585
347. Miyazawa T, Nakajo S, Nishikawa M, Imagawa K, Yanagihara R, Yamada T (1996) J. Chem. Soc. Perkin Trans. 1: 2867
348. Gololobov MY, Petrauskas A, Pauliukonis R, Koske V, Borisov IL, Svedas V (1990) Biochim. Biophys. Acta 1041: 71
349. Sekizaki H, Itoh K, Toyota E, Tanizawa K (1998) Chem. Pharm. Bull. 46: 846
350. Isowa Y, Ohmori M, Ichikawa T, Mori K, Nonaka Y, Kihara K, Oyama K (1979) Tetrahedron Lett. 20: 2611
351. Oyama K (1992) The industrial production of aspartame. In: Collins AN, Sheldrake GN (eds) Chirality in Industry. Wiley, New York, p 237
352. Montalbetti CAGN, Falque V (2005) Tetrahedon 61: 10827.
353. Stoineva IB, Petkov DD (1985) FEBS Lett. 183: 103
354. Di Maio J, Nguyen TMD, Lemieux C, Schiller PW (1982) J. Med. Chem. 25: 1432
355. Inouye K, Watanabe K, Morihara K, Tochino K, Kanaya T, Emura J, Sakakibara S (1979) J. Am. Chem. Soc. 101: 751
356. Rose K, Gladstone J, Offord RE (1984) Biochem. J. 220: 189
357. Obermeier R, Seipke G (1984) In: Voelter W, Bayer E, Ovchinnikov YA, Wünsch E (eds) Chemistry of Peptides and Proteins, vol 2. de Gruyter, Berlin, p 3
358. Masaki T, Nakamura K, Isono M, Soejima M (1978) Agric. Biol. Chem. 42: 1443
359. Morihara K, Oka T, Tsuzuki H, Tochino Y, Kanaya T (1980) Biochem. Biophys. Res. Commun. 92: 396
360. Cerovsky V, Kockskämper J, Glitsch HG, Bordusa F (2000) ChemBioChem 1: 126.
361. Sun L, Coy DH (2016) Curr. Drig Targets 17: 529.
362. Rizo J, Gierasch LM (1992) Annu. Rev. Biochem. 61: 387.
363. Hwang BK, Gu QM, Sih CJ (1993) J. Am. Chem. Soc. 115: 7912
364. Bernhardt P, Hult K, Kazlauskas RJ (2005) Angew. Chem. Int. Ed. 44: 2742
365. Ankudey EG, Olivo H F, Peeples TL (2006) Green Chem. 8: 923

366. Lemoult SC, Richardson PF, Roberts SM (1995) J. Chem. Soc. Perkin Trans. 1: 89
367. Chavez G, Rasmussen J-A, Janssen M, Mamo G, Hatti-Kaul R, Sheldon RA (2014) Top. Catal. 57: 349.
368. Björkling F, Frykman H, Godtfredsen SE, Kirk O (1992) Tetrahedron 48: 4587
369. Björkling F, Godtfredsen S E, Kirk O (1990) J. Chem. Soc. Chem. Commun. 1301
370. Warwel S, Rüsch gen. Klaas M (1995) J. Mol. Catal. B: Enzym. 1: 29.
371. de Zoete MC, van Rantwijk F, Maat L, Sheldon RA (1993) Recl. Trav. Chim. Pays-Bas 112: 462
372. Grunwald J, Wirz B, Scollar MP, Klibanov AM (1986) J. Am. Chem. Soc. 108: 6732
373. Adlercreutz P (1996) Biocatalysis Biotrans. 14: 1
374. Gorrebeeck C, Spanoghe M, Lanens D, Lemiere GL, Domisse RA, Lepoivre JA, Alderweireldt FC (1991) Recl. Trav. Chim. Pays-Bas 110: 231
375. Snijder-Lambers AM, Vulfson EN, Doddema H (1991) Recl. Trav. Chim. Pays-Bas 110: 226
376. Adlercreutz P (1991) Eur. J. Biochem. 199: 609
377. Itoh S, Terasaka T, Matsumiya M, Komatsu M, Ohshiro Y (1992) J. Chem. Soc. Perkin Trans. 1: 3253
378. Duine JA, van der Meer RA, Groen BW (1990) Ann. Rev. Nutr. 10: 297
379. Malmstrom BG, Ryden L (1968) The copper containing oxidases. In: Singer TP (ed) Biological Oxidations. Wiley, New York, p 419
380. Wood BJB, Ingraham LL (1965) Nature 205: 291
381. Siegel SM, Roberts K (1968) Space Life Sci. 1: 131
382. Carrea G, Ottolina G, Riva S (1995) Trends Biotechnol. 13: 63
383. Margolin AL, Tai DF, Klibanov AM (1987) J. Am. Chem. Soc. 109: 7885
384. Faber K, Ottolina G, Riva S (1993) Biocatalysis 8: 91
385. Secundo F, Carrea G (2003) Chem. Eur. J. 9: 3194
386. Jongejan J A, van Tol J B, Duine J A (1994) Chim. Oggi 12: 15
387. Wescott C R, Klibanov A M (1994) Biochim. Biophys. Acta 1206: 1
388. Carrea G, Riva S (2000) Angew. Chem. Int. Ed. 39: 2226
389. Tawaki S, Klibanov AM (1992) J. Am. Chem. Soc. 114: 1882
390. Wu SH, Chu FY, Wang KT (1991) Bioorg. Med. Chem. Lett. 1: 339
391. Ueji S, Fujino R, Okubo N, Miyazawa T, Kurita S, Kitadani M, Muromatsu A (1992) Biotechnol. Lett. 14: 163
392. Kitaguchi H, Itoh I, Ono M (1990) Chem. Lett. 1203
393. Secundo F, Riva S, Carrea G (1992) Tetrahedron Asymmetry 3: 267
394. Wernerova M, Hudlicky T (2010) Synlett 2701.
395. Dixon M (1948) Multi-Enzyme Systems, Univ. Press, London.
396. Mayer S F, Kroutil W, Faber K (2001) Chem. Soc. Rev. 30: 332.
397. Tietze L F (ed.) (2014) Domino Reactions: Concepts for Efficient Organic Synthesis, Wiley, Weinheim.
398. Hayashi Y (2016) Chem. Sci. 7: 866.
399. Simon R, Richter N, Busto E, Kroutil W (2014) ACS Catal. 4: 129.
400. Ricca E, Brucher B, Schrittwieser J (2011) Adv. Synth. Catal. 353: 2239.
401. Bruggink A, Schoevaart R, Kieboom T (2003) Org. Proc. Res. Dev. 7: 622.
402. Muschiol J, Peters C, Oberleitner N, Mihovilovic M D, Bornscheuer U T, Rudroff F (2015) Chem. Commun. 51: 5798.
403. Köhler V, Turner N J (2015) Chem. Commun. 51: 450.
404. Hamid M H S, Slatford P, Williams J M (2007) Adv. Synth. Catal. 349: 1555.
405. Martzen M R, McCraith S M, Spinelli S L, Torres F M, Fields S, Grayhack E J, Phizicky E M (1999) Science 286: 1153.
406. Riva S, Fessner W-D (2014) Cascade Biocatalysis, Wiley-VCH, Weinheim.
407. Fessner W-D (2015) New Biotechnol. 32: 658.
408. Tessaro D, Pollegioni L, Piubelli L, D'Arrigo P, Servi S (2015) ACS Catal. 5: 1604.
409. Kroutil W, Rueping M (2014) ACS Catal. 4: 2086.

410. Schober M, Toesch M, Knaus T, Strohmeier G A, van Loo B, Fuchs M, Hollfelder F, Macheroux P, Faber K (2013) Angew. Chem. Int. Ed. 52: 3277.
411. Fessner W-D, Walter C (1992) Angew. Chem. Int. Ed. 31: 614.
412. Wu S, Zhou Y, Wang T, Too H-P, Wang D I C, Li Z (2016) Nature Commun. 7: 11917.
413. Chmura A, Rustler S, Paravidino M, van Rantwijk F, Stolz A, Sheldon R A (2013) Tetrahedron: Asymmetry, 24:1225.
414. Osprian I, Fechter M H, Griengl H (2003) J. Mol. Catal. B: Enzym. 24-25: 89.
415. van Pelt S, van Rantwijk F, Sheldon R (2009) Adv. Synth. Catal. 351: 397.
416. Steinreiber J, Schürmann M, Wolberg M, van Assema F, Reisinger C, Fesko K, Mink D, Griengl H (2007), Angew. Chem. Int. Ed. 46: 1624.
417. Sehl T, Hailes H C, Ward J M, Wardenga R, von Lieres E, Offermann H, Westphal R, Pohl M, Rother D (2013) Angew. Chem. Int. Ed. 52: 6772.
418. Sehl T, Hailes H C, Ward J M, Meynes U, Pohl M, Rother D (2014) Green Chem. 16: 3341.
419. Bäckvall J-E (2002) J. Organomet. Chem. 652: 105.
420. Edwards M G, Jazzar R F C, Paine B M, Shermer D J, Whittlesey M K, Williams J M J, Edney D D (2004) Chem. Commun. 90.
421. Wandrey C, Fiolitakis E, Wichmann U, Kula M-R (1984) Ann. N. Y. Acad. Sci. 434: 91.
422. Sattler J H, Fuchs M, Tauber K, Mutti F G, Faber K, Pfeffer J, Haas T, Kroutil W (2012) Angew. Chem. Int. Ed. 51: 9156.
423. Mutti F G, Knaus T, Scrutton N S, Breuer M, Turner N J (2015) Science 349: 1525.
424. Willetts A J, Knowles C J, Levitt M S, Roberts S M, Sandey H, Shipston N F (1991) J. Chem. Soc. Perkin Trans. 1, 1608.
425. Mallin H, Wulf H, Bornscheuer U T (2013) Enzyme Microb. Technol. 53: 283.
426. Bornadel A, Hatti-Kaul R, Hollmann F, Kara S (2015) ChemCatChem 7: 2442.
427. Song J-W, Jeon E-Y, Song D-H, Jang H-J, Bornscheuer U T, Oh D-K, Park J-B (2013) Angew. Chem. Int. Ed. 52: 2534.
428. Srere PA (1987) Annu. Rev. Biochem. 56: 89.
429. Wheeldon I, Minteer SD, Banta S, Barton SC, Atanassov P, Sigman M (2016) Nat. Chem. 8: 299.
430. You C, Myung S, Zhang Y-H P (2012) Angew. Chem. Int. Ed. 51: 8787.
431. Can M, Armstrong FA, Ragsdale SW (2014) Chem. Rev. 114: 4149.
432. Perham RN (2000) Ann. Rev. Biochem. 69: 961.
433. Castellana M, Wilson MZ, Xu Y, Joshi P, Cristea IM, Rabinowitz JD, Gitai Z, Wingreen NS (2014) Nat. Biotechnol. 32: 1011.
434. Napora-Wijata K, Strohmeier GA, Winkler M (2014) Biotechnol. J. 9: 822.
435. Venkitasubramanian P, Daniels L, Rosazza JPN (2007) J. Biol. Chem. 282: 478.
436. Schwendenwein D, Fiume G, Weber H, Rudroff F, Winkler M (2016) Adv. Synth. Catal. 358: 3414.
437. France SP, Hussain S, Hill AM, Hepworth LJ, Howard RM, Mulholland KR, Flitsch SL, Turner NJ (2016) ACS Catal. 6: 3753.
438. Akhtar MK, Turner NJ, Jones PR (2013) Proc. Natl. Acad. Sci. USA 110: 87.
439. Suckling CJ, Suckling KE (1974) Chem. Soc. Rev. 3: 387
440. Sharma BP, Bailey LF, Messing RA (1982) Angew. Chem. Int. Ed. 21, 837
441. Tischer W, Wedekind F (1999) Top. Curr. Chem. 200: 95
442. Rosevaer A (1984) J. Chem. Technol. Biotechnol. 34: 127
443. Zaborsky OR (1973) Immobilized Enzymes. CRC, Cleveland
444. Trevan MD (1980) Immobilized Enzymes: Introduction and Applications in Biotechnology. Wiley, New York
445. Hartmeier W (1986) Immobilisierte Biokatalysatoren. Springer, Berlin, Heidelberg, New York
446. Suckling CJ (1977) Chem. Soc. Rev. 6: 215
447. Cao L (2005) Curr. Opinion Chem. Biol. 9: 217

448. Cao L (2005) Carrier-Bound Immobilised Enzymes – Principles, Applications and Design. Wiley-VCH, Weinheim
449. Bornscheuer U T (2003) Angew. Chem. Int. Ed. 42: 3336
450. Adamczak M, Krishna S H (2004) Food Technol. Biotechnol. 42: 251
451. Krajewska B (2004) Enzyme Microb. Technol. 35: 126
452. Franssen MCR, Steunenberg P, Scott EL, Zuilhof H, Sanders JPM (2013) Chem. Soc. Rev. 42: 6491.
453. Christen M, Crout DHG (1987) Enzymatic reduction of β-ketoesters using immobilized yeast. In: Moody GW, Baker PB (eds) Bioreactors and Biotransformations. Elsevier, London, p 213
454. Cabral JMS, Kennedy JF (1993) In: Gupty MN (ed) Thermostability of Enzymes. Springer, Berlin, p 163
455. Martinek K, Klibanov AM, Goldmacher VS, Tchernysheva AV, Mozhaev VV, Berezin IV, Glotov BO (1977) Biochim. Biophys. Acta 485: 13
456. Klibanov AM (1983) Science 219: 722
457. Sheldon R A (2007) Adv. Synth. Catal. 349: 1289
458. Miyawaki O, Wingard jr LB (1984) Biotechnol. Bioeng. 26: 1364
459. Krakowiak W, Jach M, Korona J, Sugier H (1984) Starch 36: 396
460. Petri A, Marconcini P, Salvadori P (2005) J. Mol. Catal. B 32: 219
461. Takahashi H, Li B, Sasaki T, Myazaki C, Kajino T, Inagaki S (2001) Micropor. Mesopor. Mater. 44-45: 755
462. Yan AX, Li XW, Ye YH (2002) Appl. Biochem. Biotechnol. 101: 113
463. Wiegel J, Dykstra M (1984) Appl. Microbiol. Biotechnol. 20: 59
464. Kato T, Horikoshi K (1984) Biotechnol. Bioeng. 26: 595
465. Sugiura M, Isobe M (1976) Chem. Pharm. Bull. 24: 72
466. Akita H (1996) Biocatalysis 13: 141
467. Balcao VM, Paiva AL, Malcanta FX (1996) Enzyme Microb. Technol. 18: 392
468. Lavayre J, Baratti J (1982) Biotechnol. Bioeng. 24: 1007
469. Boudrant J, Ceheftel C (1975) Biotechnol. Bioeng. 17: 827
470. Tosa T, Mori T, Fuse N, Chibata I (1967) Enzymologia 31: 214
471. Tosa T, Mori T, Chibata I (1969) Agric. Biol. Chem. 33: 1053
472. Bryjak J, Kolarz BN (19998) Biochemistry 33: 409
473. Janssen MHA, van Langen LM, Pereita SRM, van Rantwijk F, Sheldon RA (2002) Biotechnol. Bioeng. 78: 425
474. Weetall HH, Mason RD (1973) Biotechnol. Bioeng. 15: 455
475. Cannon JJ, Chen LF, Flickinger MC, Tsao GT (1984) Biotechnol. Bioeng. 26: 167
476. Ibrahim M, Hubert P, Dellacherie E, Magdalou J, Muller J, Siest G (1985) Enzyme Microb. Technol. 7: 66
477. Monsan P, Combes D (1984) Biotechnol. Bioeng. 26: 347
478. Chipley JR (1974) Microbios 10: 115
479. Marek M, Valentova O, Kas J (1984) Biotechnol. Bioeng. 26: 1223
480. Vilanova E, Manjon A, Iborra JL (1984) Biotechnol. Bioeng. 26: 1306
481. Miyama H, Kobayashi T, Nosaka Y (1984) Biotechnol. Bioeng. 26: 1390
482. Burg K, Mauz S, Noetzel S, Sauber K (1988) Angew. Makromol. Chem. 157: 105
483. Katchalski-Katzir E, Kraemer DM (2000) J. Mol. Catal. B 10: 157
484. Galaev I Y, Mattiasson B (1999) Trends Biotechnol. 17: 335
485. Roy I, Sharma S, Gupta MN (2004) Adv. Biochem. Eng. Biotechnol. 86: 159
486. Galaev IY, Mattiasson B (eds) (2004) Smart Polymers for Bioseparation and Bioprocessing. Taylor & Francis, London
487. Wong SS, Wong LJC (1992) Enzyme Microb. Technol. 14: 866
488. Cao L, van Langen L, Sheldon RA (2003) Curr. Opinion Biotechnol. 14: 387
489. Khan SS, Siddiqui AM (1985) Biotechnol. Bioeng. 27: 415
490. Kaul R, D'Souza SF, Nadkarni GB (1984) Biotechnol. Bioeng. 26: 901

491. Mateo C, Palomo J M, van Langen L M, van Rantwijk F, Sheldon R A (2004) Biotechnol. Bioeng. 86: 273
492. Quiocho FA, Richards FM (1964) Proc. Natl. Acad. Sci. USA 52: 833
493. Persichetti RA, St. Clair NL, Griffith JP, Navia MA, Margolin AL (1995) J. Am. Chem. Soc. 117: 2732
494. Häring D, Schreier P (1999) Curr. Opinion Chem. Biol. 3: 35
495. Zelinski T, Waldmann H (1997) Angew. Chem. Int. Ed. 36: 722
496. Margolin AL (1996) Trends Biotechnol. 14: 223
497. Roy JJ, Abraham TE (2004) Chem. Rev. 104: 3705
498. Sheldon RA, Schoevaart R, van Langen LM (2005) Biocatal. Biotrans. 23: 141
499. Cao L, van Rantwijk F, Sheldon RA (2000) Org. Lett. 2: 1361
500. St. Clair NL, Navia MA (1992) J. Am. Chem. Soc. 114: 7314
501. Karube I, Kawarai M, Matsuoka H, Suzuki S (1985) Appl. Microbiol. Biotechnol. 21: 270
502. Qureshi N, Tamhane DV (1985) Appl. Microbiol. Biotechnol. 21: 280
503. Umemura I, Takamatsu S, Sato T, Tosa T, Chibata I (1984) Appl. Microbiol. Biotechnol. 20: 291
504. Reetz MT (1997) Adv. Mater. 9: 943
505. Braun S, Rappoport S, Zusman R, Avnir D, Ottolenghi M (1990) Mater. Lett. 10: 1
506. Avnir D, Braun S, Lev O, Ottolenghi M (1994) Chem. Mater. 6: 1605
507. Avnir D (1995) Acc. Chem. Res. 28: 328
508. Gill I (2001) Chem. Mater. 13: 3404
509. Pierre A C, Pajonk G M (2002) Chem. Rev. 102: 4243
510. Fukui S, Tanaka A (1984) Adv. Biochem. Eng. Biotechnol. 29: 1
511. Mori T, Sato T, Tosa T, Chibata I (1972) Enzymologia 43: 213
512. Martinek K, Klibanov AM, Goldmacher VS, Berezin IV (1977) Biochim. Biophys. Acta 485: 1
513. Hoar TP, Schulman JH (1943) Nature 152: 102
514. Bonner FJ, Wolf R, Luisi PL (1980) Solid Phase Biochem. 5: 255
515. Poon PH, Wells MA (1974) Biochemistry 13: 4928
516. Wells MA (1974) Biochemistry 13: 4937
517. Martinek K, Levashov AV, Klyachko NL, Khmelnitsky YL, Berezin IV (1986) Eur. J. Biochem. 155: 453
518. Luisi PL (1985) Angew. Chem. Int. Ed. 24: 439
519. Luisi PL, Laane C (1986) Trends Biotechnol. 4: 153
520. Meier P, Luisi PL (1980) Solid Phase Biochem. 5: 269
521. Okahata Y, Niikura K, Ijiro K (1995) J. Chem. Soc. Perkin Trans. 1: 919
522. Barbaric S, Luisi PL (1981) J. Am. Chem. Soc. 103: 4239
523. Grandi C, Smith RE, Luisi PL (1981) J. Biol. Chem. 256: 837
524. Martinek K, Semenov AN, Berezin IV (1981) Biochim. Biophys. Acta 658: 76
525. Martinek K, Levashov AV, Khmelnitsky YL, Klyachko NL, Berezin IV (1982) Science 218: 889
526. Hilhorst R, Spruijt R, Laane C, Veeger C (1984) Eur. J. Biochem. 144: 459
527. Flaschel E, Wandrey C, Kula MR (1983) Adv. Biochem. Eng. Biotechnol. 26: 73
528. Kragl U, Vasic-Racki D, Wandrey C (1993) Indian J. Chem. 32B: 103
529. Biselli M, Kragl U, Wandrey C (1995) Reaction engineering for enzyme-catalyzed biotransformations. In: Drauz K, Waldmann H (eds) Enzyme Catalysis in Organic Synthesis. Verlag Chemie, Weinheim, p 89
530. Bednarski MD, Chenault HK, Simon ES, Whitesides GM (1987) J. Am. Chem. Soc. 109: 1283
531. Grimes MT, Drueckhammer DG (1993) J. Org. Chem. 58: 6148
532. Thiem J, Stangier P (1990) Liebigs Ann. Chem. 1101
533. van Eikeren P, Brose DJ, Muchmore DC, West JB (1990) Ann. NY Acad. Sci. 613: 796
534. Bückmann AF, Carrea G (1989) Adv. Biochem. Eng. Biotechnol. 39: 97

535. Bückmann AF, Kula MR, Wichmann R, Wandrey C (1981) J. Appl. Biochem. 3: 301
536. Vasic-Racki DJ, Jonas M, Wandrey C, Hummel W, Kula MR (1989) Appl. Microbiol. Biotechnol. 31: 215
537. Wykes JR, Dunnill P, Lilly MD (1972) Biochim. Biophys. Acta 286: 260
538. Malinauskas AA, Kulis JJ (1978) Appl. Biochem. Microbiol. 14: 706
539. Breslow R (ed) (2005) Artificial Enzymes. Wiley-VCH, Weinheim
540. Penning T M, Jez JM (2001) Chem. Rev. 101: 3027
541. Lundblad RL (1991) Chemical Reagents for Protein Modification, 2nd edn. CRC, London
542. Glazer AN (1976) The chemical modification of proteins by group- specific and site-specific reagents. In: Neurath H, Hill RL (eds) The Proteins, vol II, p 1, Academic Press, London
543. Inada Y, Yoshimoto T, Matsushima A, Saito Y (1986) Trends Biotechnol. 4: 68
544. Inada Y, Matsushima A, Hiroto M, Nishimura H, Kodera Y (1995) Adv. Biochem. Eng. Biotechnol. 52: 129
545. Inada Y, Furukawa M, Sasaki H, Kodera Y, Hiroto M, Nishimura H, Matsushima A (1995) Trends Biotechnol. 13: 86
546. Kodera Y, Takahashi K, Nishimura H, Matsushima A, Saito Y, Inada Y (1986) Biotechnol. Lett. 8: 881
547. Takahashi K, Ajima A, Yoshimoto T, Okada M, Matsushima A, Tamaura Y, Inada Y (1985) J. Org. Chem. 50: 3414
548. Takahashi K, Ajima A, Yoshimoto T, Inada Y (1984) Biochem. Biophys. Res. Commun. 125: 761
549. Uemura T, Fujimori M, Lee HH, Ikeda S, Aso K (1990) Agric. Biol. Chem. 54: 2277
550. Ferjancic A, Puigserver A, Gaertner H (1988) Biotechnol. Lett. 10: 101
551. Pasta P, Riva S, Carrea G (1988) FEBS Lett. 236: 329
552. Bremen U, Gais HJ (1996) Tetrahedron Asymmetry 7: 3063
553. Beecher JE, Andrews AT, Vulfson EN (1990) Enzyme Microb. Technol. 12. 955
554. Takahashi K, Nishimura H, Yoshimoto T, Saito Y, Inada Y (1984) Biochem. Biophys. Res. Commun. 121: 261
555. Yoshimoto T, Takahashi K, Nishimura H, Ajima A, Tamaura Y, Inada Y (1984) Biotechnol. Lett. 6: 337
556. Yoshimoto T, Mihama T, Takahashi K, Saito Y, Tamaura Y, Inada Y (1987) Biochem. Biophys. Res. Commun. 145: 908
557. Nishio T, Takahashi K, Yoshimoto T, Kodera Y, Saito Y, Inada Y (1987) Biotechnol. Lett. 9: 187
558. Heiss L, Gais HJ (1995) Tetrahedron Lett. 36: 3833
559. Ruppert S, Gais HJ (1997) Tetrahedron Asymmetry 8: 3657
560. Matsushima A, Kodera Y, Takahashi K, Saito Y, Inada Y (1986) Biotechnol. Lett. 8: 73
561. Cambou B, Klibanov AM (1984) J. Am. Chem. Soc. 106: 2687
562. Ajima A, Yoshimoto T, Takahashi K, Tamaura Y, Saito Y, Inada Y (1985) Biotechnol. Lett. 7: 303
563. Lee H, Takahashi K, Kodera Y, Ohwada K, Tsuzuki T, Matsushima A, Inada Y (1988) Biotechnol. Lett. 10: 403
564. Babonneau MT, Jaquier R, Lazaro R, Viallefont P (1989) Tetrahedron Lett. 30: 2787
565. Matsushima A, Okada M, Inada Y (1984) FEBS Lett. 178: 275
566. Okahata Y, Mori T (1997) Trends Biotechnol. 15: 50
567. Mori T, Kobayashi A, Okahata Y (1998) Chem. Lett. 921
568. Okahata Y, Mori T (1998) J. Mol. Catal. B 5: 119
569. Jene Q, Pearson JC, Lowe CR (1997) Enzyme Microb. Technol. 20: 69
570. Kaiser ET (1988) Angew. Chem. Int. Ed. 27: 902
571. Letondor C, Ward T R (2006) ChemBioChem 7: 1845
572. van der Velde F, Könemann L, van Rantwijk F, Sheldon RA (1998) Chem. Commun. 1891
573. Nakatsuka T, Sasaki T, Kaiser ET (1987) J. Am. Chem. Soc. 109: 3808
574. Polgár L, Bender MC (1966) J. Am. Chem. Soc. 88: 3153

575. Neet KE, Koshland DE (1966) Proc. Natl. Acad. Sci. USA 56: 1606
576. Hilvert D (1989) Design of enzymatic catalysts. In: Whitaker JR, Sonnet PE (eds) Biocatalysis in Agricultural Biotechnology, ACS Symposium Series 389. ACS, Washington, p 14
577. Häring D, Herderich M, Schüler E, Withopf B, Schreier P (1997) Tetrahedron Asymmetry 8: 853
578. Kaiser ET, Lawrence DS (1984) Science 226: 505
579. Levine HL, Kaiser ET (1980) J. Am. Chem. Soc. 102: 343
580. Slama JT, Radziejewski C, Oruganti SR, Kaiser ET (1984) J. Am. Chem. Soc. 106: 6778
581. Radziejewski C, Ballou DP, Kaiser ET (1985) J. Am. Chem. Soc. 107: 3352
582. Aitken DJ, Alijah R, Onyiriuka SO, Suckling CJ, Wood HCS, Zhu L (1993) J. Chem. Soc. Perkin Trans. 1: 597
583. Hilvert D, Hatanaka Y, Kaiser ET (1988) J. Am. Chem. Soc. 110: 682
584. Okrasa K, Kazlauskas R J (2006) Chem. Eur. J. 12: 1587
585. Ward TR (2010) Acc. Chem. Res. 44: 47.
586. Wilson ME, Whitesides GM (1978) J. Am. Chem. Soc. 100: 306.
587. Qi D, Tann C-M, Haring D, Distefano MD (2001) Chem. Rev. 101: 3081.
588. Lu Y, Yeung N, Sieracki N, Marshall NM (2009) 460: 855.
589. Bos J, Garcia-Herraiz A, Roelfes G (2013) Chem. Sci. 4: 3578.
590. Chatterjee A, Mallin H, Klehr J, Vallapurackal J, Finke AD, Vera L, Marsh M, Ward TR (2016) Chem. Sci. 7: 673.
591. Jeschek M, Reuter R, Heinisch T, Trindler C, Klehr J, Panke S, Ward TR (2016) Nature 537: 661.
592. Reetz MT, Jaeger KE (1999) Top. Curr. Chem. 200: 31
593. Reetz MT, Becker MH, Kühling KM, Holzwarth A (1998) Angew. Chem. Int. Ed. 37: 2647
594. Arnold FH, Volkov AA (1999) Curr. Opinion Chem. Biol. 3: 54
595. Arnold FH, Moore JC (1999) Optimizing industrial enzymes by directed evolution. In: Scheper T (ed) New Enzymes for Organic Synthesis. Springer, Berlin Heidelberg New York, p 1
596. Woodley JM (2013) Curr. Opin. Chem. Biol. 17: 310.
597. Sun Z, Wikmark Y, Baeckvall J-E, Reetz MT (2016) Chem. Eur. J. 22: 5046.
598. Renata H, Wang ZJ, Arnold FH (2015) Angew. Chem. Int. Ed. 43: 3351.
599. Brustad EM, Arnold FH (2011) Curr. Opin. Chem. Biol. 15: 201.
600. Cronin CN (1998) J. Biol. Chem. 273: 24465
601. Vellom DC, Radic Z, Li Y, Pickering NA, Camp S, Taylor P (1993) Biochemistry 32: 12
602. Cantu C, Huang W, Palzkill T (1997) J. Biol. Chem. 272: 29144
603. Tanaka T, Matsuzawa H, Ohta T (1998) Biochemistry 37: 17402
604. Mei HC, Liaw Y C, Li YC, Wang DC, Takagi H, Tsai YC (1998) Protein Eng. 11: 109
605. Zhu Z, Sun D, Davidson VL (2000) Biochemistry 39: 11184
606. Ma H, Penning TM (1999) Proc. Natl. Acad. Sci. USA 96: 11161
607. Oue S, Okamoto A, Yano T, Kagamiyama H (1999) J. Biol. Chem. 274: 2344
608. Vacca RA, Giannattasio S, Graber R, Sandmeier E, Marra E, Christen P (1997) J. Biol. Chem. 272: 21932
609. Scrutton NS, Berry A, Perhan RN (1990) Nature 343: 38
610. Bohren KM, Bullock B, Wermuth B, Gabbay KH (1989) J. Biol. Chem. 264: 9574
611. Ratnam K, Ma H, Penning TM (1999) Biochemistry 38: 7856
612. Jiang RT, Dahnke T, Tsai MD (1991) J. Am. Chem. Soc. 113: 5485
613. Sakowicz R, Gold M, Jones JB (1995) J. Am. Chem. Soc. 117: 2387
614. Kuroki, R, Weaver LH, Matthews BW (1999) Proc. Natl. Acad. Sci. USA 96: 8949
615. van den Heuvel RH, Fraaije MW, Ferrer M, Mattevi A, van Berkel WJ (2000) Proc. Natl. Acad. Sci. USA 97: 9455
616. Terao Y, Iijima Y, Miyamoto K, Ohta H (2007) J. Mol. Catal. B: Enzym. 45: 15
617. Liebeton BW, Reetz MT, Jaeger K (2000) Chem. Biol. 7: 709
618. Fong S, Machajewski TD, Mak CC, Wong CH (2000) Chem. Biol. 7: 873

619. Mugford PF, Wagner UG, Jiang Y, Faber K, Kazlauskas RJ (2008) Angew. Chem. Int. Ed. 47: 8782.
620. Bornscheuer UT, Kazlauskas RJ (2004) Angew. Chem. Int. Ed. 43: 6032
621. Hult K, Berglund P (2007) Trends Biotechnol. 25: 231
622. Lopez-Iglesias M, Gotor-Fernandez V (2015) Chem. Rec. 15: 743.
623. Miao Y, Rahimi M, Geertsema EM, Poelarends GJ (2015), Curr. Opin. Chem. Biol. 25: 115.
624. Dufour E, Storer AC, Menard R (1995) Biochemistry 34: 16382
625. Millard CB, Lockridge O, Broomfield CA (1998) Biochemistry 37: 237
626. Newcomb RD, Campbell PM, Ollis DL, Cheah E, Russell RJ, Oakeshott JG (1997) Proc. Natl. Acad. Sci. USA 94: 7464
627. Jez JM, Penning T M (1998) Biochemistry 37: 9695
628. Jochens H, Stiba K, Savile C, Fujii R, Yu JG, Gerassenkov T, Kazlauskas RJ, Bornscheuer UT (2009) Angew. Chem. Int. Ed. 48: 3532
629. Branneby C, Carlqvist P, Magnusson A, Hult K, Brinck T, Berglund P (2003) J. Am. Chem. Soc. 125: 874
630. Carlqvist P, Svedendahl M, Branneby C, Hult K, Brinck T, Berglund P (2005) ChemBioChem 6: 331
631. Svedendahl M, Jovanovic B, Fransson L, Berglund P (2009) ChemCatChem 1: 252
632. Yow GY, Watanabe A, Yoshimura T, Esaki N (2003) J. Mol. Catal. B 23: 311
633. Graber R, Kasper P, Malashkevich VN, Strop P, Gehring H, Jansonius JN, Christen P (1999) J. Biol. Chem. 274: 31203
634. Mutter M (1985) Angew. Chem. Int. Ed. 24: 639
635. Murakami Y, Kikuchi J, Hisaeda Y, Hayashida O (1996) Chem. Rev. 96: 721
636. Kirby AJ (1996) Angew. Chem. Int. Ed. 35: 705
637. Allen JV, Roberts SM, Williamson NM (1999) Adv. Biochem. Eng. Biotechnol. 63: 125
638. Chadwick DJ, Marsh J (1991) Catalytic Antibodies. Ciba Foundation Symposium, vol 159. Wiley, New York
639. Schultz PG (1989) Angew. Chem. Int. Ed. 28: 1283
640. Kirby AJ (1996) Acta Chem. Scand. 50: 203
641. Reymond JL (1999) Top. Curr. Chem. 200: 59
642. Lerner RA (1990) Chemtracts – Org. Chem. 3: 1
643. Kiss G, Celebi-Ölcüm N, Moretti R, Baker D, Houk KN (2013) Angew. Chem. Int. Ed. 52: 5700.
644. French DL, Laskov R, Scharff MD (1989) Science 244: 1152
645. Pollack SJ, Schultz PG (1989) J. Am. Chem. Soc. 111: 1929
646. Janjic N, Tramontano A (1989) J. Am. Chem. Soc. 111: 9109
647. Janda KD, Schloeder D, Benkovic SJ, Lerner RA (1988) Science 241: 1188
648. Tramontano A, Janda KD, Lerner RA (1986) Science 234: 1566
649. Pollack SJ, Jacobs JW, Schultz PG (1986) Science 234: 1570
650. Janda KD et al. (1991) J. Am. Chem. Soc. 113: 291
651. Napper AD, Benkovic SJ, Tramontano A, Lerner RA (1987) Science 237: 1041
652. Janda KD, Benkovic SJ, Lerner RA (1989) Science 244: 437
653. Shokat KM, Leumann CJ, Sugasawara R, Schultz PG (1989) Nature 338: 269
654. Uno T, Schultz PG (1992) J. Am. Chem. Soc. 114: 6573
655. Nakayama GR, Schultz PG (1992) J. Am. Chem. Soc. 114: 780
656. Hilvert D (1992) Pure Appl. Chem. 64: 1103
657. Hoffmann T, Zhong G, List B, Shabat D, Anderson J, Gramatikova S, Lerner RA, Barbas III CF (1998) J. Am. Chem. Soc. 120: 2768
658. Iverson BL, Cameron KE, Jahangiri GK, Pasternak DS (1990) J. Am. Chem. Soc. 112: 5320
659. Jackson DY, Schultz PG (1991) J. Am. Chem. Soc. 113: 2319
660. Cochran AG, Sugasawara R, Schultz PG (1988) J. Am. Chem. Soc. 110: 7888
661. Hilvert D, Nared KD (1988) J. Am. Chem. Soc. 110: 5593

662. Jackson DY, Jackson DY, Jacobs JW, Sugasawara R, Reich SH, Bartlett PA, Schultz PG (1988) J. Am. Chem. Soc. 110: 4841
663. Braisted AC, Schultz PG (1990) J. Am. Chem. Soc. 112: 7430
664. Meekel AAP, Resmini M, Pandit UK (1996) Bioorg. Med. Chem. 4: 1051
665. Hasserodt J, Janda KD, Lerner RA (1997) J. Am. Chem. Soc. 119: 5993
666. Hasserodt J, Janda KD, Lerner RA (1996) J. Am. Chem. Soc. 118: 11654
667. Zhong G, Hoffmann T, Lerner RA, Danishefsky S, Barbas CF (1997) J. Am. Chem. Soc. 119: 8131
668. Tramontano A, Ammann AA, Lerner RA (1988) J. Am. Chem. Soc. 110: 2282
669. Kitazume T, Lin JT, Takeda M, Yamazaki T (1991) J. Am. Chem. Soc. 113: 2123
670. Sinha SC, Keinan E, Reymond JL (1993) J. Am. Chem. Soc. 115: 4893
671. Sinha SC, Keinan E (1995) J. Am. Chem. Soc. 117: 3653
672. Janda KD, Shevlin CG, Lerner RA (1993) Science 259: 490
673. Levitt M (2007) Proc. Natl. Acad. Sci. USA 104: 3183.
674. Röthlisberger D, Khersonsky O, Wollacott AM, Jiang L, DeChancie J, Betker J, Gallaher JL, Althoff EA, Zanghellini A, Dym O, Albeck S, Houk KN, Tawfik DS, Baker D (2008) Nature 453: 190.
675. Siegel JB, Zanghellini A, Lovick HM, Kiss G, Lambert AR, St. Clair JL, Gallaher JL, Hilvert D, Gelb MH, Stoddart BL, Houk KN, Michael FE, Baker D (2010) Science 329: 309.
676. Steinkellner G, Gruber CC, Pavkov-Keller T, Binter A, Steiner K, Winkler C, Lyskowski A, Schwamberger O, Oberer M, Schwab H, Faber K, Macheroux P, Gruber K (2014) Nat. Commun. 5: 4150; DOI: 10.1038/ncomms5150.

Chapter 4
State of the Art and Outlook

The biotransformations described in this book demonstrate that this area is in an active state of development and that enzymes have attained an important position in contemporary organic synthesis both on the academic and the industrial level [1–10]. The increased interest of synthetic chemists in the biotransformation of nonnatural compounds has spurred the commercial availability of enzymes in various forms and grades of purity [11, 12]. To date, about ~8 % of all papers published in the area of synthetic organic chemistry contain elements of biotransformations [13] and >300 industrial-scale processes are documented worldwide [14–24]. As shown in Fig. 4.1, the frequency of use of a particular biocatalyst is not evenly distributed among the various types of enzymes and it is obvious that the more 'simple' hydrolytic reactions are dominant.[1] In this book, attention has been focused on those methods which are most useful for and accessible to synthetic chemists and the rating of methods according to their general applicability and reliability is intended!

In the following paragraphs a brief summary on the state of the art of biotransformations is given (Table 4.1) with an outlook on future developments.

Hydrolytic enzymes such as proteases, esterases and lipases are ready-to-use catalysts for the preparation of optically active carboxylic acids, amino acids, alcohols, and amines. This reliable technology is applicable to a broad range of synthetic problems and is widely used in industry. The broad knowledge on biotransformations from these areas has evolved a large number of commercially available proteases and lipases in conjunction with their substrate tolerance and techniques for the improvement of their selectivities and predictive models for the stereochemical outcome of a given reaction. A search for novel esterases to enrich the limited number of available enzymes and for lipases showing 'anti-Kazlauskas' stereo-specificities would be a worthwhile endeavor.

---

[1]Data from database Kroutil/Faber, ~18,000 entries, 2016.

© Springer International Publishing AG 2018
K. Faber, *Biotransformations in Organic Chemistry*,
DOI 10.1007/978-3-319-61590-5_4

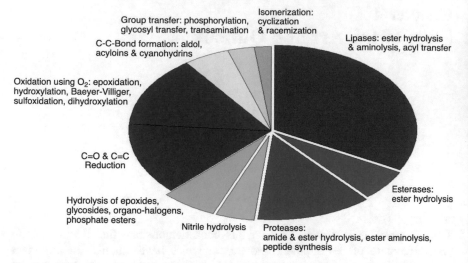

Fig. 4.1 Frequency of use of particular biocatalysts in biotransformations

The asymmetric hydrolysis of epoxides offers a valuable method for the synthesis of nonracemic epoxides and *vic*-diols on a multigram-scale. The enzymatic hydrolysis of nitriles to amides or carboxylic acids often does not proceed with high enantioselectivity, but it offers a mild alternative to the harsh reaction conditions required using traditional methodology. As a consequence, nitrile-hydrolysing enzymes have outcompeted chemical catalysts on industrial scale.

The chemo- and regio-selective synthesis of phosphate esters used as flavor-enhancers, active pharmaceutical ingredients and as high-energy metabolites required for pathway engineering is now possible using cheap inorganic pyrophosphate circumventing the need of ATP.

In contrast to the dominating hydrolases, about one fifth of the research in biotransformations involves reduction reactions. Alcohol dehydrogenases and/or whole microbial cells enable the stereo-selective reduction of ketones to furnish the corresponding *sec*-alcohols on industrial scale. Enzymes possessing opposite stereospecificity are available and the recycling of NAD(P)H-cofactors is state-of-the-art. Recent breakthroughs in the cloning of oxygen-stable ene-reductases allows the asymmetric bioreduction of activated carbon–carbon double bonds for multi-gram scale applications. Access to amines in nonracemic form has recently be enabled by development of imine reductases and amine dehydrogenases.

Biocatalytic oxygenation processes are becoming increasingly important since traditional methodology is either not feasible or makes use of hypervalent metal oxides, which are ecologically undesirable when used on a large scale. As the use of isolated mono- and di-oxygenases will presumably be always impeded by their multicomponent nature and their requirement for NAD(P)H recycling, the majority of oxygenation reactions, such as mono- and dihydroxylation, epoxidation, sulfoxidation, and Baeyer-Villiger reactions will continue to be performed using engineered whole cell systems. Nonheme oxygenases and oxidases may help to circumvent these limitations. The regio- and stereoselective oxidation of (poly) hydroxy compounds, which is notoriously difficult with nicotinamide-dependent dehydrogenases, is more efficiently performed with Cu- or flavin-dependent alcohol oxidases. The deracemization of pharmacologically important (cyclic) *sec*-amines can be achieved by combining enantioselective oxidation using amine oxidases combined with non-stereoselective (chemical) reduction in a cyclic fashion on industrial scale.

The stereoselective formation of carbon–carbon bonds by means of aldolases and transketolases is a well-researched standard technique for the synthesis of carbohydrate-like α- and β-hydroxy ketones. It is noteworthy that the full set of stereo-complementary aldolases are now available, which accept cheap (non-phosphorylated) donors, such as dihydroxy acetone, pyruvate, acetaldehyde and glycin. Along these lines, PLP-dependent lyases for acyloin and benzoin condensations are gaining importance due to their lack of sensitive phosphorylated cosubstrates.

Glycosyl transfer reactions using glycosidases and glycosyl transferases are gaining ground as more of these enzymes are made available by genetic engineering. Non-phosphorylated cheap high-energy sugar donors, such as sucrose, currently supersede the expensive glycosyl phosphates previously employed. Within only a few years, synthesis of enantiopure amines became feasible on industrial scale by the advent of (R)- and (S)-selective transaminases.

The synthesis of optically active (R)- and (S)-cyanohydrins by hydroxynitrile lyases is now well established. Several of these enzymes are commercially available and were successfully implemented in industry. The lyase-catalyzed addition of water and ammonia onto activated C=C bonds has little counterpart in traditional chemical catalysis and provides a highly desirable atom efficiency of 100%. Unfortunately, the presently used hydratases and ammonia lyases are impeded by a narrow substrate tolerance, which possibly might be circumvented by the design of suitable mutants.

Although halogenation and dehalogenation reactions can be catalyzed by enzymes, it is doubtful whether these reactions will be used widely, mainly because the corresponding conventional chemical methods are highly competitive.

Emerging fields exploit the applicability of lesser-used enzymes, such as cyclases and racemases [25], sulfatases, diol dehydratases, aldoxime dehydratases and carboxylate reductases. Notoriously 'dirty' and inefficient Friedel-Crafts-alkylation and -acylations are feasible by employing alkyl- and acyl-transferases and enzymatic carboxylation reactions are currently being investigated for the selective formation of carbon–carbon bonds using CO_2 as a raw material for organic synthesis [26].

The methodology concerning the employment of enzymes in nonaqueous organic solvents with respect to enzyme *activity* is well understood and highly predictable. Thus, the synthesis of esters, lactones, amides, peptides, and peracids by using enzymes is standard methodology in industry. On the other hand, the influence of the nature of organic solvents on an enzyme's *selectivity* is still insufficiently understood.

Table 4.1 Pros and cons of biotransformations according to enzymes types

Enzyme type	Reaction catalyzed	Strength	Weakness	Solution
Lipase	Ester hydrolysis and aminolysis, acyl transfer	Many stable enzymes, organic solvents	—	Anti-Kazlauskas lipases
Esterase	Ester hydrolysis and formation	Pig liver esterase, esterase-activity of proteases	Few esterases, organic solvents	Novel esterases
Protease	Ester and amide hydrolysis, ester aminolysis, peptide synthesis	Many stable proteases	No D-proteases	Engineered D-proteases
Nitrilase, nitrile hydratase	Nitrile hydrolysis	Chemo- and regioselectivity	Stereoselectivity, enzyme stability	Stable stereoselective nitrilases
Epoxide hydrolase	Hydrolysis of epoxides	Chemocatalysis weak	Few enzymes, enzyme stability	New stable enzymes?
Phosphatase/Kinase	Phosphate ester hydrolysis and formation	ATP-Independent phosphatases	ATP-Dependent kinases, equilibrium	Engineered phosphatases with reduced hydrolytic activity
Glycosidase/ Glycosyl transferase	Oligosaccharide formation	Anomeric selectivity, non-phosphorylated disaccharides as donors	Regio- and diastereoselectivity, equilibrium	More selective enzymes accepting non-phosphorylated donors
Dehalogenase	Hydrolysis of haloalkanes	Chemocatalysis weak, haloacid and halohydrin dehalogenases	Limited catalytic activity of haloalkane dehalogenases	More active haloalkane dehalogenases
Dehydrogenase/ Carbonyl reductase	Reduction of aldehydes and ketones	Prelog-selectivity, NADH-enzymes	Anti-Prelog-enzymes, NADPH-enzymes	NADH-dependent anti-Prelog enzymes
Ene-reductase	Reduction of activated C=C bonds	trans-Reduction, stereocontrol	Large-scale applications	Stable enzymes
Mono-oxygenase	Hydroxylation, Baeyer-Villiger, epoxidation	Chemocatalysis fails	Sensitive multi-component NADPH-enzymes	Non-heme enzymes
Di-oxygenase	Dihydroxylation of aromatics	Whole-cell systems	Isolated enzymes, non-aromatic substrates	Enzymes for non-aromatic substrates
Aldolase	Aldol reaction in H_2O	Stereo-complementary DHAP-enzymes	Non-phosphorylated donors	Enzymes accepting non-phosphorylated donors
Transketolase	Acyloin/benzoin condensation	Non-phosphorylated donors	Few enzymes	More enzymes accepting ketones and CO_2 as donors
Hydroxynitrile lyase	Cyanohydrin formation	(R)- and (S)-enzymes	Equilibrium	Acid-stable enzymes
Fumarase, aspartase	Addition of H_2O, NH_3	Chemocatalysis fails	Narrow substrate tolerance	New enzymes
Alcohol oxidase	Oxidation of alcohols	O_2 as oxidant, prim-alcohols	sec-alcohols	sec-Alcohol oxidases
Transaminase	Amino transfer	High selectivity, (R)- and (S)-enzymes	Equilibrium	State of the art

Over the past decade, the combination of multiple (bio)catalytic steps onto each other in a cascade [27, 28] has resulted in the design of 'artificial synthetic pathways', which show drastically improved efficiency compared to traditional step-wise synthesis, because decomposition of sensitive intermediates is minimized and loss-causing purification of intermediates is omitted.

After all, enzymes have been optimizing their skills for more than 3×10^9 years so as to develop a lot of sophisticated chemistry, whereas organic chemists and genetic engineers have a track record of less than a century. There is strong evidence that the number of distinctly different enzyme mechanisms is finite, since we know examples of protein molecules that are unrelated by evolution but possess almost identically arranged functional groups. In many cases, Nature has obviously faced the same biochemical problem and has found the same optimum solution. In parallel to the development of novel enzymes by evolution through natural selection, tremendous advances in the field of genetic engineering provide biocatalysts at drastically decreased costs, and also allow the design of enzymes possessing enhanced stabilities and altered (stereo)specificities.

Most of the environmental contamination of greatest concern involves xenobiotic materials which have been produced and distributed over our globe with increasing diversity and volume in industrial and agricultural processes during the past century. These include hydrocarbons, heavy metals, neurotoxic pesticides, halogenated (aromatic) compounds, explosives, and carcinogens, which often do not have natural counterparts. Although their introduction into the ecosphere is lamentable, and the effects they can have on the environment is often devastating, numerous microbial systems have already acquired the ability to degrade most of these xenobiotics. The likelihood that totally new enzymes with specific detoxification activities could evolve naturally in such a brief period of time is unrealistic, it is more likely that enzymes already present in microbial communities – used for defense and maintenance functions – were evolutionary adapted within a few decades through chemical stress to address the new challenge. These novel biocatalytic activities represent a large potential to enable the transformation of functional groups, which are unknown to biochemistry.

The enormous amount of ~9.5 Gt[2] of carbon derived from non-renewable fossil resources – oil, coal and natural gas – are currently converted annually worldwide. Only a small fraction (~5–9%) of this precious raw material is used as starting material for the synthesis of organic compounds, such as polymers, dyes, detergents, pharmaceuticals, vitamins, etc., which make our high-tech life safe, healthy and comfortable. The vast majority of it (>90 %) is burnt to generate energy going in hand with an undesired accumulation of CO_2 in the atmosphere, where it accelerates global warming [29].[3]

---

[2]BP Statistical Report 2017

[3]World primary energy patterns are changing only slowly: Currently, oil constitutes 35% of global energy consumption, coal 29%, natural gas 24%, nuclear power 5.5% and hydroelectricity 6.6%.

Traditional industrial synthesis of organic chemicals predominantly proceeds via steam-cracking of crude oil to furnish a small number of high-volume primary platform chemicals, from which all major synthetic routes of industrial organic chemicals are branching off. The primary intermediates are composed of:

(i) *Olefins*, i.e. ethylene, propene and butenes (1- and 2-butene, 1,3-butadiene, *i*-butene),

(ii) *aromatics* (benzene, toluene, xylenes), and

(iii) *one alkane* (methane).

The latter have a low oxidation state of carbon (from -4 in methane to -1 in aromatics), which renders the introduction of functional groups bearing elements with high electronegativity (such as O, N, S, halogen, etc.) a 'hidden oxidation', which is thermodynamically favored. This – together with the high efficiency of steam-cracking (~98%) – is the main reason for the unparalleled success of traditional synthetic organic chemistry on industrial scale.

As non-renewable carbon feedstocks are slowly depleted, alternative sources are sought with increasing intensity from biomass. So far, three main groups of compounds have been proposed as substitute for non-renewable carbon feedstocks for the synthesis of organic chemicals: [30]

(i) Carbohydrates In a simplistic version, they are oligomeric forms of [—CH-OH—] and as such contain carbon at comparably high oxidation state (0), which renders the introduction of functional groups thermodynamically less favorable. Carbohydrates are anything but ideal as starting materials for synthesis, because they are *over-decorated* by hydroxy-groups, which prevent direct functionalization of the carbon framework without protecting and/or activating strategies. In addition, they are *under-functionalized*, because they contain only a single functional group – the alcohol moiety – besides the occasional carbonyl group in aldoses or ketoses, which is usually masked as (hemi)acetal. Hence, carbohydrates appear more useful as feedstocks for fermentation or synthesis gas production than as starting material for organic synthesis.

Advances in biotechnology have enabled us to produce complex fermentations products, such as vitamins (e.g. B12, niacin), antibiotics (penicillins), steroids and terpenoids (taxadiene, artemisinic acid), which are either used directly as bioactive (pharmaceutical) ingredient, or as indispensable building block for the (semi) synthesis of natural products for therapy of hormone disorders, cancer (taxol) or parasites (artemisinin), which could never be achieved economically by chemical synthesis alone. In addition to these highly complex structures, fermentation yields a range of low molecular metabolites, which are mainly composed of carboxylic acids (citric, succinic, fumaric, itaconic, lactic acid) or alcohols (ethanol, 1,2- and 1,3-propanediol, 1-butanol, 1,4-butanediol). On a first glimpse, this looks appealing, but it has to be kept in mind that fermentation is far less efficient than steam-cracking and after all, the overall energy efficiency of the initial step in bio-synthesis – photosynthesis – is $\leq 2\%$. Consequently, the use of biofuels for the conversion of sunlight into chemical energy – in contrast to photovoltaics – might massage our ecological ego, but is scientific nonsense [31].

(ii) Lignin breakdown products Lignin (simplified as 'bark') is a very heterogeneous polymeric material composed of phenolic aromatics and C_3-units, whose
main role is to shield the cellulose from microbial attack. It can be chemically
converted via partial oxidative degradation to mono-ligninols (sinapyl, coniferyl
and hydroxycinnamoyl alcohol). Common to these compounds is their electron-rich
(phenolic) aromatic core, which renders them susceptible to auto-oxidation, which
creates serious problems upon storage. The color of lignin degradation product
streams range from dark yellow to tarry black – not an ideal starting point for
selective organic synthesis. No efficient use of these compounds for synthesis has
been found so far.

(iii) Oils and fats are predominantly composed of triglycerides. The alcohol moiety –
glycerol – resembles a sugar alcohol and thus has limited value for synthesis[4], but
fatty acids – composed of a dominant hydrocarbon chain with a terminal carboxylic
acid group – can be synthetically utilized to some extent. However, the most efficient
production of triglycerides is palm oil, which has turned Indonesia into the third
largest CO_2-producer worldwide – after the US and China – due to (intended and
illegal) large-scale deforestation. Not a long-term solution.

Since current technology for the synthesis of organic chemicals on industrial
scale is mainly based on the use of alkenes and aromatics, it needs a lot of research
(and optimism) for the development of highly efficient processes, which draw on
'natural' starting materials – carboxylic acids, alcohols and the like – to bypass
traditional industrial routes starting from hydrocarbons. In this context, it can be
expected that biotransformations will most likely play a dominant role, because
since natural carbon sources are *made* by enzyme catalysis, they can also be
expected to be *converted* by them.

If such pathways cannot be developed in an efficient manner, we have to resort to
the old-fashioned, reliable – and relatively efficient – generation of synthesis gas, in
short a mixture of CO and H_2. The latter can be made from almost any carbona-
ceous material, including natural gas, coal, peat, wood and agricultural residues,
which is converted into hydrocarbons by Fischer-Tropsch synthesis.

An immediate action to be taken is the de-carbonization of energy-production, in
order to reserve our non-renewable carbon feedstocks for chemical synthesis of
long-lasting products instead of burning them for short-term energy generation.
This would extend their availability[5] for more than one order of magnitude and
would help to limit global warming as a highly desirable side effect.

There is a general need for cleaner ('green') and sustainable chemistry, and
although it is not the prerogative of biocatalysis, it is nevertheless an important
feature of it. Social pressure forces the chemical industry to pay more attention to

---

[4]About 4 Mt of glycerol are annually obtained as (low-value) by-product from biodiesel production worldwide

[5]Projected availability according to www.eia.doe.gov: oil 43–67 years, gas 64–150 years, coal
200–1500 years.

these issues, as the public becomes aware that the resources are limited and have to be used with utmost efficiency.

References

1. Faber K (1997) Pure Appl. Chem. 69: 1613
2. Drauz K, Gröger H, May O (eds.) (2012) Enzyme Catalysis in Organic Synthesis, 3rd edn., Wiley, Hoboken, NJ.
3. Faber K, Fessner W-D, Turner N J (eds.) (2014) Science of Synthesis, Biocatalysis in Organic Synthesis, 3 vols., Thieme, Stuttgart.
4. Patel R N (ed.) (2016) Green Biocatalysis, Wiley, Hoboken, NJ.
5. Grunwald P (ed.) (2015) Industrial Biocatalysis, Pan Stanford Publishing, Singapore.
6. Hoyos P, Pace V, Hernaiz M J, Alcantara A R (2014) Curr. Green Chem. 1: 155.
7. Turner N J (2013) Nat. Chem. Biol. 9: 285.
8. Nestl B M, Hammer C S, Nebel B A, Hauer B (2014) Angew. Chem. Int. Ed. 53: 3070.
9. Goswami A, Stewart J D (eds.) (2015) Organic Synthesis Using Biocatalysis, Elsevier, Amsterdam.
10. Hilterhaus L, Liese A, Kettling U, Antranikian G (eds.) (2016) Applied Biocatalysis: From Fundamental Science to Industrial Applications, Wiley, Hoboken, NJ.
11. White JS, White DC (1997) Source Book of Enzymes. CRC Press, Boca Raton
12. Lauchli R, Rozzell D (2015) Enzyme Sources and Selection of Biocatalysts, in: Science of Synthesis, Biocatalysis in Organic Synthesis, vol. 1, pp. 75-93, Thieme, Stuttgart.
13. Crosby C (1992) Chirality in industry – an overview. In: Collins, AN, Sheldrake GN, Crosby J (eds) Chirality in Industry. Wiley, Chichester, pp 1–66
14. Liese A, Seelbach K, Wandrey C (2006) Industrial Biotransformations, 2nd edn. Wiley-VCH, Weinheim
15. Breuer M, Ditrich K, Habicher T, Hauer B, Kesseler M, Stürmer R, Zelinski T (2004) Angew. Chem. Int. Ed. 43: 788
16. Straathof AJJ, Panke S, Schmid A (2002) Curr. Opinion Biotechnol. 13: 548
17. Hasan F, Shah AA, Hameed A (2006) Enzyme Microb. Technol. 39: 235
18. Schmid A, Hollmann F, Park JB, Bühler B (2002) Curr. Opinion Biotechnol. 13: 359
19. Thomas SM, DiCosimo R, Nagarajan V (2002) Trends Biotechnol. 20: 238
20. Schmid A, Dordick JS, Hauer B, Kiener A, Wubbolts M, Witholt B (2001) Nature 409: 258
21. Rasor JP, Voss E (2001) Appl. Catal. A: Gen. 221: 145
22. Wandrey C, Liese A, Kihumbu D (2000) Org. Proc. Res. Dev. 4: 286
23. Schoemaker HE, Mink D, Wubbolts MG (2003) Science 299: 1694
24. OECD (2011) Future Prospects for Industrial Biotechnology, OECD Publishing, 10.1787/9789264126633-en; ISBN: 9789264126633 (PDF); 9789264119567(print). p. 24.
25. Schnell B, Faber K, Kroutil W (2003) Adv. Synth. Catal. 345: 653
26. Glueck SM, Gümüs S, Fabian WMF, Faber K (2010) Chem. Soc. Rev. 39: 313
27. Mayer SF, Kroutil W, Faber K (2001) Chem. Soc. Rev. 30: 332
28. Riva S, Fessner W-D (eds.) (2014) Cascade Biocatalysis. Wiley, Hoboken, NJ.
29. Boden TA, Marland G, Andres RJ (2017) Global, Regional, and National Fossil-Fuel CO_2 Emissions. doi 10.3334/CDIAC/00001_V2017.
30. IEA Bioenergy, Bio-based Chemicals – Value Added Products from Biorefineries (2012) http://www.ieabioenergy.com/wp-content/uploads/2013/10/Task-42-Biobased-Chemicals-value-added-products-from-biorefineries.pdf.
31. Michel H (2012) Angew. Chem. Int. Ed. 51: 2516.

Chapter 5
Appendix

5.1 Basic Rules for Handling Biocatalysts

Like with chemical catalysts, a few guidelines should be observed with biocatalysts to preserve activity and to handle them safely. Depending on the formulation – liquid, lyophilized, spray-dried, immobilized – different rules apply. Enzymes are often perceived as unstable and delicate entities, however, when treated in the right way, they can be as sturdy as almost any chemical catalyst. For optimal stability, enzyme preparations are usually stored best in their original commercial form, either as lyophilizate, spray-dried powder or (stabilized) liquid formulation.

Safety
Direct contact with enzymes should be avoided. This is particularly important with respect to the inhalation of aerosols (from liquid formulations) or protein dust from lyophilized or spray-dried protein preparations. Depending on the person, the tendency to develop allergies (itching, red stains) can be very different. However, the development of health problems due to contact with cell-free protein formulations is extremely unlikely.

When working with whole (viable) cells, make sure you find out to which safety class the organism belongs *before* you start. Microorganisms are classified into four safety categories (Classes 1–4): Class 1 organisms, such as baker's yeast, and the vast majority of microorganisms described in this book are generally regarded as safe and special precautions are not required. On the other hand, cells belonging to Class 2 may cause infections (particularly if the immune system is damaged due to an already present infection) and should be handled with care in a laminar-flow cabinet. Every good culture collection provides reasonably reliable information on the safety classification of their microorganisms. When working with your own isolates, you should definitely obtain some information on the health hazards of your strains. For instance, this can be performed at reasonable cost at a culture collection, e.g., at DSMZ (see Sect. 5.5).

© Springer International Publishing AG 2018
K. Faber, *Biotransformations in Organic Chemistry*,
DOI 10.1007/978-3-319-61590-5_5

Occasionally, the use of an enzyme inhibitor may be necessary to suppress undesired side reactions due to further metabolism. You should be aware that the majority of them (in particular serine hydrolase inhibitors) are *extremely toxic* and must be handled with the utmost care, like (for example) osmium tetraoxide.

Preservation of Enzyme Stability
The following processes lead to rapid enzyme deactivation and thus should be avoided:

*   Dilution of enzyme solutions: Enzymes are not happy in very dilute solutions and are therefore more stable at higher protein concentrations.
*   Additives serving as enzyme stabilizers (such as polyhydroxy compounds, carbohydrates, inactive filler proteins, etc.) should not be removed.
*   Enzyme purification: It is an unfortunate, but commonly encountered fact that pure proteins are generally less stable than crude preparations.
*   Freezing and thawing: If required, freezing should be performed *fast* via shock-freezing (by dropping the protein solution or cell suspension into liquid nitrogen). Slow freezing leads to the formation of large ice crystals, which tend to damage proteins. The addition of polyhydroxy compounds (e.g., glycerol, sorbitol) serving as cryo-protectant is strongly recommended.
*   Very high concentrations of substrate and/or product.
*   Highly charged (inorganic) ions, such as Cr^{3+}, Cr^{6+}, Co^{2+}, Ni^{2+}, Fe^{3+}, Al^{3+} should be avoided. Alkali and halide ions are usually harmless if present at low to moderate concentrations. Some enzymes are stabilized by Mg^{2+}, Ca^{2+} or Zn^{2+} ions at low concentrations.
*   High ionic strength: Buffer concentrations should be low, i.e., 0.05–0.1 M.
*   Extreme temperatures: The best temperature range for performing a biocatalytic reaction is between 20 and 30 °C. Proteins and whole cells should be stored at 0 to +4 °C. Most lyophilized enzyme preparations and whole cells can be stored at this temperature for many months without a significant loss of activity.
*   Extreme pH: The optimal pH-range for the most commonly used enzymes is 7.0–7.5. If acid or base is emerging during the reaction, a proper pH must be maintained by, e.g., addition of an acid or base scavenger; if possible, this should occur in a continuous mode via an autotitrator.
*   High shear forces, caused by sharp edges or rapid stirring, cause deactivation. Thus gentle shaking on a rotary shaker is recommended. If stirring is required, use an overhead stirrer with a glass paddle with round edges at low to moderate speed rather than a magnetic stirring bar, which acts as mortar and grinds your enzymes or cells to pieces.
*   Metal surfaces: Some alloys can liberate traces of harmful metal ions; reactors and components thereof made from glass or plastic are preferable.
*   Precipitation of proteins – e.g., by addition of $(NH_4)_2SO_4$ or a water-miscible organic solvent, such as acetone or *i*-propanol – often leads to a significant degree of deactivation. On the contrary, lyophilization is the milder method for the isolation of proteins from aqueous solutions.
*   Complete dehydration causes irreversible deactivation. Proteins should therefore be able to keep their minimum of 'structural water' to retain their activity. In general, this is safely achieved via lyophilization.

- Large surfaces at the liquid–gas interface of bubbles and foams can lead to deactivation; foaming should therefore be avoided by, e.g., addition of an antifoam agent.
- Some agents, such as sodium azide and metal chelators (EDTA) are known to (irreversibly) deactivate enzymes.
- When you use a solid (lyophilized or spray-dried) enzyme formulation, allow the protein to rehydrate in the reaction buffer (at its pH optimum) for ca. 30 min *before* you add the reagents.

Liquids Liquid enzyme formulations usually contain a significant amount of stabilizing agents, such as carbohydrates, polyols and inactive (filler) proteins. In addition, side-products from the fermentation may be present. In general, they tend to stabilise enzymes and thus their removal (e.g., by dialysis) is not recommended. The same applies to dilution, as concentrated protein solutions are usually more stable. Liquid formulations should be stored in the cold (0 to +4 °C), but avoid freezing! If long-term storage is required, two options are possible (if in doubt, test both methods on a sample first and check for any loss of activity):

- Split the liquid formulation into aliquots and shock-freeze them by dropping into liquid nitrogen. The frozen samples can be stored at −25 °C for an (almost) unlimited period of time.
- Subject the liquid formulation to lyophilization. The latter can be conveniently stored at 0 to +4 °C.

Lyophilizates and Powders Most technical-grade enzyme preparations are shipped as solid lyophilizates or powders. In general, they are more stable than liquid formulations and they can be conveniently stored in the cold (0 to +4 °C), but should not be frozen! Before you open an enzyme container (from the refrigerator) make sure that it has reached room temperature *before* you open it. Otherwise moisture will condense onto the (hygroscopic) enzyme powder and will gradually turn it into a sticky paste which goes in hand with a loss of activity. Be aware that protein dust is more susceptible to electrostatic interaction than solid organic compounds. Thus, enzymes tend to stick onto plastic surfaces and to spread as dust, particularly when you collect static electricity by shuffling along a plastic lab floor. Dissolve any solid enzymes in water before disposal.

Carrier-Fixed Immobilized Enzymes The risk of formation of dust with immobilized enzymes is minimal. Storage at 0 to +4 °C is recommended, but avoid freezing! Any vials from the refrigerator should be brought to room temperature before opening. Due to the presence of a macroscopic carrier particle, immobilized enzymes are particularly susceptible to mechanical disintegration. Thus, avoid rapid stirring and agitate only by gentle shaking. Soak immobilized enzymes in water before disposal.

5.2   Make Your Own Enzymes: Some Guidelines for the Non-specialist

While basic knowledge of molecular biology is necessary to allow designing and performing cloning, the corresponding theoretical background and most standard protocols can be learnt rapidly by non-specialists. Provided the enzyme of interest is neither particularly exotic nor resilient, any researcher with minimal equipment (molecular biology kits, PCR machines and incubators are of advantage) can obtain a first sample of protein from scratch within few weeks.

Accessing protein and gene sequence information

In case the enzyme of interest has been already characterized and (at least) its amino acid sequence is known, a search in a protein database (most comprehensive one being UniprotKB) can directly grant access to the peptide sequence. Back-translation using online bioinformatics tools (numerous and freely accessible) delivers a possible nucleic acid sequence, i.e. the gene, which codes for the protein of interest.

For most known protein sequences, the encoding gene is available directly in the protein database (ID card of the protein) or the genetic information can be retrieved from gene databases (GenBank, EMBL).

From the gene to the protein

With the gene of interest at hand, the next step involves decision on the host organism to be used for protein expression. Typically, and for most standard enzymes, the bacterium *Escherichia coli* is the workhorse. The gene can be ordered from commercial vendors, who also offer complementary codon usage optimization, i.e. the gene sequence will be synthesized using the most frequently used codon for each amino acid for a given host organism. In order to facilitate protein purification, you may add a tag to the protein sequence.

Once an appropriate expression vector for gene cloning has been chosen, the gene of interest is inserted in the plasmid through ligation and the recombinant DNA product thus obtained can be transferred to the selected host (transformation). Common commercial vectors are the pET vectors.

Finally, the recombinant protein is obtained by over-expression through growing a cell culture of the host. Typically, IPTG is used as inducer of protein expression. In case the protein forms inclusion bodies, is misfolded or turns out to be toxic towards the host, ask your friends at the molecular biology department for trouble-shooting strategies.

Several options for the use of this biocatalyst are available:

- Whole recombinant (*E. coli*) host cells can be used as such in a non-fermenting (resting) state, particularly for preparative-scale biotransformations. In this case, control experiments are necessary to ensure the absence of disturbing background activity by 'empty' host cells.
- Host cells are lysed to liberate the enzyme of interest.
- The enzyme of interest is purified after cell lysis, which is greatly facilitated by an affinity tag.

5.3 Abbreviations

ACE	Acetylcholine esterase
ADH	Alcohol dehydrogenase
ADP	Adenosine diphosphate
AMP	Adenosine monophosphate
Ar	Aryl
ATP	Adenosine triphosphate
Bn	Benzyl
Boc	*tert*-Butyloxycarbonyl
BPO	Bromoperoxidase
CAL	*Candida antarctica* lipase
Cbz	Benzyloxycarbonyl
CEH	Cytosolic epoxide hydrolase
CLEA	Cross-linked enzyme aggregate
CLEC	Crosslinked enzyme crystal
CPO	Chloroperoxidase
CRL	*Candida rugosa* lipase
CSL	*Candida* sp. lipase
CTP	Cytosine triphosphate
Cyt P-450	Cytochrome P-450
DAHP	3-Deoxy-D-*arabino*-heptulosonate-7-phosphate
d.e.	Diastereomeric excess
DER	2-Deoxyribose-5-phosphate
DH	Dehydrogenase
DHAP	Dihydroxyacetone phosphate
DOPA	3,4-Dihydroxyphenyl alanine
E	Enantiomeric ratio
e.e.	Enantiomeric excess
EH	Epoxide hydrolase
Enz	Enzyme
e.r.	Enantiomer ratio
FAD	Flavine adenine dinucleotide
FDH	Formate dehydrogenase
FDP	Fructose-1,6-diphosphate
FMN	Flavine mononucleotide
Gal	Galactose
GDH	Glucose dehydrogenase
Glc	Glucose
GluDH	Glutamate dehydrogenase
G6P	Glucose-6-phosphate
G6PDH	Glucose-6-phosphate dehydrogenase
GTP	Guanosine triphosphate
HLADH	Horse liver alcohol dehydrogenase
HSDH	Hydroxysteroid dehydrogenase
HLE	Horse liver esterase
IPTG	Isopropyl-β-D-thiogalactopyranoside
KDO	3-Deoxy-D-*manno*-2-octulosonate-8-phosphate
LDH	Lactate dehydrogenase
LG	Leaving group
MEEC	Membrane-enclosed enzymatic catalysis
MEH	Microsomal epoxide hydrolase

(continued)

ACE	Acetylcholine esterase
MOM	Methoxymethyl
MSL	*Mucor* sp. lipase
NAD$^+$/NADH	Nicotinamide adenine dinucleotide
NADP$^+$/NADPH	Nicotinamide adenine dinucleotide phosphate
NDP	Nucleoside diphosphate
NeuAc	*N*-acetylneuraminic acid
Nu	Nucleophile
O-5-P	Orotidine-5-phosphate
PEP	Phosphoenol pyruvate
PEG	Polyethylene glycol
PLE	Porcine liver esterase
PPL	Porcine pancreatic lipase
PQQ	Pyrroloquinoline quinone
PRPP	5-Phospho-D-ribosyl-α-1-pyrophosphate
PSL	*Pseudomonas* sp. lipase
PYR	Pyruvate
RAMA	Rabbit muscle aldolase
Sub	Substrate
TBADH	*Thermoanaerobium brockii* alcohol dehydrogenase
TEPP	Tetraethyl pyrophosphate
Thex	Thexyl = 1,1,2-trimethylpropyl
TOF	Turnover frequency
TON	Turnover number
Tos	*p*-Toluenesulfonyl
TPP	Thiamine pyrophosphate
TTN	Total turnover number
UDP	Uridine diphosphate
UMP	Uridine monophosphate
UTP	Uridine triphosphate
XDP	Nucleoside diphosphate
XTP	Nucleoside triphosphate
YADH	Yeast alcohol dehydrogenase
Z	Benzyloxycarbonyl

5.4 Suppliers of Enzymes

Almac	UK
Amano	Japan
American Biosystems	USA
ASA Spezialenzyme	Germany
BASF	Germany
Biocatalysts	UK
Biozym	Germany
Biozyme Laboratories	UK
Calzyme	USA
Cambrex	USA
Chiralvision	The Netherlands

(continued)

CHR Hansen	Denmark
CLEA Technologies	The Netherlands
c-LEcta	USA
DSM	The Netherlands
Enz Bank	Korea
Enzymicals	Germany
Eucodis	Austria
Evocatal	Germany
Evonik	Germany
Syncozymes	China
Libragen	France
Innotech MSU	Russia
Iris Biotech	Germany
Maxygen	USA
Meito Sangyo	Japan
MP Biomedicals	UK
Nagase ChemteX	Japan
Novozymes	Denmark
Nzomics Biocatalysis	UK
Oriental Yeast	Japan
Proteus	France
Prozomix	UK
Sanofi Genzyme	UK
Sigma-Aldrich (Enzyme Explorer)	USA
Strem (Codexis Enzymes)	USA
Sustainable Chemistry Solutions (Enzyme Company Guide)	USA
SyncoZymes	China
Toyobo	Japan
Wako Chemicals	Germany
Worthington	USA
X-Zyme (Johnson Matthey)	Germany
ZA Biotech	South Africa

5.5 Major Culture Collections

ARS (NRRL)	Agricultural Research Service Culture Collection, USA
ATCC	American Type Culture Collection, USA
CBS	Centraalbureau voor Schimmelcultures, The Netherlands
DSMZ	Deutsche Sammlung von Mikroorganismen und Zellkulturen, Germany
ECACC	European Collection of Authenticated Cell Cultures, UK
IFO	Institute of Fermentation, Japan
NCIMB (NCIB)	National Collections of Industrial and Marine Bacteria, UK
NCYC	National Collection of Yeast Cultures, UK
NITE (NBRC)	Biological Resource Center, Japan
WDCM	World Data Centre for Microorganisms, China

5.6   Pathogenic Bacteria and Fungi

Dangerous Pathogens

Bacteria	
Bacillus anthracis	*Mycobacterium bovis*
Bordetella pertussis	*Mycobacterium leprae*
Clostridium bifermentans	*Mycobacterium tuberculosis*
Clostridium botulinum	*Neisseria gonorrhoeae*
Clostridium fallax	*Neisseria meningitidis*
Clostridium histolyticum	*Pasteurella pestis*
Clostridium oedematiens	*Pseudomonas pseudomallei*
Clostridium septicum	*Salmonella typhi*
Clostridium welchii (perfringens)	*Streptococcus pneumoniae*
Corynebacterium diphtheriae	*Treponema pallidum*
Flavobacterium meningosepticum	*Treponema pertenue*
Leptospira icterohaemorrhagiae	*Vibrio cholerae*

Fungi	
Aspergillus fumigatus	*Histoplasma capsulatum*
Blastomyces dermatitidis	*Histoplasma farcimimosum*
Coccidioides immitis	*Paracoccidioides braziliensis*

Readily Infectious Pathogens

Bacteria	
Borrelis sp.	*Moraxella lacunata*
Brucella abortus	*Pasteurella tularensis*
Brucella melitensis	*Pseudomonas aeruginosa*
Brucella suis	*Pseudomonas pyocyanae*
Fusobacterium fusiforme	*Shigella dysenteriae*
Haemophilus aegypticus	*Shigella flexneri*
Haemophilus ducreyi	*Shigella sonnei*
Haemophilus influenzae	*Staphylococcus aureus*
Klebsiella pneumoniae	*Staphylococcus pyogenes*
Klebsiella rhinoscleromatis	

Fungi	
Candida albicans	*Sporotrichum schenkii*
Epidermophyton flossosum	*Trychophyton verrucosum*
Microsporum sp.	

Index

© Springer International Publishing AG 2018
K. Faber, *Biotransformations in Organic Chemistry*,
DOI 10.1007/978-3-319-61590-5

Printed in the United States
By Bookmasters